# Lecture Notes in Computer Science 1178

Edited by G. Goos, J. Hartmanis and J. van Leeuwen

Advisory Board: W. Brauer   D. Gries   J. Stoer

**Springer**
*Berlin*
*Heidelberg*
*New York*
*Barcelona*
*Budapest*
*Hong Kong*
*London*
*Milan*
*Paris*
*Santa Clara*
*Singapore*
*Tokyo*

Tetsuo Asano   Yoshihide Igarashi
Hiroshi Nagamochi   Satoru Miyano
Subhash Suri   (Eds.)

# Algorithms and Computation

7th International Symposium, ISAAC '96
Osaka, Japan, December 16-18, 1996
Proceedings

 Springer

Series Editors

Gerhard Goos, Karlsruhe University, Germany
Juris Hartmanis, Cornell University, NY, USA
Jan van Leeuwen, Utrecht University, The Netherlands

Volume Editors

Tetsuo Asano
Dept. of Engineering Informatics, Osaka Electro-Communication University
Hatsu-cho, Neyagawa 572, Japan

Yoshihide Igarashi
Dept. of Computer Science, Gunma University
Kiryu 376, Japan

Hiroshi Nagamochi
Dept. of Applied Mathematics and Physics, Kyoto University
Sakyo-ku, Kyoto 606-01, Japan

Satoru Miyano
Institute of Medical Science, University of Tokyo
Shirokanedai 4-6-1, Minato-ku, Tokyo 108, Japan

Subhash Suri
Dept. of Computer Science, Washington University
St. Louis, MO 63130, USA

Cataloging-in-Publication data applied for

Die Deutsche Bibliothek - CIP-Einheitsaufnahme

**Algorithms and computation** : 7th international symposium ;
proceedings / ISAAC '96, Osaka, Japan, December 16 - 18,
1996. Tetsuo Asano ... (ed.). - Berlin ; Heidelberg ; New York ;
Barcelona ; Budapest ; Hong Kong ; London ; Milan ; Paris ;
Santa Clara ; Singapore ; Tokyo : Springer, 1996
  (Lecture notes in computer science ; Vol. 1178)
  ISBN 3-540-62048-6
NE: Asano, Tetsuo [Hrsg.]; ISAAC <7, 1996, Osaka>; GT

CR Subject Classification (1991): F.1-2, G.2-3, I.3.5, C.2

ISSN 0302-9743
ISBN 3-540-62048-6 Springer-Verlag Berlin Heidelberg New York

Typesetting: Camera-ready by author
SPIN 10550196    06/3142 – 5 4 3 2 1 0    Printed on acid-free paper

# Preface

The first International Symposium on Algorithms was held in Tokyo in 1990 and the second one in Taipei in 1991. The title was changed to International Symposium on Algorithms and Computation in 1992 when it was held in Nagoya. Later the symposiums were held in Hong Kong, Beijing, and Cairns in 1993, 1994, and 1995 respectively. The papers in this volume were presented at the seventh symposium in Osaka, Japan, during December 16-18, 1996. The symposium was sponsored by Osaka Electro-Communication University, The Telecommunications Advancement Foundation, and The Asahi Glass Foundation in cooperation with the Special Interest Group on Algorithms of the Information Processing Society of Japan, the Technical Group on Theoretical Foundation of Computing of the Institute of Electronics, Information and Communication Engineers of Japan, the ACM Special Interest Group on Automata and Computability Theory, and the European Association for Theoretical Computer Science.

In response to the Call for Papers, 119 extended abstracts were submitted. The papers in this volume were selected from them by the program committee consisting of S. Miyano, S. Suri, T. Akutsu, V. Chandru, D. Dobkin, D.-Z. Du, H. ElGindy, M. Golin, T. Hirata, M. Golin, A.K. Lenstra, D. Kirkpatrick, W. Rytter, M. Ogihara, A. Srinivasan, and M. Sharir. Each submitted paper was reviewed by about four program committee members. Thus, the job of the program committee is quite challenging.

We wish to thank all who submitted papers for consideration and all program committee members for their excellent work. We also thank our colleagues who contributed to the success of the symposium and the sponsors for their assistance and supports.

September, 1996

Tetsuo Asano
Yoshihide Igarashi
Hiroshi Nagamochi
Satoru Miyano
Subhash Suri

# Conference Organization

**Symposium Chairs**    T. Asano (Osaka Electro-Communication U., Japan)
Y. Igarashi (Gunma U., Japan)

**Program Committee**    T. Akutsu (U. of Tokyo, Japan)
V. Chandru (Ind. Inst. Sci., India)
D. Dobkin (Princeton U., USA)
D. -Z. Du (U. of Minnesota, USA)
H. ElGindy (U. of Newcastle, Australia)
M. Golin (HK UST, Hong Kong)
T. Hirata (Nagoya U., Japan)
D. Kirkpatrick (UBC, Canada)
A. K. Lenstra (Bellcore, USA)
S. Miyano (U. of Tokyo, Japan, Co-Chair)
M. Ogihara (Rochester, USA)
W. Rytter (Warsaw, Poland)
M. Sharir (Tel Aviv U., Israel and NYU, USA)
S. Suri (Washington U., USA, Co-Chair)
A. Srinivasan (Nat. U. Singapore, Singapore)

**Local Arrangement**    N. Katoh (Kobe U. of Commerce, Japan)
H. Nakano (Osaka City U., Japan)

**Publicity**    D. Z. Chen (U. of Notre Dame, USA)
T. Tokuyama (Tokyo Research Lab., IBM Japan)

**Finance**    H. Suzuki (Ibaraki U., Japan)

**Publication**    H. Nagamochi (Kyoto U., Japan)

**Sponsors**
Osaka Electro-Communication University
The Telecommunications Advancement Foundation
The Asahi Glass Foundation
Special Interest Group on Algorithms,
    Information Processing Society of Japan
Technical Group on Theoretical Foundation of Computing,
    Institute of Electronics, Information and Communication Engineers of Japan
ACM Special Interest Group on Automata and Computability Theory
European Association for Theoretical Computer Science

# Contents

# Applications of a Numbering Scheme for Polygonal Obstacles in the Plane*

Mikhail J. Atallah[†]     Danny Z. Chen[‡]

## Abstract

We present efficient algorithms for the problems of matching red and blue disjoint geometric obstacles in the plane and connecting the matched obstacle pairs with mutually nonintersecting paths that have useful geometric properties. We first consider matching $n$ red and $n$ blue disjoint isothetic rectangles and connecting the $n$ matched rectangle pairs with nonintersecting monotone rectilinear paths; each such path consists of $O(n)$ segments and is not allowed to touch any rectangle other than the matched pair that it is linking. Based on a numbering scheme for certain geometric objects and on several useful geometric observations, we develop an $O(n \log n)$ time, $O(n)$ space algorithm that produces a desired matching for isothetic rectangles. If an explicit printing of all the $n$ paths is required, then our algorithm takes $O(n \log n + \lambda)$ time and $O(n)$ space, where $\lambda$ is the total size of the desired output. We then extend these matching algorithms to other classes of red/blue polygonal obstacles. The numbering scheme also finds applications to other problems.

## 1 Introduction

The problem of computing paths that avoid obstacles is fundamental in computational geometry and has many applications. It has been studied in both sequential and parallel settings and using various metrics. The rectilinear version of the problem, which assumes that each of a path's constituent segments is parallel to a coordinate axis, is motivated by applications in areas such as VLSI wire layout, circuit design, plant and facility layout, urban transportation, and robot motion. There are many efficient sequential algorithms that compute various shortest rectilinear paths avoiding different classes of obstacles [11, 12, 13, 14, 15, 17, 18, 19, 21, 24, 25, 27, 28, 29, 38, 39, 40, 41], and some parallel algorithms as well [4, 5].

In this paper, we present efficient algorithms for the problems of matching red and blue disjoint geometric obstacles in the plane and connecting the matched obstacle pairs with mutually nonintersecting paths that have certain useful geometric properties. The first problem we consider has the following input: $n$ of the given $2n$ pairwise disjoint isothetic rectangles are

---

*This work was carried out in part at the Center for Applied Science and Engineering and Institute of Information Science. Academia Sinica, Nankang, Taiwan, during its 1996 Summer Institute on Computational Geometry and Applications.

†Department of Computer Sciences, Purdue University, West Lafayette, Indiana 47907, USA. mja@cs.purdue.edu. This author gratefully acknowledges support from the National Science Foundation under Grant DCR-9202807, and the COAST Project at Purdue University and its sponsors, in particular Hewlett Packard.

‡Department of Computer Science and Engineering, University of Notre Dame, Notre Dame, Indiana 46556, USA. chen@cse.nd.edu. The research of this author was supported in part by the National Science Foundation under Grant CCR-9623585.

Figure 1: Example of a matching of red and blue rectangles with monotone paths.

colored red (think of them as sources of something, e.g., electric power in a VLSI circuit), and
the other $n$ are colored blue (think of them as consumers of power). By *isothetic* objects, we
mean that each edge of such an object is parallel to a coordinate axis. We are interested in
matching each red rectangle with one and only one blue rectangle, and *vice versa*. Specifically,
we would like to find such a matching and connect each matched pair of red/blue rectangles
with a planar rectilinear path in such a way that (i) each path is monotone with respect to a
coordinate axis, (ii) each path does not touch any rectangle other than the matched pair that it
is supposed to connect, (iii) no two such paths intersect each other, and (iv) each path consists
of $O(n)$ segments. Figure 1 shows an example of such a matching.

Several geometric algorithms have been developed for solving various problems of finding
obstacle-avoiding pairwise disjoint paths that connect certain geometric objects [26, 32, 36], because of their relevance to VLSI layout applications [16, 26, 35] (e.g., VLSI single-layer routing).
Lee *et al.* [26] designed an $O((k^2!)n \log n)$ time algorithm for computing $k$ shortest non-crossing
rectilinear paths in a *plane region*. Takahashi, Suzuki, and Nishizeki [36] studied the problem
of finding shortest non-crossing rectilinear paths in a plane region that is bounded by an outer
box and an inner box and that contains a set of disjoint isothetic rectangle obstacles, giving
an $O(n \log n)$ time algorithm for computing $k$ such paths whose endpoints are all on the two
bounding boxes (with $k \leq n$). Papadopoulou [32] very recently obtained an $O(n + k)$ time
algorithm for computing $k$ shortest non-crossing paths in a simple polygon whose endpoints are
all on the polygon boundary. However, these problems are different from the one we study here
since they often assume that a specification on which object matches with which other object is
already given (hence, these problems require only to compute a set of non-crossing paths that
realize the specified matching).

We develop an $O(n \log n)$ time, $O(n)$ space algorithm that produces a desired matching for
red/blue isothetic rectangles. If an explicit printing of all the $n$ paths for such a matching is
required, then our algorithm takes $O(n \log n + \lambda)$ time and $O(n)$ space, where $\lambda$ is the total size
of the desired output.

We then extend these matching algorithms to a more general geometric setting which consists of disjoint red/blue polygonal obstacles that are all monotone to a coordinate axis (say, the $y$-axis). The matching paths we compute for this more general setting have similar structures to those for isothetic rectangles, except that in this case their monotonicity has to be weaker: Each such matching path can be partitioned into at most two subpaths, each of which is monotone to the $y$-axis. Our matching algorithms for $y$-monotone polygonal obstacles have the same complexity bounds as those for isothetic rectangles.

We also prove that all the matching problems studied in this paper have an $\Omega(n \log n)$ lower bound in the algebraic computation tree model [8]. Our matching algorithms are based on a numbering scheme for certain geometric objects and on several useful geometric observations. This numbering scheme also finds applications to other problems [7].

Our algorithms can also be viewed as proofs that such matchings always exist, a fact that, to the best of our knowledge, was not previously established. We should point out that without the requirement that all matching paths must satisfy a monotonicity constraint, the existence of nonintersecting paths for any red/blue polygonal obstacle matching is trivial to prove: For every matched pair of geometric objects in turn, draw a direct rectilinear path $P$ between them, ignoring all previously drawn paths and obstacles; at each place where path $P$ intersects a previously drawn path or an obstacle, "deform" $P$ so that $P$ goes around that previously drawn path or the obstacle.

Section 2 gives some preliminary definitions, Section 3 presents one of the ingredients needed by the matching algorithms for isothetic rectangles, Section 4 describes the data structures that our matching algorithms will use, Section 5 gives the algorithm for computing a desired matching for isothetic rectangles, Section 6 extends this algorithm to also producing the $n$ actual monotone paths that link the matched rectangle pairs, Section 7 generalizes these algorithms to matching $y$-monotone polygonal obstacles, Section 8 proves $\Omega(n \log n)$ lower bounds for the matching problems we consider, and Section 9 makes further remarks on several consequences and possible extensions of this work.

## 2 Preliminaries

A geometric object in the plane is *rectilinear* if each of its constituent segments is parallel to either the $x$-axis or the $y$-axis. Without loss of generality (WLOG), we assume that no two boundary edges of the input obstacles are collinear. We use $R = \{R_1, R_2, \ldots, R_{2n}\}$ to denote the set of $2n$ input isothetic rectangles.

A path consists of a contiguous sequence of line segments in the plane. The number of line segments in a path $P$ is called the *size* of $P$, denoted by $|P|$, and the *length* of $P$ is the sum of

the distances of its edges in a certain metric. A path is said to be *monotone* with respect to the
$x$-axis (resp., $y$-axis) if and only if its intersection with every vertical (resp., horizontal) line is
either empty or a contiguous portion of that line. A path is said to be *monotone* if and only
if it is monotone to the $x$-axis or to the $y$-axis. A rectilinear path is *convex* if it is monotone
to both the $x$-axis and the $y$-axis. In general, a convex (rectilinear) path has the shape of a
staircase, and in fact we shall henceforth use the word "staircase" as a shorthand for "convex
path". Staircases can be either *increasing* or *decreasing*, depending on whether they go up or
down as we move along them from left to right. A staircase is *unbounded* if it starts and ends
with a semi-infinite segment, i.e., a segment that extends to infinity on one end. A staircase is
said to be *clear* if it does not intersect the interior of any input obstacle.

A polygon $G$ is said to be *monotone* to the $x$-axis (resp., $y$-axis) if and only if its intersection
with any vertical (resp., horizontal) line $L$ is either empty or a contiguous segment of $L$; the
boundary of such a monotone polygon $G$ can be partitioned into two paths each of which is
monotone to the $x$-axis (resp., $y$-axis). In fact, the notion of monotonicity of a polygon or a
path is in general with respect to an arbitrary line [34]. Note that it is possible to find out in
linear time whether there is a line (in an arbitrary direction) to which all polygons in a polygon
set are monotone, by using Preparata and Supowit's monotonicity test algorithm [34].

A point $p$ in the plane is defined by its $x$-coordinate $x(p)$ and $y$-coordinate $y(p)$. A point $p$
is strictly *below* (resp., to the *left* of) a point $q$ if and only if $x(p) = x(q)$ and $y(p) < y(q)$ (resp.,
$y(p) = y(q)$ and $x(p) < x(q)$); we can equivalently say that $q$ is strictly *above* (resp., to the *right*
of) $p$. A rectangle $r$ is *below* (resp., to the *left* of) an unbounded staircase $S$ if no point of $r$ is
strictly above (resp., to the right of) a point of $S$; we can equivalently say that $S$ is *above* (resp.,
to the *right* of) $r$.

Unless otherwise specified, all geometric objects in the rest of this paper (e.g., paths, rays,
lines, polygons, obstacles, etc) are assumed to be rectilinear in the plane.

## 3   Partitioning Isothetic Rectangles with a Staircase

Given a set $R = \{R_1, R_2, \ldots, R_{2n}\}$ of $2n$ pairwise disjoint isothetic rectangles in the plane
and an integer $k$ with $1 \leq k < 2n$, we present an algorithm for partitioning the set $R$ into two
subsets of respective sizes $k$ and $2n - k$, such that the two resulting subsets are separated by an
increasing staircase. This algorithm runs in $O(n \log n)$ time, or in $O(\min\{k, 2n - k\})$ time if $R$
is given in a suitably preprocessed form. The algorithm can also be implemented optimally in
parallel (see Section 9). A key idea of this partition algorithm is a useful numbering scheme for
certain geometric objects, which also finds applications to other problems [7].

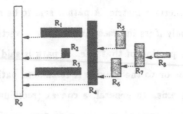

Figure 2: Illustrating the tree $T$ of the rectangles in $R$.

Not only is the result of this section needed as a key ingredient in the algorithms for matching isothetic rectangles given later, but it also implies simpler algorithms for a number of unrelated divide-and-conquer sequential and parallel algorithms for various rectilinear shortest path problems among rectangles, in which such a staircase is needed for bipartitioning the problem before recursively solving the two subproblems defined by the staircase [4, 5, 11, 29].

We begin by describing the $O(n \log n)$ time preprocessing. The first step of the preprocessing algorithm consists of computing a horizontal trapezoidal decomposition of $R$ [33], in $O(n \log n)$ time. Recall that this gives, among other things, the following *Parent* information (actually, it gives more than what follows, but we only need what follows): For each rectangle $R_i$ of $R$, *Parent*(i) is the first rectangle $R_j$ encountered by shooting a leftwards-moving horizontal ray from the *bottom-left corner* of $R_i$ (see Figure 2). If no such rectangle exists for $R_i$, then the ray goes to infinity, a fact that we denote by saying that *Parent*(i) is empty. Note that the rectangles in $R$ and their *Parent* information together define a forest of the rectangles. The trapezoidal decomposition algorithm [33] also produces a sorted list of each subset of rectangles having the same *Parent* (including the "empty" parent). Every rectangle $R_j$ maintains an adjacency list of all the rectangles whose *Parent* is $R_j$, sorted by the *decreasing* $y$-coordinates of their leftwards-moving horizontal rays. For example, the sorted adjacency list of $R_4$ in Figure 2 is $\{R_5, R_6\}$.

The second step of the preprocessing algorithm is now given. To simplify the presentation, we assume that we have added to the given collection $R$ of input rectangles an extra "dummy" rectangle $R_0$ to the left of all the other rectangles in $R$, such that the horizontal projection of $R_0$ on the $y$-axis properly contains the horizontal projections of all the other rectangles (see Figure 2). This amounts to replacing every empty *Parent*(i) by $R_0$, effectively making $R_0$ the root of a tree each of whose nodes corresponds to exactly one rectangle in $R$. We use $T$ to denote this tree. Figure 2 shows an example of such a tree $T$. The preprocessing algorithm then computes the preorder numbers of $T$ in $O(n)$ time [1], and re-labels the rectangles of $R$ (which are the nodes of $T$) so that rectangle $R_i$ now denotes the one whose preorder number in $T$ is $i$. The

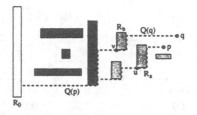

Figure 3: An example of the paths $Q(p)$ and $Q(q)$.

preorder numbers of $T$ start from 0. Hence the dummy rectangle, the root, retains the name $R_0$. This completes the description of the preprocessing.

This preprocessing algorithm clearly takes altogether $O(n \log n)$ time and $O(n)$ space. In the rest of this section, we assume that the rectangles have been re-labeled as explained above.

For any set $R'$ of disjoint rectangles, we henceforth use $CH(R')$ to denote the rectilinear convex hull of $R'$ in the plane (see [30] for a study of rectilinear convex hulls of planar rectilinear geometric objects). For every point $p$ in the plane that is to the right of the root rectangle $R_0$ and is not in the interior of any obstacle, we define a path $Q(p)$ from $p$ to $R_0$, as follows:

$Q(p)$ starts at $p$ and follows the leftwards-moving horizontal ray $r(p)$ from $p$; if the ray $r(p)$ first hits a rectangle $R_i \neq R_0$, then $Q(p)$ goes downwards along the boundary of $R_i$ to its bottom-right vertex and then leftwards to its bottom-left vertex, from which $Q(p)$ continues as it did at $p$, until it reaches $R_0$.

Note that for every such point $p$, the path $Q(p)$ is uniquely defined, and in fact is always an increasing obstacle-avoiding staircase. Also, note that every vertical segment of $Q(p)$ is completely on the right edge of a rectangle and the lower vertex of such a vertical segment is at the bottom-right vertex of that rectangle. Figure 3 gives an example of such paths.

The following lemmas are useful to proving the theorem on staircase separators.

**Lemma 1** *Let $p$ and $q$ be two points in the plane such that they both are to the right of $R_0$, and $x(p) \leq x(q)$. If $p$ is below (resp., above) some point of $Q(q)$, then no point of $Q(p)$ is strictly above (resp., below) any point of $Q(q)$ (Figure 3).*

**Proof.** The proof is straightforward and will be given in the full version of the paper. □

**Lemma 2** *Let $p$ and $q$ be two points in the plane such that they both are to the right of $R_0$ and that $x(p) \leq x(q)$. Let $u$ (resp., $v$) be the bottom-left vertex of a rectangle $R_a$ (resp., $R_b$), such that $u$ (resp., $v$) is on $Q(p)$ (resp., $Q(q)$) but not on $Q(q)$ (resp., $Q(p)$). If $p$ is strictly below (resp., above) some point of $Q(q)$, then the preorder number of $R_a$ in the tree $T$ of rectangles is larger (resp., smaller) than that of $R_b$, i.e., $a > b$ (resp., $a < b$).*

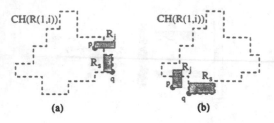

Figure 4: Illustrating the proof of the staircase separator theorem.

**Proof.** This follows from Lemma 1 and from the definition of the tree $T$. An example illustrating the lemma is given in Figure 3. □

We are now ready to give the staircase separator theorem.

**Theorem 1 (Staircase Separator Theorem)** *Given a preprocessed set $R$ of $2n$ disjoint isothetic rectangles, the subsets $\{R_1, R_2, \ldots, R_k\}$ and $\{R_{k+1}, R_{k+2}, \ldots, R_{2n}\}$, for any integer $k$ with $1 \leq k < 2n$, form a partition of the set $R$ that has the desired property, that is, there exists a rectangle-avoiding increasing staircase of size $O(n)$ that separates these two subsets. Furthermore, such a staircase separator can be computed in $O(\min\{k, 2n - k\})$ time.*

**Proof.** Let $R(a, b)$ denote the subset $\{R_a, R_{a+1}, \ldots, R_b\}$ of $R$. For the existence of such a staircase separator, we first show that for any $i < j$, the following holds: (1) $CH(R(1, i))$ does not intersect $R_j$, and (2) $CH(R(j, 2n))$ does not intersect $R_i$. We give the proof only for (1), that for (2) being similar. We prove (1) by contradiction: Suppose to the contrary that for some $j > i$, $R_j$ intersects $CH(R(1, i))$. Then one of the following two cases must occur:

- $CH(R(1, i))$ contains some point $p$ on the bottom edge of $R_j$ ($CH(R(1, i))$ possibly contains $R_j$ completely). Note that there can be no rectangles $R_s$ and $R_l$, $s \leq i < l$, such that the leftwards-moving horizontal ray from the bottom-left vertex of $R_s$ first hits $R_l$ (otherwise, this would make $R_l$ the parent of $R_s$, contradicting the fact that $R_l$ has a larger preorder number than $R_s$ in the tree $T$). Since the point $p$ of $R_j$ is inside $CH(R(1, i))$, there must be a rectangle $R_s$, $s \leq i < j$, such that the bottom edge of $R_s$ contains a point $q$ that satisfies both $x(p) \leq x(q)$ and $y(p) > y(q)$ (see Figure 4(a)). But then the path $Q(p)$ (resp., $Q(q)$) contains the bottom-left vertex of $R_j$ (resp., $R_s$) and by Lemma 2, the preorder number of $R_j$ in $T$ is smaller than that of $R_s$, a contradiction.

- $CH(R(1, i))$ contains some point of $R_j$ but the bottom edge of $R_j$ is completely outside $CH(R(1, i))$. Then $R_j$ must intersect the lower hull of $CH(R(1, i))$ (see Figure 4(b)). Again there can be no rectangles $R_s$ and $R_l$, $s \leq i < l$, such that the leftwards-moving horizontal ray from the bottom-left vertex of $R_s$ first hits $R_l$. But then, there must be a point $q$ on the bottom edge of a rectangle $R_s$, $s \leq i < j$, such that $x(p) \leq x(q)$ and $y(p) > y(q)$ for some point $p$ on the bottom edge of $R_j$ (Figure 4(b)). Again by Lemma 2, this implies that the preorder number of $R_j$ in $T$ is smaller than that of $R_s$, a contradiction.

We can now let such a desired staircase separator $S$ for the subsets $R(1,k)$ and $R(k+1,2n)$ consist of (say) the portion of the boundary of $CH(R(1,k))$ from its rightmost edge clockwise to its lowest edge, augmented by two semi-infinite segments, one extended leftwards horizontally from its lowest edge and the other extended upwards vertically from its rightmost edge. By using the same arguments as above, we can show that for every $j$ with $j > k$, $S$ is above or to the left of $R_j$. Hence $S$ so constructed is an obstacle-avoiding increasing staircase and consists of $O(k)$ segments.

WLOG, assume $k = \min\{k, 2n-k\}$. We now show how to compute such a staircase separator $S$ in $O(k)$ time. In fact, we will compute $CH(R(1,k))$, which is a little more than the above staircase $S$, in $O(k)$ time. Note that the boundary of $CH(R(1,k))$ can be obtained from four staircase paths, each of which corresponds to an ordered sequence of certain suitably defined elements of maximal domination [33] for the $4k$ rectangle vertices of $R(1,k)$. WLOG, we only show the procedure for computing one such sequence of maximal elements.

Our procedure is based on a simple divide-and-conquer strategy. First, partition the set $R(1,k)$ into two subsets $R(1,k/2)$ and $R(k/2,k)$ (WLOG, assume $k$ is an even integer greater than 1). Then, recursively compute the sequence of maximal elements for each such subset, represented by a balanced search tree, such as a 2-3 tree [1]. Finally, compute the sequence of maximal elements for the vertices of $R(1,k)$ from the two sequences for the two subsets. By the above discussion, these two sequences are respectively contiguous portions of the boundaries of two disjoint convex hulls. Hence by performing $O(1)$ standard 2-3 tree operations, the sequence of maximal elements for $R(1,k)$ can be obtained, also maintained by a 2-3 tree. The recurrence relation for the time complexity of this divide-and-conquer procedure is

$T(k) = 2T(k/2) + O(\log k)$, for $k > 1$

$T(1) = O(1)$

Hence it follows that $T(k) = O(k)$. After the above divide-and-conquer procedure terminates, it is easy to obtain the sequence of maximal elements for $R(1,k)$ from its 2-3 tree in $O(k)$ time. The space used for computing $CH(R(1,k))$ is clearly $O(k)$.

This completes the proof of the staircase separator theorem. □

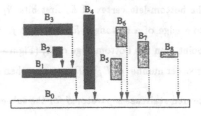

Figure 5: Illustrating the definition of the tree $T'$.

# 4    Data Structures

In this section, we describe the data structures that the algorithm in the next section will use. Since that algorithm from time to time will delete some rectangles from the collection $R = \{R_1, R_2, \ldots, R_{2n}\}$, we use $L_+$ to denote the current list of rectangles sorted by their preorder numbers in $T$. The list $L_+$ is initially $\{R_1, R_2, \ldots, R_{2n}\}$, but may change as the algorithm proceeds. However, the following invariants must hold:

1. The list $L_+$ must contain as many red as blue rectangles.

2. $CH(L_+)$ does not intersect any of the rectangles in $R - L_+$. This invariant insures that we can solve the problem on $L_+$ without having to worry about interfering with the solution of $R - L_+$, so long as our solution paths for $L_+$ (resp., $R - L_+$) do not wander outside (resp., inside) of $CH(L_+)$. Note that if the algorithm decides to match the pair of rectangles $R', R''$ and delete $R', R''$ from $L_+$, then this invariant requires that the resulting new list $L_+ - \{R', R''\}$ should also satisfy the invariant, i.e., that $CH(L_+ - \{R', R''\})$ must intersect neither $R'$ nor $R''$.

We define another list $L_-$ which contains exactly the same set of rectangles as $L_+$ but is ordered differently (as explained next). $L_-$ initially contains all the input rectangles, but they are sorted according to their preorder numbers in a tree $T'$ rather than $T$, where $T'$ is defined just like $T$ except for the following differences:

- Instead of the "leftwards-shooting horizontal ray emanating from the bottom-left corner of each rectangle" that we used in the definition of $T$, in $T'$ we use "downwards-shooting vertical ray emanating from the bottom-right corner of each rectangle" (see Figure 5).

- Instead of sorting adjacency lists by the decreasing $y$-coordinates of the horizontal shooting rays, in $T'$ the adjacency lists are sorted by the *increasing* $x$-coordinates of the vertical shooting rays.

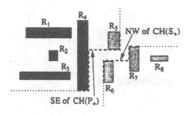

Figure 6: An example for Lemma 3, with $P_+ = \{R_1, \ldots, R_5\}$ and $S_+ = \{R_6, R_7, R_8\}$.

- The "dummy" rectangle corresponding to the root of $T'$ is *below* all the input rectangles (whereas for $T$ it was to their left).

Figure 5 illustrates the tree $T'$ in which the rectangles are named $B_i$'s (for boxes) instead of $R_i$'s.

The $L_-$ list is not explicitly maintained by our algorithm. But, the order in which the elements of $L_+$ would appear in this hypothetical list $L_-$ is conceptually important, and will be exploited by our algorithm; we henceforth use the shorthand "$T'$ preorder" to refer to this order.

Because $L_+$ (hence $L_-$) satisfies Invariant 2 above, the proofs of the following lemmas are very similar to the proof of Theorem 1 and are therefore omitted. (Note how the proof falls apart without Invariant 2, specifically at the place where we deduce that $R_l$ must be the parent of $R_s$ — this need not hold if Invariant 2 is violated, and indeed we cannot even claim that $R_l$ is an ancestor of $R_s$.)

**Lemma 3** *Let $P_+$ be a prefix of the list $L_+$, and $S_+$ be the remaining suffix of $L_+$, i.e., $S_+ = L_+ - P_+$. Then the increasing staircase defined by the South-East portion of $CH(P_+)$ is (geometrically) above all of the rectangles in $S_+$. Equivalently, the increasing staircase defined by the North-West portion of $CH(S_+)$ is below all of the rectangles in $P_+$.*

Figure 6 illustrates Lemma 3.

**Lemma 4** *Let $P_-$ be a prefix of the list $L_-$, and $S_-$ be the remaining suffix of $L_-$, i.e., $S_- = L_- - P_-$. Then the decreasing staircase defined by the North-East portion of $CH(P_-)$ is (geometrically) below all of the rectangles in $S_-$. Equivalently, the decreasing staircase defined by the South-West portion of $CH(S_-)$ is above all of the rectangles in $P_-$.*

Figure 7 illustrates Lemma 4.

When the algorithm to be described in the next section is solving a problem corresponding to the rectangles in $L_+$, it is not given just the list $L_+$ but rather a tree structure $S(L_+)$ built

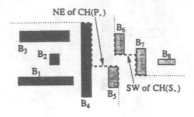

Figure 7: An example for Lemma 4, with $P_- = \{B_1, \ldots, B_5\}$ and $S_- = \{B_6, B_7, B_8\}$.

"on top" of $L_+$. Specifically, $S(L_+)$ is a 2-3 tree structure [1] whose leaves contain the rectangles in $L_+$, in the same order as in $L_+$; these leaves are doubly linked together. Each internal node $v$ of $S(L_+)$ contains a *label* equal to the smallest $T'$ preorder number (i.e., according to the $L_-$ ordering) of the rectangles stored in the subtree of $S(L_+)$ rooted at $v$. In addition, there are cross-links between every internal node $v$ of $S(L_+)$ and the leaf in the subtree of $S(L_+)$ rooted at $v$ corresponding to the label of $v$. We will perform only deletion and split operations on $S(L_+)$, both of which can be done in logarithmic time using standard techniques [1]. The deletions will take place after we have matched a pair of rectangles — we then delete them from $S(L_+)$ and recurse on the resulting $S(L_+)$. The split operations will take place when we process $L_+$ by solving recursively two pieces of $L_+$: A prefix $L'$ of $L_+$, and the remaining suffix $L'' = L_+ - L'$ (of course $L'$ and $L''$ must satisfy the required invariants mentioned earlier). Splitting $S(L_+)$ allows us to create $S(L')$ and $S(L'')$ in logarithmic time.

## 5   The Matching Algorithm for Rectangles

The goal of this procedure is to compute a desired matching for the rectangles in $R$, without worrying about describing the actual paths that join the matched pairs of red/blue rectangles (the next section explains how this procedure can be modified to also produce the actual paths connecting the matched pairs).

The procedure is recursive, and takes as input the 2-3 tree data structure $S(L_+)$ described in the previous section.

**Procedure MATCH($L_+$)**

*Input:* $S(L_+)$, where $L_+ = (R'_1, R'_2, \ldots, R'_m)$.

*Output:* A matching of the red and blue rectangles in $L_+$.

1. If $m = 2$, then the two rectangles in $L_+$ surely have different colors (by Invariant 1): Match them and return. If $m > 2$, then proceed to the next step.

   *Comment:* The path that will join the pair just matched will be along the boundary of $CH(L_+)$.

2. Find the first leaf (for $R'_1$) and the last leaf (for $R'_m$) of $S(L_+)$, in $O(\log m)$ time. If $R'_1$ and $R'_m$ have different colors, then proceed to the next step. Otherwise $R'_1$ and $R'_m$ have the same color (say, it is red). For each integer $s$, $1 \leq s \leq m$, let $f(s)$ be the number of red elements minus the number of blue elements in the set $\{R'_1, R'_2, \ldots, R'_s\}$; observe that $|f(s+1) - f(s)| = 1$ and that in this case $f(1) = 1$ whereas $f(m-1) = -1$. This implies, by a simple "continuity" argument, that there is some integer $\ell$, $1 < \ell < m-1$, for which $f(\ell) = 0$. (A somewhat similar continuity argument was used in [3] in the context of matching points.) Next, we shall search for such an $\ell$ in time $O(\min\{\ell, m-\ell\})$ rather than in time $O(m)$, as follows. We linearly search for it along the leaf sequence of $S(L_+)$, by two interleaved searches: One starting from the beginning of $L_+$, from $R'_1$ up, and the other starting from the end of $L_+$, from $R'_{m-1}$ down, where we alternate between the two searches until one of them first hits a desired value $\ell$ which we know must exist. Hence, we find an $\ell$ value for which $f(\ell) = 0$ in $O(\min\{\ell, m-\ell\})$ time, rather than in $O(m)$ time. This defines two subproblems $L'$ and $L''$: $L' = \{R'_1, R'_2, \ldots, R'_\ell\}$ and $L'' = \{R'_{\ell+1}, R'_{\ell+2}, \ldots, R'_m\}$. In $O(\log m)$ time, we split $S(L_+)$ into $S(L')$ and $S(L'')$. Then we recursively call **MATCH**$(L')$ and **MATCH**$(L'')$.

*Analysis:* This step has a cumulative total cost of $O(n \log n)$ time rather than $O(n^2)$ even though the two subproblems so generated and solved recursively can be very "unbalanced", e.g., $|L'|$ could be $O(1)$. The analysis is as follows: We spend only $O(\log m + \min\{\ell, m-\ell\})$ time in generating the two subproblems; we can "charge" the $\log m$ term of this cost to the recursive call itself (i.e., to the node of that recursive call in the recursion tree), and the $\min\{\ell, m-\ell\}$ term to the rectangles of the smaller subproblem ($O(1)$ time per rectangle). A rectangle that is so "charged" ends up in a subproblem of no more than half the size of its previous subproblem, and hence cannot be charged more than $\log n$ times, for a total (over all the $2n$ rectangles of $R$) of $O(n \log n)$. The total number of nodes in the recursion tree is $O(n)$, and hence the overall cost of the charges to the nodes of that recursion tree ($\log m$ per node) is $O(n \log n)$.

3. $R'_1$ and $R'_m$ have different colors. Obtain, from the label at the root of $S(L_+)$, the smallest rectangle of $L_+$ according to the $L_-$ ordering. Let $R''$ be this rectangle. Rectangle $R''$ must have the same color as one of $\{R'_1, R'_m\}$, so suppose WLOG that it has the same color as $R'_1$. Then we (i) match $R'_1$ and $R''$, (ii) delete $R'_1$ and $R''$ from $S(L_+)$ in $O(\log m)$ time, and (iii) recursively solve the problem on the resulting $L_+$.

*Comment:* The path that will join the pair just matched will be along the boundary of $CH(L_+ - \{R'_1, R''\})$. The justification for the monotonicity of this path follows from Lemmas 3 and 4, which ensure that the path from $R'_1$ to $R''$ along the boundary of $CH(L_+ - \{R'_1, R''\})$ consists of at most two subpaths: An increasing staircase followed by a decreasing staircase. This step also has a cumulative total cost of $O(n \log n)$ time, because each of the $n$ matched pairs is charged a cost of $O(\log n)$ time by the step.

As analyzed above, algorithm **MATCH** computes $n$ matched pairs of red/blue rectangles of $R$ in $O(n \log n)$ time and $O(n)$ space.

# 6  Reporting the Actual Paths

This section shows how to output the actual monotone paths between all the $n$ matched red/blue rectangle pairs in $O(n \log n + \lambda)$ time, where $\lambda$ is the total number of segments that make up these $n$ paths.

Recall the comments we made after a rectangle pair was matched by the algorithm of the previous section (specifically, following Steps 1 and 3). These comments described the desired path between the pair just matched in terms of a rectilinear convex hull $CH(v)$ of a subproblem associated with a particular place (i.e., a node) $v$ in the recursion tree of algorithm **MATCH** at which this subproblem occurred. We postponed the actual computation of these $CH(v)$ convex hulls, because once we have the overall structure of the recursion tree, we can traverse it and compute these $CH(v)$ hulls bottom up, with insertion operations only (since the subproblem of a child node in the recursion tree is that of its parent node *minus* some rectangles). Thus, this enables us to use the fact that maintaining rectilinear convex hulls, in the face of insertions only, is possible in logarithmic time per insertion [31].

Hence, the idea is to run the matching algorithm of Section 5 and make sure that, after that algorithm has executed, it leaves behind the skeleton of its recursion tree, which we call *RecTree*, together with certain information describing how a path between a matched rectangle pair is related to $CH(v)$ (i.e., the description in the "comments" of algorithm **MATCH**). This description information uses $O(1)$ space per matched pair. This skeleton just gives the overall structure of *RecTree*. It does not store directly the rectangles of the subproblem associated with each node $v$ of *RecTree* (that would be too expensive in terms of the space complexity), but rather how the rectangles of $v$ are related to those of $v$'s children:

1. If $v$ has only one child in *RecTree*, then its associated rectangles are those of its only child plus two rectangles that are matched by algorithm **MATCH** at $v$: It is these two rectangles that are explicitly stored at $v$ in *RecTree*.

2. If $v$ has two children in *RecTree*, then its associated rectangles are the union of the rectangles of both its children.

In either case, we store $O(1)$ information at each node $v$, so that *RecTree* uses altogether $O(n)$ space. The problem of computing the actual monotone path (if any) associated with each node $v$ in *RecTree* clearly reduces to computing $CH(v)$ in turn and using it to print that path. The computation of the $CH(v)$'s associated with all the nodes $v$ of *RecTree* is done by a simple traversal of *RecTree* during which the $CH(v)$'s are computed according to the postorder numbers [1] of the nodes $v$ in *RecTree*. Of course, at a node $v$ of *RecTree* that has two children (say, $u$ and $w$), we do not create $CH(v)$ by individually inserting the vertices of $CH(u)$ into $CH(w)$, but rather we obtain $CH(v)$ by "merging" $CH(u)$ and $CH(w)$ in logarithmic time [31]. After $CH(v)$ is computed, the actual path between the matched rectangle pair of node $v$ is computed by walking along $CH(v)$, in time proportional to the size of the path plus a logarithmic additive term. We assume that if two such matching paths share some common portions on certain convex hulls so computed, then the two paths are apart by at least a positive distance that can be made arbitrarily small. The overall time of this algorithm is therefore $O(n \log n)$ plus the time needed to print all the output paths, i.e., $O(\lambda)$.

# 7  Extensions to Monotone Polygonal Obstacles

In this section, we extend our techniques for matching red/blue isothetic rectangle obstacles to matching red/blue polygonal obstacles in the plane that are all monotone with respect to a coordinate axis (say, the $y$-axis). Let $W$ be a set of $r$ red and $r$ blue disjoint polygonal obstacles in the plane, with a total of $n$ vertices. We assume that all the polygonal obstacles in $W$ are monotone to the $y$-axis, and call them $y$-*monotone polygons*. We show that it is possible to match all the red and blue polygons in $W$, by connecting the $r$ matched red/blue polygon pairs with $r$ mutually disjoint paths. The properties of the matching paths are similar to those for isothetic rectangles, except for the monotonicity: In this case, a path can be used for the matching if it can be partitioned into at most *two* subpaths, each of which is monotone to the $y$-axis. Our algorithms for computing such a matching have the same complexity bounds as the matching algorithms for isothetic rectangles in the previous sections.

One consequence of considering $y$-monotone polygonal obstacles (whose structures are less nice than those of isothetic rectangles) is that we must use a weaker monotonicity constraint on the matching paths. This is because even with a geometric setting consisting of disjoint *convex* polygonal obstacles in the plane, there is in general no obstacle-avoiding path between two arbitrary points that is monotone to the $x$-axis or to the $y$-axis. But in such a setting, a

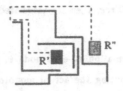

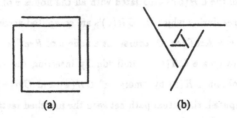

Figure 8: A path with two $y$-monotone subpaths among rectilinear convex obstacles.

(a)                              (b)

Figure 9: There is no staircase separator for rectilinear and non-rectilinear convex obstacles.

path consisting of at most two $y$-monotone subpaths always exists between any two points (see. Figure 8 for an example). Another consequence of considering $y$-monotone polygonal obstacles is that there is in general no staircase separator for partitioning such geometric object sets. In the two examples of Figure 9, there exists no staircase (even with respect to *any* two orthogonal lines) that partitions each *convex* obstacle set into two subsets, such that every subset contains more than one obstacle. However, as we will show, there exist $y$-monotone paths that partition $y$-monotone polygons. Note that a key difference between staircases and $y$-monotone paths is that staircases are monotone to *both* the $x$-axis and $y$-axis, while $y$-monotone paths need not be monotone to the $x$-axis.

It turns out that the matching algorithms based on the geometric structures of $y$-monotone polygonal obstacles are similar to and in fact simpler than the matching algorithms for isothetic rectangles. Also, although we have chosen in this section to focus our discussion on rectilinear geometric objects (obstacles, paths, etc), it is actually not difficult to modify our algorithms so that they will work with non-rectilinear objects under the $y$-monotonicity constraint.

Let the obstacle set $W = \{W_0, W_1, \ldots, W_{2r}\}$, where $W_0$ is the extra "dummy" rectangle $R_0$ to the left of all the other obstacles in $W$ (as introduced in Section 3). We first preprocess $W$ as in Section 3. From the left vertex of the lowest edge of every $W_i$, shoot a leftwards-moving horizontal ray $r_i$; let $Parent(i)$ be $W_j$, where $W_j$ is the first obstacle in $W$ hit by the ray $r_i$. Maintain for every $W_j$ an adjacency list of all the obstacles in $W$ whose $Parent$ is $W_j$, sorted by the *decreasing* $y$-coordinates of their leftwards-moving horizontal rays. This gives a tree structure whose nodes are the obstacles in $\dot{W}$ (as the tree $T$ in Section 3) and which we again denote

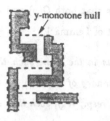

Figure 10: An example of the y-monotone hull of a set of obstacles.

by $T$. Label the nodes of $T$ by their preorder numbers in $T$, and re-label the obstacles in $W$ by their corresponding preorder numbers in $T$. This preprocessing can be done by a horizontal trapezoidal decomposition of $W$ [33] and a preorder traversal of $T$ [1], in altogether $O(n \log n)$ time and $O(n)$ space. WLOG, let $i$ be the label of $W_i$ in the preprocessed form. In addition, we also construct, as part of the preprocessing, the planar subdivision [33] that is defined by the horizontal trapezoidal decomposition of $W$. The construction of this planar subdivision also takes $O(n \log n)$ time and $O(n)$ space.

For any consecutive subset $W' = \{W_i, W_{i+1}, \ldots, W_j\}$ of $W$, where $i > 0$, we define the y-monotone hull of $W'$, denoted by $CH_y(W')$, to be the region with the smallest area that contains all the obstacles in $W'$ and that is y-monotone (see Figure 10 for an example). Note that the region $CH_y(W')$ so defined may be disconnected. If this is the case, we assume that we link the connected components of $CH_y(W')$ together with some paths of zero width, so that $CH_y(W')$ becomes connected and is still y-monotone.

Note that the boundary of every y-monotone polygon can be easily partitioned into two y-monotone paths, which we call the *left boundary* and *right boundary* of such a polygon. For every point $p$ in the plane that is to the right of the root obstacle $W_0$ of $T$ and is not in the interior of any obstacle, we define the path $Q(p)$ from $p$ to $W_0$ as in Section 3, with one small exception: When $Q(p)$ follows a leftwards-moving horizontal ray and hits an obstacle $W_i \neq W_0$, $Q(p)$ goes to the left vertex of the lowest edge of $W_i$ along a downwards y-monotone path on the right boundary of $W_i$. $Q(p)$ so defined is clearly a unique y-monotone path, although it need not be $x$-monotone simultaneously.

The following observations are analogous to those of Lemmas 1 and 2 and Theorem 1. The differences in these observations and their proof arguments stem from the structural differences between the convex hulls of isothetic rectangles and the y-monotone hulls of y-monotone polygons in our matching problems.

**Lemma 5** *For an obstacle $W_i$ in $W - \{W_0\}$, let $p$ and $q$ be two points such that $p$ is on the left boundary of $W_i$ and $q$ is on the right boundary of $W_i$. Then no point of $Q(p)$ is strictly below any point of $Q(q)$.*

**Proof.** A crucial fact to the proof is that both $Q(p)$ and $Q(q)$ are planar $y$-monotone paths. The proof argument is similar to that of Lemma 1. □

**Lemma 6** *Let $p$ and $q$ be two points in the plane such that $p$ is on the left boundary of an obstacle $W_i$ and $q$ is on the right boundary of $W_i$, with $i > 0$. Let $u$ (resp., $v$) be the left vertex of the lowest edge of an obstacle $W_a$ (resp., $W_b$), such that $u$ (resp., $v$) is on $Q(p)$ (resp., $Q(q)$) but not on $Q(q)$ (resp., $Q(p)$). Then the preorder number of $W_a$ in the tree $T$ of obstacles is smaller than that of $W_b$, i.e., $a < b$.*

**Proof.** This follows from Lemma 5 and from the definition of the tree $T$. □

**Theorem 2** *Given a preprocessed set $W$ of $2r$ disjoint $y$-monotone polygonal obstacles with $n$ vertices in total, the subsets $\{W_1, W_2, \ldots, W_k\}$ and $\{W_{k+1}, W_{k+2}, \ldots, W_{2r}\}$, for any integer $k$ with $1 \leq k < 2r$, form a partition of the set $W$ that has the desired property, that is, there exists an obstacle-avoiding $y$-monotone path of size $O(n)$ that separates these two subsets. Furthermore, such a $y$-monotone path can be computed in $O(n)$ time.*

**Proof.** Let $W(a, b)$ denote the subset $\{W_a, W_{a+1}, \ldots, W_b\}$ of $W$. For the existence of such a $y$-monotone path, we first show that for any $i < j$, the following holds: (1) $CH_y(W(1, i))$ does not intersect $W_j$, and (2) $CH_y(W(j, 2r))$ does not intersect $W_i$. We give the proof only for (1), that for (2) being similar.

We prove (1) by contradiction: Suppose to the contrary that for some $j > i$, $W_j$ intersects $CH_y(W(1, i))$. Then for a point $w \in CH_y(W(1, i)) \cap W_j$, there must be a point $z$ of a $W_s$, $s \leq i < j$, such that $y(w) = y(z)$ and $x(w) < x(z)$, (i.e., $z$ is strictly to the right of $w$). (If such a point $z$ did not exist, then $w$ would have not belonged to $CH_y(W(1, i))$ by the definition of $y$-monotone hulls, a contradiction.) WLOG, let $z \in W_s$ be the leftmost such point. Then $z$ must be on the left boundary of $W_s$ and the leftwards-moving horizontal ray from the left vertex of the lowest edge of $W_s$ cannot first hit $W_j$ (otherwise, we would have a contradiction). Let $z'$ be a point on the right boundary of $W_s$ such that $y(z) > y(z')$. Then by Lemma 6, the preorder number of $W_j$ in $T$ is smaller than that of $W_s$, a contradiction.

We can compute a desired $y$-monotone path by letting the path first go along the right boundary of $CH_y(W(1, k))$ as much as possible, then along the left boundary of $CH_y(W(k + 1, 2r))$ (if necessary), and finally extend vertically upwards and downwards to infinity. The $y$-monotone path so obtained clearly has a size of $O(n)$. Given the planar subdivision based on the horizontal trapezoidal decomposition of the obstacle set $W$ (this subdivision is part of the preprocessing result), it is possible to obtain such a $y$-monotone path in $O(n)$ time. This is done by examining the $O(n)$ cells of the planar subdivision to identify those cells that separate the two subsets $W(1, k)$ and $W(k + 1, 2r)$, i.e., the cells whose left (resp., right) boundaries are on the right (resp., left) boundaries of the polygons in $W(1, k)$ (resp., $W(k + 1, 2r)$). □

Note that in a fashion similar to Theorem 2, we can also partition the preprocessed set $W$ into two subsets based on the total sizes of the polygons in the resulting subsets. That is, for an integer $j$ with $1 \le j < n$, we can partition the preprocessed obstacle set $W$ into two subsets $W(1, k)$ and $W(k+1, 2r)$ with a $y$-monotone path, such that the total number of polygon vertices of $W(1, k)$ is no bigger than $j$ but the total number of polygon vertices of $W(1, k + 1)$ is strictly larger than $j$. This partitioning can also be done in $O(n)$ time.

Theorem 2 enables us to obtain efficient algorithms for computing a desired matching for $y$-monotone polygons, as did Theorem 1 for isothetic rectangles. In fact, the matching algorithms for $y$-monotone polygons are similar to and actually simpler than the ones for isothetic rectangles.

Like the matching algorithms for isothetic rectangles, the algorithms here also maintain the list $L_+$. However, unlike the algorithms for isothetic rectangles, $L_+$ here is always a consecutive sublist of the original list $W(1, 2r)$ and is maintained only as a doubly linked list. Further, the algorithms here do not need to use the tree $T'$ and hence the list $L_-$, and do not use the 2-3 tree $S(L_+)$. We only sketch below the computation of these algorithms, since they are very similar to those of Sections 5 and 6.

To specify the matching pairs of the red/blue polygons in a list $L_+ = (W_1', W_2', \ldots, W_m')$ (without computing the actual paths), the algorithm simply does the following:

> If $W_1'$ and $W_m'$ are of different colors, then match $W_1'$ and $W_m'$ (by letting the $W_1'$-to-$W_m'$ path go along first the left boundary of $CH_y(L_+)$ and then the right boundary of $CH_y(L_+)$), and recursively solve the problem on $L_+ - \{W_1', W_m'\}$ if $L_+ - \{W_1', W_m'\}$ is non-empty; otherwise, partition $L_+$ into two consecutive sublists (as in Step 2 of algorithm **MATCH**) and recursively solve the two subproblems.

A matching path so specified consists of at most two $y$-monotone subpaths because it follows first the left boundary and then the right boundary of a $y$-monotone hull. As analyzed in Section 5 for algorithm **MATCH**, the matching algorithm here takes $O(r \log r)$ time after the ordered list $W(1, 2r)$ is made available by the $O(n \log n)$ time preprocessing.

The algorithm for computing the $r$ actual paths of a matching here is similar to the one for isothetic rectangles in Section 6: It maintains the recursion tree $RecTree$ of the above matching algorithm, and computes the $y$-monotone hull $CH_y(v)$ for the subproblem on every node $v$ of $RecTree$. Each of the left and right boundaries of $CH_y(v)$ can be maintained by a

2-3 tree. The geometric structures of the $y$-monotone hulls of the input polygons in $RecTree$ can be exploited by our computation in the following way: When we need to "merge" two $y$-monotone hulls $CH_y(u)$ and $CH_y(w)$ to obtain $CH_y(v)$ (with $u$ and $w$ being the left and right children of $v$, respectively), we replace the corresponding portions of the (say) left boundary of $CH_y(w)$ by the left boundary of each connected component of $CH_y(u)$ (if $CH_y(u)$ indeed consists of more than one connected component). This can be done by using $O(1)$ split and concatenation operations of 2-3 trees for each component of $CH_y(u)$, in logarithmic time. Since we can charge the time for "merging" each such connected component to a horizontal line segment of the horizontal trapezoidal decomposition and since there are $O(n)$ such line segments in the trapezoidal decomposition, the total time for our algorithm to output all the $r$ actual paths between the matched red/blue polygon pairs is $O(n \log n + \lambda)$, where $\lambda$ is the total number of segments that make up these $r$ paths. The space bounds of the matching algorithms in this section are $O(n)$.

# 8  Lower Bounds for the Matching Problems

In this section, we prove $\Omega(n \log n)$ lower bounds in the algebraic computation tree model [8] for the matching problems studied in this paper.

First, we show that the problem of matching $2n$ disjoint red/blue isothetic rectangles with nonintersecting monotone rectilinear paths in the plane requires $\Omega(n \log n)$ time in the worst case. Actually, we will show an $\Omega(n \log n)$ lower bound for the following (simpler) problem **P**: Giving $n$ red and $n$ blue disjoint isothetic rectangles in the plane, find a monotone rectilinear obstacle-avoiding path from a *specified* red rectangle (say, $R_1$) to some (unspecified) blue rectangle $V_i$. The reason for considering problem **P** is that this problem can be easily reduced to our matching problem since any solution to the matching problem definitely contains such a monotone path between the red rectangle $R_1$ and some blue rectangle $V_i$. The key to our proof is a reduction from the problem of sorting $O(n)$ pairwise distinct positive integers (in an arbitrary range) to problem **P**. Note that based on Yao's $\Omega(n \log n)$ lower bound result for the element uniqueness problem on $n$ arbitrary integers [42], Chen, Das, and Smid [10] showed that sorting $O(n)$ pairwise distinct positive integers in the worst case requires $\Omega(n \log n)$ time in the algebraic computation tree model.

The reduction goes as follows. Consider a set $K$ of $n$ pairwise distinct positive integers $I_1$, $I_2$, ..., $I_n$. Let $I_a$ (resp., $I_b$) be the smallest (resp., largest) integer in the set $K$ (it is easy to find $I_a$ and $I_b$ in $O(n)$ time). WLOG, assume that $I_a > 2$. For every integer $I_j \in K$, map $I_j$ to a set $U_j$ of four *red* isothetic rectangles $R_l^j$, $R_r^j$, $R_u^j$, and $R_d^j$ in the plane, as follows (see Figure

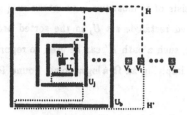

Figure 11: Illustrating the reduction of the lower bound proofs.

11): The shorter edges of all the four red rectangles in $U_j$ have the same length of 0.5 units; the right (resp., left) edge of $R_r^j$ (resp., $R_l^j$) has the point $(I_j, 0)$ (resp., $(-I_j, 0)$) as its middle point and has a length of $2I_j$, while the top (resp., bottom) edge of $R_u^j$ (resp., $R_d^j$) has the point $(0, I_j)$ (resp., $(0, -I_j)$) as its middle point and has a length of $2I_j - 1 - 2\epsilon$, for a very small fixed $\epsilon > 0$. Let $R_1$ be a red isothetic unit box whose center is at the origin of the coordinate system. We then have $4n + 1$ red rectangles. We next create $4n + 1$ isothetic blue rectangles $V_l$'s in the following way: These blue rectangles are all isothetic unit boxes whose centers are all on the $x$-axis; every two consecutive blue boxes are one unit distance apart, and the leftmost blue box is at least one unit distance to the right of $U_b$ (see Figure 11). It is clear that the $O(n)$ red/blue isothetic rectangles so obtained are pairwise disjoint (since the input integers are pairwise distinct), and that the construction of this rectangle set takes $O(n)$ time.

Now, it is an easy matter to observe that (1) an $R_1$-to-$V_i$ path in this setting can be monotone only to the $x$-axis (but not to the $y$-axis), and (2) any such monotone $R_1$-to-$V_i$ path must get around every red rectangle set $U_j$ in the sorted order of the corresponding $I_j$ values of the $U_j$'s (Figure 11). Let $H$ be a monotone rectilinear $R_1$-to-$V_i$ path computed by any algorithm for problem **P**, with $|H| = O(n)$. We assume that when the path $H$ is getting around a particular rectangle set $U_j$, it picks up the index $j$ and associates $j$ with the horizontal edge of $H$ that contains the $x$-coordinate of the rightmost edge of $U_j$. Then given such a path $H$, we can output the sorted sequence of the input integers in $K$ by tracing $H$ and picking up the indices of the integers $I_j$ from their associated horizontal edges of $H$ along the path order of $H$. Such a tracing of $H$ can be easily done in $O(n)$ time. This completes the lower bound proof for problem **P**.

Our lower bound proof for the matching problem on $y$-monotone polygons uses the same reduction construction as for that on isothetic rectangles, except that we now compute a path which consists of at most *two* $y$-monotone subpaths instead of one monotone path. That is, we use any algorithm for computing such an $R_1$-to-$V_i$ path among $y$-monotone polygons to build a geometric sorting device for integer input; the reduction is the same as the one illustrated in Figure 11 and takes $O(n)$ time. This reduction works because any $R_1$-to-$V_i$ obstacle-avoiding rectilinear path $H'$ that consists of at most two $y$-monotone subpaths in the setting of Figure 11 must get around every red rectangle set $U_j$ in the sorted order of the corresponding $I_j$ values of the $U_j$'s. Therefore, such a path $H'$ can be used to report the sorted sequence of the input integers in $O(n)$ time, implying an $\Omega(n \log n)$ lower bound for the matching problem on $y$-monotone polygons.

## 9   Further Remarks

As mentioned earlier, Theorem 1 implies an efficient parallel bound for equipartitioning a set of disjoint isothetic rectangles. This fact is potentially useful in the parallel algorithmics of other, not necessarily red/blue, rectangle problems (as is clear from [4, 5], where tremendous simplifications follow from the next theorem). Therefore, this useful side-effect of Theorem 1 is summarized below.

**Theorem 3** *Let $R$ be a set of $2m$ disjoint isothetic rectangles (not given in any particular order). Then an $m$-processor CREW PRAM can compute, in $O(\log m)$ time, an increasing staircase $S$ that does not intersect the interior of any rectangle in $R$ and partitions $R$ into two equal parts, with $|S| = O(m)$.*

**Proof.** This follows from Theorem 1 and the fact that a trapezoidal decomposition [6] as well as the preorder numbers in a tree [37] can all be computed in parallel within these bounds.   $\Box$

In fact, the preprocessed form of $R$ required by Theorem 1 can be obtained as a by-product of Theorem 3, in $O(\log m)$ time using $m$ CREW PRAM processors. Once this form is available, we can do a little more than Theorem 3: We can partition the set $R = \{R_1, R_2, \ldots, R_{2m}\}$ into two subsets $\{R_1, R_2, \ldots, R_k\}$ and $\{R_{k+1}, R_{k+2}, \ldots, R_{2m}\}$, for any integer $k$ with $1 \leq k < 2m$, in $O(\log t)$ time using $t/\log t$ processors in the CREW PRAM or even the EREW PRAM model [22], where $t = \min\{k, 2m - k\}$. This is done by using, instead of the two-way divide-and-conquer algorithm given in the proof of Theorem 1, a many-way divide-and-conquer approach as in [9, 20]. The details of this parallel algorithm are very similar to (and in fact even simpler than) those of [9, 20], and hence are omitted.

The following partition result may also be useful to designing parallel algorithms for certain geometric problems.

**Theorem 4** *Let $W$ be a set of $2r$ disjoint $y$-monotone polygons (not given in any particular order) with a total of $m$ vertices. Then an $m$-processor CREW PRAM can compute, in $O(\log m)$ time, a $y$-monotone path $P$ that does not intersect the interior of any polygon in $W$ and partitions $W$ into two subsets of $r$ polygons each, with $|P| = O(m)$.*

**Proof.** This follows from Theorem 2 and the fact that a trapezoidal decomposition and the planar subdivision based on it [6] as well as the preorder numbers in a tree [37] can all be computed in parallel within these bounds. □

Again, we can also preprocess $W$ in $O(\log m)$ time using $m$ CREW PRAM processors. After that, such a $y$-monotone path $P$, as defined in Theorem 4, can be obtained in $O(\log m)$ time using $m/\log m$ CREW PRAM processors. This is done by first examining the cells of the planar subdivision (to identify those cells that separate the two subsets of the polygons in $W$) and then using parallel list ranking [22] to find the path $P$. Note that it is also possible to modify Theorem 4 to partition $W$ into two subsets based on the total sizes of the polygons in the resulting subsets.

We conclude with an implementation note about our algorithms. If we are to program the matching algorithms for isothetic rectangles, we would modify them by creating (in Step 2) $S(L')$ and $S(L'')$ only as a last resort, by inserting before Step 2 a Step 1' in which we check whether $R'_1$ and $R'_2$ are of different colors — if so we match them, delete them, etc, and if not we check whether $R'_{m-1}$ and $R'_m$ are of different colors — if so we match them, delete them, etc, and if not we go to Step 2. Thus, we go to Step 2 only if we are unable to match the pair $\{R'_1, R'_2\}$ and the pair $\{R'_{m-1}, R'_m\}$. Performing such a Step 1' before Step 2 gives preference to short paths over long ones, since an $R'_1$-to-$R'_m$ path is likely to be longer than an $R'_1$-to-$R'_2$ (or $R'_{m-1}$-to-$R'_m$) path. For $y$-monotone polygons, an efficient heuristic that may produce short paths for a matching we desire is to use a modification of the so called red/blue matching approach [2, 23] for matching red/blue elements in an ordered list (in our situation, the ordered list is $W(1, 2r)$). Of course, this assumes that short paths are practically better than long ones.

The above discussion suggests the obvious open problems of finding matchings that satisfy some additional length criteria, such as:

- Minimum sum of lengths of all $n$ paths, or

- Minimum maximum length of all $n$ paths, or

- Versions of the above two where "length" means number of links rather than the usual $L_1$ length (hence this version of the sum-of-lengths problem amounts to minimizing what we earlier called $\lambda$).

# References

[1] A. V. Aho, J. E. Hopcroft, and J. D. Ullman, *The Design and Analysis of Computer Algorithms*, Addison-Wesley, Reading, Mass., 1974.

[2] M. G. Andrews, M. J. Atallah, D. Z. Chen, and D. T. Lee, "Parallel algorithms for maximum matching in interval graphs," *Proc. 9th IEEE International Parallel Processing Symp.*, 1995, pp. 84–92.

[3] M. J. Atallah, "A matching problem in the plane," *J. of Computer and Systems Sciences*, 31 (1985), pp. 63–70.

[4] M. J. Atallah and D. Z. Chen, "Parallel rectilinear shortest paths with rectangular obstacles," *Computational Geometry: Theory and Applications*, 1 (1991), pp. 79–113.

[5] M. J. Atallah and D. Z. Chen, "On parallel rectilinear obstacle-avoiding paths," *Computational Geometry: Theory and Applications*, 3 (1993), pp. 307–313.

[6] M. J. Atallah, R. Cole, and M. T. Goodrich, "Cascading divide-and-conquer: A technique for designing parallel algorithms," *SIAM J. Computing*, 18 (1989), pp. 499–532.

[7] M. J. Atallah, S. E. Hambrusch, and L. E. TeWinkel, "Parallel topological sorting of features in a binary image," *Algorithmica*, 6 (1991), pp. 762–769.

[8] M. Ben-Or, "Lower bounds for algebraic computation trees," *Proc. 15th Annual ACM Symp. on Theory of Computing*, 1983, pp. 80–86.

[9] D. Z. Chen, "Efficient geometric algorithms on the EREW PRAM," *IEEE Trans. on Parallel and Distributed Systems*, 6 (1) (1995), pp. 41–47.

[10] D. Z. Chen, G. Das, and M. Smid, "Lower bounds for computing geometric spanners and approximate shortest paths," *Proc. 8th Canadian Conf. on Computational Geometry*, 1996, pp. 155–160.

[11] D. Z. Chen and K. S. Klenk, "Rectilinear short path queries among rectangular obstacles," *Information Processing Letters*, 57 (6) (1996), pp. 313–319.

[12] D. Z. Chen, K. S. Klenk, and H.-Y. T. Tu, "Shortest path queries among weighted obstacles in the rectilinear plane," *Proc. 11th Annual ACM Symp. Computational Geometry*, 1995, pp. 370–379.

[13] J. Choi and C.-K. Yap, "Rectilinear geodesics in 3-space," *Proc. 11th Annual ACM Symp. Computational Geometry*, 1995, pp. 380–389.

[14] K. L. Clarkson, S. Kapoor, and P. M. Vaidya, "Rectilinear shortest paths through polygonal obstacles in $O(n(\log n)^2)$ time," *Proc. 3rd Annual ACM Symp. Computational Geometry*, 1987, pp. 251–257.

[15] K. L. Clarkson, S. Kapoor, and P. M. Vaidya, "Rectilinear shortest paths through polygonal obstacles in $O(n \log^{3/2} n)$ time," manuscript.

[16] W. Dai, T. Asano, and E. S. Kuh, "Routing region definition and ordering scheme for building-block layout," *IEEE Trans. on Computer-Aided Design*, CAD-4 (3) (1985), pp. 189–197.

[17] M. de Berg, M. van Kreveld, and B. J. Nilsson, "Shortest path queries in rectangular worlds of higher dimension," *Proc. 7th Annual Symp. Computational Geometry*, 1991, pp. 51–59.

[18] P. J. de Rezende, D. T. Lee, and Y. F. Wu, "Rectilinear shortest paths in the presence of rectangles barriers," *Discrete & Computational Geometry*, 4 (1989), pp. 41–53.

[19] H. ElGindy and P. Mitra, "Orthogonal shortest route queries among axes parallel rectangular obstacles," *International J. of Computational Geometry and Applications*, 4 (1) (1994), pp. 3–24.

[20] M. T. Goodrich, "Finding the convex hull of a sorted point set in parallel," *Information Processing Letters*, 26 (1987/88), pp. 173–179.

[21] M. Iwai, H. Suzuki, and T. Nishizeki, "Shortest path algorithm in the plane with rectilinear polygonal obstacles" (in Japanese), *Proc. of SIGAL Workshop*, July 1994.

[22] J. JáJá, *An Introduction to Parallel Algorithms*, Addison-Wesley, Reading, MA, 1992.

[23] S. K. Kim, "Optimal parallel algorithms on sorted intervals," *Proc. 27th Annual Allerton Conf. Communication, Control, and Computing*, 1989, pp. 766–775.

[24] R. C. Larson and V. O. Li, "Finding minimum rectilinear distance paths in the presence of barriers," *Networks*, 11 (1981), pp. 285–304.

[25] D. T. Lee, T. H. Chen, and C. D. Yang, "Shortest rectilinear paths among weighted obstacles," *International J. of Computational Geometry and Applications*, 1 (2) (1991), pp. 109–124.

[26] D. T. Lee, C. F. Shen, C. D. Yang, and C. K. Wong, "Non-crossing paths problems," manuscript, Dept. of EECS, Northwestern University, 1991.

[27] J. S. B. Mitchell, "An optimal algorithm for shortest rectilinear path among obstacles," *First Canadian Conf. on Computational Geometry*, 1989.

[28] J. S. B. Mitchell, "$L_1$ shortest paths among polygonal obstacles in the plane," *Algorithmica*, 8 (1992), pp. 55–88.

[29] P. Mitra and B. Bhattacharya, "Efficient approximation shortest-path queries among isothetic rectangular obstacles," *Proc. 3rd Workshop on Algorithms and Data Structures*, 1993, pp. 518–529.

[30] T. M. Nicholl, D. T. Lee, Y. Z. Liao, and C. K. Wong, "On the X-Y convex hull of a set of X-Y polygons," *BIT*, 23 (4) (1983), pp. 456–471.

[31] M. H. Overmars and J. van Leeuwen, "Maintenance of configurations in the plane," *J. of Computer and Systems Sciences*, 23 (1981), pp. 166–204.

[32] E. Papadopoulou, "$k$-Pairs non-crossing shortest paths in a simple polygon," to appear in the *7th Annual International Symp. on Algorithms and Computation*, 1996, Osaka, Japan.

[33] F. P. Preparata and M. I. Shamos, *Computational Geometry: An Introduction*, Springer-Verlag, New York, 1985.

[34] F. P. Preparata and K. J. Supowit, "Testing a simple polygon for monotonicity," *Information Processing Letters*, 12 (1981), pp. 161–164.

[35] J. Takahashi, H. Suzuki, and T. Nishizeki, "Algorithms for finding non-crossing paths with minimum total length in plane graphs," *Proc. 3rd Annual International Symp. on Algorithms and Computation*, 1992, pp. 400–409.

[36] J. Takahashi, H. Suzuki, and T. Nishizeki, "Finding shortest non-crossing rectilinear paths in plane regions," *Proc. 4th Annual International Symp. on Algorithms and Computation*, 1993, pp. 98–107.

[37] R. E. Tarjan and U. Vishkin, "Finding biconnected components and computing tree functions in logarithmic parallel time," *SIAM J. Computing*, 14 (1985), pp. 862–874.

[38] P. Widmayer, Y. F. Wu, and C. K. Wong, "On some distance problems in fixed orientations," *SIAM J. Computing*, 16 (4) (1987), pp. 728–746.

[39] Y. F. Wu, P. Widmayer, M. D. F. Schlag, and C. K. Wong, "Rectilinear shortest paths and minimum spanning trees in the presence of rectilinear obstacles," *IEEE Trans. on Computers*, C-36 (1987), pp. 321–331.

[40] C. D. Yang, T. H. Chen, and D. T. Lee, "Shortest rectilinear paths among weighted rectangles," *Journal of Information Processing*, 13 (4) (1990), pp. 456–462.

[41] C. D. Yang, D. T. Lee, and C. K. Wong, "Rectilinear path problems among rectilinear obstacles revisited," *SIAM J. Computing*, 24 (3) (1995), pp. 457–472.

[42] A. C.-C. Yao, "Lower bounds for algebraic computation trees with integer inputs," *SIAM J. Computing*, 20 (1991), pp. 655–668.

# Multicast Communication in High Speed Networks

Jonathan S. Turner

Department of Computer Science

Washington University

Campus Box 1045

One Brookings Drive

St. Louis, MO 63130-4899, U.S.A.

Email: jst@cs.wustl.edu

# Incremental Convex Hull Algorithms Are Not Output Sensitive[*]

David Bremner

McGill University[**]

**Abstract.** A *polytope* is the bounded intersection of a finite set of half-spaces of $\mathbb{R}^d$. Every polytope can also be represented as the convex hull conv $\mathcal{V}$ of its vertices (extreme points) $\mathcal{V}$. The *convex hull* problem is to convert from the vertex representation to the halfspace representation or (equivalently by geometric duality) vice-versa. Given an ordering $v_1 \dots v_n$ of the input vertices, after some initialization an incremental convex hull algorithm constructs halfspace descriptions $\mathcal{H}_{n-k} \dots \mathcal{H}_n$ where $\mathcal{H}_i$ is the halfspace description of conv$\{ v_1 \dots v_i \}$. Let $m_i$ denote $|\mathcal{H}_i|$, and let $m$ denote $m_n$. In this paper we give families of polytopes for which $m_{n-1} \in \Omega(m^{\sqrt{d/2}})$ for *any* ordering of the input. We also give a family of 0/1-polytopes with a similar blowup in intermediate size. Since $m_{n-1}$ is not bounded by any polynomial in $m$, $n$, and $d$, incremental convex hull algorithms cannot in any reasonable sense be considered output sensitive. It turns out the same families of polytopes are also hard for the other main types of convex hull algorithms known.

## 1 Introduction

A *polytope* is the bounded intersection of a finite set of halfspaces of $\mathbb{R}^d$. A fundamental theorem of convexity is that every polytope can be represented as the convex hull of its extreme points or vertices. Converting from the vertex representation to the halfspace representation is called *facet enumeration*. Converting from the halfspace representation to the vertex representation is called *vertex enumeration*. Since these problems are equivalent under point-hyperplane duality, where the distinction is unimportant we use the term *convex hull problem* to refer to either or both problems.

The term *output sensitive* is used to describe algorithms with performance guarantees in terms of the output size as well as the input size. For problems such as the convex hull problem, where the output size can range from exponential to logarithmic in the input size, such a bound is highly desirable. Implicit in describing an algorithm as output sensitive is that the dependence on the output size is "reasonable", usually bounded by a small polynomial. Here we accept any polynomial bound. For a $d$-dimensional polytope (or $d$-polytope) $P$ with vertices

[*] This research supported by FCAR Québec and NSERC Canada

[**] School of Computer Science, #318-3480 University St., Montréal Canada H3A 2A7, bremner@cs.mcgill.ca

$\mathcal{V}$ and facet defining halfspaces $\mathcal{H}$, we define the summed input and output size as $\text{size}\,P = (|\mathcal{V}| + |\mathcal{H}|)d$. It is an open problem whether there is a convex hull algorithm polynomial in $\text{size}\,P$. In the rest of this paper we call such an algorithm polynomial, where the dependence on $\text{size}\,P$ is implicit.

The only known class of algorithms not previously known to be superpolynomial [12,3], or to have an NP-complete subproblem [15] are incremental algorithms based on the *double description method* of Motzkin et al. [21]. In this paper we show that any incremental algorithm is superpolynomial in the worst case.

Geometric algorithms in high dimensions are sometimes analyzed under the assumption that the number of input objects grows much faster than the dimension $d$. In this context, $d$ is considered a constant for the purposes of analysis. The algorithm of Chazelle [9] has an upper bound of $O(n^{\lfloor d/2 \rfloor})$ in each fixed dimension $d > 3$ and is thus worst case optimal, up to factors that depend only on $d$. With $d$ fixed, a satisfactory output-sensitive time bound may be $O(\text{size}(P)^c)$ for some $c$ independent of the dimension. We show here that even this weaker condition cannot be met by incremental algorithms.

Convex hull computations are one area where assumptions of general position or non-degeneracy seem to make a great deal of difference to the tractability of the problem. While vertex enumeration algorithms based on pivoting that are polynomial on simple input have been known for over forty years (see [8], and refinements in e.g. [24,23,4,7]), finding an algorithm polynomial for non-simple (so-called degenerate) input has proved a much more difficult undertaking. The (unmodified) pivoting method searches all feasible bases of the polytope. In general the number of feasible bases can be superpolynomial in $\text{size}\,P$. Neither of the two well known modifications — perturbation and recursion on the dimension — yields a polynomial algorithm (see [3]).

Incremental algorithms are not necessarily affected by degeneracy, but have a fundamental weakness of their own. In an incremental algorithm, after some initialization, we insert points one by one, maintaining the convex hull of the points inserted so far at every step. A necessary condition for such algorithms to be polynomial is that the size of each intermediate polytope be polynomial in $\text{size}\,P$. It turns out the order the points are inserted can make a huge difference in the size of the intermediate polytopes. This is analogous to the simplex method of linear programming where a family such as the Klee-Minty [19] cubes can be superpolynomial for one pivoting rule but easily solvable using a different pivoting rule. Dyer [12] gave a family of polytopes for which inserting the points in the order given yields superpolynomial intermediate polytopes. Avis, Bremner, and Seidel [3] showed that several more sophisticated insertion orders used in practice also produce superpolynomial intermediate polytopes in the worst case. In this paper, we show that are families for which there is no polynomial insertion order. This may be contrasted with the situation of the simplex method, where the existence of a polynomial pivoting rule remains an open problem.

## 2   Preliminaries

In this section we introduce some notation and fundamental results from the theory of convex polytopes that will be useful in the sequel. We also make precise our measures of complexity.

Given a set of points $X = \{ x_1 \ldots x_n \}$, a combination $\sum_{i=1}^{n} \lambda_i x_i$ is called *affine* if $\sum \lambda_i = 1$. An affine combination is called *convex* if $\lambda_i \geq 0$. The *affine hull* aff $X$ is the set of all affine combinations of $X$, or equivalently the smallest affine subspace containing $X$. The *dimension* dim $X$ is the dimension of aff $X$. The *convex hull* conv $X$ is the set of all convex combinations of $X$. Point $p$ is *extreme* for $X$ if $p$ is not a convex combination of $X \setminus \{ v \}$.

A hyperplane $h$ *supports* a polytope $P$ if $h \cap P \neq \emptyset$ and $P$ is contained in one of the closed halfspaces (the *supporting halfspace*) induced by $h$. A *face* of a convex polytope $P$ is the intersection of one or more supporting hyperplanes of $P$, or $P$ itself. For a face $F$ with supporting hyperplane $h$, we write $F^+$ or $h^+$ for the corresponding supporting halfspace and $F^-$ or $h^-$ for the other (open) halfspace induced by $h$. Faces of dimension $k$ are called $k$-faces; $f_k(P)$ denotes the number of $k$-faces of $P$. The names *vertices*, *edges*, and *facets* refer to 0, 1, and $(d-1)$-faces respectively. The *face lattice* of a polytope is the poset of its faces partially ordered by inclusion. Polytopes $P$ and $Q$ are *combinatorially equivalent* (respectively *dual*) if their face lattices are isomorphic (respectively anti-isomorphic). For any point set $X$, the *polar* $X^*$ of $X$ is defined as $\{ y \mid yx \leq 1 \ \forall x \in X \}$. It is known (see e.g. [5]) that if $P$ is a polytope containing the origin in its interior, $P^*$ is a polytope dual to $P$, containing the origin in its interior.

A *family* of polytopes is simply an infinite set of polytopes. Usually, but not necessarily, families arise in some natural way from a problem such as the traveling salesman problem [13], or a construction such as those described below. Given a family of polytopes $\mathcal{F}$, a function $g : \mathcal{F} \to \mathbb{R}$ is called *polynomial* for $\mathcal{F}$ if there exists some univariate polynomial $p(x)$ such that for every $P \in \mathcal{F}$, $g(P) \leq p(\text{size } P)$. A function $g : \mathcal{F} \to \mathbb{R}$ is called *weakly polynomial* for $\mathcal{F}$ if there exists positive function $f(x)$ and polynomial $p(x)$ such that $\forall P \in \mathcal{F}$, $g(P) \leq f(\dim P) \cdot p(\text{size } P)$; this corresponds to the notion of considering $d$ to be a constant. If $g$ is not (weakly) polynomial for $\mathcal{F}$ we say that $g$ is *(strongly) superpolynomial* for $\mathcal{F}$.

The convex hull of $d$ points on the moment curve $c(t) = (t, t^2, \ldots t^d)$ is called the *cyclic polytope* on $n$ vertices. Here we assume that the $d$ vertices are $\{ c(0), c(1), \ldots c(n-1) \}$. It follows from "Gale's evenness condition" [17,18,26] that:

$$\gamma(n,d) = \begin{cases} \dfrac{n}{n - d/2} \dbinom{n - d/2}{n - d} & d \text{ even}; \\[2mm] 2 \dbinom{n - (d+1)/2}{n - d} & d \text{ odd} \end{cases} \tag{1}$$

The famous Upper Bound Theorem of McMullen [20] says that no $d$-polytope with $n$ vertices has more than $\gamma(n,d)$ facets.

A very useful construction for building new "more complicated" families of polytopes from known families is the *Cartesian product of polytopes* construction.

Let $P$ be a subset of $\mathbb{R}^k$ and let $Q$ be a subset of $\mathbb{R}^l$. Let $P \times Q$, called the *product* of $P$ and $Q$, denote $\{(p, q) \mid p \in P, q \in Q\}$. We regard $\mathbb{R}^l \times \mathbb{R}^k$ as naturally embedded in $\mathbb{R}^{k+l}$. The following lemma (whose proof we omit) summarizes the basic properties of the construction.

**Lemma 1.** *Let $P$ be a $k$-polytope and $Q$ an $l$-polytope.*

*(a) $P \times Q$ is a $(k + l)$-polytope.*

*(b) For $i \geq 0$, the $i$-faces of $P \times Q$ are precisely $F_p \times F_q$ where $F_p$ is a $j$-face of $P$, $j \geq 0$ and $F_q$ is an $(i - j)$ face of $Q$.*

For polytopes $P \subset \mathbb{R}^k$ and $Q \subset \mathbb{R}^l$, we define the *orthogonal sum* $P \oplus Q$ as

$$P \oplus Q \equiv \operatorname{conv}(\{(p, 0^l) \mid p \in P\} \cup \{(0^k, q) \mid q \in Q\}).$$

If $P$ and $Q$ are full dimensional and contain the origin as an interior point, $P \oplus Q = (P^* \times Q^*)^*$.

## 3  Polytopes without good insertion orders

In the most general sense, insertion orders are procedures to determine at each step of an incremental algorithm, what input element should be processed next. In some cases, such as lexicographic or random ordering, all of the choices can be made before the input is processed. In other cases, such as the maxcutoff rule (where we choose the next element which causes the largest drop in intermediate size), the insertion order is inherently dynamic. In either case, for every input and for every insertion order there are one or more possible permutations of the input generated. We will say that $\pi$ is a *good* insertion order for $\mathcal{F}$ if the size of intermediate polytopes created by $\pi$ is polynomial for $\mathcal{F}$ (obviously a much stronger bound is necessary for an insertion order to be "good" in practice). A good insertion order does not by itself guarantee a polynomial algorithm: in particular the use of triangulation can still cause an incremental algorithm to be superpolynomial (see [3]). On the other hand, a naive implementation of the double description method will be polynomial given a good insertion order. We show that there are polytopes for which every permutation is bad; in fact we show the slightly stronger result that every permutation is bad at the last step.

At each step of an incremental facet enumeration algorithm, we maintain (at minimum) the vertex description $\mathcal{V}_i$ and the halfspace description $\mathcal{H}_i$ of the current intermediate polytope (the "double description" of Motzkin et al. [21]). We are interested here in the drop in the size of the halfspace description caused by inserting the last vertex. This is the same as the increase in the number of facets caused by removing one vertex of a polytope and recomputing the convex hull of the remaining vertices. In the following definitions, let $P$ be a $d$-polytope and let $v$ be a vertex of $P$. Let $\mathcal{V}(P)$ denote the vertices of $P$. Let $P - v$ denote $\operatorname{conv}(\mathcal{V}(P) \setminus \{v\})$. Let $\mathcal{F}(P)$ denote the facets of $P$, $\mathcal{F}_v(P)$ the facets of $P$ containing $v$, and let $\widetilde{\mathcal{F}}_v(P)$ be defined as follows:

$$\widetilde{\mathcal{F}}_v(P) \equiv \begin{cases} \{F \in \mathcal{F}(P - v) \mid v \in F^-\} & \text{if } \dim(P - v) = d; \\ \{P - v\} & \text{otherwise.} \end{cases}$$

We define $\mathrm{loss}(v, P) \equiv |\tilde{\mathcal{F}}_v(P)|$ as the number of halfspaces deleted by inserting $v$ last[1]. Similarly, we define $\mathrm{gain}(v, P) \equiv |\mathcal{F}_v(P)|$ as the number of halfspaces created by inserting $v$ last. The net drop in intermediate size is then $\mathrm{drop}(v, P) = \mathrm{loss}(v, P) - \mathrm{gain}(v, P)$. Finally we define $\mathrm{gain}(P) \equiv \max_v \mathrm{gain}(v, P)$, $\mathrm{loss}(P) \equiv \min_v \mathrm{loss}(v, P)$, and $\mathrm{drop}(P) \equiv \min_v \mathrm{drop}(v, P)$. $\mathrm{drop}(P)$ could be negative if there is a vertex whose removal decreases the number of facets (as in for example a stacked polytope).

The lower bounds in this paper follow from using the product and sum of polytopes constructions defined above. The central geometric observation is that while the number of facets of the final polytope sums under the product of polytopes operation, the number of intermediate facets multiplies. A polytope $P$ is called *robust* if $\dim(P - v) = \dim P$ for every vertex $v$ of $P$.

**Theorem 1.** *Let $P$ and $Q$ be polytopes with dimension at least 2. $P \times Q$ is robust and $\mathrm{drop}(P \times Q) = \mathrm{loss}(P) \cdot \mathrm{loss}(Q)$.*

Before proving Theorem 1, we state some consequences for particular families of polytopes. To construct families of polytopes hard for incremental convex hull algorithms, it suffices to take products of families with large loss functions. We now present three such families. For continuity we defer the proofs of lemmas in this section until Section 3.2. Recall that $C_d(n)$ denotes the cyclic $d$-polytope with $n$ vertices.

**Lemma 2.** *For even $d$, $n \geq d + 2$, for any vertex $v$ of $C_d(n)$,*

$$\mathrm{loss}(v, C_d(n)) = \binom{n - d/2 - 2}{n - d - 1} = \left[ \frac{d(n - d)}{(2n - d - 2)n} \right] \gamma(n, d)$$

$$\in \Theta(n^{d/2 - 1}) \qquad d \text{ fixed.}$$

A polytope $P$ is called *centered* if for every $v \in \mathcal{V}(P)$, $P - v$ contains the origin as an interior point. The following lemma gives a general method for constructing families of polytopes with large loss functions.

**Lemma 3.** *Let $P$ be a centered polytope. Let $Q$ be a polytope with $m$ facets containing the origin in its interior. Let $v = (p, \bar{0})$ be a vertex of $P \oplus Q$.*

$$\mathrm{loss}(v, P \oplus Q) = \mathrm{loss}(p, P) \cdot m.$$

For a polytope $P$, let $\bigoplus[P; k]$ denote the $k$-fold sum of $P$ with itself, i.e. $\bigoplus[P; 1] \equiv P$, $\bigoplus[P; k] \equiv P \oplus \bigoplus[P; k - 1]$. Let $P_n$ be a centered convex polygon with $n$ vertices. Let $\Pi_d(n)$ denote $\bigoplus[P_n; d/2]$. By careful choice of $P_n$, both descriptions of $\Pi(n)$ can be given using small integers independent of the dimension.

---

[1] We make the notation simplifying assumption that in the case where $\dim(P - v) = \dim(P) - 1$, the affine hull of $P - v$ is stored as two halfspaces.

**Corollary 4.** *For even* $d$, $n \geq 5$, loss $\Pi_d(n) = n^{d/2-1}$.

Following [3], for any even dimension $2d \geq 4$ define $a = \lceil \sqrt{d} \rceil$, $b = \lfloor d/a \rfloor$, $c = d \bmod a$, and

$$K_{2d}(n) = Q_{2a}^b(n) \times Q_{2c}(n)$$

Where $Q_k(n)$ is either a cyclic polytope or a dual product of polygons. The following theorem follows from Theorem 1, Lemma 2 and Corollary 4.

**Theorem 2.** *For* $d \geq 2$ *held fixed,*

 (a) $s \equiv \text{size } K_{2d}(n) \in O(n^{\lceil \sqrt{d} \rceil})$, *and*

 (b) $\text{drop } K_{2d}(n) \in \Omega(s^{d/\lceil \sqrt{d} \rceil - 1})$.

From this theorem we can see that incremental convex hull algorithms are strongly superpolynomial in the worst case, irrespective of the insertion order used. Since these same families have previously been shown to be hard for perturbation/triangulation and for face lattice producing algorithms, it follows that no method consisting on running several of the well known methods in parallel will be output sensitive either.

An important class of polytopes for facet enumeration is the 0/1-*polytopes*, whose vertices are a subset of $\{0,1\}^d$. Because of their importance in combinatorial optimization, a facet enumeration algorithm polynomial for 0/1-polytopes would be significant result; unfortunately incremental algorithms fail here also, at least from a theoretical point of view. We consider the equivalent case of polytopes whose vertices are a subset of $\{+1,-1\}^d$. Let $H_d$ denote the hypercube with vertices $\{+1,-1\}^d$. The reader can verify that $H_d$ is centered for $d \geq 3$. Define $U_{3d}$ as $(\bigoplus[H_3; \sqrt{d}])^{\sqrt{d}}$ Since $\text{loss}(H_d) = 1$, by Theorem 1 and Lemma 3 we have the following:

**Theorem 3.**

 (a) $s \equiv \text{size}(U_{3d}) \in O(d^{\sqrt{d}}) = O(2^{\sqrt{d}\log d})$

 (b) $\text{drop } U_{3d} \in \Omega(2^d) = \Omega(s^{\sqrt{d}/\log d})$

Since duality preserves the size of the face lattice, $U_{3d}$ has face lattice size $3^{3d}$, hence is also difficult for face lattice generating convex hull algorithms. In [3], the authors observe that if $P$ has $\bar{f}$ non-empty faces, then any triangulation of the boundary of $P$ contains at least $(\bar{f}-1)/2^{\dim P}$ maximal simplices. It follows that the family $U_{3d}$ is also difficult for algorithms based on triangulation and perturbation. By a construction similar to the one in Theorem 2 we can obtain families with members in every sufficiently large dimension with about the same behaviour.

### 3.1 Proof of Theorem 1

We prove Theorem 1 via several lemmas. Since $\text{gain}(v, P)$ is non-negative, the following holds:

$$\text{loss}(P) - \text{gain}(P) \leq \text{drop}(P) \leq \text{loss}(P).$$

Let $P$ and $Q$ be robust polytopes with dimension at least 2. Theorem 1 follows from the following two facts.

$$\text{gain}(P \times Q) = 0 \tag{2}$$
$$\text{loss}(P \times Q) = \text{loss}(P) \cdot \text{loss}(Q), \tag{3}$$

We start with (2), which is equivalent to saying that every facet of $(P \times Q)$ is robust. Since each facet of $P \times Q$ is the product of a facet of $P$ (respectively $P$) and $Q$ (respectively a facet of $Q$), this follows from the next lemma.

**Lemma 5.** *Let $P$ and $Q$ be polytopes with $\dim P \geq 1$ and $\dim Q \geq 1$. $P \times Q$ is robust.*

We now turn our attention to (3).

**Lemma 6.** *Let $P$ and $Q$ be polytopes containing the origin as a vertex.*

$$\tilde{\mathcal{F}}_{\bar{0}}(P \times Q) = \{\, F_p \oplus F_q \mid F_p \in \tilde{\mathcal{F}}_{\bar{0}}(P), F_q \in \tilde{\mathcal{F}}_{\bar{0}}(Q) \,\}$$

Since by change of coordinates we can assume without loss of generality that an arbitrary vertex of $P \times Q$ lies on the origin, (3) and hence Theorem 1 follows. From Lemma 6 we also get a complete characterization of the facets of $(P \times Q) - v$ since $\mathcal{F}((P \times Q) - v) = \{\, F \in \mathcal{F}(P \times Q) \mid v \in F^+ \,\} \cup \tilde{\mathcal{F}}_v(P \times Q)$.

### 3.2 Proofs of Lemmas

**Proof of Lemma 2:** Let the dimension $d = 2k$ for some positive integer $k$. Let $v$ be a vertex of $P = C_d(n)$. It is known (see e.g. [2], p. 102) that each vertex of an even dimensional cyclic polytope is contained in $\gamma(n - 1, d - 1)$ facets. Since $P$ is simplicial, $\text{gain}(v, P) = \gamma(n - 1, d - 1)$. Since $P - v$ is full dimensional, $f_{d-1}(P - v) - f_{d-1}(P) = \text{loss}(v, P) - \text{gain}(v, P)$. It follows that

$$\text{loss}(v, P) = f_{d-1}(P - v) - f_{d-1}(P) + \text{gain}(v, P)$$
$$= \gamma(n - 1, d) - \gamma(n, d) + \gamma(n - 1, d - 1).$$

Substituting in the appropriate values of $\gamma(n, d)$ from (1) for odd and even $d$

$$\text{loss}(v, P) = \left[\frac{n - 1}{k} - \frac{(n - k - 1)n}{(n - 2k)k} + \frac{2n - 2k - 2}{n - 2k}\right]\frac{(n - k - 2)!}{(n - 2k - 1)!\,(k - 1)!}$$

$$= \frac{(n - k - 2)!}{(n - 2k - 1)!\,(k - 1)!} \qquad \square$$

**Proof of Lemma 3:** Note that $(P \oplus Q) - (p, \bar{0}) = (P - p) \oplus Q$. Since $P$ is centered, by Lemma 1 and polarity, $ax + by \leq 1$ defines a facet of $(P - p) \oplus Q$ iff $ax \leq 1$ (respectively $by \leq 1$) defines a facet of $P - p$ (respectively $Q$). Vertex $(p, \bar{0})$ is infeasible for $ax + by \leq 1$ iff $p$ is infeasible for $ax \leq 1$. $\qquad\square$

**Proof of Lemma 5:** Let $v = (p, q)$ be a vertex of $P \times Q$. Let $P'$ denote $(P \times Q) - v$:

$$P' = \text{conv}(\{\,(P - p) \times Q\,\} \cup \{\,P \times (Q - q)\,\}).$$

If $\dim(Q - q) = \dim Q$, then the lemma follows by Lemma 1. Otherwise, $q \notin \text{aff}(Q - q)$. Let $p'$ be some vertex of $P$ other than $p$. If $(x, y) \in \text{aff}(X \times Y)$, by manipulation of sums we see that $x \in \text{aff}\, X$ and $y \in \text{aff}\, Y$. It follows that $(p', q) \notin \text{aff}(P \times (Q - q))$. But $(p', q) \in P'$, so

$$\dim P' > \dim [P \times (Q - q)] \geq \dim P + \dim Q - 1 \qquad\qquad\square$$

**Proof of Lemma 6:** Suppose we have facets $F_p \in \tilde{\mathcal{F}}_{\bar{0}}(P)$, $F_q \in \tilde{\mathcal{F}}_{\bar{0}}(Q)$. We can write linear constraints $ax \geq 1$ and $by \geq 1$ that support $F_p$ and $F_q$ respectively. The constraint $ax + by \geq 1$ supports $P$. The vertices of $P \times Q$ that lie on $ax + by = 1$ are precisely $F_p \oplus F_q$. It is known (see e.g. [22]) that if $\text{aff}\, P_1 \cap \text{aff}\, P_2 = \emptyset$ then $\dim(P_1 \oplus P_2) = \dim P_1 + \dim P_2 + 1$.

Now suppose we have some $F \in \tilde{\mathcal{F}}_{\bar{0}}(P \times Q)$. Let $h$ denote $\text{aff}\, F$. Every vertex of $P \times Q$ in $h$ must have at least one adjacent edge $e$ in $h^-$, since otherwise $P \times Q \subset h^+$. The other vertex of $e$ must be $\bar{0}$, since otherwise $h$ is not a supporting hyperplane for $(P \times Q) - \bar{0}$. By Lemma 1 the vertices defining $h$ must be of the form $(p, \bar{0})$ or $(\bar{0}, q))$ for $p \in \mathcal{V}(P)$, $q \in \mathcal{V}(Q)$. It follows that any basis $B$ for $h$ must have the form

$$B = \begin{bmatrix} B_p & 0 \\ 0 & B_q \end{bmatrix}$$

where $B_p$ (respectively $B_q$) is a basis of some $F_p \in \tilde{\mathcal{F}}_{\bar{0}}(P)$ (respectively $F_q \in \tilde{\mathcal{F}}_{\bar{0}}(Q)$). $\qquad\square$

## 4  Conclusions

In a previous paper Avis, Bremner, and Seidel [3] showed that products of cyclic polytopes (and products of sums of polytopes) form a family hard for algorithms based on triangulation, perturbation, and on computing the face lattice. Here we have extended their results to show they are also hard for incremental algorithms, regardless of insertion order. We mention in closing a minor extension of [3] with regard to algorithms that compute the face lattice. Swart [24] suggested computing the "abbreviated face lattice" as follows. Rather than recursively computing the entire face lattice, stop the recursion at the first simplicial face (i.e. a $k$-face with $k+1$ vertices) encountered. While this avoids counterexamples based on "large dimensional" simplicial faces, any product of $d$-polytopes will have no simplicial faces of dimension larger than $d$. Thus the families given here are also hard for convex hull algorithms that compute the abbreviated face lattice.

The question of the practical usefulness of the double description method is hardly settled by the existence of families without good insertion orders. There have been many practical success stories using this technique (see e.g. [6,16]) and its simplicity and "immunity" to degeneracy make it the method of choice for most implementors (see e.g. [1,10,14,25]). Nonetheless, the results of this paper show that rather than searching for a universally good insertion order, a more fruitful approach for families that defeat the usual heuristics (and pivoting) is to try and use knowledge about the combinatorial structure of the family to compute a good insertion order. Remarkable success in this vein has recently been reported by Deza, Deza, and Fukuda [11].

## Acknowledgments

The author would like to thank David Avis, Komei Fukuda, and Raimund Seidel for useful and interesting conversations about convex hulls and for comments on a previous version of this paper. The polytopes of Theorem 3 were suggested to the author by David Avis. Raimund Seidel reminded the author that products of sums of polygons could stand in for products of cyclic polytopes.

## References

[1] D. Alevras, G. Cramer, and M. Padberg. *DODEAL*. ⟨ftp://elib.zib-berlin.de/pub/mathprog/polyth/dda⟩

[2] A. Altshuler and M. Perles. Quotient polytopes of cyclic polytopes. Part I: Structure and characterization. *Israel J. Math.*, 36(2):97–125, 1980.

[3] D. Avis, D. Bremner, and R. Seidel. How good are convex hull algorithms? To appear in Comput. Geom.: Theory and Appl., 1996.

[4] D. Avis and K. Fukuda. A pivoting algorithm for convex hulls and vertex enumeration of arrangements and polyhedra. *Disc. Comput. Geom.*, 8:295–313, 1992.

[5] A. Brøndsted. *Introduction to Convex Polytopes*. Springer-Verlag, 1981.

[6] G. Ceder, G. Garbulsky, D. Avis, and K. Fukuda. Ground states of a ternary lattice model with nearest and next-nearest neighbor interactions. *Physical Review B*, 49:1–7, 1994.

[7] T. M. Chan. Output-sensitive results on convex hulls, extreme points, and related problems. In *Proc. 11th ACM Symp. Comp. Geom.*, pages 10–19, 1995.

[8] A. Charnes. The simplex method: optimal set and degeneracy. In *An introduction to Linear Programming*, Lecture VI, pages 62–70. Wiley, New York, 1953.

[9] B. Chazelle. An optimal convex hull algorithm in any fixed dimension. *Disc. Comput. Geom.*, 10:377–409, 1993.

[10] T. Christof and A. Loebel. porta v1.2.2. April 1995.
⟨http://www.iwr.uni-heidelberg.de/iwr/comopt/soft/PORTA/porta.tar⟩

[11] A. Deza, M. Deza, and K. Fukuda. On Skeletons, Diameters and Volumes of Metric Polyhedra. Technical report, Laboratoire d'Informatique de l'Ecole Supérieure, January 1996. To appear in Lecture Notes in Computer Science, Springer-Verlag.

[12] M. Dyer. The complexity of vertex enumeration methods. *Math. Oper. Res.*, 8(3):381–402, 1983.

[13] R. Euler and H. Le Verge. Complete linear descriptions of small asymetric travelling salesman polytopes. *Disc. App. Math.*, 62:193–208, 1995.

[14] K. Fukuda. cdd+ Reference manual, version 0.73. ETHZ, Zurich, Switzerland. ⟨ftp://ifor13.ethz.ch/pub/fukuda/cdd⟩

[15] K. Fukuda, T. M. Liebling, and F. Margot. Analysis of backtrack algorithms for listing all vertices and all faces of a convex polyhedron. To appear in Comput. Geom.: Theory and Appl., 1996.

[16] K. Fukuda and A. Prodon. The double description method revisited. In *Lecture Notes in Computer Science*, volume 1120. Springer-Verlag, 1996. To appear.

[17] D. Gale. Neighborly and cyclic polytopes. In V. Klee, editor, *Proc. 7th Symp. Pure Math.*, 1961.

[18] B. Grünbaum. *Convex Polytopes*. Interscience, London, 1967.

[19] V. Klee and G. Minty. How good is the simplex method? In O. Shisha, editor, *Inequalities-III*, pages 159–175. Academic Press, 1972.

[20] P. McMullen. The maximal number of faces of a convex polytope. *Mathematika*, 17:179–184, 1970.

[21] T. S. Motzkin, H. Raiffa, G. Thompson, and R. M. Thrall. The double description method. In H. Kuhn and A. Tucker, editors, *Contributions to the Theory of Games II*, volume 8 of *Annals of Math. Studies*, pages 51–73. Princeton University Press, 1953.

[22] J. A. Murtha and E. R. Willard. *Linear Algebra and Geometry*. Holt Rinehart and Winston, New York, 1969.

[23] R. Seidel. *Output-size sensitive algorithms for constructive problems in computational geometry*. Ph.D. thesis, Dept. Comput. Sci., Cornell Univ., Ithaca, NY, 1986. Technical Report TR 86-784.

[24] G. Swart. Finding the convex hull facet by facet. *J. Algorithms*, pages 17–48, 1985.

[25] D. K. Wilde. A library for doing polyhederal applications. Technical Report 785, IRISA, Campus Universitaire de Beaulieu – 35042 Rennes CEDEX France, 1993. ⟨ftp://ftp.irisa.fr/local/API⟩

[26] G. M. Ziegler. *Lectures on Polytopes*. Graduate Texts in Mathematics. Springer-Verlag, 1994.

# Separating and Shattering Long Line Segments*

Alon Efrat[1] and Otfried Schwarzkopf[2]

[1] School of Mathematical Sciences, Tel Aviv University, Tel-Aviv 69982, Israel.
*alone@cs.tau.ac.il*
[2] Dept. of Computer Science, Pohang University of Science and Technology,
Pohang 790-784, South Korea. *otfried@postech.ac.kr*

**Abstract.** A line $l$ is called a *separator* for a set $S$ of objects in the plane if $l$ avoids all the objects and partitions $S$ into two non-empty subsets, lying on both sides of $l$. A set $L$ of lines is said to *shatter* $S$ if each line of $L$ is a separator for $S$, and every two objects of $S$ are separated by at least one line of $L$. We give a simple algorithm to construct the set of all separators for a given set $S$ of $n$ line segments in time $O(n \log n)$, provided the ratio between the diameter of $S$ and the length of the shortest line segment is bounded by a constant. We also give an $O(n \log n)$-time algorithm to determine a set of lines shattering $S$, improving (for this setting) the $O(n^2 \log n)$ time algorithm of Freimer, Mitchell and Piatko.

**Keywords:** Computational Geometry, BSP-trees, line-separation.

## 1 Introduction

Given a set $S$ of $n$ objects in the plane, we say that a line $l$ *avoids* $S$ if $l$ avoids every element in $S$. We call a line $l$ a separator for $S$ if $l$ avoids $S$ and partitions $S$ into two non-empty subsets. In this paper we consider the problem of finding separators for a set of line-segments. Clearly this is sufficient to treat the case of general polygonal objects as well.

For a set $S$ of line segments, a separator can be found using duality. Under duality, the segments transform to double wedges, and a separator line transforms to a point lying in the complement of the union of these double wedges, but not above all or below all of them. In other words, we can construct the set of all possible separators by computing the union of these double wedges in dual space. This can be done by constructing the arrangement of the $2n$ lines that are dual to the endpoints of the segments in $S$, and then determining for every face of the arrangement whether it is inside one of the double wedges. All this can be done in $O(n^2)$ time using for example the topological sweep technique of Edelsbrunner and Guibas [5].

---

* Work on this paper by the second author has been supported by the Netherlands' Organization for Scientific Research (NWO), and partially by Pohang University of Science and Technology Grant P96005, 1996.

On the other hand, it is easy to see that there are sets of $n$ line segments for which the union of double wedges—and hence the set of all possible separators—has complexity $\Omega(n^2)$. So it seems that the algorithm sketched above is already the best one can do.

In fact, even if we do not need to compute the set of all possible separators, but are just interested in finding some arbitrary separator (if one exists at all), it is not known whether the problem can be solved in subquadratic time. In fact, it can be shown that the problem belongs to the class of so-called *three-sum hard problems*. The problems in this class are suspected to admit no solution in subquadratic time [10].

In this paper we investigate the special case in which the segments of $S$ are *long*. By this, we mean that the ratio of the diameter of $S$ and the length of the shortest segment is bounded by a constant. Or, rescaling the problem, we will assume that the segments are contained in the unit disk and their length is at least a constant $\lambda > 0$.

As Efrat, Rote, and Sharir [7] observe, the dual wedge of a line segment whose slope is bounded away from $90°$ is *fat*, meaning that its interior angle is bounded from below by a constant $\delta = \delta(\lambda) > 0$. It can be shown [7, 12], that the overall complexity of the union of $n$ fat double wedges is linear in $n$, with the constant of proportionality depending on $\delta$, and this union can also be computed in close to linear time. Efrat, Rote, and Sharir used this fact to obtain an algorithm to find a separator for a set of long segments: They partition the set of segments into two subsets depending on their slope, use the above observation to conclude that the dual wedges for both subsets have linear complexity, and finally overlay the two unions until they find a point corresponding to a separating line, or determine that no such point exists. The running time of the algorithm is $O(n \log n)$, or, more precisely, $O(n/\lambda^2 \log(1/\lambda) \log(n/\lambda))$.

In this paper we will show that the union of the dual wedges of a set of long segments actually has only linear complexity. As a consequence, we obtain an algorithm that computes *all* separators for the set of segments (instead of just one), and that is considerably simpler than the algorithm by Efrat et al. Furthermore, our algorithm can be extended to find a separator that partitions the segments as equally as possible, or even to find a line that stabs only a small number of segments. This is, for instance, attractive for the construction of binary space partition trees (BSP-trees).

We say that a set of segments $S$ is *shatterable* if there exists a set of lines $L$ avoiding $S$ and such that every pair of segments is separated by at least one line in $L$. If such a set $L$ exists, we say that $L$ shatters $S$. Freimer, Mitchell and Piatko [8] gave an $O(n^2 \log n)$ time algorithm to find a linear-cardinality set $L$ that shatters $S$, or to determine that no such $L$ exists. We show that for a set of long segments this can be done in time $O(n \log n)$. We also show that $\Omega(n)$ lines are always needed to shatter a set of $n$ long segments (in contrast to the case of arbitrary segments).

# 2  The combinatorial bound

Let $S$ be a set of $n$ line segments in the plane. We assume that they are all contained in the unit disk centered at the origin, and that their length is bounded from below by a constant $\lambda > 0$. We also assume (without loss of generality) that no segment in $s$ has slope $30° + i \cdot 60°$, for $i = 0, 1, 2$.

We dualize every segment $s =$ to a double wedge $s^*$ that is the union of all lines $p^*$, where $p$ is a point on $e$. (We use the duality transformation that maps a point $(a, b)$ to a line $y = ax + b$, and a line $y = cx + d$ to the point $(c, d)$.) A non-vertical line $l$ meets $s$ if and only if its dual point $l^*$ lies inside the double wedge $s^*$. It is not difficult to see that a non-vertical line $l$ is a separator for $S$ if and only if $l^*$ lies in the complement of the union of the double wedges for the elements of $S$, but not above or below *all* the double wedges.

We will denote the union of a set $S^*$ of double wedges by $\mathcal{U}(S^*)$, and prove the following little lemma.

**Lemma 1.** *For a set $S$ of line segments, let $S_\theta$ denote $S$ rotated by an angle $\theta$ around the origin. Then the complexity of $\mathcal{U}(S^*)$ is the same as the complexity of $\mathcal{U}(S_\theta^*)$.*

*Proof.* Note that vertex $v$ of $\mathcal{U}(S^*)$ corresponds to a line in the primal plane that avoids $S$ and is tangent to two of the segments of $S$. There is a line partitioning $S_\theta$ in the same manner, and hence there is a one-to-one correspondence between the vertices of $\mathcal{U}(S^*)$ and those of $\mathcal{U}(S_\theta^*)$.

We can now prove the main theorem.

**Theorem 2.** *Let $S$ be a set of $n$ long line segments. Then the complexity of the union of the double wedges $s^*$, for $s \in S$, is $O(n)$.*

*Proof.* We partition the set $S$ of segments into three subsets, depending on the angle that the segments make with the $x$-axis: segments in $S_1$ make an angle between $0°$ and $60°$, segments in $S_2$ make an angle between $60°$ and $120°$, and segments in $S_3$ make an angle between $120°$ and $180°$, see Figure 2.

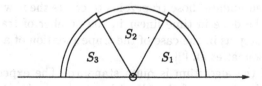

**Fig. 1.** Dividing $S$ into $S_1$, $S_2$ and $S_3$, based on the angle they create with the $X$-axis

A vertex of $\mathcal{U}(S^*)$ is either the center of a double wedge, or is the intersection between the boundaries of two double wedges. There is only a linear number of vertices of the first kind, so it is sufficient to count only the second kind of vertices.

Consider a vertex $v$ of $\mathcal{U}(S^*)$ that is formed by the boundary of the double wedges $s^*$ and $t^*$. Let $i, j \in \{1, 2, 3\}$ be such that $s \in S_i$, $t \in S_j$. Clearly, $v$ is also a vertex of $\mathcal{U}(S_i^* \cup S_j^*)$. It therefore suffices to show that the complexity of $\mathcal{U}(S_i^* \cup S_j^*)$ is $O(n)$ for all choices of $i$ and $j$.

By rotating by $60°$ in either direction, we can reduce all cases to the case of $S_1 \cup S_3$. By Lemma 1, it therefore suffices to prove that the complexity of $\mathcal{U}(S_1^* \cup S_3^*)$ is $O(n)$.

The segments of $S_1 \cup S_3$ have slope between $-60°$ and $+60°$. As observed before, their dual wedges therefore have an interior angle larger than some constant $\delta = \delta(\lambda) > 0$. Using the result on the union of fat wedges [7, 12], it follows that the complexity of $\mathcal{U}(S_1^* \cup S_3^*)$ is $O(n)$, completing the proof.

## 3 Computing all separators and other applications

The set of all separators is the complement of the union of the double wedges $S^*$ (minus the two regions above and below all the double wedges). By Theorem 2, the complexity of this set is linear. We can compute it quite easily using a randomized incremental algorithm, inspired by the algorithm by Miller and Sharir [12, 13] for the computation of the union of fat-triangles.[3]

Our algorithm treats the wedges in random order, maintaining the trapezoidation (vertical decomposition) of the complement of the union of the wedges so far. We also maintain, in what is nowadays a quite standard fashion, a history graph of the trapezoidation as defined by Boissonnat et al. [2]. The nodes of this directed acyclic graph are the trapezoids created during the course of the computation, its leaves are the trapezoids of the current trapezoidation. A child trapezoid in the graph intersects the parent trapezoid, all children of a trapezoid together cover its parent completely. Any node has at most a constant number of children (at most six, to be precise).

In every step, we have to identify the trapezoids intersected by or contained in the new double wedge. This can be done in a standard way by a traversal of the history graph until we reach the leaf nodes (which correspond to the trapezoids in question). We then update those trapezoids to create the new trapezoidation. This updating can be done in time linear in the number of trapezoids deleted and created in this step as in the case of the trapezoidation of a set of segments considered by Boissonnat et al. [2].

The analysis of the algorithm is quite standard: The expected number of trapezoids created in stage $r$ of the algorithm is proportional to the average number of trapezoids incident to a double wedge in a set of $r$ double wedges. It follows immediately from Theorem 2 that this is constant. It follows that the expected number of trapezoids created by the algorithm, and therefore the expected size of the history graph, is $O(n)$. The remaining running time of the

---

[3] A triangle is called $\delta$-*fat* if all its angles are at least $\delta$.

algorithm is dominated by the time necessary for the graph traversals. The standard analysis for randomized incremental construction [2, 4] for first-order moments can be applied to show that the expected time for this step is $O(n \log n)$.

**Theorem 3.** *Given a set $S$ of $n$ long segments, the set of all lines separating $S$ can be computed in expected time $O(n \log n)$.*

It is easy to extend the algorithm to compute a separator that splits $S$ as equally as possible. This is useful for the computation of *binary space partition trees* (BSP-trees) in the plane.

We will not go into details about the construction of BSP-trees. Let's just remark that it is based on repeatedly finding lines that partition a set of segments while intersecting only a few of them. If a set of segments admits no separator, or if all its separators split off only a small number of segments, it is therefore useful to find a line that splits it as equally as possible while intersecting not more than $k$ of the segments, where $k$ is a parameter.

Our argument can be extended to deal with this situation. A line $l$ intersects at most $k$ segments when it lies in the $\leq k$-level of the set of double wedges dual to the segments of $S$. Using Clarkson and Shor's analysis [3, 4], it can be shown that under our conditions on the set $S$, the complexity of the $\leq k$-level is $O(kn)$. The algorithm can be extended to compute it in time $O(nk \log n)$, and we can find the line that splits $S$ as equally as possible while intersecting only $k$ segments in the same time.

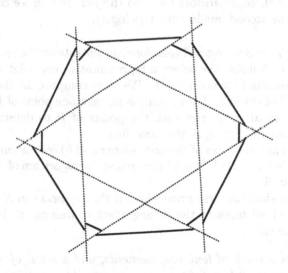

**Fig. 2.** The set of 12 segments $S$ (plotted by solid lines) is shatterable, but no separator has more than 3 segments on one of its sides.

A somewhat surprising fact, observed by Sariel Har-Peled [9], is that a set of segment $S$ can be shatterable, and still have no separator $l$ that divides $S$ in a *balanced* way, that is, at least constant fraction of $S$ lies at each side of $l$. This is demonstrated in Figure 2. We do not know yet whether such an example can be found for long segments.

## 4   The shattering problem

A related problem is the problem of shattering a set $S$ of line segments. We consider two problems:

1. Given a set $S$ of $n$ line segments, find a set $L$ of $O(n)$ lines shattering $S$.
2. Given a set $S$ of $n$ line segments and a set $L$ of $O(n)$ lines, determine whether $L$ shatters $S$.

A general algorithm (that is, not assuming that the line segments are long) for the first problem was given by Freimer, Mitchell and Piatko [8]; its running time is $O(n^2 \log n)$. The first step in this algorithm is to create a set of lines that are suspected to be separators for $S$. These lines are the lines containing the visibility edges of the visibility graph of $S$. Their number is $\Theta(n^2)$ in the worst case.

For the case that the segments are long, we can limit our interest to a much smaller set: As we have seen above, there are only $O(n)$ combinatorially distinct separators for the set of line segments. Clearly, if the set $S$ can be shattered at all, then the set of these separators must do the job. Hence, we can reduce the first problem to the second one in time $O(n \log n)$.

In the rest of this section we will therefore concentrate on the second problem. It can be solved as follows: We select a representative endpoint for every line segment in $S$, resulting in a point set $P$. We then compute all the faces in the arrangement $\mathcal{A}(L)$ of the lines $L$ that contain at least one point of $P$. After that, we perform point location queries with the points of $P$ to determine whether there are two points in $P$ lying in the same face.

In the general case, the faces of the arrangement $\mathcal{A}(L)$ containing the points $P$ can be computed in time $O(n^{4/3} \log n)$ time using the algorithm of Edelsbrunner, Guibas and Sharir [6].

To improve on that, we first prove that, if the segments in $S$ are long, the total complexity of all faces in the arrangement containing at least one long segment is only linear.

**Lemma 4.** *Given a set $S$ of long line segments, and a set $L$ of $m$ lines. Then the total complexity of all faces of the arrangement $\mathcal{A}(L)$ containing at least one segment of $S$ is $O(m)$.*

*Proof.* We cover the unit circle with a grid of $O(1/\lambda)$ horizontal and vertical lines $H$ such that every cell of the arrangement of $H$ inside the unit circle has diameter less than $\lambda$. That implies that every segment in $S$ must intersect a line of $H$. Consequently, every face of $\mathcal{A}(L)$ that contains a segment of $S$ lies in the zone of one of the lines $H$. By the zone theorem, the total complexity of all these faces is $O(m/\lambda)$.

The idea of the proof immediately leads to an algorithm for our problem: We create a set $H$ of $O(1/\lambda)$ lines as in the proof of the lemma, and compute all faces of $\mathcal{A}(L)$ that intersect at least one line of $H$. That can be done in time $O(n \log n)$ using the lazy randomized incremental algorithm by de Berg et al. [1]. We then compute a point location structure for the subdivision we have obtained so far, and perform $2n$ point location queries with the endpoints of the segments $S$. This takes time $O(n \log n)$ in total. If the two endpoints of a segment in $S$ do not lie in the zone of $H$, or do not lie in the same face of the zone, then the set $L$ contains lines that do not separate $S$. If there are two endpoints of different line segments that lie in the same face of the subdivision, then the two line segments are not separated by $L$. Otherwise, $L$ shatters $S$. To summarize, we have shown the following theorem.

**Theorem 5.** *Given a set $S$ of $n$ long segments, and a set $L$ of $\Theta(n)$ lines, we can determine in expected time $O(n \log n)$ if $L$ shatters $S$.*

**Corollary 6.** *Given a set $S$ of $n$ long segments, we can determine in expected time $O(n \log n)$ a set $L$ of $O(n)$ lines shattering $S$, or determine that no such set exists.*

## 5  The number of lines needed to shatter a set

The algorithm described above doesn't try to minimize the number of shattering lines. The following theorem proves that such a minimization cannot decrease the size of the shattering set by more than a constant factor.

**Theorem 7.** *Let $S$ be a set of long line segments, and assume $L$ shatters $S$. Then $|L| = \Omega(n)$.*

*Proof.* By our assumption, each element $s \in S$ is contained inside the unit disk $D$. Let $\mathcal{A}(L)$ denote the arrangement formed by $L$. The perimeter of the cell $r_s$ of $\mathcal{A}(L)$ containing $s$ is at least $2\lambda$, and hence, as is easily shown using the triangle inequality, the perimeter of $(\partial r_s) \cap D$, the region of the boundary of $r_s$ which is contained inside $D$, is at least $\lambda$, see Figure 3. Summing this perimeter over all segments $s \in S$, we see that the total length of the segments $\{l_i \cap D \mid l_i \in L\}$ is at least $(\lambda/2)|S|$, since both sides of such a segment $l_i \cap D$ can contribute to the boundaries of two cells $r_s$ and $r_{s'}$, for some $s$, $s' \in S$.

On the other hand, for each $l_i \in L$ the length of the segment $l_i \cap D$ is at most 2. Hence

$$2|L| \geq \frac{\lambda}{2}|S|$$

yielding $|L| = \Omega(|S|)$

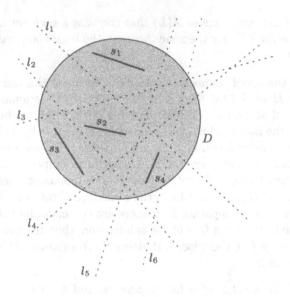

**Fig. 3.** The set of lines $L \equiv \{l_1, \ldots l_6\}$ shatters $S \equiv \{s_1, \ldots, s_4\}$

*Acknowledgment.* Discussions with Alon Itai contributed significantly to this paper, and we are grateful to him.

# References

1. M. de Berg, K. Dobrindt, and O. Schwarzkopf. On Lazy Randomized Incremental Construction. *Proceedings 26 Annual ACM Symposium on Theory of Computing*, 1994, pages 105–114.
2. J.-D. Boissonnat, O. Devillers, R. Schott, M. Teillaud, and M. Yvinec. Applications of random sampling to on-line algorithms in computational geometry. *Discrete and Computational Geometry* 8 (1992), 51–71.
3. K. L. Clarkson. New applications of random sampling in computational geometry. *Discrete and Computational Geometry* 2 (1987), 195–222.
4. K. L. Clarkson and P. W. Shor. Applications of random sampling in computational geometry, II. *Discrete and Computational Geometry* 4 (1989), 387–421.
5. H. Edelsbrunner and L.J. Guibas. Topologically sweeping an arrangement. *J. Comput. Syst. Sci.* 38 (1989), 165–194.
6. H. Edelsbrunner, L. Guibas, and M. Sharir. The complexity and construction of many faces in arrangements of lines and of segments. *Discrete and Computational Geometry* 5 (1990), 161–196.
7. A. Efrat, M. Sharir, and G. Rote. On the union of fat wedges and separating a collection of segments by a line. *Computational Geometry: Theory and Applications* 3 (1994), 277–288.
8. R. Freimer, J. S. B. Mitchell, and C. D. Piatko. On the complexity of shattering using arrangements. *Proceedings 2 Canadian Conf. Computational Geometry*, 1990, pages 218–222.

9. Sariel Har-Peled. Private communication.
10. A. Gajentaan and M. H. Overmars. On a class of $O(n^2)$ problems in computational geometry. *Computational Geometry: Theory and Applications*, 5 (1995), 165–185.
11. J. Hershberger and S. Suri. Applications of a semi-dynamic convex hull algorithm. *Proceedings 2 Scan. Workshop on Algorithms Theory, Lecture Notes in Computer Science*, 1990, vol. 447, Springer-Verlag, New York, pages 380–392.
12. J. Matoušek, N. Miller, J. Pach , M. Sharir, S. Sifrony, and E. Welzl. Fat triangles determine linearly many holes. *Proceedings 32 Annual ACM Symposium on Theory of Computing*, 1991, pages 49–58.
13. N. Miller and M. Sharir. Efficient randomized algorithms for constructing the union of fat triangles and of pseudodiscs. Manuscript, 1991.

# Optimal Line Bipartitions of Point Sets*

Olivier Devillers[1], Matthew J. Katz[2]

[1] INRIA, BP 93, 06902 Sophia-Antipolis cedex
Olivier.Devillers@sophia.inria.fr
[2] Department of Computer Science, Utrecht University, P.O.Box 80.089, 3508 TB
Utrecht, The Netherlands   matya@cs.ruu.nl

**Abstract.** Let $S$ be a set of $n$ points in the plane. We study the following problem: Partition $S$ by a line into two subsets $S_a$ and $S_b$ such that $max\{f(S_a), f(S_b)\}$ is minimal, where $f$ is any monotone function defined over $2^S$. We first present a solution to the case where the points in $S$ are the vertices of some convex polygon and apply it to some common cases — $f(S')$ is the perimeter, area, or width of the convex hull of $S' \subseteq S$ — to obtain linear solutions (or $O(n \log n)$ solutions if the convex hull of $S$ is not given) to the corresponding problems. This solution is based on an efficient procedure for finding a minimal entry in matrices of some special type, which we believe is of independent interest. For the general case we present a linear space solution which is in some sense output sensitive. It yields solutions to the perimeter and area cases that are never slower and often faster than the best previous solutions.

## 1 Introduction

Let $S$ be a set of $n$ points in the plane. We wish to partition $S$ by a line into two subsets $S_a$ and $S_b$, such that $max\{f(S_a), f(S_b)\}$ is minimal, where $f$ is some real-valued monotone function that is defined over the collection of subsets of $S$. ($f$ is *monotone* if $S_1 \subseteq S_2 \Rightarrow f(S_1) \leq f(S_2)$, for $S_1, S_2 \subseteq S$.) The problem of efficiently finding an optimal bipartition in respect to some criterion is the simplest form of the more general $k$-partition problem, and it has been studied extensively, see e.g. [1, 2, 3, 7, 8, 9, 11, 12, 13, 15]. The problem considered here was studied by Mitchell and Wynters [14] who present solutions for two instances of the problem, in which $f(S')$ is either the perimeter or area of the convex hull of $S'$. Their solutions require $O(n^3)$ time and $O(n)$ space. They also present an $O(n^2)$ time $O(n^2)$ space solution to the perimeter case. (In their paper they also solve the *minsum* versions of these cases, i.e., find a line bipartition for which $f(S_a) + f(S_b)$ is minimal.) Rokne et al. [18] present an $O(n^2 \log n)$ time and $O(n)$ space algorithm for the four cases considered in [14]. Both Mitchell and Wynters and Rokne et al. claim that an optimal bipartition is necessarily a line bipartition. However, Glozman et al. [6] show that this claim is false except for the minsum perimeter case.

---

* Part of this work was done while the first author was visiting INRIA Sophia-Antipolis.

Here, we first consider the restricted problem where the points in $S$ are in convex position, i.e., they are the vertices of some convex polygon $P$. For a portion $P'$ of (the boundary of) $P$ whose corresponding set of points is $S'$, we often write $f(P')$ instead of $f(S')$. For the restricted problem we present a solution of complexity $O(g(n) + n \cdot h(n))$, where $g(n)$ is the complexity of the preprocessing (if required), and $h(n)$ is the complexity of maintaining the value $f(P')$ under insertions and deletions of extreme vertices to/from $P'$, where $P'$ is a portion of (the boundary of) $P$. We apply this solution to some common cases, and obtain the following results. (The width case is especially interesting since it is not obvious how to maintain the width of $P'$ in constant time per operation.)

**Theorem 1.1.** *Let $S$ be a set of $n$ points in convex position in the plane, and assume that the convex hull of $S$ is known. It is possible to compute an optimal line bipartition of $S$ with respect to the perimeter, area, or width functions in linear time. For the width function, an optimal line bipartition is also an optimal (general) bipartition.*

Our solution is based on a procedure for searching for a minimal entry in matrices $M$ of the following kind, which we believe is of independent interest. Let $A = (a_{i,j}), B = (b_{i,j})$ be two $m \times n$ matrices of real values, such that the rows and columns of $A$ (resp. $B$) define increasing (resp. decreasing) sequences. The corresponding matrix $M$ is a $m \times n$ matrix $M = (m_{i,j})$ where $m_{i,j} = max\{a_{i,j}, b_{i,j}\}$. If $A$ and $B$ are, for example, discrete representations of two surface patches in 3-D lying above a rectangular region in the plane, then the procedure computes the lowest point lying on the upper envelope of these surfaces. In [19], Sharir considers this problem for 0-1 matrices $A$ and $B$. His solution examines $O((m + n) \log(m + n))$ entries of $M$, while our solution examines only $O(m + n)$ entries.

We now return to the non-restricted problem. Clearly it is enough to consider all the lines passing through a pair of points $a, b$ of $S$. Every such line $l$ defines two possible bipartitions ($a$ belongs to the right half plane and $b$ to the left half plane, or vice versa). We present a solution to the non-restricted problem which is sensitive to some value $k$ defined in Section 4. The problem of establishing tight bounds for $k$ is still open; $k$ is clearly at most quadratic but the authors believe that $k$ is subquadratic. In any case, in practical situations $k$ is almost always linear or nearly linear. (For some very specialized point sets $k$ is larger, but still far from quadratic.) More precisely the complexity of the algorithm is $O((n + k)h(n))$, where $h(n)$ is typically a polylogarithmic factor. We apply this solution to some common cases, and obtain the following results.

**Theorem 1.2.** *Let $S$ be a set of $n$ points in the plane. It is possible to compute an optimal line bipartition of $S$ with respect to the perimeter or area functions in $O((n+k)\log^2 n)$ time (or in randomized $O((n+k)\log n)$ time) and linear space, where $k$ is as above.*

Thus, whenever $k$ is less than $O(n^2)/\log n$ (or $O(n^2)$ for the randomized version), our algorithm is more efficient than the previous algorithms. (As mentioned, it is possible that $k$ is always less than $O(n^2)$; we were unable to prove or disprove this claim. However, in many examples $k$ is clearly less than $O(n^2)$.)

The paper is organized as follows. In Section 2 we describe the matrix searching procedure. In Section 3 we present the (general) solution to the restricted problem, that is based on the procedure of Section 2. The solution to the non-restricted problem is sketched in Section 4.

Recently Hurtado et al. [10] have independently studied the restricted problem of Section 3 (and some other related problems), and have obtained through an alternative method that is more direct but less general some similar results.

## 2 A Matrix Searching Procedure

Let $A = (a_{i,j}), B = (b_{i,j})$ be two $m \times n$ matrices of real values, such that the rows and columns of $A$ (resp. $B$) define increasing (resp. decreasing) sequences. Define a third $m \times n$ matrix $M = (m_{i,j})$ by $m_{i,j} = max\{a_{i,j}, b_{i,j}\}$.

In this section we describe an efficient procedure for finding an entry $m_{k,l}$ of $M$ such that $m_{k,l} \leq m_{i,j}$, for $1 \leq i \leq m$ and $1 \leq j \leq n$. The procedure finds such an entry of $M$ after inspecting only $O(m+n)$ entries of $M$. The sequence of entries that are inspected by the procedure forms a continuous path in $M$. This latter property is crucial for some of our applications.

We first observe that for any $1 \leq j \leq n$, the $j$'th column satisfies exactly one of the following conditions.

$a_{1,j} \geq b_{1,j}$ : The "best" entry in this column (i.e., the entry for which the maximum between the $a$-value and the $b$-value is minimal) is therefore $m_{1,j}$. (Since the sequence $a_{1,j}, \ldots, a_{m,j}$ is increasing while the sequence $b_{1,j}, \ldots, b_{m,j}$ is decreasing.) We set $i_j$ to 0.

$a_{m,j} \leq b_{m,j}$ : The best entry in this column is therefore $m_{m,j}$. We set $i_j$ to $m$.

$a_{1,j} < b_{1,j}$ and $a_{m,j} > b_{m,j}$ : The best entry in this column is therefore either the entry corresponding to the meeting point of the curves

$$A = \begin{pmatrix} a_{1,1} \leq \cdots \leq a_{1,n} \\ \wedge \qquad\quad \wedge \\ \vdots \qquad\quad \vdots \\ \wedge \qquad\quad \wedge \\ a_{m,1} \leq \cdots \leq a_{m,n} \end{pmatrix}$$

$$B = \begin{pmatrix} b_{1,1} \geq \cdots \geq b_{1,n} \\ \vee \qquad\quad \vee \\ \vdots \qquad\quad \vdots \\ \vee \qquad\quad \vee \\ b_{m,1} \geq \cdots \geq b_{m,n} \end{pmatrix}$$

$$M = \begin{pmatrix} b_{1,1} \geq \cdots \geq b_{1,j} & \cdots \\ \vee \qquad\quad \vee \\ \vdots \qquad\quad \vdots \\ \vee \qquad\quad \vee \\ b_{i_j,1} \geq \cdots \geq b_{i_j,j} \\ \qquad\qquad\qquad a_{i_j+1,j} \leq \cdots \leq a_{i_j+1,n} \\ \qquad\qquad\qquad \wedge \qquad\qquad \wedge \\ \vdots \qquad\quad \vdots \\ \wedge \qquad\qquad \wedge \\ \cdots \qquad a_{m,j} \leq \cdots \leq a_{m,n} \end{pmatrix}$$

**Figure 1.** The matrix $M$

defined by the sequences $a_{1,j}, \ldots, a_{m,j}$ and $b_{1,j}, \ldots, b_{m,j}$, if such an entry exists (i.e., if the meeting point corresponds to a sample point), or one of the two entries adjacent to the meeting point of these curves. We set $i_j$ to the index such that $m_{i_j,j} = b_{i_j,j}$ and $m_{i_j+1,j} = a_{i_j+1,j}$; the best entry is thus either $m_{i_j,j}$ or $m_{i_j+1,j}$ (see Figure 1).

Thus, if we begin our search for the best entry of the $j$'th column by inspecting some arbitrary entries $a_{i,j}$, $b_{i,j}$ of the $j$'th columns, then we can determine easily whether $m_{i,j}$ is the desired entry (possibly by inspecting one of its vertically adjacent entries). If $m_{i,j}$ is not the desired entry, we can immediately say which direction (i.e., upwards or downwards) leads to the desired entry.

We next observe that

**Lemma 2.1.** $i_j$ *is decreasing.*

*Proof.* We will prove that $i_j \geq i_{j+1}$. The claim is trivial if $i_j = 0$ or $i_{j+1} = m$. (Assume, for example, that $i_j = 0$. Then $m_{1,j} = a_{1,j} \geq b_{1,j}$, and since $a_{1,j+1} \geq a_{1,j}$ and $b_{1,j} \geq b_{1,j+1}$, we have $a_{1,j+1} \geq b_{1,j+1}$, and therefore $i_{j+1} = 0$.) Assume therefore that $1 \leq i_j < m$ and thus $m_{i_j,j} = b_{i_j,j}$ and $m_{i_j+1,j} = a_{i_j+1,j}$. The monotonicity of row $i_j + 1$ of matrices $A$ and $B$ ensures that $m_{i_j+1,j+1} = a_{i_j+1,j+1}$, since $a_{i_j+1,j+1} \geq a_{i_j+1,j} \geq b_{i_j+1,j} \geq b_{i_j+1,j+1}$. Thus the transition between the $b$-values and the $a$-values in column $j + 1$ cannot appear below this transition in the $j$-th column, which implies that $i_j \geq i_{j+1}$.

We are now ready to describe the procedure for finding a best entry of $M$. First compute the best entry $m_{i,1}$ of the first column of $M$, by inspecting the entries of the first column one by one beginning at the bottom entry and stopping once the best entry is encountered. According to the remark just above Lemma 2.1 this involves at most one inspection

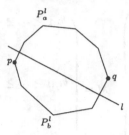

**Figure 2.** The line bipartition defined by $p$ and $q$

above the best entry. Now we inspect the entry $m_{i,2}$ and its adjacent entries in the second column. If either $m_{i,2}$ or $m_{i+1,2}$ is the best entry of the second column, then we move on to the third column (to $m_{i,3}$). Otherwise, we know by Lemma 2.1 that the best entry of the second column is above $m_{i,2}$, and we search for it (beginning at $m_{i,2}$ and moving upwards in the second column). We continue in this way until we reach the top entry of the last column. The best entry of $M$ is the best of the column-best entries that were found.

The following theorem summarizes the result of this section.

**Theorem 2.2.** *A minimal entry of $M$ can be found in $O(m + n)$ time. The search for such an entry requires only $O(m + n)$ entry inspections, and the sequence of entries that are inspected by the procedure forms a continuous path in $M$.*

**Remark:** For our purposes we are only interested in the case where $m = n$, and, by the theorem above, the desired entry of $M$ can be found in linear time in this case. However, in the general case where, say, $m \gg n$, it is still possible to compute the desired entry of $M$ efficiently; more precisely in $O(n \log m)$ time. This follows from the remark just above Lemma 2.1 which implies that the best entry of the $j$'th column can be found in $O(\log m)$ time.

## 3   The Solution to the Restricted Problem

In this section we assume that the points in $S$ are the vertices of some convex polygon $P$. A pair of distinct points $p, q \in S$ define a line bipartition of $S$ as follows (see Figure 2). Consider a line $l$ passing through a point on (the boundary of) $P$ immediately following $p$ in clockwise direction and a point on $P$ immediately following $q$. Then $l$ partitions $P$ into two portions and $S$ into two subsets. Denote by $P_a^l$ (resp. $P_b^l$) the portion of (the boundary of) $P$ lying above (resp. below) $l$. Denote by $S_a^l$ (resp. $S_b^l$) the subset of points of $S$ lying above (resp. below) $l$. Clearly every bipartition of $S$ by a line into two non empty subsets is obtained by a pair of points

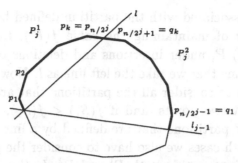

**Figure 3.** Step 1 - the recursive division of $P_{j-1}$

of $S$ in this way. Thus the total number of such bipartitions is $\binom{n}{2}$, and we want to find such a bipartition $\{S_a, S_b\}$ such that $max\{f(S_a), f(S_b)\}$ is minimal, where $f$ is a monotone function defined over $2^S$.

First we observe that if $f(P_a^l) \geq f(P_b^l)$ for some line $l$, then any line that cuts $P$ below $l$ defines a worse partition. Thus the best line bipartition is either defined by a line crossing $l$ inside $P$, by $l$ itself, or by a line that cuts $P$ above $l$. The latter case is treated by the algorithm below by recursively exploring portions of $P$, and the first case is treated by searching in an appropriate matrix. We now describe the $j$-th stage of our algorithm which consists of $O(\log n)$ stages. After the $j$-th stage we are left with a portion $P_j$ of $P$ of size (number of vertices) $n/2^j$, such that the bipartitions that we still need to consider at this point are only those that are defined by a line that intersect $P_j$ at two points.

**The $j$-th stage**

Let $P_{j-1}$ be the portion of $P$ that is left after the $(j-1)$'th stage ($P_0 = P$), and assume the vertices of $P_{j-1}$ in clockwise order are $p_1, \ldots, p_{n/2^{j-1}}$. The $j$-th stage consists of two steps.

**Step 1**

Consider any bipartition of $S$ that splits $P_{j-1}$ into two portions $P_j^1$ and $P_j^2$, such that the vertices of $P_j^1$ are $p_1, \ldots, p_{n/2^j}$ and those of $P_j^2$ are $p_{n/2^j+1}, \ldots, p_{n/2^{j-1}}$ (see Figure 3). Let $l$ be the line defining this partition. Compute the values $f(S_1)$ and $f(S_2)$, where $S_1$ (resp. $S_2$) is the subset of $S$ defined by $l$ containing the vertices of $P_j^1$ (resp. $P_j^2$). If $f(S_1) \geq f(S_2)$, then for any bipartition of $S$ by a line $l'$ that intersects $P_j^2$ at two points, we have $f(S_1') \geq f(S_1)$, where $S_1'$ is the subset of the bipartition defined by $l'$ that contains $S_1$. Thus we can discard all these partitions from further consideration. An analogue argument applies when $f(S_2) \geq f(S_1)$. As the line $l$ we take one of the two lines intersecting $P$ between the vertices $p_{n/2^j}$ and $p_{n/2^j+1}$ and at the left or right endpoint of $P_{j-1}$. The appropriate values of $f$ for these two lines can be computed in $O(n/2^j h(n))$ time from the values associated with the partition defined by $l_{j-1}$, where $h(n)$ is the complexity of maintaining the value $f(P')$, for a portion $P'$ of (the boundary of) $P$, under insertions and deletions of extreme vertices to/from $P'$. Assume that we take the left line as $l$. Now, if $f(S_1) \geq f(S_2)$, then we still have to consider all the partitions that are defined by a line intersecting $P_j^1$ at two points, and if $f(S_1) < f(S_2)$, then we still have to consider all the partitions that are defined by a line intersecting $P_j^2$ at two points. In both cases we also have to consider the partitions that are defined by a line intersecting both $P_j^1$ and $P_j^2$. In the former case, we set $l_j$ to the line $l$, i.e, the line that is defined by the endpoints of $P_j^1$ and we set $P_j$ to $P_j^1$, and in the latter case, we set $l_j$ to the line which was the alternative choice for $l$, i.e., the line that is defined by the endpoints of $P_j^2$ and we set $P_j$ to $P_j^2$.

**Step 2**

In this step we consider the partitions that are defined by a line intersecting both $P_j^1$ and $P_j^2$. Let $p_1, \ldots, p_k$ be the points in $P_j^1$ in clockwise order, and let $q_1, \ldots, q_k$ be the points in $P_j^2$ in counterclockwise order ($k = n/2^j$). We define a $k \times k$ matrix $M = (m_{i,j})$ as follows. Consider the bipartition of $S$ that is defined by the points $p_i$ and $q_j$. The entry $m_{i,j}$ is the maximum between the pair of real values $a_{i,j}, b_{i,j}$, where $a_{i,j}$ is the value of $f$ for the subset of the bipartition that has vertices of $P$ that do not belong to $P_{j-1}$, and $b_{i,j}$ is the value of $f$ for the second subset. At this point, we would like to use the procedure described in the previous section. For this we need to check that the corresponding matrices $A$ and $B$ have the required properties. Indeed, it is easy to see that the rows and columns of $A$ form increasing sequences while the rows and columns of $B$ form decreasing sequences. For example, if we fix $i$ and let $j$ change from 1 to $k$, then the subset of the bipartition that has vertices of $P$ that do not belong to $P_{j-1}$ only grows, and thus the value of $f$ only grows. Applying the procedure of the previous section to the matrix $M$ yields the best partition that is defined by a mixed pair of points. The cost of applying the procedure is $O(kh(n)) = O(n/2^j h(n))$. Note that we do not need to compute the entire matrix $M$, we only need to compute $O(k)$ entries as required by the procedure of the previous section. Moreover, the property of the procedure concerning the sequence of entries that is generated allows us to update the values that are computed dynamically, since the partition of the current entry is obtained from the previous one by moving a single extreme point from one of the sets to the other.

Thus at the end of this stage we are left with the set $P_j$ whose size is only $n/2^j$. The total running time of the $j$-th stage is $O(n/2^j h(n))$, and thus the total running time of the solution is $O(g(n) + n \cdot h(n))$. The term $g(n)$ is the cost of the preprocessing that is required for the dynamic updates.

The following theorem summarizes the result of this section.

**Theorem 3.1.** *Let $S$ be a set of $n$ points in the plane, and assume that the points in $S$ are the vertices of some convex polygon $P$. Let $f$ be some monotone function that is defined over the collection of subsets of $S$. A line bipartition for which the maximal value (between the two corresponding values of $f$) is minimal can be computed in $O(g(n) + n \cdot h(n))$, where $g(n)$ is the cost of the preprocessing that is required, and $h(n)$ is the complexity of maintaining the value $f(P')$ under insertions and deletions of extreme vertices to/from $P'$, where $P'$ is a portion of (the boundary of) $P$.*

In the full version of this paper [5] this theorem is applied to a few common functions including the perimeter, area, width, and diameter functions to obtain Theorem 1.1.

## 4 The Solution to the Non-Restricted Problem

In this section, we outline how some of the ideas above can be used in the case where the points of $S$ are not necessarily in convex position. More details are available in the full version of this paper [5].

As in the previous section, we will search in some matrix whose rows are bitonic (first decreasing and then increasing), but now the columns are no longer bitonic.

It is clear that any line bipartition can also be defined by a line passing through a pair of points of $S$, so we may restrict ourselves to the set of such lines. However, we prefer to search in a larger set of lines consisting of the lines that pass through a point of $S$ and are parallel to a line that passes through two points of $S$.

Denote by $N = \binom{n}{2}$ the number of pairs of points of $S$, and let $\theta_1 < \theta_2 < \ldots < \theta_N$ be the slopes defined by these pairs. For a fixed slope $\theta_i$, denote by $\sigma_i$ the permutation of the points of $S$ corresponding to the direction $\theta_i - \frac{\pi}{2}$, that is, a line of slope $\theta_i$ sweeping $S$ from left to right will first encounter $p_{\sigma_i(1)}$ and then $p_{\sigma_i(2)}, p_{\sigma_i(3)} \cdots p_{\sigma_i(n)}$. Finally, denote by $l_{i,j}$ the line of slope $\theta_i$ passing through point $p_{\sigma_i(j)}$, and by $m_{i,j}$ the maximum between the two values of $f$ for the bipartition defined by $l_{i,j}$ (see Figure 4). ($l_{i,0}$ is to the left of $S$, and $p_{\sigma_i(j)}$ is considered a member of the left subset.)

The $N \times (n+1)$ matrix $M = (m_{i,j})$ has some interesting properties. First its rows are bitonic. When the line of direction $\theta_i$ sweeps $S$, it splits

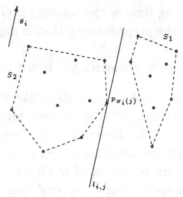

**Figure 4.** Definition of $l_{i,j}$

$S$ into a subset $S_1$ which is decreasing from $S$ to $\emptyset$ and a subset $S_2$ which is increasing from $\emptyset$ to $S$. Thus $m_{i,j} = \max\{f(S_1), f(S_2)\}$, $j = 0, \ldots, n$, is bitonic.

A second interesting property is that the difference between two consecutive rows is always small. For a given $i$, $m_{i,j} \neq m_{i+1,j}$ for at most 4 consecutive values of $j$. Moreover, for many values of $i$ the optimal bipartition for slope $\theta_{i+1}$ is the same as the one for slope $\theta_i$.

Thus the main idea is to explore the (implicit) matrix $M$ by examining only the rows where the optimum changes. Given the optimal bipartition for slope $\theta_i$ it is possible to determine geometrically the next slope $\theta_{i'}$, $i < i'$, where the optimum might change. We thus obtain an algorithm that is sensitive to the number of rows $k$ that were examined. Unfortunately no tight bound for $k$ has been established. However, notice that (i) $N = O(n^2)$ is a trivial upper bound for $k$, and therefore the *proven* complexity of our algorithm is similar to the best previous results, and (ii) in many examples $k$ is much smaller.

The following theorem is proven in [5].

**Theorem 4.1.** *Let $S$ be a set of $n$ points in the plane and let $f$ be some monotone function that is defined over the collection of subsets of $S$. A line bipartition for which the maximal value (between the two corresponding values of $f$) is minimal can be computed in $O((n+k)h(n))$, where $h(n)$ is the complexity of maintaining the convex hull of $S'$ and the value $f(S')$, where $S'$ is a subset of $S$, under insertions and deletions to/from $S'$, and of some basic queries concerning the convex hulls of the two sets consisting of the current partition, and $k = O(n^2)$ is as above.*

If $f(S')$ is the perimeter or the area of the convex hull of $S'$, then the basic problem is to maintain the convex hull under insertions and deletions. After the convex hull has been updated, it is easy to update $f$.

The convex hull can be maintained dynamically in $O(\log^2 n)$ time [16, 17] or in randomized $O(\log n)$ time [4] per operation. Both methods allow the maintenance of the perimeter and the area in logarithmic time, and also the common tangents of the two convex polygons can be computed in logarithmic time. This leads to Theorem 1.2.

# References

1. Pankaj K. Agarwal and M. Sharir. Planar geometric location problems. *Algorithmica*, 11:185–195, 1994.
2. Te. Asano, B. Bhattacharya, J. M. Keil, and F. Yao. Clustering algorithms based on minimum and maximum spanning trees. In *Proc. 4th Annu. ACM Sympos. Comput. Geom.*, pages 252–257, 1988.
3. D. Avis. Diameter partitioning. *Discrete Comput. Geom.*, 1:265–276, 1986.
4. K. L. Clarkson, K. Mehlhorn, and R. Seidel. Four results on randomized incremental constructions. *Comput. Geom. Theory Appl.*, 3(4):185–212, 1993.
5. O. Devillers and M. Katz. Optimal line bipartitions of point sets. Research Report 2871, INRIA, BP93, 06902 Sophia-Antipolis, France, 1996. http://www.inria.fr/rapports/sophia/RR-2871.html.
6. A. Glozman, K. Kedem, and G. Shpitalnik. Finding optimal bipartitions of points, 1994. manuscript.
7. Alex Glozman, Klara Kedem, and Gregory Shpitalnik. On some geometric selection and optimization problems via sorted matrices. In *Proc. 4th Workshop Algorithms Data Struct.*, volume 955 of *Lecture Notes in Computer Science*, pages 26–37. Springer-Verlag, 1995.
8. J. Hagauer and G. Rote. Three-clustering of points in the plane. In T. Lengauer, editor, *Proc. 1st Annu. European Sympos. Algorithms (ESA '93)*, volume 726 of *Lecture Notes in Computer Science*, pages 192–199. Springer-Verlag, 1993.
9. J. Hershberger and S. Suri. Finding tailored partitions. *J. Algorithms*, 12:431–463, 1991.
10. F. Hurtado, M. Noy, and S. Whitesides. Finding optimal $k$-partitions for points in convex position. Technical Report MA2-IR-95-0010, Departament de Matemàtica Aplicada II, Universitat Politècnica de Catalunya, Barcelona, Spain, 1995.
11. J. W. Jaromczyk and M. Kowaluk. An efficient algorithm for the Euclidean two-center problem. In *Proc. 10th Annu. ACM Sympos. Comput. Geom.*, pages 303–311, 1994.
12. M. J. Katz. Improved algorithms in geometric optimization via expanders. In *Proc. 3rd Israel Symposium on Theory of Computing and Systems*, pages 78–87, 1995.
13. M. J. Katz and M. Sharir. An expander-based approach to geometric optimization. In *Proc. 9th Annu. ACM Sympos. Comput. Geom.*, pages 198–207, 1993.
14. J. S. B. Mitchell and E. L. Wynters. Finding optimal bipartitions of points and polygons. In *Proc. 2nd Workshop Algorithms Data Struct.*, volume 519 of *Lecture Notes in Computer Science*, pages 202–213. Springer-Verlag, 1991.
15. C. Monma and S. Suri. Partitioning points and graphs to minimize the maximum or the sum of diameters. In *Graph Theory, Combinatorics and Applications (Proc. 6th Internat. Conf. Theory Appl. Graphs)*, volume 2, pages 899–912, New York, NY, 1991. Wiley.
16. M. H. Overmars and J. van Leeuwen. Maintenance of configurations in the plane. *J. Comput. Syst. Sci.*, 23:166–204, 1981.
17. F. P. Preparata and M. I. Shamos. *Computational Geometry: An Introduction.* Springer-Verlag, New York, NY, 1985.
18. J. Rokne, S. Wang, and X. Wu. Optimal bipartitions of point sets. In *Proc. 4th Canad. Conf. Comput. Geom.*, pages 11–16, 1992.
19. Micha Sharir. A near-linear algorithm for the planar 2-center problem. In *Proc. 12th Annu. ACM Sympos. Comput. Geom.*, pages 106–112, 1996.

# Interval Finding and Its Application to Data Mining

Takeshi Fukuda, Yasuhiko Morimoto, Shinich Morishita and Takeshi Tokuyama

IBM Japan, Tokyo Research Laboratory, Yamato-shi 242, Japan

**Abstract.** In this paper, we investigate inverse problems of the interval query problem in application to *data mining*. Let $\mathcal{I}$ be the set of all intervals on $U = \{0, 1, 2, .., n\}$. Consider an objective function $f(I)$, conditional functions $u_i(I)$ on $\mathcal{I}$, and define an optimization problem of finding the interval $I$ maximizing $f(I)$ subject to $u_i(I) > K_i$ for given real numbers $K_i$ $(i = 1, 2, .., h)$. We propose efficient algorithms to solve the above optimization problem if the objective function is either *additive* or *quotient*, and the conditional functions are additive, where a function $f$ is additive if $f(I) = \sum_{i \in I} \hat{f}(i)$ extending a function $\hat{f}$ on $U$, and quotient if it is represented as a quotient of two additive functions. We use computational-geometric methods such as convex hull, range searching, and multidimensional divide-and-conquer.

## 1 Introduction

Let us consider the set $\mathcal{I}$ of intervals of a set $U$ of $n$ ordered items. For simplicity, we assume $U = \{1, 2, .., n\} = [1, n]$. A function $f$ on $\mathcal{I}$ is called an *additive* interval function (or additive function, in short) if $f(I) = \sum_{i \in I} \hat{f}(i)$, extending a function $\hat{f}$ on $U$. From now on, we use the same notation $f$ for $\hat{f}$ for simplicity. We assume that the values of $f(i)$ for each $i \in U$ is known. If $f(I) \geq 0$ (resp. $f(I) \leq 0$) for all $I \in \mathcal{I}$, we say that $f$ is *nonnegative* [1] (resp. *nonpositive*).

If $f(I) > 0$ for all $I \neq \emptyset \in \mathcal{I}$, we say that $f$ is *positive*. A function $f$ is called *pure* if it is either nonnegative or nonpositive. A function $r(I)$ on intervals is called a *quotient function* if there exist an additive function $f(I)$ and a positive additive function $g(I)$ such that $r(I) = f(I)/g(I)$.

For a fixed function $f$ on $\mathcal{I}$, a basic operation in data base searching (SQL query) is the *interval query*.

**Interval query:** "For a given query interval $I$, compute $f(I)$."

If $f$ is either additive or quotient, the interval query can be trivially done in $O(n)$ time. Moreover, after linear-time preprocessing, $f(I)$ can be queried in $O(1)$ time with $O(n)$ space for an additive interval function $f$; Indeed, we can precompute and store $f([1, t])$ for $t = 1, 2, .., n$ in a table, and answer the value of $f$ at $I = [s, t]$ as $f([s, t]) = f([1, t]) - f([1, s - 1])$. Several data structures for a dynamic interval query permitting local change on both $f$ and $U$ are known in literature [14].

---

[1] Nonnegative additive interval function is a kind of *measure* on intervals.

However, instead of giving an interval as a query input, we sometimes want to perform a query that outputs an interval. In this paper, we consider the following inverse problem of interval query:

**Optimal-Interval Finding (OIF):** "Compute the interval $I$ that maximizes a function $v(I)$ under the condition that $u_i(I) > K_i$ for given additive functions $u_i$ and real numbers $K_i$ for $i = 1, 2, .., h$."

The function $v(I)$ is called the *objective functions*, while the other $u_i$ ($i = 1, 2, .., h$) are called *conditional functions*. OIF with an additive objective function is called AIF (additive interval finding), and that with a quotient objective function is called QIF (quotient interval finding).

We consider $h$ as a constant if we discuss computational complexity. Since there are $n(n-1)/2$ different intervals, it is almost trivial that AIF and QIF can be solved in $O(n^2)$ time. Our interest is to devise more efficient algorithms.

If $h = 0$, a linear time algorithm (Section 3.1), for AIF can be found in [6], and QIF can be reduced to the convex-hull computation (Section 4.2). If $h = 1$, linear time algorithms for AIF for a nonnegative objective function and a special QIF have been designed and implemented by Fukuda et al. [7] in application to data mining (automatic rule finding from large databases, see Section 2). This paper gives extension of the paper of Fukuda et al.[7]. We show the following results:

AIF can be solved in $O(n)$-space and $O(n \log^h n)$-time. If all conditional functions are pure, it can be solved in $O(n)$ time. If the objective function $v$ is nonnegative, the time complexity becomes $O(n \log^{h-1} n)$. QIF can be solved in $O(n \log^h n)$ time and space. The time complexity can be further reduced in several special cases.

## 2  Application to Data Mining

Recent progress in data input technologies made it much easier for finance and retail organizations to collect massive amounts of data and to store them on disk at a low cost. Such organizations are interested in extracting from these huge databases unknown information (in the form of rules) that inspires new marketing strategies. Database systems are the primary means of achieving this purpose, and there has been a growing interest in *data mining*, that is, efficient discovery of useful rules, [2, 3, 4, 7, 8, 9, 12, 13, 15].

We consider a database consisting of data vectors called *tuples*. For example, imagine a database containing data on a bank's customers. The records associated with a customer (or a cluster of customers, if the data is compressed) form a vector called a *tuple*. Each tuple has many *attributes* (entries of the vector). If the value of an attribute is a real number, we call it a *numeric attribute*. An example of a numeric attribute is *Balance*(), which is the total balance of accounts of a customer corresponding to a tuple.

Suppose we have a set $Y$ of $n$ tuples. We consider additive functions on $Y$, which usually are linear combinations of attributes.

57

*Example 1.* Consider a bank's data on customers. Each tuple contains the bank balance and a service (credit card loan, say) for one customer. Suppose that among customers whose balance is between $10^4$ and $10^5$, more than a threshold ratio (say, 20%) of customers use credit card loans, and less than 10% of cardloan users overdue: then, we have the following rule, $(Balance \in [10^4, 10^5]) \Rightarrow (CardLoan = yes)\&(Overdue = no).$ ∎

This kind of rule (called *association rule*) is interesting only if there is a sufficiently large number of cardloan users in the interval (in this case, $[10^4, 10^5]$) of ballance. Hence, we want to find an interval (called *optimized support interval*) of ballance that contains the maximum number of cardloan users under the conditions that the ratio of the card loan users should be more than 20% and the ratio of the number of overdue cardloan users to the number of cardloan users should be less than 10%.

For each customer, $Balance(p)$ is his bank balance. The set consisting of $Balance(p)$ for all customers $p$ in the database corresponds to the ordered set $U$ in the introduction. The function $Size(p) = 1$ for each $p$ is called the size indicator. $CardLoan(p)$ is an attribute such that $Cardloan(p) = 1$ if the customer makes credit card loans, and 0 otherwise. Similarly, $Overdue(p) = 1$ if and only if the customer overdues the loan payment. For the set of intervals of balance, let us define additive interval functions $Cardloan(I) = \sum_{Balance(p)\in I} Cardloan(p)$, $Size(I) = \sum_{Balance(p)\in I} Size(I)$, $u_1(I) = Cardloan(I) - 0.2Size(I)$, and $u_2(I) = 0.1Cardloan(I) - Overdue(I)$. Then, the problem of finding the optimized support interval is transformed into the following AIF problem, where the objective function $Size$ is a positive additive function:

"Compute the interval that maximizes $Cardloan(I)$ under the condition that $u_i(I) > 0$ for $i = 1, 2$."

Next, suppose we want to find an interval $I$ of $Balance$ that contains at least 10,000 customers and maximizes the ratio $Cardloan(I)/Size(I)$ of card loan users, under the condition that $Overdue(I)/Cardloan(I) < 0.1$. Then, the problem becomes the following QIF:

"Compute the interval maximizing $Cardloan(I)/Size(I)$ under the condition that $Size(I) > 10,000$ and $u_2(I) > 0$."

Since finding association rules is a fundamental function of a data mining system, efficient methods for querying the optimal intervals are crucial. For $h = 1$, linear time algorithms for AIF for a nonnegative objective function and QIF (where the objective function $r(I) = f(I)/g(I)$ and $u_1(I) \equiv g(I)$, precisely speaking) have been given and implemented by Fukuda et al [7]. They reported that the elapsed times for these special AIF and QIF were 0.014 and 0.09 seconds for $n = 10000$, respectively, whereas a naive $O(n^2)$ time method took 90 seconds. The efficiency enables to realize a real-time interactive system [9] for handling a database where each tuple has more than 100 different attributes. Our algorithms in this paper extend the ability of the system so that it can handle more general OIF problems which has more than one conditional functions.

# 3 Algorithms for AIF

## 3.1 AIF without constraint (back to *Programming Pearls*)

AIF with no conditional function was discussed in "Programming Pearls" (column by J. Bentley in CACM [6]), where Bentley used it as an example to demonstrate how algorithm design techniques are important for writing an efficient program.

> "Compute the interval $I$ that maximizes $f(I)$ for an additive interval function $f$."

Bentley presented four algorithms, which have time complexities $O(n^3)$, $O(n^2)$, $O(n \log n)$, and $O(n)$, respectively.

## 3.2 If conditional functions are pure

We next give a solution of AIF when conditional functions are pure. Let $u_1, ..., u_h$ are pure conditional functions, and $u_i > K_i$ are conditions (for $i = 1, 2, .., h$). For each $s$, we define $first_i(s)$ (resp. $last_i(s)$) to be the minimum (resp. maximum) index $t$ such that $u_i([s, t]) > K_i$. If there is no such index, we define $last_i(s) = first_i(s) = n + 1$ if $u_i$ is nonnegative, and $last_i(s) = first_i(s) = s - 1$ if $u_i$ is nonpositive. Note that, if $u_i$ is nonnegative, $last_i(s) = n$ or $n + 1$, and if it is nonpositive, $first_i(s) = s$ or $s - 1$. The following lemma is an easy observation:

**Lemma 1.** *For each* $i = 1, 2, .., h$, $first_i(s) \geq first_i(s')$ *and* $last_i(s) \geq last_i(s')$ *if* $s \geq s'$.

From the above lemma, we can compute $first_i(s)$ and $last_i(s)$ for $s = 1, 2, .., n$ in $O(n)$ time. We define $first(s) = \max_{1 \leq i \leq h} first_i(s)$ and $last(s) = \min_{1 \leq i \leq h} last_i(s)$. From the lemma, $first(s) \geq first(s')$ and $last(s) \geq last(s')$ if $s \geq s'$. For each $s$, $opt(s)$ is the index $t \in [first(s), last(s)]$ that maximizes $v([s, t])$, where $v$ is the objective function. The solution of AIF is $\max_{s=1, 2, .., n} v([s, opt(s)])$.

**Theorem 2.** *AIF can be solved in* $O(n)$ *time if all conditional functions are pure. It can also be computed in* $O(\log n)$ *time by using* $O(n)$ *processors in the CRCW model.*

*Proof.* It suffices to compute all the indices $opt(s)$ ($s = 1, 2, .., n$). We define a partially-defined matrix $M$ by $M(s, t) = v([s, t])$ if $first(s) < t < last(s)$. The row maximum of the $s$-th row of $M$ is $v([s, opt(s)])$. The function $v(I)$ satisfies the linearity $v(I) + v(I') = v(I \cup I') + v(I \cap I')$ for every pair of $I$ and $I'$ if $I \cap I' \neq \emptyset$. From the linearity, we can see that either $opt(s) \in [last(s - 1) + 1, last(s)]$, $opt(s - 1) \in [first(s - 1), first(s) - 1]$, or $opt(s) = opt(s - 1)$. This enables us to compute all the row maxima by using $O(n)$ entries of $M$, which takes $O(n)$ time, since each entry can be queried in $O(1)$ time. To create a parallel version, we apply a parallel matrix-searching algorithm in a *totally monotone* matrix (Aggarwal and Park [1]).

## 3.3 If conditional functions are mixed

Removing the purity of the conditional functions drastically alters the solution. We first assume that the objective function $v(I)$ is nonnegative, and then remove the assumption later.

Since each $s \in U$, we consider an interval $[1, s]$, and $h + 1$ values $v([1, s])$ and $u_i([1, s])$ $(i = 1, 2, .., h)$. This generates a point $q_s = (v([1, s]), u_1([1, s]), ..., u_h([1, s]))$ in $(h + 1)$-dimensional space. Let $S = \{q_s : s = 1, 2, .., n\}$. We therefore can consider AIF a computational-geometric problem.

We denote $x^{(i)}$ for the $i$-th coordinate of the $(h + 1)$-dimensional space. We say that a point $q = (x^{(0)}, .., x^{(h)})$ in the $(h + 1)$-dimensional space dominates another point $q = (x'^{(0)}, .., x'^{(h)})$ if $x^{(i)} \geq x'^{(i)}$ for each $i = 0, 1, ..., h$. Similarly, a point $q$ is dominated by a set $A$ of points if there exists a point $q' \in A$ that dominates $q$. A point $q_s$ of $S$ is a maximal (resp. minimal) point of $S$ if there is no other point $q_{s'}$ of $S$ dominating (resp. dominated by) $q_s$. The set of maximal (resp. minimal) points of $S$ is called the maxima (resp. minima).

From a point $z$ in space, let $F(z)$ be the unique point of $S$ (if it exists) that has the maximum $x^{(1)}$-coordinate value among the points dominating $q$. By definition, if there is a point of $S$ that dominates $z$, $F(z)$ exists, and is included in the maxima of $S$. We define $\mathbf{K} = (0, K_1, K_2, ..., K_h)$, and $q + \mathbf{K}$ is the point associated with the vector sum of $q$ and $\mathbf{K}$. We denote $S + \mathbf{K}$ for $\{q + \mathbf{K} : q \in S\}$.

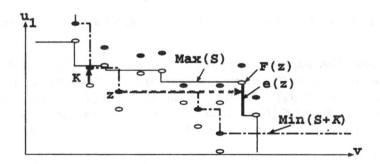

**Fig. 1.** Maxima of $S$ (white points), minima of $S + \mathbf{K}$ (black points), and ray shooting from a point $z$

**Lemma 3.** *There exists a solution* $I = (s, t]$ *of the AIF with a nonnegative objective function, such that* $q_s$ *is a minimal point, and* $q_t$ *is a maximal point. Moreover,* $q_t = F(q_s + \mathbf{K})$.

*Proof.* Because $v(I)$ is nonnegative, if $q_t$ dominates $q_s$, $s \leq t$ holds. Thus, the condition that $q_t$ dominates $q_s + \mathbf{K}$ is equivalent to the interval $(s, t]$ satisfies the conditions $u_i((s, t]) > K_i$ for $i = 1, 2, .., h$. If $q_s$ is not a minimal points, then there is a minimal point $q_l$ dominated by $q_s$. It is clear that $q_l + \mathbf{K}$ is dominated by $q_t$, and $v((l,t]) \geq v((s,t])$. Hence, we can replace $q_s$ with a minimal point $q_l$. The other statements can be similarly proven.

Thus, it is sufficient to compute $F(q_s + \mathbf{K})$ for all $s = 1, 2, .., n$ in order to solve AIF. In order to improve readability, we first demonstration an algorithm for a special case where $h = 1$. Note that the point set $S$ and $S + \mathbf{K}$ are sorted with respect to the first coordinate, since $v(I)$ is nonnegative. It is known that the two-dimensional maxima and minima of a sorted point set (with respect to the first coordinate) in linear time [14]. We construct the maxima $Max(S)$ of $S$ and the minima $Min(S + \mathbf{K})$ of $S + \mathbf{K}$. Two-dimensional maxima and minima have geometric representations as rectilinear monotone chains shown in Figure 1.

For each point $z$ on the minima $Min(S + \mathbf{K})$, we draw a horizontal ray. Let $e(z)$ be the intersecting vertical edge of the chain associated with $Max(S)$. Then $F(z)$ is the upper endpoint of $e(z)$ (Figure 1). The computation of $e(z)$ is a special case of *ray-shooting*, which is a popular notion in computational geometry. If we compute $e(z)$ in the increasing order of $z$ with respect to the second coordinate, we can find all $F(z)$ in $O(n)$ total time. Note that the sorted order of $Min(S)$ with respect to the second coordinate is given in linear time, although computation of the sorted order of $S$ requires $\Omega(n \log n)$.

**Lemma 4.** *If $h = 1$ and the objective function is nonnegative. AIF can be solved in $O(n)$ time.*

If $h \geq 2$, we use the following multidimensional divide-and-conquer method, which we call *group ray shooting*.

**Lemma 5.** $F(q_s + \mathbf{K})$ *for all* $s = 1, 2, .., n$ *can be computed in* $O(n \log^{h-1} n)$ *time and* $O(n)$ *space. Moreover, they can be computed in* $O(\log^2 n)$ *time by using* $O(n \log^{h-2} n)$ *processors.*

*Proof.* For simplicity, we assume that $n = 2^m$ for an integer $m$. We define a subset $S_{L,b} = \{q_i : i \in (2^b L, 2^b (L+1))\}$ of $S$ for $b = 0, 1, .., m$ and $L = 0, 1, .., 2^{m-b} - 1$. Note that, for each $b$, $S_{L,b} = S_{2L,b-1} \cup S_{2(L+1),b-1}$.

We consider the orthogonal projection $pr$ defined by $pr(q_j) = (x_i^{(1)}, ..., x_i^{(h)})$ to the $h$-dimensional subspace $R^h$ spanned by $x^{(1)}, .., x^{(h)}$. Then, we can compute $F(q_s + \mathbf{K})$ for $s = 1, 2, .., n$ as follows: First, we compute the subset $U(1)$ consisting of points of $S + \mathbf{K}$ whose projection images by $pr$ are dominated by $pr(S_{1,m-1})$. We define $U(2) = S - U(1)$ (set difference).

Even if be a point $q_i$ in $S_{0,m-1}$ dominates a point $z = q_j + \mathbf{K}$ in $U(1)$, $q_i$ cannot be $F(z)$; Indeed, there exists a point $q_l$ in $S_{1,m-1}$ dominating $z + \mathbf{K}$, and the $x^{(1)}$ coordinate value of $q_l$ is larger than that of $q_i$ from definition of $S_{0,m-1}$. Thus, $F(z)$ for $z \in U(1)$ is in $S_{1,m-1}$, whereas that for $z \in U(2)$ is in $S_{0,m-1}$, if there exists a point of $S$ that dominates $z$.

We next compute the subset $U(1, 1)$ consisting of points of $U_1$ whose projection images are dominated by $pr(S_{3,m-2})$, and define $U(1, 2) = U(1) - U(1, 1)$. Similarly, we compute the subset $U(2, 1)$ consisting of points of $U(2)$ whose projection images are dominated by $pr(S_{1,m-2})$, and define $U(2, 2) = U(2) -$

$U(2,1)$. We continue this operation until we have computed $U(e_1, ..., e_m)$ for all $e_i \in \{1, 2\}$. Then, $F(z)$ for a point $z$ in $U(e_1, ..., e_m)$ is $q_{M+1}$, where $M = e_1 2^{m-1} + e_2 2^{m-2} + ... + e_m$, if it dominates $z$ (otherwise, no point of $S$ dominates $z$).

It is known that for point sets $A$ and $B$ in $R^d$ ($d \geq 2$) of sizes $n_A$ and $n_B$, respectively, we can find the subset consisting of points of $A$ dominated by $B$ in $T_d(n_A, n_B) = O((n_A + n_B) \log^{d-2}(n_A + n_B))$ time [10, 14], if the point sets are sorted with respect to a coordinate. This operation is called *filtering* in Preparata and Shamos [14], and used as a subroutine of computing the maxima of a point set. More generally, it is a multidimensional divide-and-conquer algorithm [5]. If we examine the multidimensional divide-and-conquer algorithm, we find that its space complexity $s_d(n_A, n_B)$ satisfies $s_d(n_A, n_B) \leq n_A + n_B + s_{d-1}(m, n_B/2)$. Thus, $s_d(n_A, n_B) = O(dn) = O(n)$ if $d$ is a constant. Hence, $U(1)$ can be computed in $O(n \log^{h-2} n)$ time, and for each fixed $l \leq m = \log n$, $U(e_1, e_2, ..., e_{l-1}, 1)$ can be computed from $U(e_1, e_2, ..., e_{l-1})$ in $O((2^{m-l+1} + |U(e_1, e_2, ..., e_{l-1})|) \log^{h-2} n)$ time. Hence, the time complexity of our algorithm is $O(n \log^{h-1} n)$. Since $\sum_{e_i \in \{1,2\}} |U(e_1, e_2, ..., e_{l-1})| = n$, the space complexity of the algorithm is $O(n)$.

In the parallel model, the filtering operation can be done in $O(\log n)$ time by using $O((n_A + n_B) \log^{d-2} n)$ processors. Hence, in total, parallel complexity of the group shooting becomes $O(\log^2 n)$ time and $O(n \log^{h-2} n)$ processors.

When the objective function $v$ is not nonnegative, $i \leq j$ need not hold even if $q_i$ is dominated by $q_j$. Here, we add a new conditional function $w(I)$ generated by $w(i) = 1$ for all $i \in U$. Then, $i \leq j$ holds if $q_i$ is dominated by $q_j$ after adding the condition $w(I) > 0$.

**Theorem 6.** *If the objective function is nonnegative, AIF with $h$ conditional functions can be solved in $O(n \log^{h-1} n)$ time and $O(n)$ space. Moreover, it can be solved in $O(\log^2 n)$ time by using $O(n \log^{h-2} n)$ processors. If the objective function is not nonnegative, the sequential time complexity and the number of processors used in parallel computation are increased by a factor of $\log n$.*

If the conditional functions has linear dependence, we can improve the time complexity (we omit proof):

**Theorem 7.** *If conditional functions $u_i()$ are linear combinations of $d$ additive interval functions $a_1(), .., a_d()$ for $d \ll h$, AIF can be solved in $O(h^{\lfloor d/2 \rfloor} n \log^{d+1} n)$ time and space.*

## 4  Algorithms for QIF

### 4.1  Parametric searching method

One possible solution of the QIF problem is as follows: The objective function $r(I)$ is written as $f(I)/g(I)$ using a pair of additive functions. Suppose that we guess the optimal value $\alpha_{opt}$ of $r(I)$, and consider an additive objective function $f(I) - g(I)\alpha$, where $\alpha$ is guessed value. Then, solve the AIF problem "Compute the interval $I$ that maximizes function $f(I) - g(I)\alpha$ under the condition that $u_i(I) > K_i$ for additive functions $u_i, i = 1, 2, .., h$."

If $\alpha$ is the real optimal value of the original QIF problem, the objective function of the AIF problem takes the optimal value 0, and the corresponding interval $I$ is the solution of the QIF problem. If the optimal value of the AIF problem is positive (resp. negative), the optimal value $\alpha_{opt}$ of the QIF is larger (resp. smaller) than $\alpha$. Hence, we can binary search the optimal value $\alpha_{opt}$ by solving the AIF $\log L$ time, where $L$ is the problem precision.

Moreover, applying parametric searching paradigm by using the $O(\log^2 n)$ time $O(n \log^{h-1} n)$ processors parallel algorithm for AIF. Applying the general scheme of Meggido [11], it suffices to use the $O(n \log^h n)$ time sequential AIF algorithm $O(\log^3 n)$ times; hence, we obtain an $O(n \log^{h+3} n)$ time algorithm for QIF. We give a simpler $O(n \log^h n)$ time solution below.

## 4.2 Algorithms using convex hulls

First, we give a linear time algorithm for $h = 0$. We define $x_s = g([1, s])$ and $y_s = f([1, s])$ for $s = 1, 2, .., n$. We define $x_0 = y_0 = 0$. Since $g$ is positive, $x_s < x_t$ if $s < t$. We consider the planar point set $P = \{(x_s, y_s) : s = 0, 1, 2, .., n\}$. For each $t \leq n + 1$, we define $P(< t) = \{(x_i, y_i) \in P : i < t\}$. For $I = (s, t]$, $f(I)/g(I) = (y_t - y_s)/(x_t - x_s)$ coincides with the slope of the segment between $(x_s, y_s)$ and $(x_t, y_t)$.

For a planar point set $A$, let $LConv(A)$ be the lower chain of the convex hull of $A$. We can easily observe the following lemma:

**Lemma 8.** If $I = (s, t]$ maximizes $r(I) = f(I)/g(I)$, then $(x_s, y_s)$ is on $LConv(P < t)$, and they form the tangent segment from $(x_t, y_t)$ to $LConv(P < t)$, equivalently, the rightmost edge of $Lconv(P < t + 1)$.

Hence, it suffices to compute $LConv(P < t)$ for $t = 1, 2, .., n + 1$. This can be done by using the incremental convex hull algorithm [14] (starting from $\{0\} = LConv(P < 1)$ increasing $t$ by one in each step) on a planar point set sorted with respect to the $x$-coordinate values. It is well-known that this can be done in linear time. We extend this idea to a general case.

**Theorem 9.** QIF can be solved in $O(n \log^h n)$ time and $O(n \log^h n)$ space.

*Proof.* Let $q_i = (g([1, i]), f([1, i]), u_1([1, i]), ..., u_h([1, i]))$ for $i = 1, 2, .., n$, $q_0 = \mathbf{0}$, and $S = \{q_i : i = 0, 1, 2, .., n\}$. Let $S(< t) = \{q_0, ..., q_{t-1}\}$ for each $t$. Let $\Phi$ be a projection map on $S$ to a two dimensional plane defined by $\Phi(q_i) = (g([1, i]), f([1, i]))$. For $I = (s, t]$, $f(I)/g(I) = (y_t - y_s)/(x_t - x_s)$ is the slope of the segment between $\Phi(q_s)$ and $\Phi(q_t)$. We also define $q_i + \mathbf{K} = (g([1, i]), f([1, i]), u_1([1, i]) + K_1, ..., u_h([1, i]) + K_h)$. Let $W(t) = \{q_j \in S(< t) : q_j \text{ is dominated by } q_t + \mathbf{K}\}$. By definition, $s \in W(t)$ if and only if $I = (s, t]$ satisfies the conditions (except the maximality of $r(I)$) of QIF.

$W(t) \subset S(< t)$ is defined as an orthogonal region in the $(h + 2)$-dimensional space. Hence, we represent $W(t)$ as a decomposition into some subsets stored in an *orthogonal range searching* data structure $\mathcal{D}(t)$ of $S(< t)$. We can reduce the number of effective dimensions from $h + 2$ to $d = h$, since the dominance relation between $q_t$ and any point in $S(< t)$ with respect to the second coordinate (associated with the positive function $g$) automatically holds, and that with respect to the first coordinate (associated with $f$) comes from the maximality of $r$, and hence can be neglected.

We construct an orthogonal range search tree data structure $\mathcal{D}(n + 1)$ of $S(< n + 1) = S$, and other data structure can be dynamically obtained by simply erasing data in the structure associated with deleted points. $\mathcal{D}(n + 1)$ stores *primal subsets* of $S = S(< n + 1)$, so that any orthogonal region can be reported as a union of $O(\log^d n)$ primal subsets. For a primal subset $Z$ in $\mathcal{D}(n + 1)$, we denote $Z(t)$ for the associated primal subset of $\mathcal{D}(t)$ obtained by deleting data associated with $q_t, q_{t+1}, .., q_n$ from $Z$. For each primal subset $Z(t)$, we precompute the two-dimensional lower convex hull $LConv(\Phi(Z(t)))$.

When we construct $\mathcal{D}(t-1)$ from $\mathcal{D}(t)$, $O(\log^d n)$ primal subsets are updated, and an update of the lower convex hull in each primal subset can be done in $O(1)$ amortized time. Hence, the update of the whole data structure can be done in $O(\log^d n)$ time.

A way of solving QIF by using above data structure is as follows. Starting from $t = n$, we decrease $t$. For each $t$, we query $W(t)$ using $\mathcal{D}(t)$, and compute the tangent from $\Phi(q_t)$ to $LConv(Z(t))$ for each $Z$ appeared in the decomposition of $W(t)$. The computation of the tangent from $\Phi(q_t)$ to $LConv(\Phi(Z(t)))$ needs $O(\log n)$ time for each $Z$, that makes the total computation time $O(n \log^{d+1} n)$. The space complexity is $O(n \log^d n)$.

We can reduce the time complexity by a factor of $\log n$ as follows: For each $Z$, suppose that $Z(t_1), ... Z(t_k)$ appears in the query, so that $t_k > ... > t_1$. We have $k$ tangents, and only need the one with the maximum slope. It is easy to see that $t_i$ is in the upper convex hull of the point set $\{t_i, t_{i+1}, ..., t_k\}$ if the associated tangent attains the maximum slope. Then, the maximum slope can be computed in $O(|Z| + k)$ in total (we omit details). Hence, the total computation time is reduced to $O(n \log^d n) = O(n \log^h n)$ .

## 5 Concluding Remarks

The data mining system of Fukuda et al. [9] is designed so that users often execute many OIF queries for slightly modified objective and conditional functions. Typically, we often want to compute optimal-intervals many times changing the values of $K_i$. It would be useful if we could design a more efficient query system for this kind of series of OIFs.

We feel that data mining involves many research problems, that can be solved by using techniques in computational geometry, such as a higher dimensional extension of the OIF problem to compute an optimal region (instead of interval) of the space of tuples in a database that indicates a valuable rule. An initial approach can be found in [8].

It is an open problem to design of a dynamic data structure for OIF where $U$ and values of $f$ are changed dynamically. An $O(1)$ query and $O(\log n)$ update time data structure with $O(n)$ space can be constructed for AIF without constratint (omitted in this version because of space limitation). However, no such efficient data structure is known even for the case where only one conditional function is given.

# References

1. A. Aggarwal and J. Park, "Notes on Searching in Multidimensional Monotone Arrays," Proc. 29th IEEE FOCS (1988) 497-512.
2. R. Agrawal, S. Ghosh, T. Imielinski, B. Iyer, and A. Swami: "An Interval Classifier for Database Mining Applications," *Proceedings of the 18th VLDB Conference,* (1992) 560-573.
3. R. Agrawal, T. Imielinski and A. Swami, "Mining Association Rules between Sets of Items in Large Databases," *Proceedings of the ACM SIGMOD Conference on Management of Data,* (1993) 207-216.
4. R. Agrawal and R. Srikant, "Fast Algorithms for Mining Association Rules," *Proceedings of the 20th VLDB Conference,* (1994), 487-499.
5. J. L. Bentley, "Multidimensional divide-and-conquer," *CACM* 23-4 (1980) 214–229.
6. J. L. Bentley, "Programming Pearls – Algorithm Design Techniques," *CACM* 27-9 (1984), 865-871.
7. T. Fukuda, Y. Morimoto, S. Morishita, and T. Tokuyama, "Mining Optimized Association Rules for Numeric Attributes," Proc. 15th ACM SIGACT-SIGMOD-SIGART Principle of Data Systems (1996), 182-191.
8. T. Fukuda, Y. Morimoto, S. Morishita, and T. Tokuyama, "Data Mining Using Two-Dimensional Optimized Association Rules: Scheme, Algorithms, and Visualization," Proc. ACM SIGMOD Conference of Management of Data (1996), 13-23.
9. T. Fukuda, Y. Morimoto, S. Morishita, and T. Tokuyama, "SONAR: System for Optimized Numeric Association Rules (demonstration)", Proc. ACM SIGMOD Conference on Management of Data (1996) p. 553.
10. H. T. Kung, F. Luccio, and F. P. Preparata, "On Finding the Maxima of a Set of Vectors," *JACM* 22-4 (1975), 469-476.
11. N. Megiddo, Applying Parallel Computation Algorithms in the Design of Serial Algorithms, *J. ACM* **30** (1983), 852-865.
12. J. S. Park and M.-S. Chen and P. S. Yu, "An Effective Hash-Based Algorithm for Mining Association Rules," *Proceedings of the ACM SIGMOD Conference on Management of Data,* (1995), 175-186.
13. G. Piatetsky-Shapiro, "Discovery, Analysis, and Presentation of Strong Rules," *Knowledge Discovery in Databases,* AAAI Press, (1991) 229-248.
14. F. P. Preparata and M. I. Shamos, *Computational Geometry, an Introduction,* 2nd edition (1988).
15. M. Stonebraker, R. Agrawal, U. Dayal, E. J. Neuhold and A. Reuter, "DBMS Research at a Crossroads: The Vienna Update," *Proceedings of the 19th VLDB Conference,* (1993) 688-692.

# On the Approximimability of the Steiner Tree Problem in Phylogeny

David Fernández-Baca[1]*, Jens Lagergren[1]**

[1] Department of Computer Science, Iowa State University, Ames, IA 50011.
E-mail: fernande@cs.iastate.edu
[2] Department of Numerical Analysis and Computing Science, Royal Institute of
Technology, S-100 44 Stockholm, Sweden. Email: jensl@nada.kth.se

**Abstract.** Three results on the Steiner tree problem are presented: (i)
Computing optimum $k$-restricted Steiner tree is APX-complete for $k \geq 4$,
(ii) the minimum-cost $k$-restricted Steiner tree problem in phylogeny is
APX-complete for $k \geq 4$, and (iii) the $k$-Steiner ratio for the Steiner tree
problem in phylogeny matches the corresponding ratio for metric spaces
defined on graphs. The results are significant because $k$-restricted trees
are used in various approximation algorithms for the Steiner tree problem, and (i) and (ii) suggest that there is a limit to the approximimability
of the optimum solution. The $k$-Steiner ratio, which establishes a relation between the size of the optimum Steiner tree and the optimum
$k$-restricted tree, arises in the analysis of many of the same algorithms.

## 1   Introduction

A fundamental problem in biology and linguistics is that of inferring the evolutionary history of a set of taxa, each of which is specified by the set of *traits* or
*characters* that it exhibits [9, 10, 20]. In mathematical terms, the problem can be
expressed as follows. Let $C$ be a set of *characters*, and for every $c \in C$, let $A_c$ be
the set of allowable *states* for character $c$. Let $m = |C|$ and $r_c = |A_c|$. A *species*
(or, more generally, a taxon) $s$ is an element of $A_1 \times \cdots \times A_m$; $s(c)$ is referred to
as the *state of character $c$ for $s$*. A *phylogeny* for a set of $n$ distinct species $S$ is a
tree $T$ whose leaves are all elements of $S$ and where $S \subseteq V(T) \subseteq A_1 \times \cdots \times A_m$.
Vertices of $T$ that are not in $S$ are called *Steiner vertices*. The *length* of $T$ is
given by $L(T) = \sum_{(u,v) \in T} d(u, v)$, where $d(u, v)$ denotes the Hamming distance
between two species $u$ and $v$; i.e., $d(u, v)$ is the number of character states in
which $u$ and $v$ differ. The *Steiner tree problem in phylogeny* (STP) is to find a
phylogeny $T$ of minimum length for a given set of species $S$. STP and many of
its variants are NP-hard even when characters are binary [3, 5, 11, 19]. Jiang et
al. have noted that STP is MAXSNP-complete [15].

In order to state our results, we need to introduce some terminology. The
Steiner tree problem over a metric space $E$ is: given a finite set of points $S \subseteq E$,

* Supported in part by the National Science Foundation under grants CCR-9211262
and CCR-9520946.
** Supported by grants from NFR and TFR.

find a tree of minimum total length spanning the elements of $S$. STP is a Steiner tree problem over a metric space whose points set is $\mathcal{A}_1 \times \cdots \times \mathcal{A}_m$ and where interpoint distance is given by Hamming distance. The special case where characters are binary ($\mathcal{A}_i = \{0, 1\}$ for $i = 1, \ldots, m$) is equivalent to a Steiner tree problem on the metric space associated with the $m$-dimensional hypercube $Q_m$ [5, 11]. Other metric spaces of interest to us here are those associated with weighted graphs. In this case, points in the space are vertices of a graph and the distance between two points is the length of the path of least total weight between them in the graph. The associated Steiner tree problem is referred to as the *Steiner tree problem in networks*.

Let us consider the Steiner tree problem for a set of points $U$ in a metric space $E$. A tree $T$ is *full* if no internal node of $T$ is a point of $U$. A *full component* of $T$ is a maximal full subtree of $T$. A *k-restricted* Steiner tree is one where no full component has more than $k$ vertices of $U$ (thus, a 2-restricted Steiner tree is a spanning tree). The *k-Steiner ratio* in a metric space $E$ is defined as

$$\rho_k(E) = \inf_{S \subseteq E} \frac{L_{ST}(S)}{L_{kST}(S)} \tag{1}$$

where $L_{ST}(S)$ is the length of the minimum Steiner tree for $S$, and $L_{kST}(S)$ is the length of the minimum $k$-restricted Steiner tree for $S$.

We have three results:

- Computing a minimum-length $k$-restricted steiner tree on a graph is APX-complete for $k \geq 4$.
- Computing a minimum-length $k$-restricted phylogeny is APX-complete for $k \geq 4$, even when characters are binary.
- The $k$-Steiner ratio $\rho_k(Q)$ for the metric space associated with STP with binary characters equals the bound established for graphs by Borchers and Du [4]. That is, for any $k$ with $k = 2^r + s$, where $0 \leq s < 2^r$,

$$\rho_k(Q) = \frac{r2^r + s}{(r+1)2^r + s}$$

The results are significant because of the use of $k$-restricted trees in approximation algorithms for the Steiner tree problem in various metric spaces. In particular, Berman and Ramaiyer have shown that there exists a polynomial-time approximation algorithm with performance ratio

$$\rho_2^{-1} - \sum_{i=3}^{k} \frac{\rho_i^{-1} - \rho_{i-1}^{-1}}{i-1}$$

for any metric space where a minimum Steiner tree is polynomially-computable for any fixed-size set of points [2]; the latter condition can be shown to hold for STP.

While the close relationship between $k$-restricted binary STP and the graph version of the problem is, perhaps, not surprising, it must be pointed out that there are metric spaces such as the Euclidean and rectilinear planes, where the $k$-Steiner ratio is lower than on graphs [2, 7]. Thus, it is not clear a priori whether the $k$-Steiner ratio for phylogeny should match that for graphs (Gusfield's observation that $\rho_2(Q) = \rho_2 = 1/2$ [12], does, however, suggest that this might be true). Furthermore, there is an important difference between the Steiner tree problem in networks and STP. In the former, we are given a weighted graph and a subset of its vertices as input, and the goal is to obtain a minimum-weight tree connecting all the vertices. In STP, the underlying graph is given *implicitly* and can be much larger than the input.

## 2 The complexity of computing $k$-restricted Steiner trees

We will now show that the $k$-restricted Steiner tree problem on networks is APX-complete for $k \geq 4$, thereby giving evidence that there is a limit to the approximability of the problem. We will use this proof as a basis for arguing that the phylogeny version of the problem is also APX-complete (see Khanna et al. [16] for definitions regarding the class APX).

That the $k$-restricted Steiner tree problem belongs to APX is already known. We will give an L-reduction from a known APX-complete problem, Max 3-Sat-3 (Max 3-Sat where each variable has either two or three occurrences) [17, 13], to the $k$-restricted Steiner tree problem for any $k \geq 4$. In what follows, an instance of the Steiner tree problem on networks will be denoted by $(G, U)$, where $G$ is the edge-weighted graph and $U \subseteq V(G)$ is the set of vertices to be spanned. We will write $\text{opt}_A$ to denote the cost of an optimum solution to an instance $A$ of a problem and we will write $c_A(z)$ to denote the cost of a feasible solution $z$ of $A$. We will often drop the subscript from $c$ when the identity of $A$ is clear from the context.

To prove L-reducibility, we must show there exists a polynomial-time transformation $f$ from instances of Max 3-Sat-3 to instances of the Steiner tree problem on graphs, a polynomial-time mapping $g$ from solutions to the Steiner tree problem to solutions of Max 3-Sat-3, and constants $\alpha$ and $\beta$ such that, for any instance $\psi$ of Max 3-Sat-3:

i) $\text{opt}_{f(\psi)} \leq \alpha \cdot \text{opt}_\psi$,
ii) for every feasible solution $y$ of $f(\psi)$, $g(y)$ is a solution for $\psi$ and

$$|\text{opt}_\psi - c_\psi(g(y))| \leq \beta \,|\text{opt}_{f(\psi)} - c_{f(\psi)}(y)|.$$

The transformation $f$ maps an instance $\psi$ of Max 3-Sat-3 with $m$ clauses and $n$ variables to an instance $(G, U)$ of the Steiner tree problem defined as follows. For each clause $C$, $V(G)$ contains a vertex $C$. For each variable $x$, $V(G)$ contains three vertices $x$, $x_0$, and $x_1$. Finally $V(G)$ contains a vertex $r$. For each variable $x$, there are edges $(r, x_0)$, $(r, x_1)$, $(x, x_0)$, and $(x, x_1)$. Moreover, for each clause $C$, if the literal $x$ appears in $C$ then there is an edge $(x_1, C)$ and if the literal $\overline{x}$ appears in $C$ then there is an edge $(x_0, C)$. The graph $G$ has no other vertices or edges; all edges are of length one. Finally, the set of *given* vertices is

$$U = \{x|\ x \text{ variable}\} \cup \{C|\ C \text{ clause}\} \cup \{r\}.$$

**Lemma 2.1.** $opt_{(G,U)} \leq 2(n+m) - opt_\psi$ and hence $opt_{(G,U)} \leq \frac{33}{7} opt_\psi$.

*Proof.* Let $A$ be an assignment of truth values to the variables in $\psi$ that maximizes the number of satisfied clauses. Let $F$ be the set of edges consisting of (a) all edges $(r, x_i)$ and $(x_i, x)$ such that $A(x) = i$, (b) all edges $(C, x_i)$ such that $A(x) = i$ and $x$ is the first variable whose literal in $C$ is made true by $A$, and (c) all edges $(C, x_i)$ and $(x_i, r)$ such that $C$ is not satisfied by $A$ and $x$ is the first variable in $C$.

It is straightforward to verify that $F$ induces a subtree $T$ of $G$ that contains all the given vertices. Since, moreover, no Steiner vertex in $G$ has degree greater than 4, $T$ is a $k$-restricted Steiner tree for any $k \geq 4$. Hence $c(T) \leq 2n + c(A) + 2(m - c(A)) = 2(n+m) - opt_\psi$. Since we have two or three occurrences of each variable and $opt_\psi \geq 7m/8$, it follows that

$$c(T) \leq 3m + 2m - opt_\psi \leq \frac{40}{7} opt_\psi - opt_\psi = \frac{33}{7} opt_\psi.$$

*Remark:* If edges between clauses and literals have length 2, then we obtain $opt_{(G,U)} \leq 2n + 3m - opt_\psi$. Hence, we will still have $opt_{(G,U)} = O(opt_\psi)$.

**Definition 1.** A Steiner tree $T$ is *canonical* if and only if for any variable $x$ and any clause $C$ (a) $(x_i, C) \in E(T) \implies (r, x_i) \in E(T)$ (b) $(x_i, x) \in E(T) \implies (r, x_i) \in E(T)$.

Note that each clause $C$ has degree 1 in a canonical Steiner tree.

**Lemma 2.2.** *Given a $k$-restricted Steiner tree $S$ for $(G, U)$, a canonical no costlier $k$-restricted Steiner tree $S$ for $(G, U)$ can be found in polynomial time.*

*Proof.* Let $F_x$ be the set of edges that are incident to either $x_0$ or $x_1$. We first prove that we can transform $S$ into a no costlier $k$-restricted Steiner tree $T$ such that: if $C$ is in the same connected component as $r$ in $T \backslash F_x$, then $(x_i, C) \notin E(T)$ (for $i = 0$ and $i = 1$). Note that when this condition is satisfied, for each variable $x$, either $(r, x_0) \in E(T)$ or $(r, x_1) \in E(T)$.

Assume that if $C$ is in the same connected component as $r$ in $S \backslash F_x$, and $(x_i, C) \in E(S)$. Then we can delete $(x_i, C)$ from $S$ and add $(x_i, r)$ instead, whithout increasing the cost. This can be done recursively.

Assume that $(x, x_0) \in E(S)$ and $(x, x_1) \in E(S)$. Assume also that $(x_i, r) \in E(S)$. Then we can delete $(x, x_{1-i})$ from $S$ and add $(r, x_{1-i})$ instead, whithout increasing the cost. Also this can be done recursively. It follows that the tree obtained is canonical and no costlier.

For a Steiner tree $S$, we define $g(S)$ to be the truth assignment $A$ for variables in $\psi$ defined as follows. Let $T$ be a no costlier $k$-restricted canonical Steiner tree obtained from $S$ (using a fixed polynomial time algorithm). For each variable $x$, (1) if $(r, x_i) \in E(T)$ and $(r, x_{1-i}) \notin E(T)$ then $A(x) = i$ (2) if $(r, x_i) \in E(T)$,

$(r, x_{1-i}) \in E(T)$, and the literal $x$ appears twice in $\psi$ then $A(x) = 1$ (3) if $(r, x_i) \in E(T)$, $(r, x_{1-i}) \in E(T)$, and the literal $\bar{x}$ appears twice in $\psi$ then $A(x) = 0$.

**Lemma 2.3.** *For any canonical Steiner tree* $T$, $c(T) = 2(n + m) - c(g(T))$.

*Proof.* Let $A = g(T)$. We will now count the edges of $T$

- There are exactly $n$ edges of the form $(x, x_i)$ in $T$, since for each variable $x$, either $(x, x_0)$ or $(x, x_1)$ is an edge of $T$, but not both.
- There are exactly $n$ edges $(r, x_i)$ in $T$ such that $A(x) = i$
- For each clause $C$ that is satisfied by $A$, there is an $x_i$ such that $(C, x_i)$ is an edge of $T$. There are exactly $c(A)$ such edges in $T$.
- For each clause $C$ which is not satisfied by $A$ there is an $x_i$ such that both $(C, x_i)$ and $(x_i, r)$ are edges of $T$. Since no two different clauses use the same $x_i$, there are exactly $2(m - c(A))$ such edges.

Note that all edges have been enumerated and that no edge has been enumerated twice above. Thus, $c(T) = 2n + c(A) + 2(m - c(A)) = 2(n + m) - c(g(T))$. The lemma follows.

Combining our results we get the following corollary.

**Corollary 2.4.** For any $k \geq 4$ and any solution $T$ of the $k$-restricted Steiner tree problem instance $(G, U) = f(\psi)$, we have $|\text{opt}_\psi - c(g(T))| \leq |\text{opt}_{(G,U)} - c(T)|$.

*Remark:* If edges between clauses and literals have length 2, then we can prove that $c(T) = 2n + 3m - c(g(T))$. By the same reasoning as that used to prove the above result, we will have $|\text{opt}_\psi - c(g(T))| \leq |\text{opt}_{(G,U)} - c(T)|$.

This result and Lemma 2.1 yield

**Theorem 2.5.** *For any* $k \geq 4$, *the minimum* $k$-*restricted Steiner tree problem on graphs is APX-complete.*

**$k$-restricted STP.** We will now show that the $k$-restricted Steiner tree problem ($k$-STP) in phylogeny is APX-complete for $k \geq 4$. The proof is based on a construction similar to the one we have just presented for graphs. It will suffice to consider binary characters.

As before, we will exhibit an L-reduction from Max 3-Sat-3 to $k$-STP; we start by showing the mapping $f$ that goes from the first problem to the second. The sets of species we will consider will be given by vectors with many zeroes and few ones. To streamline the notation, we will write $\langle i_0, i_1, \ldots, i_l \rangle$, $0 \leq l \leq M-1$, $0 \leq i_j \leq M-1$, to denote the species where characters $i_0, \ldots, i_l$ have state 1, and all other characters have state 0. We write $\mathbf{0}$ to denote the species where all states are zero.

Let $\psi$ be any instance of Max 3-Sat-3. We will assume without loss of generality that every two clauses in $\psi$ differ in at least two variables [13, 17]. Assume that the variables of $\psi$ are $y_0, \ldots, y_{n-1}$. Write $y_{i0}$ to denote $\overline{y_i}$; $y_{i1}$ is just $y_i$. Associate with $\psi$ a set $S$ of species on $M$ characters, where $M$ is $2n$ plus the number of two-literal clauses.

Literal $y_{ir}$ is associated with character $2i+r$; we call $\langle 2i+r \rangle$ the *literal point* of $y_{ir}$. Each of the remaining characters corresponds to a different 2-literal clause. The set $S$ contains a species for each variable and clause of $\psi$:

- For each variable $y_i$, there is a *variable point* $x_i = \langle 2i, 2i+1 \rangle$.
- For each three-literal clause $C$, there is a *clause point* $z_c$:
  - If $C = (y_{i_1,r_1}, y_{i_2,r_2}, y_{i_3,r_3})$, $z_C = \langle 2i_1+r_1, 2i_2+r_2, 2i_3+r_3 \rangle$.
  - If $C = (y_{i_1,r_1}, y_{i_2,r_2})$, $z_C = \langle 2i_1+r_1, 2i_2+r_2, j_C \rangle$, where $j_C$ is the index of the position unique to $C$.

Then,

$$S = \{x_i : y_i \text{ is a variable in } \psi\} \cup \{z_C : C \text{ is a clause in } \psi\} \cup \{0\}$$

A Steiner tree for $S$ is *vertex-nice* if it only contains vertices that are variable, clause, or literal points. Let the *natural mapping* be the mapping from variable, clause, literal points, and $0$ that maps a variable point of $y_i$ to $y_i$, a literal point of $y_{ir}$ to $y_{ir}$, a clause point of $C$ to $C$, and $0$ to $r$. A Steiner tree for $S$ is *nice* if it is vertex nice and the natural mapping maps it to a a subtree of $f(\psi)$, using the reduction $f$ defined in the last section. By the remarks following Lemma 2.1 and Corollary 2.4, the results proved for graphs earlier in this section apply to nice trees as well. Thus, all we have to prove is that any $k$-restricted Steiner tree for $S$ can be transformed in polynomial-time into a no costlier nice $k$-restricted Steiner tree for $S$. This follows from the next two lemmas.

**Lemma 2.6.** *Any $k$-restricted Steiner tree for $S$ can be transformed in polynomial time into a no costlier vertex-nice $k$-restricted Steiner tree for $S$.*

*Proof.* Let us call a full component of a Steiner tree *good* if it is either a single edge or a star whose center is a literal point; all other full components are *bad*. The total number of leaves in all bad components of a Steiner tree $T$ is its *bad leaf sum*. Let $T$ be any $k$-restricted Steiner tree for $S$. It is enough to show that, unless the bad leaf sum of $T$ is 0, we can, in polynomial time, perform an operation on $T$ that yields a $k$-restricted Steiner tree for $S$ with smaller bad leaf sum than $T$ and whose length is at most that of $T$; we say that such an operation *improves* $T$. It is clear that we can assume that no Steiner vertex of $T$ has degree two in $T$.

Assume that $F$ is a bad full component of $T$. Let $v$ be an internal vertex of $F$ of largest distance from $0$ in $T$. Let $u$ be the neighbor of $v$ of smallest distance from $0$ in $T$. Note that all vertices except one in the neighborhood of $v$ in $T$ are leaves of $F$ different from $u$.

Let $N_T(w)$ be the set of neighbors of a node $w$ of $T$; $d(a,b)$ will denote the (Hamming) distance between the points $a$ and $b$ in the hypercube; $d_T(a,b)$ will denote the distance between the points $a$ and $b$ in the tree $T$. Obviously, $d_T(a,b) \geq d(a,b)$ for all points $a,b$.

The following two observations are crucial to our proof. If there is a clause point $z \in N_T(v) \setminus \{u\}$ such that $d(z, v) \geq 3$, then removing the edge $(v, z)$ and adding the edge $(z, 0)$ improves $T$. If there is a variable point $x \in N_T(v) \setminus \{u\}$ in $T$ such that $d(x, v) \geq 2$, then removing the edge $(v, x)$ and adding the edge $(x, 0)$ improves $T$.

We now conclude the proof by considering three exhaustive cases. We will find that, in each case, we can improve $T$.

**Case 1:** $N_T(v) \setminus \{u\}$ contains two variable points $x$ and $x'$. Since $d(x, x') = 4$, we have $d(x, v) \geq 2$ or $d(x', v) \geq 2$. In both cases, we can improve $T$.

**Case 2:** $N_T(v) \setminus \{u\}$ contains two clause points $z$ and $z'$. Since $d(z, z') \geq 4$, either both $d(z, v)$ and $d(z', v)$ equal 2 or one of these distances at least 3. If one of distances is at least 3, we can improve $T$. If both distances equal 2, then there must be a literal $l$ shared by the clauses corresponding to $z$ and $z'$, respectively. Let $w$ be the literal point for $l$. Clearly, $d(w, z) = 2$ and $d(w, z') = 2$. Thus, we can improve $T$ by removing $v$ (along with all incident edges) and adding the literal point $w$ together with edges $(w, 0)$, $(w, z)$, and $(w, z')$.

**Case 3:** $N_T(v) \setminus \{u\} = \{x, z\}$ where $x$ is a variable point and $z$ is a clause point. Note that $d(x, z)$ is either 3 or 5. Furthermore, if it is 3 then the clause corresponding to $z$ contains a literal of the variable corresponding to $x$.

If $d(x, z) = 5$, then either $d(x, v) \geq 2$ or $d(z, v) \geq 3$. In both cases, we can improve $T$ as above.

If $d(x, z) = 3$, we can improve $T$ as follows. Remove $v$ (together with incident edges) and add a literal point $v'$ for the literal of the variable corresponding to $x$ that is contained in the clause corresponding to $z$. Furthermore, add the edges $(v', 0)$, $(v', x)$, and $(v', z)$.

**Lemma 2.7.** *Any vertex-nice $k$-restricted Steiner tree for $S$ can be transformed into a no costlier nice $k$-restricted Steiner tree $T$ for $S$ in polynomial time.*

The preceding lemma, together with earlier results, implies the following.

**Theorem 2.8.** *The minimum $k$-restricted Steiner tree problem in phylogeny is APX-complete.*

## 3 Bounds on the Steiner ratio in phylogeny

Borchers and Du proved matching lower and upper bounds for $\rho_k$. The lower bound holds for any metric space, including the one associated with the hypercube; we will now show that the upper bound holds for the hypercube as well. For this we will need to examine a certain metric space $M_N$ introduced by Borchers and Du [4].

Let $B_N$ be a binary tree with $N + 2$ levels, numbered from 0 to $N + 1$. The first $N + 1$ levels of $B_N$ form a complete binary tree; edges between levels $i - 1$ and $i$, $1 \leq i \leq N$ have length $2^{N-i}$. Level $N + 1$ consists of $2^N$ nodes, the $j$th of which is attached to the $j$th node at level $N$ by an edge of length 1. The metric space $M_N$ has the nodes of $B_N$ as its points; the distance between two points is the length of the path between them in $B_N$. The leaves make up the set $U_N$ of given points. Borchers and Du [4] showed that for any $k$ with $k = 2^r + s$, where $0 \leq s < 2^r$,

$$\lim_{N \to \infty} \frac{L_{ST}(U_N)}{L_{kST}(U_N)} = \frac{r2^r + s}{(r+1)2^r + s}, \tag{2}$$

where, as in the Introduction, $L_{ST}(U_N)$ and $L_{kST}(U_N)$ denote the length of the optimum (unrestricted) Steiner tree and the length of the optimum $k$-restricted Steiner tree for $U_N$, respectively. We will prove an upper bound on $\rho_k(Q)$ by showing that the Borchers-Du construction can be embedded in the hypercube. Before proceeding, we will prove some preliminary results. For any two points $a$ and $b$ in a metric space $E$, and for any tree $T$ in $E$ containing $a$ and $b$, $d(a, b)$ will denote the distance between $a$ and $b$ in $E$, while $d_T(a, b)$ will be the total length of the path joining $a$ and $b$ in $T$.

**Theorem 3.1.** *Let $T$ be a tree such that the set of leaves of $T$ is $S$ and $d_T(a, b) = d(a, b)$ for each $a, b \in S$. Then $T$ is a minimum Steiner tree for $S$.*

The metric space $M_N$ will be embedded within a hypercube of dimension $2^N(N+1)$. The embedding is defined by a mapping $p$ that assigns a vector of character states to each node of $M_N$ so that the distance between nodes in $B_N$ equals the Hamming distance between the corresponding character vectors. In order to describe the state assignments, we will number the nodes of $B_N$ level by level, from left to right within a level. The $j$th node at level $l$, $0 \leq l \leq N$, $0 \leq j \leq 2^l - 1$, will be denoted by $u_{lj}$, while the $j$th node at level $N + 1$ will be denoted by $v_j$. Given this numbering scheme, the root is $u_{00}$ and an internal node $u_{lj}$ with $l \geq 1$ has parent $u_{l-1,\lfloor j/2 \rfloor}$. The $j$th leaf, $0 \leq j \leq 2^N - 1$, is denoted $v_j$; the parent of $v_j$ is $u_{N,j}$.

Since we will be dealing with species that, unlike those in the previous section, have many non-zero character states, we will use a different notation to describe character vectors (i.e., points on the hypercube). Each point will be denoted by a 0-1 string whose $i$th element equals the state of character $i$.

In the embedding we are about to describe, it may be useful to view each string of length $2^N(N+1)$ as being composed of blocks of length $2^N$, where the $l$th block corresponds to level $l$. The block for level $l$ is itself divided into equally-sized sub-blocks, where the $j$th sub-block corresponds to the $j$th node at level $l$. The mapping $p$ is such that the $j$th node at level $l$ has a unique sequence of 1's, stored in its sub-block, which appears in all of that node's descendants, but in no other nodes. The number of 1's is sufficiently large to guarantee the required distance from the node's parent. We now describe the construction precisely.

Let us write $c^b$ to denote the symbol $c$ repeated $b$ times, and "+" to denote the bitwise "or" of two 0-1 strings. Let $p(u_{00}) = 0^{2^N(N+1)}$. For $1 \leq l \leq N$ and and $0 \leq j \leq 2^l - 1$, we have

$$p(u_{lj}) = p(u_{l-1,\lfloor j/2 \rfloor}) + 0^{2^N(l-1)+2^{N-l}j} 1^{2^{N-l}} 0^{2^N(N-l+2)-2^{N-l}(j+1)}$$

For $0 \leq j \leq 2^N$, let

$$p(v_j) = p(u_{N,j}) + 0^{2^N N + j} 10^{2^N - j - 1}.$$

Let $d$ denote Hamming distance and let $d_B(a, b)$ denote the distance between two points $a$ and $b$ in the tree $B_N$ (which is also their distance in $M_N$). Observe that, for $l \geq 1$,

$$d(p(u_{lj}), p(u_{l-1, \lfloor j/2 \rfloor})) = d_B(u_{lj}, u_{l-1, \lfloor j/2 \rfloor}) = 2^{N-l}. \tag{3}$$

and

$$d(p(v_j), p(u_{N,j})) = d_B(v_j, u_{N,j}) = 1. \tag{4}$$

The preceding equations state that distances between neighbors in $B_N$ equal the distances between the corresponding points in the embedding. Indeed, since $p$ is such that the root of any subtree is assigned a unique sequence of 1's, which appears in all the descendants, but is not present in any other point, one can show the following.

**Lemma 3.2.** *For any two nodes $x$, $y$ in $B_N$, $d(p(x), p(y)) = d_B(x, y)$.*

By Theorem 3.1, $B_N$ is a minimum Steiner tree for $U = \{v_0, \dots, v_{2^N - 1}\}$ in $M_N$. By the same Theorem and Lemma 3.2, $B_N$ is also a minimum Steiner tree, of the same length, for $S = \{p(v_0), \dots, p(v_{2^N - 1})\}$ in the hypercube of dimension $2^N(N + 1)$. Moreover, we have the following result.

**Lemma 3.3.** *For every $k \geq 2$, the length of the optimum $k$-restricted tree for $M_N$ is the same as that of the optimum $k$-restricted tree for $S$ in the hypercube of dimension $2^N(N + 1)$.*

*Proof.* Let $T_Q$ be an optimum $k$-restricted tree for $S$ in the hypercube, and let $T_M$ be an optimum $k$-restricted tree for $S$ in $M_N$. By Lemma 3.2, $T_M$ is a $k$-restricted tree of the same cost in $Q$; thus, $L(T_Q) \leq L(T_M)$. We will prove the lemma by showing that $L(T_Q) \geq L(T_M)$. For this we shall argue that $T_Q$ can be transformed into a $k$-restricted Steiner tree in $M_N$ with at most the same cost. Let $K$ be any full component of $T_Q$ and let $K'$ be the minimal subtree of $B_N$ that contains the leaves of $K$. By Theorem 3.1, $K'$ must be a minimum Steiner tree for the leaves of $K$; moreover $K'$ is also a Steiner tree in the hypercube and $L(K') \leq L(K)$. Thus, we can replace $K$ with $K'$ and the result is a $k$-restricted tree of at most the same length as $T_Q$. Doing this recursively, we obtain the desired tree.

By the preceding discussion and equation (2),

**Theorem 3.4.** *For any $k$, with $k = 2^r + s$, where $0 \leq s < 2^r$,*

$$\rho_k(Q) = \frac{r 2^r + s}{(r + 1) 2^r + s}. \tag{5}$$

# References

1. S. Arora, C. Lund, R. Motwani, M. Sudan, and M. Szegedy. Proof verification and hardness of approximation problems. In *33rd FOCS*, pages 14–23, 1992.

2. P. Berman and V. Ramaiyer. Improved approximations for the Steiner tree problem. In *3rd SODA*, pages 1–10, 1992.

3. H. Bodlaender, M. Fellows, and T. Warnow. Two strikes against perfect phylogeny. In *Proceedings of the 19th International Colloquium on Automata, Languages, and Programming*, pp. 273–283, Springer Verlag, Lecture Notes in Computer Science, 1992.

4. A. Borchers and D.-Z. Du. The $k$-Steiner ratio in graphs. In *Proceedings 27th Annual Symposium on Theory of Computing*, pp. 641–249.

5. W.H.E. Day, D.S. Johnson, and D. Sankoff. The computational complexity of inferring rooted phylogenies by parsimony. *Mathematical Biosciences*, 81:33–42, 1986.

6. A. Dress and M. Steel. Convex tree realizations of partitions. *Appl. Math. Letters*, 5(3):3–6, 1992.

7. D.-Z. Du and F.K. Hwang. An approach for proving lower bounds: solution of Gilbert-Pollack's conjecture on the Steiner ratio. *Proceedings of the 31st Annual Symposium on Foundations of Computer Science*, pp. 76–85, 1990.

8. G. F. Estabrook, C. S. Johnson Jr., and F. R. McMorris. An idealized concept of the true cladistic character. *Mathematical Biosciences*, 23:263–272, 1975.

9. G. F. Estabrook. Cladistic methodology: A discussion of the theoretical basis for the induction of evolutionary history. *Annual Review of Ecology and Systematics*, 3:427–456, 1972.

10. W. M. Fitch. Aspects of Molecular Evolution. *Annual Reviews of Genetics*, 7:343–380, 1973.

11. L.R. Foulds and R.L. Graham. The Steiner tree problem in phylogeny is NP-complete. *Advances in Applied Mathematics*, 3:43–49, 1982.

12. D. Gusfield. The Steiner tree problem in phylogeny. *Unpublished manuscript.*

13. J. Håstad, J. Lagergren, and V. Kann. *Personal communication.*

14. F.K. Hwang, D.S. Richards, and P. Winter. *The Steiner Tree Problem.* North-Holland, Amsterdam, 1992.

15. T. Jiang, E.L. Lawler, and L. Wang. Aligning sequences via an evolutionary tree: Complexity and approximation (extended abstract). In *Proceedings of the Twenty-Sixth Annual ACM Symposium on Theory of Computing*, pp. 760–769, Montréal, Québec, Canada, 1994.

16. S. Khanna, R. Motwani, M. Sudan, and U. Vazirani. On Syntactic versus Computational Views of Approximability. In *35th Annual Symposium on Foundations of Computer Science*, pp. 819–830, Santa Fe, New Mexico, 1994.

17. C.H. Papadimitriou. *Computational Complexity.* Addison-Wesley, Reading, MA,1994.

18. C.H. Papadimitriou and M. Yannakakis. Optimization, approximation, and complexity classes. *Journal of Computer and System Sciences*, 43. Also in STOC '88.

19. M.A. Steel. The complexity of reconstructing trees from qualitative characters and subtrees. *Journal of Classification*, 9:91–116, 1992.

20. Tandy Warnow, Donald Ringe, and Ann Taylor. Reconstructing the evolutionary history of natural languages. To appear in *Proc. SODA 96.*

# Approximation and Special Cases of Common Subtrees and Editing Distance*

Magnús M. Halldórsson[1] and Keisuke Tanaka[2]

[1] Science Institute, University of Iceland
IS-107 Reykjavik, Iceland
mmh@rhi.hi.is
[2] School of Information Science,
Japan Advanced Institute of Science and Technology – Hokuriku
Tatsunokuchi, Ishikawa 923-12, Japan
tanaka@jaist.ac.jp

**Abstract.** Given two rooted, labeled, unordered trees, the *common subtree* problem is to find a bijective matching between subsets of vertices of the trees of maximum cardinality which preserves labels and ancestry relationship. The *tree editing distance* problem is to determine the least cost sequence of additions, deletions and changes that converts a tree into another given tree. Both problems are known to be hard to approximate within some constant factor in general.

We present polynomial algorithms for several special classes of trees as well as a tighter approximation hardness proof, which together comes close to characterizing the complexity of both problems on all interesting special classes of trees. We also present the first approximation algorithm with non-trivial approximation ratios. In particular, we achieve a ratio of $\log^2 n$, where $n$ is the number of vertices in the trees.

## 1 Introduction

An important part of computer science is in detecting and efficiently recognizing similarities among data sets or changes thereof. A natural measure of the difference between data sets is in the minimum cost sequence of atomic changes (or editing operations) that transform one into another.

One of the more pervasive structures of data are trees. We consider in this paper the *tree editing distance* problem and the *largest common (hereditary) subtree* problem, which model the difference and the similarity of two trees, respectively.

Given two rooted, labeled, unordered trees, the *common subtree* problem is to find a bijective matching between subsets of vertices of the trees of maximum cardinality which preserves labels and ancestry relationship.

---

* Work done in large part while the first author visited JAIST, first as PFU Visiting Chair, and later supported by JAIST Foundation. Also work done in part while the second author visited NTT Telecommunication Networks Laboratories under the supervision of Toshihiko Yamakami.

The *tree editing distance* problem is to determine the least cost sequence of additions, deletions and changes that converts one tree into the other input tree. Deletion of a vertex $v$ not only removes $v$ but makes the parent of $v$ become the new parent of the children of $v$. Addition of a vertex $v$ similarly makes $v$ become the parent of some subset of the children of the new parent of $v$. A change operation changes only the label of a vertex. Each of these operation carry a cost function that depends on the label of the respective vertices.

These problems have applications in various fields. Telecommunications networks are often in the form of rooted, labeled, unordered trees, and the rapid changes in active element have forced the introduction of automatic recognition in network design systems. Such similarity recognition problem occur also frequently in biocomputing, since different sequences (e.g. molecules, RNA) of the same type are never completely identical.

The editing distance and common subtree problems of two trees have been shown to be NP-hard, and thus no efficient algorithm are to be expected. The current paper deals with two ways of dealing with the intractability of the problem: efficient algorithms for important special classes of trees, and approximation algorithms with good performance bounds for general trees.

**Special classes of trees** Given the NP-hardness of the common subtree problem, it is valuable to study the complexity when the inputs are restricted to important special classes of trees.

We consider the following restricted classes of trees:

**Star** Trees where the root is the only internal node.

**Few internal nodes** Trees with some constant number of internal nodes. A superclass of *stars*.

**Few branching nodes** A branching node is a node of degree three or more. A superclass of *few internal nodes*, and of paths.

**Generalized Caterpillar** Trees where all branching nodes are lie on a single path. A superclass of *stars*, *paths*, and trees with at most 3 branching nodes.

**Small height** Trees whose height is bounded by a constant. A superclass of *few internal nodes*.

We present algorithms based on dynamic programming that solve the general editing distance problem for the classes of *few internal nodes, few branching nodes* and *generalized caterpillars*. For the other direction, we show that the common subtree problem and the editing problem are NP-hard, as well hard to approximate, if one tree is of height one (viz. a star) and the other is of height two.

These results characterize completely the solvability of the common subtree problem of trees from any combination of these classes. Namely, the problem is polynomial solvable iff either both trees are from one of the first four classes listed above, or if one tree has bounded number of leaves.

**Approximation** Approximation algorithms are heuristics that compute solutions that are not necessarily optimal but are guaranteed to be within some ratio from the optimal bound. The *performance ratio* of an approximation algorithm for the common subtree problem is the maximum, over all pairs of input trees, of the ratio between the size of the optimal common subtree and the size of the common subtree found by the algorithm.

We present the first approximation algorithm with sub-polynomial performance ratio. The ratio we achieve is $\log^2 n$. We also give an incomparable bound, that is twice the height of the shorter tree.

**Previous results** On computing the largest common subtree or editing distance between unordered labeled trees, the following results are known. It was shown in [3] that these problem are MAX SNP-hard for the general case, and thus cannot be approximated within some constant slightly larger than one. Previously, the problem had been shown NP-hard; even the restricted problem of deciding if the former tree was fully included in the latter tree. (Our reduction shows that the latter problem is also hard to approximate.)

A result for a special class of trees appears in [4], which contains an efficient polynomial-time algorithms for the case when one tree is a string (i.e. has only one leaf) or has a bounded number of leaves.

The only approximation result we are aware of is an incomplete manuscript of Yamaguchi [2]. It contains an algorithm for which a performance ratio of $O(OPT/\log OPT)$ for the largest common subtree problem is claimed, where $OPT$ is the size of the optimal solution. We note that this ratio is incomparable to our bounds, as it can be superior when the optimal solution is very small.

**Outline of paper** In Section 2 we present polynomial algorithms for special classes of trees, that apply both for the common subtree as well as the editing distance problems. We then give a tightened hardness result in Section 3. Section 4 contains several approximation algorithms for common subtrees of general trees. We then end with discussion of extensions to more general types of graphs.

## 2 Special classes of trees

We give in this section algorithms for computing the editing distance between two trees belonging to several special classes of trees. These can then easily be modified to output as well the largest common subtree.

We start with definitions of tree editing operations and their measures, and the equivalent concept of a mapping between subsets of nodes of the trees. We then give an algorithm based on dynamic programming for computing the editing distance between two trees with bounded number of internal nodes. These are then modified for the more general cases of trees with bounded number of branching nodes, and for generalized caterpillars.

**Editing operations and editing distance** We consider three kinds of operations: a) deleting a node $v$, where $v$ is removed from the tree while the children of $v$ become the children of the parent of $v$; b) adding a node $v$, which is the complement of deleting, as $v$ will become a parent of the subset of the children of the node that becomes the parent of $v$, and c) changing the label of a node.

These editing operations carry a cost function, that may depend on both the type of operation as well as the value of the label added/deleted/changed. We denote them by $del(u), add(v)$, and $chg(u,v)$, and assume that they satisfy a *distance metric.* The editing distance between $T_1$ and $T_2$ is defined as the total cost of a minimum cost sequence of edit operations that transform $T_1$ into $T_2$.

**Mappings** The editing distance is equivalent to finding a bijective (one-to-one) matching between subsets of the vertices of $T_1$ and $T_2$ that preserves the ancestor-descendant relationship.

For a mapping $M$, let $I$ denote the non-participating vertices in $T_1$, and $J$ the non-participating vertices in $T_2$. The cost of the mapping $M$ is defined as:

$$\sum_{(u,v)\in M} chg(u,v) + \sum_{u\in I} del(u) + \sum_{v\in J} add(v).$$

The largest common subtree problem is the natural complement of the editing distance problem, where we try to *maximize* the weight of the vertices that are not changed (or maximize the savings over deleting all vertices from $T_1$ and adding all from $T_2$).

$$\text{Weight of common subtree} = \frac{1}{2}\left(\sum_{u\in T_1} del(u) + \sum_{v\in T_2} add(v) - \sum_{(u,v)\in M} chg(u,v)\right)$$

Some of the algorithms apply only to the unweighted case, where addition and deletion cost one unit and changes cost two units.

**Notation** Let $F$ be a forest, i.e. a collection of trees, and let $|F|$ denote the number of nodes in $F$. Let $roots(F)$ denote a set of the roots of all the trees in $F$, and $tree(u)$ denote the subtree rooted at $u$. Notice that, if $T$ is a tree and $v$ is the root of $T$, let $T - v$ represent the forest obtained from $T$ by removing $v$. The editing distance between $F_1$ and $F_2$ is denoted by $dist(F_1, F_2)$.

## 2.1 The case of bounded number of internal nodes

Let the two trees be $T_1$ and $T_2$ with constant number of internal nodes. Let $F_1$ and $F_2$ be two subforest obtained from $T_1$ and $T_2$ by deleting nodes, respectively. We now introduce an observation and a lemma which assure the validity of our algorithms.

Let $M$ be one of the minimum cost mappings from $F_1$ to $F_2$ which induces the minimum length of edit operations. Let $\overline{M}$ denote the set of vertices that does not appear in the matching. The following observation is obvious. Since, for example, if $u \notin M$, then root node $u$ maps to no node and $u$ should be deleted. When $u$ is deleted from subtree $tree(u)$ in $F_1$, $tree(u)$ becomes subforest $tree(u) - u$.

**Observation 1.**

$$dist(F_1, F_2) = \begin{cases} \sum_{u\in F_1} del(u) + \sum_{v\in F_2} add(v) & \text{if } F_1 = \emptyset \text{ or } F_2 = \emptyset, \\ dist(F_1 - u, F_2) + del(u) & \text{if } \exists u, \ u \in roots(F_1) \cap \overline{M} \\ dist(F_1, F_2 - v) + add(v) & \text{if } \exists v, \ v \in roots(F_2) \cap \overline{M} \end{cases}$$

Here we consider the following condition:

$$u \text{ is a leaf or } u \in M, \text{ for all } u \in roots(F_1) \cup roots(F_2) \tag{1}$$

We can show the following (proof omitted).

**Lemma 2.** *Under (1), if* $|roots(F_1)| \leq |roots(F_2)|$, *then each root in* $F_1$ *maps to some root in* $F_2$; *otherwise, each root in* $F_2$ *is mapped from some root in* $F_1$.

**Algorithm** We now present a recursive algorithm **dist** with its sub-algorithm **distsub**, where $I = \{u|u \in roots(T_1)$ and $u$ is an internal node$\}$ and $J = \{v|v \in roots(T_2)$ and $v$ is an internal node$\}$.

Algorithm **dist** recursively calculates distances considering three major cases: one of the two forests is empty, there exists an internal root node which is not in $M$, and all internal root nodes are in $M$.

> $\text{dist}(F_1, F_2)$
>   if $(F_1 = \emptyset$ or $F_2 = \emptyset)$ then
>     return$(\sum_{u \in F_1} del(u) + \sum_{v \in F_2} add(v))$
>   else
>     Calculate $dist(F_1 - u, F_2) + del(u)$ for each $u \in I$
>     Calculate $dist(F_1, F_2 - v) + add(v)$ for each $v \in J$
>     Calculate $distsub(F_1, F_2)$
>     return the minimum of the three cases
> end

Algorithm **distsub** treats the case that all internal root nodes are in $M$. It constructs a balanced bipartite graph, with vertices for each root of the forests along with padding vertices corresponding to deleting and inserting operations. The weight of a pair of roots then depends on if both roots are matched in $M$, or if one of the roots is not matched (in which case it is not an internal node). The solution is obtained by computing a maximum weight matching on this graph.

> $\text{distsub}(F_1, F_2)$
>   $U \leftarrow roots(F_1)$
>   $V \leftarrow roots(F_2)$
>   W.l.o.g., assume $|U| \leq |V|$
>   if $|U| < |V|$ then add $|V| - |U|$ additional vertices to $|U|$.
>   $E \leftarrow \{(u, v) \mid u \in U, v \in V\}$
>   for each $u \in U$, $v \in V$ do
> 
> $$w(u,v) \leftarrow \begin{cases} dist(tree(u) - u, tree(v) - v) + chg(u,v) \\ \quad \text{if } u \in roots(F_1) \text{ and } v \in roots(F_2) \\ add(v) \quad \text{if } u \notin roots(F_1) \text{ and } v \text{ is a leaf in } F_2 \\ del(u) \quad \text{if } v \notin roots(F_2) \text{ and } u \text{ is a leaf in } F_1 \\ \infty \quad\quad \text{otherwise} \end{cases}$$
>
>   return$(MinimumWeightBipartiteMatching(U, V, E, w))$
> end

Then, we can compute the editing distance between the two trees by executing $dist(T_1, T_2)$ Correctness is obvious from Observation 1 and Lemma 2.

**Complexity** Let $\tau(t_1, t_2)$ be the time complexity of the algorithm **dist** on $T_1$ and $T_2$ with $t_1$ and $t_2$ internal nodes, respectively. Then,

$$\tau(t_1, t_2) \leq t_1 \cdot \tau(t_1 - 1, t_2) + t_2 \cdot \tau(t_1, t_2 - 1) + t_1 \cdot t_2 \cdot \tau(t_1 - 1, t_2 - 1) + O(n^3)$$
$$= O(t_1! \cdot t_2! \cdot n^3).$$

This is polynomial for $t_1, t_2$ as large as $O(\log n / \log \log n)$.

## 2.2 The case of a bounded number of branching nodes

Let $T_1$ and $T_2$ be trees with constant number of branching nodes. Let $F_1$ and $F_2$ be two subforest obtained from $T_1$ and $T_2$ by deleting nodes, respectively.

The algorithm for the case for two trees of a bounded number of internal nodes can be extended to this case.

Let $forest(U)$ denote a subforest containing all trees rooted at $u \in U$. Let $b(u)$ be the nearest branching descendant such that there is no branching node between the root $u$ and its descendant $b(u)$, and let $children(b(u))$ be a set of children of $b(u)$. If the root $u$ itself is a branching node, then $b(u) = u$. We denote all nodes between $u$ and $b(u)$ by $path(u)$.

We use the modified versions of the algorithm **dist** and the sub-algorithm **distsub**. We let $I = \{u | u \in roots(T_1) \text{ and } u \text{ has a branching descendant}\}$ and $J = \{v | v \in roots(T_2) \text{ and } v \text{ has a branching descendant}\}$, and use use $path(u)$ instead of internal node $u$. We modify a few lines of the algorithms. We use:

$$dist(path(u), \emptyset) \quad \text{and} \quad dist(\emptyset, path(v))$$

instead of $del(u)$ and $add(v)$, and we use

> Calculate
>     $dist(path(u) \cup tree(s), tree(v)) + dist(forest(children(b(u)) - \{s\}), \emptyset)$
>         for each $s \in children(b(u))$
> Calculate
>     $dist(tree(u), path(v) \cup tree(v)) + dist(\emptyset, forest(children(b(v)) - \{s\}))$
>         for each $s \in children(b(v))$
> Calculate $dist(tree(u) - path(u), tree(v) - path(v)) + dist(path(u), path(v))$
> **return** the minimum of the three cases

instead of $dist(tree(u) - u, tree(v) - v) + chg(u, v)$.

The distance between a path $S$ and a tree $T$ can be calculated in time $O(|S| \cdot |T|)$ [4]. We use this method to calculate $dist(path(u), tree(v))$ and $dist(tree(u), path(v))$ in the algorithm.

**Theorem 3.** *The algorithm correctly calculates the distance between $T_1$ and $T_2$ in polynomial time.*

*Proof.* Correctness can be checked similarly as for the previous case. Notice that the nodes in $path(u)$ of $F_1$ can map to both $path(v)$ and the nodes in one of the subtrees in $\bigcup_{s \in children(b(v))} tree(s)$ This can be handled in the modified version of the algorithm.

Let $t_1$ and $t_2$ be the number of branching nodes in $T_1$ and $T_2$, respectively. Let $\tau(t_1, t_2)$ be the time complexity of the algorithm **dist** with inputs $T_1$ and $T_2$. Then,

$$\tau(t_1, t_2) \leq t_1 \cdot \tau(t_1 - 1, t_2) + t_2 \cdot \tau(t_1, t_2 - 1) + t_1 \cdot t_2 \cdot \tau(t_1 - 1, t_2 - 1)$$
$$+ t_1 \cdot t_2 \cdot O(n^2) + t_1 \cdot n \cdot \tau(t_1, t_2 - 1) + t_2 \cdot n \cdot \tau(t_1 - 1, t_2) + O(n^3)$$
$$\leq (t_1! \cdot t_2! + t_1^{t_2} + t_2^{t_1} + O(n^2)) \cdot O(n^3) = O(n^{3 \cdot (t_1 + t_2)}).$$

Since $t_1$ and $t_2$ are constants, the running time of the algorithm is polynomial. ∎

## 2.3 The case of two generalized caterpillars

We here consider the case that each tree is a *generalized caterpillar*, which is a tree all whose branching nodes are on a unique path from the root to a leaf. Let $T_1$ and $T_2$ be generalized caterpillars. For this case, we use the same algorithm for two trees with a bounded number of branching nodes, while the analysis of the complexity is different. Correctness can be checked similarly as for the previous case.

**Theorem 4.** *The algorithm runs in time $O(n^4)$ on two caterpillars.*

*Proof.* Let $t_1$ and $t_2$ be the (unbounded) number of branching nodes in $T_1$ and $T_2$, respectively. Let $\tau(t_1, t_2)$ be the time complexity of the algorithm **dist** with inputs $T_1$ and $T_2$. Observing that there is always a unique non-path tree (i.e. generalized caterpillar) in each subforest of $F_1$ and $F_2$, it follows that

$$\tau(t_1, t_2) \leq \tau(t_1 - 1, t_2) + \tau(t_1, t_2 - 1) + \tau(t_1 - 1, t_2 - 1)$$
$$+ |roots(F_1)| \cdot |roots(F_2)| \cdot O(n_1 \cdot n_2) + O(n^3),$$

where $n_1$ and $n_2$ are the sizes of subtrees in $F_1$ and $F_2$ to be matched, respectively. Since $|F_1| = |F_2| = n$, $\tau(t_1, t_2) \leq O(n) \cdot (O(n^2) + O(n^3)) = O(n^4)$. ∎

# 3 Hardness

Zhang and Jiang [3] showed that the common subtree problem was NP-hard to approximate within some constant factor, even when one tree had a single branching node. However, the number of internal nodes and the height of the trees was not bounded. We present an approximation-preserving reduction that holds for a more restricted class of trees.

**Theorem 5.** *Let $T_1$ be restricted to stars, and $T_2$ restricted to trees of height 2. There is a fixed $\epsilon > 0$, such that it is NP-hard to distinguish between the following two cases: a) $T_1$ is a subtree of $T_2$, and b) the maximum common subtree of $T_1$ and $T_2$ contains at most $(1 - \epsilon)|T_1|$ vertices.*

**Reduction** We reduce from the problem 3-Set Packing-3:

*Given:* Finite set $S$ and a collection $C = \{C_1, C_2, \ldots, C_m\}$ of subsets of $S$ of size three, such that each element in $S$ appears at most three times in a set in $C$.

*Find:* A collection of maximum cardinality of disjoint sets from $C$.

This problem is known to be hard to approximate within some constant factor greater than one [1] (see results on 3-DM). More precisely, there exists a $\gamma > 0$, such that it is NP-hard to decide whether there is an *exact cover* of $(S, C)$, or if no set packing contains more than $n/3 - \gamma n$ sets.

Denote the elements of $C_i$ by $l_{i,1}, l_{i,2}, l_{i,3}$. Denote the elements of $S$ by $l_1, l_2, \ldots, l_n$, with $n$ being divisible by 3. Let $x$ be a new element not in $S$.

We construct two labeled trees as follows: $T_1$ is a star with $n + (m - n/3)$ rays (vertices of degree one): one for each label $l_i$, $i = 1, \ldots n$, and $m - n/3$ with the label $x$. $T_2$ is of three levels: the root which is unlabeled; $m$ internal nodes at level 2, all labeled with $x$ and representing the sets $C_i$; and, $3m$ leaves, one for each element $l_{i,j}$ of the sets $C_i$, $i = 1, 2, \ldots, m$, $j = 1, 2, 3$. A subtree rooted by a node at level 2 is called a *clause*.

An exact cover of $(S, C)$ yields a matching of all the leaves of $T_1$: using the leaves of $n/3$ subtrees of $T_2$, along with the $m - n/3$ remaining internal nodes labeled $x$.

On the other hand, suppose that we can match $n + (m - n/3) - (\gamma/2)n$ non-root vertices from $T_1$ and $T_2$. Without loss of generality, these include all the $l_1, l_2, \ldots, l_n$, along with $m - n/3 - (\gamma/2)n$ of the $x$-labeled vertices. That leaves $n/3 + (\gamma/2)n$ internal nodes in $T_2$ that are not matched and thus their leaves can be matched. Since we match all $n$ vertices, all three leaves are matched in at least $n/3 - \gamma n$ clauses. These induce a set packing in $(S, C)$, and we would be deciding an NP-hard problem.

Since $m \leq n$, the approximation ratio $\frac{n+(m-n/3)}{n+(m-n/3)-(\gamma/2)n}$ is at most $10/(10 - 3\gamma)$. Hence, if we can approximate the common subtree problem within $10/(10 - 3\gamma)$, then $P = NP$. This completes the proof of Theorem 5.

Observe that our construction has a *gap location* at 1. Namely, it is hard to determine if all the nodes can be matched, or if only a certain constant fraction of them. That implies that the complementary problem, the editing distance problem with unit costs, is also hard to approximate for the same class of trees.

## 4 Approximation Algorithms

We present two approximation algorithms for the maximum common subtree problem in this section, that are based on partitioning the input trees into easy subproblems.

**Common subforests vs. common subtrees** We shall be searching for common *subforests* in the inputs. By trying all $n^2$ choices for roots, this gives at least as good bounds for the common subtree problem as well.

**Approximation via partitioning** Suppose that we partition a forest into induced subgraphs; that is, each vertex is contained in exactly one partitioning class and the subgraph induced by the vertices within a class also forms a forest, under the ancestry relationship.

If we find solutions within each class (by matching the other input forest to the subgraph induced by each class), we can choose the largest one as the output.

This is because solutions to the common subforest problem are *hereditary* under vertex removal; subsets of solutions are also solutions. The optimal solution would be at most the union of optimal solutions of the classes. Hence, if we can find optimal solutions of each of the classes, the performance ratio attained would be at most the number of classes.

We generalize this argument to partitions of both input forests.

**Proposition 6.** *Suppose we partition input forests $F$ and $F'$ into induced subgraphs $F_1, F_2, \ldots, F_k$ and $F'_1, F'_2, \ldots, F'_l$, and compute $\rho$-approximate solutions of the maximum common subforest of $F_i$ and $F'_j$, for $i = 1, \ldots, k$, $j = 1, \ldots l$. Then the largest of these approximates the maximum weight common subforest of $F$ and $F'$ within a factor of $\rho \cdot k \cdot l$.*

*Proof.* Let $OPT(X, Y)$ denote the size of the optimal solution on trees $X$ and $Y$, and $HEU(X, Y)$ the size of the approximate solution computed. Then,

$$HEU(F, F') \geq \frac{1}{\rho} \max_{i,j} OPT(F_i, F'_j) \geq \frac{1}{\rho k l} \sum_{i,j} OPT(F_i, F'_j) \geq \frac{1}{\rho k l} OPT(F, F'). \blacksquare$$

**Approximation in terms of the height** We first give a simple greedy heuristic for the case of a zero-height forest (set of independent vertices) vs. a general forest, with a performance ratio of 2.

> Given an independent set $I$ and a forest $F$:
>> Repeat until $F$ is empty
>>> Pick any leaf $v$ of $F$ and delete it from $F$
>>> if ($v$ can be matched with a vertex $w$ of $I$) then
>>>> Add $v$ to solution
>>>> Delete $v$ and its ancestors from $F$
>>>> Delete $w$ from $I$
>> end

Each vertex $v$ added to the solution can eliminate at most two vertices from the optimal solution out of $I$: $w$, and a node that the optimal solution matched to some ancestor of $v$. Hence, the ratio attained is 2.

This yields a performance ratio for general trees that is asymptotic to the height of the shorter tree.

**Theorem 7.** *The (unweighted) common subtree problem can be approximated within a factor of $2\min(ht(F_1), ht(F_2))$.*

*Proof.* We can partition the shorter forest $F$ into $height(F)$ zero-height forests, and apply Proposition 6 to attain a ratio of $2height(F)$. $\blacksquare$

**Approximation in terms of the number of vertices** We now obtain a $\log^2 n$ performance ratio, by partitioning both input forests into at most $\log n$ different *linear forests* each, solving each pair optimally, and referring to Proposition 6. In fact, we get a more general ratio in terms of the total number of branching nodes in the trees.

A linear forest is one where each component is a path. Solving the common subforest problem of two linear forests is equivalent to the common subtree problem of trees with a single branching node. That we have shown how to do in polynomial time in the previous section.

**Lemma 8.** *A tree with $b$ branching nodes can be partitioned into $\log b$ linear forests.*

*Proof.* We apply the following bottom-up greedy strategy until the tree is empty:

Include in the class the nodes on the path from each leaf up to, but not including, the nearest branching node.

After each iteration, the new leaves are exactly the minimally lowest branching nodes. At least two leaves are descendants of each branching nodes, hence the number of leaves is halved in each step. The number of steps, and the number of classes, is therefore at most $\lfloor \log_2 b + 1 \rfloor$. This bound is tight on complete binary trees. ∎

Applying Proposition 6, Lemma 8 and the algorithm for single branching nodes, we obtain the following approximation.

**Corollary 9.** *Let $T_1$ and $T_2$ be trees with $b_1$ and $b_2$ branching nodes, respectively. The weighted common subtree of $T_1$ and $T_2$ can be approximated within a factor of $\log b_1 \cdot \log b_2$.*

We can generalize this to the case of finding common subtrees (subforests) in $t$ trees, $t$ constant.

**Theorem 10.** *The weighted common subtree problem of $t$ trees can be approximated within a factor of $t \log^t n$.*

*Proof.* Partitioning each tree into $\log n$ linear forests, we obtain $\log^t n$ subproblems. Each is an instance of a weighted $t$-dimensional matching, which can be approximated within a factor of $t$ by a greedy procedure in time $n^t$. The ratio then follows from Proposition 6. ∎

## Acknowledgments

We thank Atsuko Yamaguchi for valuable discussions.

## References

1. CRESCENZI, P., AND KANN, V. A compendium of NP optimization problems. Dynamic on-line survey available at ftp.nada.kth.se, Oct. 1995.
2. YAMAGUCHI, A. Approximation algorithm for the maximum common hereditary subtree problem. Manuscript. (In Japanese), Nov. 1994.
3. ZHANG, K., AND JIANG, T. Some MAX SNP-hard results concerning unordered labeled trees. *Inf. Process. Lett. 42* (1994), 133–139.
4. ZHANG, K., STATMAN, R., AND SHASHA, D. On the editing distance between unordered labeled trees. *Inf. Process. Lett. 42* (1992), 133–139.

# Two-Dimensional Dynamic Dictionary Matching

Ying Choi    Tak Wah Lam

Department of Computer Science, University of Hong Kong

Email: {ychoi, twlam}@cs.hku.hk

**Abstract.** This paper is concerned with the problem of managing a dynamically changing set of two-dimensional patterns (the dictionary) to support efficient searching for all occurrences of the patterns in a given text. The dictionary has to be updated efficiently when a pattern is inserted or deleted. The contribution of this paper is an improvement to the existing suffix-tree based solution [Gi93] as regards both the update and search algorithms. In comparison with the previously best scheme [AFI+93] (which is non-suffix-tree based), our new solution can perform an update more efficiently, without trading the searching time bound. Our work also gives a clue to improve the solution to the static dictionary matching problem [AF92].

## 1  Introduction

String pattern matching is a classical computation problem. It is concerned with searching for the occurrences of pattern string(s) in text string(s). In recent years, two dimensional pattern matching has also received a lot of attention due to its application in areas like image processing. This paper considers the following problem: We are given a set of patterns $D = \{P_1, P_2, \cdots, P_k\}$ (the dictionary), where each $P_i$ is a matrix of dimension $n_i \times n_i$, containing characters chosen from a bounded alphabet $\Sigma$. We want to devise a data structure to represent $D$ such that any sequence of the following operations can be executed efficiently.

- Dictionary Update: Insert a new pattern into $D$; delete a pattern from $D$.
- Text searching: Given any matrix $T$ (the text), search for all occurrences of patterns of $D$ in $T$.

It is natural to expect a data structure representing a set of patterns to integrate the similarity of patterns. That is, given a text, we do not need to match the text with each pattern separately. In the context of string matching, the method given by Amir et al. [AFG+94] can manage a dictionary $S$ of strings in such a way that insertion or deletion of a string of length $l$ takes time $O(l \log |S|)$, and searching a text of length $t$ uses time $O((t + \texttt{tocc}) \log |S|)$, where $|S|$ denotes the total length of strings in $S$ and $\texttt{tocc}$ is the total number of pattern occurrences. These time bounds were later improved by a factor of $\log \log |S|$ [AFI+93].

For the two dimensional case, it is obvious that inserting or deleting a pattern of dimension $n \times n$ requires time $\Omega(n^2)$, and searching a text of dimension $m \times m$ needs time $\Omega(m^2)$. If all patterns in a dictionary $D$ have the same dimension (say, $n_0 \times n_0$), Baker's work [Ba78] can be extended to insert and delete a pattern in time $O(n_0^2 \log \|D\|)$, where $\|D\| = \sum_{P_i \in D} n_0^2$, and search a text in time $O((m^2 + \texttt{tocc}) \log \|D\|)$.

Without the assumption of uniform dimension, the dynamic dictionary matching problem becomes non-trivial. There are currently two solutions for this problem, generalized from two typical approaches for string pattern matching, namely the failure function approach of Aho and Corasick and the suffix tree approach of McCreight. The first solution, given by Amir et al. [AFI+93], can update a dictionary $D$ with a pattern, of arbitrary dimension $n \times n$, in time $O(n^2 \log \|D\|)$, and perform a text searching in time $O((m^2 + \text{tocc}) \log \|D\|)$. The solution based on the suffix tree approach was proposed by Giancarlo [Gi93], it has slightly worse time bounds (see the table below for a comparison). Interestingly, the suffix tree has been one of the most efficient structure for many one dimensional, as well as some two dimensional, pattern matching problems; it is quite surprising that in the above setting the suffix tree approach is not as good as the other approach. It has been an interesting open problem whether the suffix tree approach can attain a better performance. In this paper, we show that the suffix tree approach is indeed no inferior, and more importantly, it gives rise to an improved solution providing the fastest update algorithm, without trading the time bound of searching. The following table compares these results. Let $|D| = \sum_{P_i \in D} n_i$.

| | Dictionary Update | Text Searching |
|---|---|---|
| Trivial lower bound | $\Omega(n^2)$ | $\Omega(m^2 + \text{tocc})$ |
| Amir *et al.* [AFI+93] | $O(n^2 \log \|D\|)$ | $O((m^2 + \text{tocc}) \log \|D\|)$ |
| Giancarlo [Gi93] | $O(n^2 \log^2 \|D\|)$ | $O((m^2 \log |D| + \text{tocc}) \log \|D\|)$ |
| This paper | $O(n^2 + n \log |D|)$ | $O((m^2 + \text{tocc}) \log |D|)$ |

Like the previous work [AFI+93], our text searching algorithm also works correctly for rectangular texts, but cannot handle rectangular patterns. Recently, Idury and Schäffer [IS95] have devised a solution for managing a dictionary of rectangular patterns. The dictionary update and text searching time increases by a factor of $\log^3 \|D\|$, though.

The remainder of this paper is organized as follows. Section 2 gives an overview of our new solution. Section 3 briefly reviews some fundamental data structures (suffix trees and dynamic trees) and several definitions used in [Gi95]. Section 4 discusses the augmented suffix tree and the constant time comparison of Lcharacters. Section 5 shows how the Lsuffix tree is adapted to support searching as fast as in the static case, while maintaining efficient updating. Section 6 sketches the usage of dynamic trees in reporting pattern occurrences.

## 2 Algorithmic Overview

The improvement stated above stems from, among others, a novel data structure for the following "substring" matching problem. Consider $S$ to be a set of strings $\{x_1, x_2, \cdots, x_m\}$, which can change over time when we insert and delete strings. Let $d_S$ denote the length of the longest string in $S$. We want a data structure to represent $S$ such that, given any string $y$ of length $l$, we can preprocess $y$ efficiently, then any substring of $y$ can be compared with any suffix of $x_i$ in constant time. Note that $y$ has as many as $\Theta(l^2)$ substrings, yet our scheme, based on a new data structure called the *augmented suffix tree*, can preprocess $y$ as fast as in time $O(l \log d_S)$.

The following sketches the way we solve the 2-dimensional dynamic dictionary matching problem.

**Definition:** Consider $T$ to be an $m \times m$ matrix. (i) Let $T[i:i', j:j']$, where $1 \le i \le i' \le m$ and $1 \le j \le j' \le m$, denote the submatrix of $T$ with the left upper corner at $(i, j)$ and right lower corner at $(i', j')$. If $i = i'$, the notation $T[i:i', j:j']$ is further simplified as $T[i, j:j']$. (ii) Let $T\langle i, j \rangle$, where $1 \le i, j \le m$, denote the largest square submatrix of $T$ with the left upper corner at $(i, j)$.

Following the work of Giancarlo, we decompose a matrix $A$ of dimension $n \times n$ into $n$ so-called Lcharacters (see Section 3.3 for definition) and represent $A$ as a string of $n$ Lcharacters (this string is called the Lstring of $A$). To manage a dictionary $D$, our scheme maintains three data structures:

I. **Two augmented suffix trees $U, U'$:** To search a text $T$ for the occurrence of patterns of $D$, we need to repeatedly compare an Lcharacter defined by $D$ with an Lcharacter defined by some square submatrix of $T$; thus, we want to preprocess $T$ so that any such comparison can be done as fast as in constant time. In section 4, we will show that, by maintaining two augmented suffix trees for $D$, we can preprocess a text $T$ of dimension $m \times m$ in $O(m^2 \log |D|)$ time, then it takes constant time to compare any two Lcharacters.

II. **An adapted version of Giancarlo's Lsuffix tree [Gi93, Gi95] $R$:** This is a compacted trie representing all Lstrings of the squares of $D$. Inside $R$, every edge is labeled with some Lcharacters and every pattern of $D$ defines a path starting from the root, of which the concatenation of the edge labels is exactly the Lstring of the pattern. The primary function of $R$ is to support the following searching operation in $O(\log |D|)$ amortized time: Given a text $T$ of dimension $m \times m$, we can match, for every position $(i, j)$ in $T$, the Lstring of the submatrix $T\langle i, j \rangle$ to the paths of the Lsuffix tree, and report the longest common prefix $\rho_{i,j}$. At first glance, $\rho_{i,j}$ may not look too meaningful. But in fact, if there exists some pattern $P$ of $D$ that matches the leftmost upper submatrix of $T\langle i, j \rangle$ then the Lstring of $P$ is also a prefix of $\rho_{i,j}$. The following data structure can reveal from $\rho$ every such $P$ efficiently.

III. **A forest of dynamic trees [ST83] $F$:** This is to represent the prefix relation of the Lstrings of the squares of $D$. In particular, it is used to report, for each $\rho_{i,j}$ found above, all the patterns of $D$ of which the Lstrings are a prefix of $\rho_{i,j}$, or equivalently, which matches the leftmost uppermost submatrix of $T\langle i, j \rangle$. The time required to report a pattern is $O(\log |D|)$. The idea of using dynamic trees [ST83] to maintain such relation is adopted from the scheme for one-dimensional dynamic dictionary matching [AFG+94].

In summary, with the above data structures, we can search the occurrences of patterns of $D$ in a matrix $T$ of size $m \times m$ in the following way. First of all, $T$ is preprocessed with respect to $U$ and $U'$. Then, for all positions $(i, j)$ in $T$, we find the longest prefix $\rho_{i,j}$ of $T\langle i, j \rangle$ that can match a path of $R$, and refer to the dynamic trees for reporting all patterns of $D$ whose Lstrings are a prefix of $\rho_{i,j}$. The total time required is $O(m^2 \log |D| + \text{tocc} \log |D|)$.

When a pattern of dimension $n \times n$ is inserted into or deleted from $D$, the augmented suffix tree can be updated in time $O(n^2)$ and the other data structures in time $O(n \log |D|)$.

# 3 Preliminary

**3.1 Suffix Tree:** Suffix tree is a data structure for representing a string. Let $x[1:n]$ be a string of $n$ characters chosen from a totally ordered alphabet $\Lambda$; let $\$ \notin \Lambda$ be a special symbol that does not match any character, including itself. Note that a suffix of the string $x\$$ cannot be the prefix of any other suffix. Let $I$ be a trie comprising the non-trivial suffixes of $x\$$, i.e. $x[1:n]\$$, $x[2:n]\$$, $\cdots$, $x[n:n]\$$. Every edge of $I$ is labeled with a character in $x\$$. A suffix tree $H$ for $x$ is a compacted version of $I$, in which the out-degree of an internal node (except the root) is at least two and every edge is labeled with a nonempty substring of $x\$$. For each node $u$, the path label of $u$ (or path-label($u$)) is defined to be the concatenation of the labels along the path from the root to $u$. $H$ has exactly $n$ leaves. The path label of each leaf is equal to a unique suffix of $x$ appended with $\$$. We store in $u$ the starting position of this suffix. When a suffix tree is built, we can put two information in each internal node (as well as each leaf) $u$, namely the suffix link and the length of path-label($u$). If path-label($u$) = $ax$ for some $a \in \Lambda$ and $x \in \Lambda^*$, the suffix link of $u$, also denoted by $SL(u)$, is a pointer to another internal node $w$ such that path-label($w$) = $x$.

The total number of nodes in a suffix tree is bounded by $2n$. In most cases $\Lambda$ is an alphabet of constant size; thus, the outdegree of a node is also a constant. When $\Lambda$ is of unbounded size, we use $n$ as an upper bound of the outdegree of a node and keep an AVL tree at each node to manage its out-going edges. Searching for the edge of a node labeled with a particular starting character takes $O(\log n)$ time. Using $H$, we can search all occurrences of $x$ in a text string $y$ of length $m$ in time $O(m)$ (or $O(m \log n)$ if $\Lambda$ is unbounded). The idea is to search every suffix of $y$ in $H$; if a suffix $y[i:m]$ can lead to the leaf representing $x\$$ then $x$ appears in $y$ starting from the position $i$. The suffix links in $H$ enable us to search each suffix in constant (or $\log n$) amortized time.

A suffix tree can be extended to represent a set of strings $S = \{x_1, x_2, \cdots, x_m\}$. Inserting or deleting a string of length $l$ requires time $O(l)$ (or $O(l \log |S|)$). Also, we can search all occurrences of strings of $S$ in a text string $y$ of length $m$ in time $O(m)$ (or $O(m \log |S|)$).

**3.2 Dynamic Trees:** In Sections 5 and 6, we will make use of a data structure called dynamic trees [ST83] to represent a forest of rooted trees, which can change over time. If the current number of nodes in the forest is $n$ then each of the following operation can be performed in time $O(\log n)$.

1. Root($v$): Return the root of the tree containing $v$;

2. Link($v, r$): Given a node $v$ of a tree and the root $r$ of another tree, we combine these two trees by adding an edge between $r$ and $v$ (i.e. $r$ becomes a child of $v$);

3. Cut($v$): $v$ is not the root of a tree, divide the tree containing $v$ into two trees by deleting the edge between $v$ and its parent;

4. New($v$): $v$ is a new node, create a new tree containing $v$ itself.

Moreover, we can assign a positive weight to every edge. Given a node $v$ and an integer $x$, the operation Ancestor($v, x$) returns a node $u$ which is the nearest ancestor of $v$ (possibly $v$ itself) such that the sum of the weights on the path from the root to $u$ is no more than $x$. This operation can also be implemented to run in time $O(\log n)$.

**3.3 Linear representation of a square matrix:** Let $P$ be an $n \times n$ matrix, with characters chosen from a bounded alphabet $\Sigma$. Below, we describe the way used by Giancarlo [Gi95] to represent $P$ as a string of $n$ special characters.

Let $L\Sigma$ be an infinite-size alphabet of which each character corresponds to a string of odd length over $\Sigma$. For example, $\Sigma = \{0, 1\}$; $L\Sigma = \{0, 1, 000, 001, 010, 011, 100, 101, \cdots\}$. Define an Lcharacter of $P$ to be a string in the form of $P[i, j:i-1]P[j:i, i]$ where $1 \leq j < i \leq n$. This Lcharacter is denoted by $Lc(P, i, j)$. The magnitude of an Lcharacter is defined to be the length of the corresponding string (over $\Sigma$). $A$ can be decomposed into $n$ Lcharacters, $a_1, a_2, \cdots, a_n$, where $a_i = P[i, 1:i-1]P[1:i, i]$ for $i \leq n$. Note that $a_i$ is in $L\Sigma$ and its magnitude is $2i - 1$. Define the Lstring of $A$, also denoted $\mathcal{L}(A)$, to be the string $a_1 a_2 \cdots a_n$, which is in $L\Sigma^*$. For any two square matrices $A_1$ and $A_2$, the $i$th Lcharacters of $\mathcal{L}(A_1)$ and $\mathcal{L}(A_2)$ must have the same magnitude.

# 4 Constant time comparison of Lcharacters

Let $D = \{P_1, P_2, \cdots, P_k\}$ be a dictionary of patterns, where each $P_i$ is a matrix of size $n_i \times n_i$, containing characters chosen from a bounded alphabet $\Sigma$. An Lcharacter of a matrix in $D$ is also called an Lcharacter of $D$. This section describes dynamic data structures for managing $D$ such that we can determine in constant time whether two Lcharacters of $D$ are equal. More importantly, the data structure supports the following operations efficiently.

- Matching: Given a matrix $T$, preprocess $T$ in such a way that we can compare in constant time an Lcharacter of any square submatrix of $T$ with an Lcharacter of $D$;

- Updating: Insert a new square $P$ into $D$, or delete an existing square $P_i$ from $D$.

As regards the time complexity, the preprocessing of an $m \times m$ text requires $O(m^2 \log B)$ time, where $B = \max_{i=1}^k n_i$; the update operation for a $n \times n$ square requires $O(n^2)$ time. Before giving the details of the data structures, we need the following definitions:

**Definition:** For any string $x[1:n]$ in $\Sigma^*$, define the upper string of $x$, denoted $U(x)$, to be the string $x[1:2^{\lfloor \log n \rfloor}]$, i.e. the longest prefix of $x$ with length equal to a power of 2. Similarly, the lower string of $x$, denoted $L(x)$, is defined to be $x[n - 2^{\lfloor \log n \rfloor} + 1:n]$, i.e. the longest suffix of $x$ with length equal to a power of 2. Obviously, for any strings $x, y \in \Sigma^*$ of equal length, $x = y$ if and only if $U(x) = U(y)$ and $L(x) = L(y)$.

By definition, an Lcharacter $Lc(P, i, j)$ of $D$ is a string in the form of $P[i, j : i - 1]P[j : i, i]$. Note that the first part of $Lc(P, i, j)$ is a suffix of $P[i, 1 : i - 1]$, or generally speaking, a suffix of a row, truncated at the diagonal, of a square in $D$. Below, we build an *augmented suffix tree* $U$ to keep track of the rows (truncated at the diagonal) of every square in $D$, i.e. the set of strings $\{P[i, 1 : i - 1] \mid P \in D; 1 < i \le n\}$. For any Lcharacter $Lc(P, i, j)$ of $D$, there always exist a pair of nodes whose paths from the root are labeled with $U(P[i, j : i - 1])$ and $L(P[i, j : i - 1])$, respectively; thus, we can represent $P[i, j : i - 1]$ by the addresses of the nodes corresponding to $U(P[i, j : i - 1])$ and $L(P[i, j : i - 1])$. $P[i, j : i - 1]$ is equal to another string $P'[i', j' : i' - 1]$ if and only if they are represented by the same pair of addresses.

Similarly, we can build another augmented suffix tree $U'$ for the columns of the squares in $D$; then, $P[j : i, i]$ can also be represented by two addresses in this tree, that correspond to $U(P[j : i, i])$ and $L(P[j : i, i])$. In summary, using two augmented suffix trees, we can represent an Lcharacter of a square in $D$ by four addresses. Any two Lcharacters are equal if and only if they are represented by the same addresses. Thus, we can check in constant time whether two Lcharacters are equal.

Below, we give the details of an augmented suffix tree for bookkeeping the rows of squares in $D$ and the way the matching and updating operations are supported. The case for the columns is similar and omitted.

| **Augmented Suffix Tree:** | Let $S = \{P[i, 1 : i - 1] \mid P \in D; 1 < i \le n\}$. Let \$ and # be two special symbols that does not match any character in $\Sigma$ including itself. Recall that the suffix tree of $S$ is a compacted trie comprising every suffix of a string $x_i \in S$ appended with \$. The augmented suffix tree of $S$, denoted by $U$, consists of two compacted tries $(U_1, U_2)$, both $U_1$ and $U_2$ comprise not only every suffix of a string $x_i \in S$ appended with \$, but also the upper string of each such suffix appended with #. Obviously, every node $\alpha$ of $U_1$ corresponds uniquely to a node $\alpha'$ of $U_2$, and they are indeed doubly linked by pointers.

In $U_1$ (or $U_2$), an internal node is said to be an ordinary one if it has two or more outgoing edges labeled with something other than #; otherwise, it is called a special one. Moreover, a leaf is said to be ordinary if its incoming edge is labeled with $; otherwise, it is also called special. Intutiviely, the ordinary internal nodes and leaves are those nodes that we can find in an ordinary suffix tree of $S$.

$U_1$ and $U_2$ keep different auxiliary information in their nodes. In $U_1$, every ordinary internal node, as well as every ordinary leaf, stores its suffix link, which points to another node of $U_1$. In $U_2$, each internal node $\alpha$ keeps a pointer, called *farthest mate*, which refers to the first node $\beta$ on the path from the root to $\alpha$ such that $\alpha$ and $\beta$ have the same upper string. The farthest mate of $\alpha$ is abbreviated as $Fm(\alpha)$.

For any Lcharacter $Lc(P, i, j)$ of $D$, there must exist a node in $U_1$ with a path label equal to the upper string of $P[i, j : i - 1]$. Moreover, the lower string of $P[i, j : i - 1]$ is equal to the upper string of some $P[i, k : i - 1]$, where $k \leq j$ (more precisely, $k = i - 2^{\lfloor \log(i-j) \rfloor}$), and it also corresponds uniquely to a node of $U_1$. Thus, we can always transform the upper string and the lower string of $P[i, j : i - 1]$ to two addresses of nodes of $U_1$. As to be shown later, whenever we insert a new square $P'$ of size $p' \times p'$ into $D$, we update $U$ and precompute a table $Ad(P', i, j)$, where $1 \leq j < i \leq p'$, to store the address corresponding to the upper string of $P'[i, j : i - 1]$.

| Matching: | Let $T$ be a matrix of size $m \times m$. An Lcharacter of a square submatrix of $T$ is a string in the form of $T[i, j : l - 1]T[k : i, l]$, where $i - k = l - j$. To compare such an Lcharacter with an arbitrary Lcharacter of $D$, we would like to transform the upper string and lower string of $T[i, j : l - 1]$ to two addresses of $U_1$, and similarly for $T[k : i, l]$. Then the comparison can be performed in constant time. To achieve constant-time transformation of $T[i, j : l - 1]$, we have the following observation: We only need to consider the case with $l - j \leq B$ (otherwise, no Lcharacter of $D$ has a first part big enough to match $T[i, j : l - 1]$), where $B$ is the dimension of the largest square in $D$. Therefore, the upper string, as well as the lower string, involved is of length $2^s$ for some integer $s \leq \lfloor \log B \rfloor$, and is in the form of $T[i, r : r + 2^s - 1]$.

We preprocess $T$ as follows: On every row $i$ of $T$, we consider all substrings of length $2^s$, where $0 \leq s \leq \lfloor \log B \rfloor$. If a substring $T[i, r : r + 2^s - 1]$ matches the path-label of a node of the trie $U_1$, we record the address of the node in an array $Ad_T$ indexed by $(i, r, s)$; otherwise, we fill in $-1$. The following procedure Mapping($i$) captures the details of preprocessing the $i$-th row of $T$.

---

**procedure Mapping($i$)**
**for** $r = 1$ **to** $m$ **do**
    1. Let $b_r$ be the length of longest prefix of $T[i, r : m]$ that matches a path of $U_1$; let $u_r$ be the node on this path, which is farthest from the root.
    /* Assume $b_0 = 0$ and $u_0$ equals the root; in general, $b_r$ and $u_r$ can be computed from $b_{r-1}$ and $u_{r-1}$ in constant amortized time using the suffix links and the array $Ad$. */
    2. $u' \leftarrow$ the node of $U_2$ doubly linked with $u_r$;   $h \leftarrow \lfloor \log B \rfloor$;
    3. **while** $u'$ is not the root **do**
        a. $v' \leftarrow$ the farthest mate of $u'$; $v \leftarrow$ the node in $U_1$ doubly linked with $v'$;
        b. Suppose $v'$ corresponds to a string of length $l$ where $2^d \leq l < 2^{d+1}$;
        c. **for** $s = d + 1$ **to** $h$ **do** $Ad_T[i, r, s] \leftarrow -1$;
        d. **if** $l = 2^d$ **then** $Ad_T[i, r, d] \leftarrow$ the address of $v$; **else** $f[i, r, d] \leftarrow -1$;
        e. $h \leftarrow d - 1$; $u' \leftarrow$ the parent of $v'$

In each iteration of Mapping($i$), step 1 uses $O(1)$ amortized time; step 3 use $O(\log B)$ time. Thus, the time required to execute Mapping($i$) is $O(m \log B)$. Preprocessing all the rows of $T$ requires a total of $O(m^2 \log B)$ time. Then, given any Lcharacter of a square submatrix of $T$, if the first part is $T[i, j : l - 1]$, we can base on the array $Ad_T$ to transform the upper string and lower string of $T[i, j : l - 1]$ accordingly in constant time. Similarly, we can preprocess every column of $T$; the result will enable us to transform the upper string and lower string of the second part of any Lcharacter accordingly in constant time.

**Insert:** When a square $P$ of dimension $p$ is inserted into $D$, we update the augmented suffix trees $U$ and $U'$ accordingly. Again, our discussion is based on $U$ only. Basically, for each row $i$ of $P$, we process the suffixes of $P[i, 1 : i - 1]$ from the longest to the shortest. For a suffix $P[i, j : i - 1]$ where $1 \le j < i$, we insert the strings $P[i, j : i - 1]\$$ and $U(P[i, j : i - 1])\#$ into $R$ as follows:

**Create a leaf for** $P[i, j : i - 1]\$$: Using a procedure similar to longest-prefix, we identify the longest prefix $\rho$ of $P[i, j : i - 1]$ that matches a path in $U_1$, as well as the node $u$, which is nearest to the root, encloses this path. If path-label($u$) is exactly $\rho$, we simply create a leaf $w$ with $u$ as its parent to represent $P[i, j : i - l]\$$. Otherwise, $\rho$ is a proper prefix of path-label($u$), we create a node $v$ between $u$ and $u$'s parent to represent $\rho$, then a leaf $w$ with $v$ as its parent to represent $P[i, j : i - l]\$$. Similarly, we create a leaf $w'$ and, if necessary, an internal node $v'$ in $U_2$ to represent $P[i, j : i - l]\$$.

**Install the additional pointers:** If the node $v$ has been created in $U_1$, $v$ is an ordinary internal node and the suffix link of $v$ will be installed as in a suffix tree (see [Mc76] for details). For $U_2$, if $v'$ has been created, we need to set the farthest mate of $v'$. Let $x'$ be the parent of $v'$ and let $u'$ be a child of $v'$ other than $w'$. **Step 1:** if path-label($v'$) $=_u$ path-label($x'$)[1] then $Fm(v')$ equals $Fm(x')$, or else $Fm(v')$ equals $v'$. **Step 2:** if $u'$ is not a leaf and path-label($u'$) $=_u$ path-label($v'$) $>_u$ path-label($x'$) then $u'$ and those descendants of $u'$ whose farthest mates are currently equal to $u'$ should update their farthest mates with $v'$. This is however very time consuming. To avoid such updating, we use the trick of reversing the roles of $v'$ and $u'$ (by exchanging the information inside $v'$ and $u'$ and updating the pointers to them), then $x'$ becomes the parent of $u'$ and $u'$ is the parent of $v'$ and $w'$. $Fm(v')$ should be set to $u'$ and $Fm(u')$ still equals $u'$.

**Create a leaf for** $U(P[i, j : i - 1])\#$: We first deal with $U_2$. Let $p(w')$ denote the current parent of $w'$ (defined above) in $U_2$. If $U(P[i, j : i - 1])$ is longer than $\rho$, we create a node between $p(w')$ and $w'$ to represent $U(P[i, j : i - 1])$. Otherwise, let $y'$ denote $Fm(p(w'))$. If $U(P[i, j : i - 1])$ is shorter than path-label($y$), we create a node as a parent of $y'$, or else $y'$ represents $U(P[i, j : i - 1])$. The farthest mate of the new node, if created, is set as described above. Finally, a new leaf is attached to this new node representing $U(P[i, j : i - 1])\#$. We also update $U_1$ accordingly. Once we have found a node in $U_1$ to represent $U(P[i, j : i - 1])$, we fill in the address of this node in the table $Ad(P, i, j)$.

**Time Complexity:** In the course of inserting each suffix of $P[i, 1 : i - 1]$ into $U$, it takes constant amortized time to find the corresponding prefix $\rho$, and constant time to create leaves and internal nodes, and install the necessary pointers. Thus, it takes $O(i)$ time to insert $P[i, 1 : i - 1]$ into $R$, and $O(n^2)$ time to insert all the rows of $P$.

The algorithm for deleting a pattern from $D$ is straightforward. The time required to delete a pattern of dimension $n \times n$ is also $O(n^2)$.

---

[1] For any string $s$ and any of its prefix $s'$, $s =_u s'$ if $s$ has the same upper string as $s'$; $s >_u s'$ if the upper string of $s$ is longer than that of $s'$.

## 5 Lsuffix Tree

Let $D = \{P_1, P_2, \cdots, P_k\}$ be a dictionary of patterns, where each $P_i$ is an $n_i \times n_i$ matrix. Below we present an adapted version of Giancarlo's Lsuffix tree [Gi95], denoted by $R$, to represent $D$. The searching algorithm for this tree is not too much different from the original one; our main contribution in this section is the algorithms for updating the tree.

$R$ consists of two parts—a compacted trie $R_t$ and a dynamic tree $R_d$ whose nodes and edges are the same as of $R_t$. $R_t$ looks like a suffix tree over the alphabet $L\Sigma$. More precisely, let $\mathcal{L}$ denote the set of Lstrings, $\{\mathcal{L}(P_i[1:n_i, 1:n_i])\$, \mathcal{L}(P_i[2:n_i, 2:n_i])\$, \cdots,$ $\mathcal{L}(P_i[n_i:n_i, n_i:n_i])\$ \mid P_i \in D\}$; $R_t$ is a compacted trie comprising the Lstrings of $\mathcal{L}$, satisfying the following properties:

1. Each internal node, except the root, in $R_t$ is of out-degree at least 2; every edge is labeled with a string of Lcharacters. Every leaf $w$ corresponds uniquely to a string $\mathcal{L}(P_i[j:n_i, j:n_i])\$ of $\mathcal{L}$ in the sense that the labels along the path from the root to $w$ is equal to $\mathcal{L}(P_i[j:n_i, j:n_i])\$. Also, $w$ stores a pointer called the suffix link of $w$ or simply $SL(w)$. Suppose $w$ corresponds to the Lstring $\mathcal{L}(P_i[j:n_i, j:n_i])\$. If $j = n_i$, $SL(w)$ points to the root; otherwise, $SL(w)$ points to the leaf corresponding to $\mathcal{L}(P_i[j+1:n_i, j+1:n_i])\$.

2. There is an association between the leaves and the internal nodes of $R_t$: Each leaf is associated with one internal node; each internal node $u$ of out-degree $d$ are associated with $d - 1$ leaves (except the root, the root is associated with $d$ leaves), which are in the subtree rooted at $u$.

A salient feature of $R_t$ is any path from the root, which is labeled with a sequence $\rho$ of Lcharacters, defines a unqiue matrix $A$, more importantly, if $\rho'$ is a prefix of $\rho$ then the matrix defined by $\rho'$ appears in $A$ as the leftmost upper submatrix. Both $R_t$ and $R_d$ contains at most $|D| = \sum_{P_i \in D} n_i$ leaves and internal nodes. The outdegree of a node is always bounded by $|D|$. In the following, we show that the adapted Lsuffix tree can support the following operations: **(1) Matching:** Given an $m \times m$ matrix $T$, we want to find, for all positions $(i, j)$ in $T$, the longest prefix of $\mathcal{L}(T\langle i, j \rangle)$ that matches a path of $R_t$. This requires a total of $(m^2 \log |D|)$ time (assuming that two Lcharacter can be compared in constant time). **(2) Update:** When an $n \times n$ matrix is inserted into $D$, $R$ can be updated in time $O(n \log |D'|)$, where $D'$ denotes the resulting dictionary. For a deletion of a pattern, $R$ can also be processed in time $O(n \log |D|)$.

**Matching:** Let $T$ be a matrix of dimension $m \times m$. We assume $T$ has been preprocessed in respect of the augmented suffix trees $U$ and $U'$. To find the longest prefix $\rho_{i,j}$ of $\mathcal{L}(T\langle i, j \rangle)$ for all positions $(i, j)$ in $T$, we process those $(i, j)$'s that are on the same diagonal together as a whole. It will be clear that each diagonal needs $O(m \log |D|)$ time, and the whole matrix $T$ requires $O(m^2 \log |D|)$ time. Let $X$ be one of the $2m - 1$ diagonals of $T$, say, $X = \{(1, d+1), (2, d+2), \cdots, (m-d, m)\}$ where $0 \le d < m$. To find $\rho_{1,d+1}$, we search $R_t$ with $\mathcal{L}(T\langle 1, d+1 \rangle)$ in a brute-force way. As a by-product, we also find the node $u_{1,d+1}$ which is farthest from the root and path-label$(u_{1,d+1})$ is a prefix of $\rho_{1,d+1}$. For the remaining positions in $X$, we search $R_t$ using the following two steps which, given $\rho_{i-1,d+i-1}$ and $u_{i-1,d+i-1}$ find $\rho_{i,d+i}$ and $u_{i,d+i}$ in $O((|\rho_{i,d+i}| - |\rho_{i-1,d+i-1}| + 1) \log |D|)$ worst-case time, or equivalently, in $O(\log |D|)$ amortized time.

1. Let $u'$ denote $u_{i-1,d+i-1}$; Let $\rho'$ denote $\rho_{i-1,d+i-1}$;
   if $u'$ is the root **then** $\alpha \leftarrow$ the root **else**

   i. **if** $u'$ is an internal node **then** $w \leftarrow$ any one of leaves associated with
      $u'$ **else** $w \leftarrow u'$;

   ii. $\alpha \leftarrow \text{Ancestor}(SL(w), |\rho'| - 1)$;

2. Let $h$ denote the length of the path label of $\alpha$. Let $\beta$ be a child of $\alpha$ such
   that the first Lcharacters labelling the edge $(\alpha, \beta)$ is equal to the $(h + 1)$-th
   Lcharacter in $\mathcal{L}(T\langle i, d + i\rangle)$. Starting from the $(|\rho'| - h)$-th Lcharcter of the
   label of $(\alpha, \beta)$, search down the tree by scanning the Lcharacters one by one
   until the search falls out of the tree. Then, $\rho_{i,d+i}$ is found and $u_{i,d+i}$ is the
   last node we have visited.

**Update:** Suppose an $n \times n$ pattern $P$ is inserted into $D$, resulting in a bigger
dictionary $D'$. Again, we assume the augmented suffix trees $U$ and $U'$ have been
updated with $P$ and an Lcharacter of $P$ can be compared with an Lcharacter of
$D$ in constant time. Then we insert the Lstrings appended with \$, $\mathcal{L}(P[1\!:\!p, 1\!:\!p])\$$,
$\mathcal{L}(P[2\!:\!p, 2\!:\!p])\$, \cdots, \mathcal{L}(P[p\!:\!p, p\!:\!p])\$$, into $R_t$ and update $R_d$ accordingly. The Lstrings
are inserted from the longest to the shortest.

For each Lstring $x_i = \mathcal{L}(P[i\!:\!p, i\!:\!p])\$$, we would like to search $R$ to match with
$x_i$ as many characters as possible, but a brute force search would require too much
time. Fortunately, except the case when $i = 1$, using the matching technique above,
we can identify in $O(\log |D'|)$ amortized time the longest prefix $\rho$ of $x_i$ that matches
a path from the root, as well as the nearest node $u$ enclosing this path. A node $z$ is
created to represent $\rho$ if it does not exist. Then, a leaf $w$ is created to represent $x_i$. In
order to maintain Condition 2, the leaf $w$ is associated with the node representing $\rho$.
If $i > 1$, the suffix link of the leaf representing $\mathcal{L}(P[i-1\!:\!p, i-1\!:\!p])\$$ should point to
the leaf $w$. If $i = p$, $SL(w)$ should point to the root. After all Lstrings $\mathcal{L}(P[i\!:\!p, i\!:\!p])\$$
have been inserted, the suffix links of all leaves created are defined correctly. Note
that whenever a node is inserted, we also update the dynamic tree $R_d$ which always
has the same structure as $R_t$.

**Time Complexity:** When the Lstring $x_i$ is inserted into $R_t$, it takes $O(\log |D'|)$
amortized time to search $R_t$ and constant time to create the necessary nodes and
maintain Condition 2; the dynamic tree $R_d$ requires $O(\log |D'|)$ time to update as at
most two edges and nodes are involved. Thus, it takes $O(n \log |D'|)$ time to process
the insertion of $P$.

The algorithm for deleting a pattern from the dictionary is quite straightforward
with the presence of Condition 2; details will be given in the full paper.

# 6    Reporting all occurrences of a pattern

Suppose we are given a string of Lcharacters $\rho$ that labels a path starting from the
root of $R_t$ (possibly ending between two nodes), and the last node $u$ on this path.
We would like to identify all patterns $P_i$ of $D$ such that $\mathcal{L}(P_i)$ is a prefix of $\rho$. A
brute force backward searching starting from $u$ would be too time consuming. Below,
we maintain a forest of dynamic trees to easy this job. Let $W$ be another dynamic
tree whose structure is the same as that of $R_t$. For a node $v$ in $R_t$, let $\hat{v}$ denote the
corresponding node in $W$. We obtain a forest of dynamic trees, denoted $F$, from $W$
as follows: A node $v$ in $R_t$ is marked if path-label$(v)$ is the Lstring of $P$ for some $P$

in $D$. $F$ is a forest of dynmaic trees obtained by deleting the edges $(p(\hat{u}), \hat{u})$ from $W$ for all marked nodes $v$ of $R_t$. Note that for all marked nodes and the root of $R$, their corresponding nodes in $F$ are roots of some dynamic trees.

Given $\rho$ and $u$, the following procedure shows how to find all patterns $P_i$ of $D$ such that $\mathcal{L}(P_i)$ is a prefix of $\rho$.

1. /* report a pattern whose Lstring is longer than the path-label of $u$ */
   if path-label($u$) is a proper prefix of $\rho$ **then**
   let $v$ be the child of $u$ such that $\rho$ is a prefix of path-label($v$); if $v$ is a leaf representing $\mathcal{L}(P)\$$ for some $P \in D$ **and** $|\rho| = |\mathcal{L}(P)|$ **then** report an occurrence of $P$;

2. $\hat{v} \leftarrow \text{Root}(\hat{u})$;

3. **while** $v$ is not the root of $R_t$ **do** /* report shorter patterns */
   Suppose $v$ represents $\mathcal{L}(P)$ for some $P \in D$; report an occurrence of $P$; $\hat{v} \leftarrow \text{Root}(p(\hat{v}))$

The time required to report a pattern is dominated by executing the Root operation and is at most $O(\log|D|)$. The algorithms for updating $F$ in accordance with the change of $R_t$ due to an insertion or deletion of patterns is basically the same as in [AFG+94]. The time bound is $O(n\log|D|)$ as there are $O(n)$ edges and nodes inserted into or deleted from $R_t$.

# References

[AC75] A.V. Aho and M.J. Corasick, Efficient string matching, *Comm. ACM* 18 (1975) 333-340.

[Ba78] T.J. Baker, A technique for extending rapid exact-match string matching to arrays of more than one dimension, *SIAM J. Comput.* 7 (1978), 533-541.

[AF92] A. Amir and M. Farach, Two-dimensional dictionary matching, *Information Processing Letters*, 44 (1992), 233-239.

[AFG+94] A. Amir, M. Farach, R. Giancarlo, Z. Galil and K. Park, Dynamic dictionary matching, *Journal of Computer and System Sciences* 49 (1994), 208-222.

[AFI+93] A. Amir, M. Farach, R.M. Idury, H.A. La Poutre and A.A. Schäffer, Improved dynamic dictionary matching, *Proc. of the 4th Annual ACM-SIAM Symposium on Discrete Algorithms*, 1993, 392-401.

[Gi93] R. Giancarlo, The suffix of a square matrix, with applications, *Proc. of the 4th Annual ACM-SIAM Symposium on Discrete Algorithms*, 1993, 402-412.

[Gi95] R. Giancarlo, A generalization of the suffix tree to square matrix, with applications, *SIAM Journal of Computing* 24 (1995), 520-562.

[IS95] R. Idury and A.A. Schäffer, Multiple matching of rectangular patterns, *Information and Computation*, 117 (1995), 78-90.

[Mc76] E.M. McCreight, A space-economical suffix tree construction algorithm, *J. ACM* 23 (1976), 262-272.

[ST83] D.D. Sleator and R.E. Tarj an, A data structure for dynamic trees, *Journal of Computer and System Sciences* 26 (1983), 362-391.

# Discovering Unbounded Unions of Regular Pattern Languages from Positive Examples

## (Extended Abstract)

Alvis Brāzma[1] Esko Ukkonen[2] Jaak Vilo[2]

[1] Institute of Mathematics and Computer Science, University of Latvia
29 Rainis Bulevard, LV-1459 Riga, Latvia
abra@cclu.lv
[2] Department of Computer Science, University of Helsinki
P.O.Box 26, FIN-00014 University of Helsinki, Finland
ukkonen,vilo@cs.helsinki.fi

**Abstract.** The problem of learning unions of certain pattern languages from positive examples is considered. We restrict to the regular patterns, i.e., patterns where each variable symbol can appear only once, and to the substring patterns, which is a subclass of regular patterns of the type $x\alpha y$, where $x$ and $y$ are variables and $\alpha$ is a string of constant symbols. We present an algorithm that, given a set of strings, finds a good collection of patterns covering this set. The notion of a 'good covering' is defined as the most probable collection of patterns likely to be present in the examples, assuming a simple probabilistic model, or equivalently using the Minimum Description Length (MDL) principle. Our algorithm is shown to approximate the optimal cover within a logarithmic factor. This extends a similar recent result for the so-called simple patterns. For substring patterns the running time of the algorithm is $O(nN)$, where $n$ is the number and $N$ the total lenght of the sequences.

## 1   Introduction

Given a set of sequences, we want to cluster them into subsets so that sequences in each subset share some common property, and find these properties. We are primarily interested in *biosequences*, i.e., sequences corresponding to biological macromolecules. Ideally, the subsets should correspond to subfamilies having some biological meaning. The set may contain noise, i.e., some sequences may be faulty or irrelevant, and we want these sequences to be left out. The level of noise and the maximal number of subfamilies are *a priori* not known.

By a shared property we simply mean a pattern of symbols that is contained in all sequences belonging to the same cluster. A natural formalism for describing such common patterns is the pattern languages of Angluin [1]. In many applications different regular subclasses of the pattern languages (such a pattern language belongs also to the set of regular languages) are reasonably sufficient and even such a simple subclass as substring patterns, i.e., a presence of a certain substring in strings, is useful in computational molecular biology

[18, 19]. In the current paper we consider regular patterns in the sense of [17] and substring patterns as a particular case.

Stated in machine learning terminology, our problem is to learn unions of unbounded number of regular pattern languages from positive examples. A similar problem for bounded unions has been considered in [2], where the learning in the limit (in the sense of Gold [7]) of such unions has been studied. The union of unbounded number of patterns cannot be learned in the limit from only positive examples [7]. Moreover, learning in the limit assumes that there are potentially infinitely many examples, while in practice the number of (even potential) examples is finite, and frequently quite small. The aim of learning for example in computational biology applications is not that much the identification in the limit of some hypotetical description of the sequences, as the discovery of interesting properties given a limited number of examples.

An approach to learning which is closer to the described situation can be based on the so called Minimum Description Length (MDL) principle [11, 15]. According to the MDL principle (in this context), the best pattern set is the one that minimizes the sum of

- the length (in bits) of the patterns; and
- the length (in bits) of data when encoded with the help of the patterns.

In practice one has to use some particular coding system, and thus to use a heuristic MDL principle. MDL based learning of pattern languages has been studied in [10]. The class of patterns considered in [10] consists of the so called simple patterns where variables correspond to unit-length substitutions (therefore all words in the generated pattern language have equal length and the language is finite). It is proved there that the best pattern according to the MDL principle using a natural coding scheme can be approximated in polynomial time within a logarithmic factor.

In this paper we extend the approach of [10] for regular patterns. Additionally, we study the Bayesian interpretation of MDL [14] in this particular case. We assume a probability distribution of patterns and sequences, and then look for the set of patterns having the highest conditional probability under the assumption that it has 'generated' the given examples. In other words, we look for a Bayesian maximum *a posteriori* estimate. Such a 'best' pattern is the same as the best according to our heuristic MDL if we select certain parameters in the MDL scheme as suggested by the probabilistic model of sequences and patterns. Unfortunately finding the best pattern set is NP-hard. Like in [10] we develop a greedy polynomial-time approximation algorithm and prove that it approximates the optimal solution within a logarithmic factor. For substring patterns the algorithm runs in $O(nN)$ time, where $n$ is the number of examples, and $N$ is the total length of examples.

The structure of the paper is the following. In the next three sections we define the problem and describe the Bayesian and MDL approaches. We also show that under some simple assumptions regarding the probability distribution the Bayesian and MDL approaches are equivalent. In Section 5 we prove the MDL approximation results and describe the respective algorithm. In the concluding section we mention some related results.

## 2 The problem

Let $\Sigma$ be a finite alphabet, and let $* \notin \Sigma$. A *regular pattern* $\pi$ is a nonempty string over the alphabet $\Sigma \cup \{*\}$. The pattern language $\mathcal{L}(\pi)$ generated by $\pi$ is the set of all words obtained from $\pi$ by replacing symbols $*$ (called the *gap characters*) by arbitrary strings from $\Sigma^*$. A *substring pattern* is a regular pattern of the type $*\varepsilon*$, where $\varepsilon \in \Sigma^+$. We say that pattern $\pi$ matches sequence[3] $\alpha$ if $\alpha \in \mathcal{L}(\pi)$.

Our aim is, given a set of sequences $A = \{\alpha_1, \ldots, \alpha_n\}$ over $\Sigma$, to find a 'good' set of patterns $\Pi = \{\pi_1, \ldots, \pi_k\}$ such that $A \subseteq \mathcal{L}(\pi_1) \cup \ldots \cup \mathcal{L}(\pi_k)$.

Sequences $\alpha_1, \ldots, \alpha_n$ are assumed to represent some experimental data (produced in a definite order and possibly having repetitions). We want to define the notion of a 'good' set of patterns in some natural way corresponding to our intuition. For instance, it should be independent of the order of the sequences and we want to exclude such trivial solutions as $k = n$ and $\pi_i = \alpha_i$, for $i = 1, \ldots, n$; or $k = 1$ and $\pi_1 = *$.

A sequence $\alpha_i$ can have several matching patterns in $\Pi$. We assume that only one of them is the correct pattern and make this explicit as follows. Let $B_1, \ldots, B_k$ be a partition of $A$ (i.e., $B_1, \ldots, B_k$ are disjoint and their union is $A$) and let $\Pi = \{\pi_1, \ldots, \pi_k\}$ be a set of patterns such that the pattern $\pi_j$ matches all the sequences of the set $B_j$. We call $\Omega = \{(\pi_1, B_1), \ldots, (\pi_k, B_k)\}$ a *cover* of $A$. We call $\Pi$ the *pattern set* of $\Omega$. Thus, the notion of the cover includes both– the partition and the patterns, and our problem can be formulated as finding the best cover.

## 3 Bayesian approach

Let us first define the best $\Pi$ using Bayesian maximum *a posteriori* estimation. By Bayes' formula, the conditional probabilities satisfy

$$P(\Pi|A) = \frac{P(A|\Pi) \cdot P(\Pi)}{P(A)}. \tag{1}$$

The 'best' $\Pi$ maximizes $P(\Pi|A)$, the conditional probability of $\Pi$ given $A$. As $P(A)$ does not depend on $\Pi$ we have to find $\Pi$ that maximizes the product $P(A|\Pi) \cdot P(\Pi)$. To this end, let us introduce a simple Bernoullian probabilistic model for sequences that gives $P(A|\Pi)$ and $P(\Pi)$.

Recall that $\Pi = \{\pi_1, \ldots, \pi_k\}$ consists of sequences over $\Sigma \cup \{*\}$. Let $\Sigma = \{a_1, \ldots, a_m\}$, and let \$ be a character (a hypothetical endmarker) such that $\$ \notin \Sigma \cup \{*\}$. To define $P(\pi)$ let us assume some probability distribution $P_1$ over the set $\Sigma \cup \{*\} \cup \{\$\}$, i.e., $P_1(a_1) + \cdots + P_1(a_m) + P_1(*) + P_1(\$) = 1$, such that $P_1(\$) = q > 0$. Let $\pi = e_1 \ldots e_h \in (\Sigma \cup \{*\})^*$ be a pattern. We define the *a priori* probability of $\pi$ as

$$P_1(\pi) = P_1(e_1) \cdots P_1(e_h) P_1(\$).$$

---

[3] 'Sequence' and 'string' are used as synonyms in this paper.

Note that the probability that $\pi$ has length $l = h$ is $(1 - q)^h q$. A similar length distribution has been used in [14] for MDL learning of decision trees. It follows from Kraft's equality (e.g., see [11]) that $\sum_{\pi \in (\Sigma \cup \{*\})^*} P_1(\pi) = 1$. In practice the probability distribution $P_1$ can be used as a tool for giving preferences to different symbols in patterns.

We define the *a priori* probability of pattern set $\Pi = \{\pi_1, \ldots, \pi_k\}$ as

$$P(\Pi) = P_1(\pi_1) \cdots P_1(\pi_k) \cdot (1 - r)^k r \tag{2}$$

for some $0 < r < 1$. Parameter $r$ gives the probability of the hypothetical endmarker of $\Pi$. Parameters $q$ and $r$ determine the *a priori* average length and number of patterns. It follows from (2) that the sum of probabilities of all pattern sets normalizes to 1.

To define $P(A|\Pi)$ we will next fix a string representation for $A$ when $\Pi$ has been given. The probability of this string then gives $P(A|\Pi)$.

If pattern $\pi$ is known to match sequence $\alpha$, then the most economical way to describe $\alpha$ is just to give the sequences $\gamma_1, \ldots, \gamma_h$ that give $\alpha$ from $\pi$ when substituted for the $h$ gap symbols $*$ of $\pi$. We introduce a probability distribution for substitutions $\gamma_1, \ldots, \gamma_h$, and define $P(\alpha|\pi)$ as the probability of the most probable substitution.

More precisely, let $P'(a_i)$ be some probability distribution over $a_i \in \Sigma$. In practice we take $P'(a_i)$ proportional to the relative frequency of $a_i$ in a sequence database (or in $A$). Let a hypothetical endmarker $\$_2 \notin \Sigma$ have probability $P_2(\$_2) = p > 0$, and let us normalize the probabilities $P'(a_i)$ to $P_2(a_i) = (1 - p)P'(a_i)$.

Let $\gamma_j = c_1 \ldots c_l \in \Sigma^*$. We define $P_2(\gamma_j) = P_2(c_1) \cdot \ldots \cdot P_2(c_l) \cdot P_2(\$_2)$, and

$$P(\alpha|\pi) = \max\{P_2(\gamma_1) \cdot \ldots \cdot P_2(\gamma_h) \mid \text{ substituting } \gamma_1, \ldots, \gamma_h \text{ for } * \text{ in } \pi \text{ gives } \alpha\}.$$

Given set $B = \{\beta_1, \ldots, \beta_m\}$ of sequences, we define $P(B|\pi) = P(\beta_1|\pi) \cdot \ldots \cdot P(\beta_m|\pi)$. Similarly for a given cover $\Omega = \{(\pi_1, B_1), \ldots, (\pi_k, B_k)\}$ of $A$, we define $P(A|\Omega) = P(B_1|\pi_1) \cdot \ldots \cdot P(B_k|\pi_k)$, and finally

$$P(A|\Pi) = \max\{P(A|\Omega) \mid \Omega \text{ is a cover of } A \text{ with pattern set } \Pi\}. \tag{3}$$

In essence this probabilistic model is similar to that of [13].

To maximize (1) we can just as well minimize $-\log Pr(\Pi|A) = -\log P(A|\Omega) - \log P(\Pi)$. This minimization can be interpreted as length-minimization problem as follows. We need some notations. Let $P_x$ be a probability distribution over some alphabet $\Sigma' \supset \Sigma$. For a sequence $\alpha = a_1 a_2 \ldots a_l \in \Sigma^*$ we define

$$\ell_{P_x}(\alpha) = -(\log P_x(a_1) + \ldots + \log P_x(a_l)).$$

The value of $\ell_{P_x}(\alpha)$ can be interpreted as the length of $\alpha$ in bits in some optimal coding in respect to the probability distribution $P_x$. For set $A = \{\alpha_1, \ldots, \alpha_n\}$, we define $\ell_{P_x}(A) = \sum_{i=1}^n \ell_{P_x}(\alpha_i)$. Let then $\Omega = \{(\pi_1, B_1), \ldots, (\pi_k, B_k)\}$ be a cover of $A$ and let the size of each $B_j$ be $l_j$. It can be shown that

$$-\log P(A|\Omega) = \ell_{P_2}(A) - \sum_{j=1}^k l_j(\ell_{P_2}(\pi_j) + n_*(\pi_j) \log p),$$

where $n_*(\pi_j)$ is the number of characters $*$ in $\pi_j$ (and $p = P_2(\$_2)$). In a similar way

$$-\log P(\Pi) = \ell_{P_1}(\Pi) - k\log(1-r) - \log r.$$

After regrouping of the terms we obtain that the most probable pattern set $\Pi$ is the pattern set of cover $\Omega$ that minimizes expression

$$M(\Omega) = M_1(\Omega) + M_2(\Omega),$$

where

$$M_1(\Omega) = \ell_{P_1}(\Pi) - \log r,$$

and

$$M_2(\Omega) = \ell_{P_2}(A) - \sum_{j=1}^{k} l_j(\ell_{P_2}(\pi_j) + n_*(\pi_j)\log p) - k\log(1-r).$$

# 4 MDL approach

The MDL principle offers another possibility to define essentially the same optimality criterion as in Section 3 for pattern set $\Pi$ as follows (c.f. [10]). Let us assume, that we want to transmit the sequences $A = \{\alpha_1, \ldots, \alpha_n\}$ over a channel. The trivial way is simply to transmit $\alpha_1, \ldots, \alpha_n$ one after another, using some optimal prefix code for the characters in $\Sigma$. Then by choosing the optimal coding according to the probability distribution $P_2$, the message length tends to $\ell_{P_2}(A)$.

If we know that set $A$ has cover $\Omega$, we can use this information for compressing the message. First we send the message '$\pi_1\$\pi_2\$ \ldots \pi_k\$:$', i.e., the patterns of $\Omega$ separated by delimiters $\$$ and some final delimiter ':'. Characters of the pattern alphabet $\Sigma \cup \{*\}$ and the delimiters $\$$ and : are encoded by variable lenght bit-strings in some prefix code depending on probabilities $P_1$. In an optimal coding the length of this part of the message tends to $M_1(\Omega)$.

After patterns we send substrings by which we have to replace the $*$ characters to obtain the strings of $A$ from the patterns. The substrings are separated by $\$_2$. Some additional delimiter is signaling the switching to the next pattern. The characters are again coded by bit-strings of different length according to the probability distribution $P_2$. The length of this part of the message tends to $M_2(\Omega)$.

The compression in respect to the trivial encoding $\ell_{P_2}(A)$ (up to some additive constant) tends to

$$C(\Omega) = \sum_{j=1}^{k}(u_j \cdot l_j - w_j), \tag{4}$$

where

$$u_j = \ell_{P_2}(\pi_j) + n_*(\pi_j)\log p \quad \text{and} \quad w_j = \ell_{P_1}(\pi_j) - \log(1-r). \tag{5}$$

Evidently minimization of $M(\Omega)$ equals to maximization of $C(\Omega)$. The best $\Pi$ according to our heuristic MDL is the pattern set of the $\Omega$ that minimizes message length $M(\Omega)$.

We have explicitly established the connection between the MDL principle for the particular coding scheme and the Bayesian inference in the particular probability model. It allows us to estimate the code lengths of pattern and sequence alphabet characters and delimiters for MDL coding from the examples and from *a priori* expectance on the number of subclasses. Thus, we can estimate the constants in expressions (4) and (5). [4]

As the substring patterns are a subclass of regular patterns, the most probable pattern set $\Pi$ for substring patterns can be obtained similarly[5].

## 5 MDL approximation algorithm

In this section we will deal with the problem of minimizing the description length, i.e., with the implementation of MDL principle. The problem that we want to solve is – given the set $A$, find a cover $\Omega = \{(\pi_j, B_j) \mid j = 1, \ldots, k\}$, $B_j = \{\beta_1^j, \ldots, \beta_{l_j}^j\}$, such that expression $M(\Omega)$ reaches minimum or $C(\Omega)$ reaches maximum. We denote the minimum value of $M(\Omega)$ by $M_{opt}$.

Because maximization of expression (4) contains the NP-hard set covering problem as a special case [10], the problem of finding $M_{opt}$ for a given set $A$ is NP-hard. Hence we will develop a fast approximation algorithm. We will prove two approximation theorems – the first for the case of substring patterns and the second for the case of regular patterns. Let us denote the length of a string $\alpha$ by $|\alpha|$. For $A = \{\alpha_1, \ldots, \alpha_n\}$ we define $\|A\| = \sum_{j=1}^{n} |\alpha_j|$, and $|A| = n$.

**Theorem 1.** *There exists an algorithm* Greedy_substrings *that, given a set of strings $A$, finds a substring pattern cover $\Omega_S$ of $A$ such that $M(\Omega_S) \leq M_{opt} \times \log |A| + O(1)$ in time $O(|A| \cdot \|A\|)$.*

In the case of more general patterns the complexity of the algorithm grows, but still is in some sense polynomial. Let $\Delta = \{\pi_1, \ldots, \pi_K\}$ be a set of patterns. By $M_{opt(\Delta)}$ we denote the minimal value of $M(\Omega)$ if the patterns are chosen only from the set $\Delta$. Let $\Theta$ be a set of pairs $\{(\pi_i, F_i) \mid i = 1, \ldots, K\}$ where each $F_i = \{\xi_1^i, \ldots, \xi_{l_i}^i\}$ is the subset of $A$ consisting of strings $\xi_j^i$ that contain pattern $\pi_i$ (i.e., $\Theta = \{(\pi_i, F_i) \mid F_i = \{\xi_1^i, \ldots, \xi_{l_i}^i\}, \xi_j^i \in L(\pi_i)\}$). We call $\Theta$ the *set of hypotheses*. Let $size(\Theta) = \|F_1\| + \cdots + \|F_K\|$, and let $\Delta(\Theta)$ be the pattern set used in $\Theta$. By $M_{opt(\Theta)}$ we denote the minimal value of $M(\Omega)$ if the patterns are chosen from $\Delta(\Theta)$ with the corresponding sets $B_i$ in $\Omega$ as subsets of $F_i$.

---

[4] By varying relative values of $P_1(\$)$, $P_1(*)$, $P_2(\$_2)$, and $r$, we can slant the optimum towards longer patterns common to fewer strings or vice-versa; or towards more or fewer gaps (i.e., characters $*$) in patterns.

[5] The only difference is that the patterns are generated as strings in the alphabet $\Sigma$ and $*$ prefixed and appended to these strings. To achieve normalization, the probability distribution $P_1$ can be assigned as $P_1(a_i) = (1 - 3P_1(\$))P(a_i)$.

**Theorem 2.** *There exists an algorithm* Greedy *that, given a set of strings $A$ and a set of hypotheses $\Theta$, finds a cover $\Omega_\Theta$ such that $M(\Omega_\Theta) \leq M_{opt(\Theta)} \times \log|A| + O(1)$ in time $O(size(\Theta) \cdot \log|A|)$.*

Note that a number of efficient heuristics are available for finding a hypotheses set $\Theta$ (e.g., [9, 16], for a survey see [3]).

**Corollary 3.** *There exists an algorithm* Greedy2 *that, given a set of strings $A$ and a set of patterns $\Delta$, finds a cover $\Omega_\Delta$ such that $M(\Omega_\Delta) \leq M_{opt(\Delta)} \times \log|A| + O(1)$ in time $O(\|A\| \cdot \|\Delta\| + |\Delta| \cdot |A| \cdot \log|A|)$.*

Corollary 3 follows from the fact that we can match all patterns of $\Delta$ against $A$ with a non-deterministic automaton of size at most $\|\Delta\|$ in time $O(\|A\| \cdot \|\Delta\|)$ and at the same time produce a hypotheses set $\Theta$ of size at most $|\Delta| \cdot |A|$.

Let $M_{opt(v)}$ be the minimal value of $M(\Omega)$ if the patterns in $\Omega$ can contain only up to $v$ variables (characters $*$). Then we obtain the following corollary.

**Corollary 4.** *There exists an algorithm* Greedy3 *that for any fixed $v$, given a set of strings $A$, finds a cover $\Omega_v$ such that $M(\Omega_v) \leq M_{opt(v)} \times \log|A| + O(1)$ in polynomial time in $\|A\|$.*

Corollary 4 follows from the fact that there exists only polynomial (in size of $A$) number of regular patterns with up to some fixed number $v$ of variables that match at least one string from the set $A$.

Next we will describe the algorithm Greedy and prove that it satisfies Theorem 2. Then we will show, how it can be implemented in a manner that it satisfies Theorem 1 in the case of substring patterns.

Let $u$, and $w$ be functions such that $u(\omega_j) = u_j$, and $w(\omega_j) = w_j$ defined by (5). Then the algorithm that computes a cover $\Omega_\Theta$ in a greedy fashion can be described as follows.

```
Greedy(A, Θ) return cover of A
        U ← A; Ω ← ∅; Γ ← Θ
        while U ≠ ∅ do
                find (πᵢ, Fᵢ) ∈ Γ maximizing  u(πᵢ)·|(Fᵢ∩U)|−w(πᵢ)
                                              ─────────────────────
                                                    |(Fᵢ∩U)|
                Ω ← Ω ∪ {(πᵢ, Fᵢ ∩ U)}
                U ← U − Fᵢ
                Γ ← Γ − {(πᵢ, Fᵢ)}
        return Ω
```

The algorithm is a variation of the greedy algorithm for the weighted set cover problem [6]. As suggested by (4), in our case the gain (in fact, the compression achieved) of selecting pattern $\pi_i$ to the cover is $u(\pi_i) \cdot |T_i| - w(\pi_i)$, where $T_i \subseteq A$ consists of strings in A that are covered by $\pi_i$ but not by any pattern that has already been selected by the algorithm to the greedy cover. Pattern $\pi_i$ that gives the maximal relative compression is selected ("greedily") next. To satisfy the complexity requirements of the theorem, the algorithm should be implemented in a slightly less straight-forward way using priority queue for managing the hypotheses set (for details see [5]).

Next, let us prove that the cover $\Omega_\Theta$ returned by the algorithm satisfies the theorems. Let $C_j(x) = u_j x - w_j$, for some reals $u_j$ and $w_j$. Thus if, $\Omega =$

$\{(\pi_j, B_j)|i = 1, \ldots, k$ and $B_j = \{\beta_1, \ldots, \beta_{l_j}\}\}$ is a cover of $A$, and $C$, $u_j$ and $w_j$ are defined by expressions (4) and (5), then $C(\Omega) = \sum_{j=i}^{k} C_j(l_j)$. Let $\Theta = \{(\pi_j, F_j)|i = 1, \ldots, K\}$ where $F_j \subset A$ and $\bigcup_{j=1}^{K} F_j = A$, be a set of hypotheses. Let $\Delta(\Theta) = \{\pi_1, \ldots, \pi_K\}$, and let $I = (i_1, \ldots, i_k)$ be a $k$-tuple of disjoint integers, such that each $i_j$ satisfies $1 \leq i_j \leq K$. We define an *ordered set cover* (of $A$) as $\Omega_o(\Theta, I) = \{(\pi_{i_1}, T_1), \ldots, (\pi_{i_k}, T_k)\}$, where sets $T_j$ are defined as follows: $T_1 = F_{i_1}, T_2 = F_{i_2} - T_1, T_3 = F_{i_3} - (T_1 \cup T_2)), \ldots, T_k = F_{i_k} - (T_1 \cup \ldots \cup T_{k-1})$. The set $\{T_1, \ldots, T_k\}$ will be called the *partition corresponding to $I$*. Thus

$$C(\Omega_o(\Theta, I)) = \sum_{j=1}^{k} C_{i_j}(|T_j|).$$

From the linearity of function $C$ we can easily obtain

**Lemma 5.** *Let $\Omega_1$ be some cover of a set $A$, and let $\Delta = \{\pi_1, \ldots, \pi_k\}$ be the set of patterns of $\Omega_1$. Let $\Theta$ be a set of hypotheses for $A$ with the pattern set $\Delta(\Theta) = \Delta$. Assume, that $|\pi_1| \geq \ldots \geq |\pi_k|$. Then $C(\Omega_o(\Theta, (1, \ldots, k))) \geq C(\Omega_1)$.*

As a corollary it follows that we can assume that the optimal cover is expressed as an ordered cover. Note that algorithm **Greedy** constructs an ordered cover.

For the next few paragraphs we will closely follow [10]. We will compare some optimal ordered cover, say $I = (i_1, \ldots, i_k)$, with the ordered cover constructed by the algorithm, say $J = (j_1, \ldots, j_p)$. Let $\{T_1, \ldots, T_k\}$ and $\{P_1, \ldots, P_p\}$ be the partitions corresponding to $I$ and $J$. Let us distribute the total compression $C(\Omega_o(\Theta, J))$ of $\Omega_o(\Theta, J)$ to elements $\alpha \in A$ as follows. If $\alpha \in P_r$, then $\alpha$ gets $c_\alpha = C_{j_r}(|P_r|)/|P_r|$ Let $H(n)$ be the harmonic number of $n$, i.e., $H(n) = \sum_{i=1}^{n}(1/i)$. We will use the fact that $H(n) = \ln n + O(1)$.

**Lemma 6.** $\sum_{\alpha \in T_r} c_\alpha \geq u_{i_r} \cdot |T_r| - w_{i_r} \cdot H(|T_r|)$.

The proof of this lemma is similar to the proof of Lemma 3 in [10], except the minus sign before $w_{i_r}$ and $\leq$ signs changed to $\geq$.

Next, from Lemma 6 we obtain that $C(\Omega_o(\Theta, J))$ equals

$$\sum_{r=1}^{k} \sum_{\alpha \in T_r} c_\alpha \geq \sum_{r=1}^{k} [u_{i_r} \cdot |T_r| - w_{i_r} \cdot H(|T_r|)] \geq \sum_{r=1}^{k} [u_{i_r} \cdot |T_r| - w_{i_r} \cdot H(|A|)].$$

According to the definition $C(\Omega_o(\Theta, I)) = \sum_{r=1}^{k} [u_{i_r} \cdot |T_r| - w_{i_r}]$. Therefore,

$$C(\Omega_o(\Theta, I)) - C(\Omega_o(\Theta, J)) \leq H(|A|) \sum_{r=1}^{k} w_{i_r}.$$

Now, using $M_{opt} = M(\Omega_o(\Theta, I)) \geq \sum_{r=1}^{k} w_{i_r}$ (follows from definition of $M$) we get $M(\Omega_o(\Theta, J)) - M(\Omega_o(\Theta, I)) = C(\Omega_o(\Theta, I)) - C(\Omega_o(\Theta, J)) \leq M(\Omega_o(\Theta, I)) \cdot (H(|A|) - 1)$, hence

$$M(\Omega_o(\Theta, J)) \leq M(\Omega_o(\Theta, I)) \cdot \ln |A| + O(1) = M_{opt} \cdot \ln |A| + O(1),$$

i.e., Theorem 2 follows.

To prove Theorem 1 we have to show, how to implement the algorithm in time $O(|A| \cdot \|A\|)$ in the case of substring patterns. This can be done by using generalized suffix trees (GST). GST generalizes the suffix trees (see for example [12]) for multiple strings. It is a trie-like data structure that represents all suffixes of multiple strings. Each leaf node contains a list of the names of the original strings that have as a suffix the string that leads from root to this leaf. All substrings of every string in GST are represented by some (may be partial) path from the root. GST can be constructed in linear time on the sum of the lengths of strings it contains [8].

For the substring pattern case the input of the algorithm is only the set $U = A$, but the pairs $(\pi_i, F_i)$ are found by the algorithm `Greedy_substring` from $A$. We first build the GST in time $O(\|A\|)$. Next for $i = 1, \ldots, |A|$ we find the longest substring $\pi_i$ present in at least $i$ strings of $A$. For each $i$ we also compute $c_i = c(\pi_i, i)$ and keep track of the best $\pi_i$ (i.e., $\pi_i$ having the highest $c_i$) so far. This can be done in time $O(\|A\|)$ by using the construction described on page 239 of [8]. After this we pick the substring pattern $\pi_i$ having the highest cost $c(\pi_i, F_i)$ among all possible subsets $F_i$ of $A$, such that $F_i = L(\pi_i) \cap A$ and also find the subset $F_i$ itself in time $O(\|A\|)$ (having $\pi_i$ this can be done by a fast string search algorithm). Simultaneously we can compute $U \leftarrow U - F_i$, and repeat the process again (starting by the building of a new GST for the updated $U$). Since in each such step $U$ becomes at least one element smaller, this can be repeated no more than $O(|A|)$ times giving the total time $O(|A| \cdot \|A\|)$ as required by Theorem 1.

# 6 Conclusions

We have studied a new approach to automatic pattern discovery in a set of sequences, based on finding a 'good' collection of patterns covering all sequences. We have shown that the 'best' cover can be efficiently approximated within a logarithmic factor. As far as we know, this is the first such approximation result regarding the learning of unbounded number of unions of regular patterns that allow variable length substitutions. For the case of substring patterns the pattern discovery algorithm is based on generalized suffix-trees and is very efficient.

We have also developed and implemented a heuristic version of our algorithm for the pattern class used in the protein sequence database PROSITE and carried out some experiments on discovering patterns and subfamilies in protein sequences [4]. The algorithm has been able to discover patterns characterizing subfamilies of biosequences recognized by biologists [4].

**Acknowledgements.** Alvis Brazma was supported by the Finnish Centre for the International Mobility and by the Latvian Council of Sciences (grant Nr. 93.593). Part of his work has been done at the Department of Computer Science, University of Helsinki.

# References

1. D. Angluin. Finding patterns common to a set of strings. *J. of Comp. and Syst. Sci.*, 21:46–62, 1980.
2. H. Arimura, T. Shinohara, and S. Otsuki. Finding minimal generalizations for unions of pattern languages and its application to inductive inference from positive data. In *Proc. of the 11th STACS, Lecture Notes in Comp. Sci., 755*, pages 649–660. Springer, 1994.
3. A. Brazma, I. Jonassen, I. Eidhammer, and D. Gilbert. Approaches to automatic discovery of patterns in biosequences. Technical Report TR-113, Department of Informatics, University of Bergen, Bergen, Norway, December 1995.
4. A. Brazma, I. Jonassen, E. Ukkonen, and J. Vilo. Discovering patterns and subfamilies in biosequences. In *Proceedings of Fourth International Conference on Intelligent Systems for Molecular Biology*, pages 34–43. AAAI Press, 1996.
5. A. Brazma, E. Ukkonen, and J. Vilo. Finding a good collection of patterns covering a set of sequences. Technical Report C-1995-60, Department of Computer Science, University of Helsinki, December 1995.
6. V. Chvátal. A greedy heuristic for the set-covering problem. *Math. Oper. Res.*, 4:233–235, 1979.
7. E. M. Gold. Language identification in the limit. *Information and Control*, 10:447–474, 1967.
8. L. C. K. Hui. Color set size problem with application to string matching. In *Proc. of Third Annual Symposium on Combinatorial Pattern Matching*, Lecture Notes in Comp. Science, 644, pages 230–243. Springer-Verlag, 1992.
9. I. Jonassen, J. F. Collins, and D. G. Higgins. Finding flexible patterns in unaligned protein sequences. *Protein Science*, 4(8):1587–1595, 1995.
10. P. Kilpeläinen, H. Mannila, and E. Ukkonen. MDL learning of unions of simple pattern languages from positive examples. In *Proceedings of the 2nd European conference EuroCOLT'95*, pages 252–260, 1995.
11. M. Li and P. Vitanyi. *An introduction to Kolmogorov complexity and its applications.* Texts and monographs in Computer Science. Springer-Verlag, 1993.
12. E. M. McCreight. A space–economical suffix tree construction algorithm. *J. ACM*, 23:262–272, 1976.
13. A. F. Neuwald, J. S. Liu, and C. E. Lawrence. Gibbs motif sampling: Detection of bacterial outer membrane protein repeats. *Protein Science*, 4:1618–1632, 1995.
14. J. R. Quinlan and R. L. Rivest. Inferring decision trees using the minimum decription length principle. *Information and Computation*, 80:227–248, 1989.
15. J. Rissanen. Modeling by the shortest data description. *Automatica-J.IFAC*, 14:465–471, 1978.
16. M.-F. Sagot, A. Viari, and H. Soldano. A distance-based block searching algorithm. In *Proc. of Third International Conference on Intelligent Systems for Molecular Biology*, pages 322–331. AAAI Press, 1995.
17. T. Shinohara. Polynomial time inference of extended regular pattern languages. In *Proceedings of RIMS Symposia on Software Science and Engineering*, Lecture Notes in Computer Science, 147, pages 115–127. Springer-Verlag, 1983.
18. R. Staden. Methods for discovering novel motifs in nucleic acid sequences. *CABIOS*, 5(4):293–298, 1989.
19. M. S. Waterman, R. Arratia, and D. J. Galas. Pattern recognition in several sequences: Consensus and alignment. *Bulletin of Mathematical Biology*, 46(4):515–527, 1984.

# Extremal Problems for Geometric Hypergraphs

Tamal K. Dey[1] and János Pach[2]

## Abstract

A *geometric hypergraph* $H$ is a collection of $i$-dimensional simplices, called *hyperedges* or, simply, *edges*, induced by some $(i+1)$-tuples of a *vertex set* $V$ in general position in $d$-space. The topological structure of geometric *graphs*, i.e., the case $d = 2, i = 1$, has been studied extensively, and it proved to be instrumental for the solution of a wide range of problems in combinatorial and computational geometry. They include the $k$-set problem, proximity questions, bounding the number of incidences between points and lines, designing various efficient graph drawing algorithms, etc. In this paper, we make an attempt to generalize some of these tools to higher dimensions. We will mainly consider extremal problems of the following type. What is the largest number of edges ($i$-simplices) that a geometric hypergraph of $n$ vertices can have without containing certain *forbidden* configurations? In particular, we discuss the special cases when the forbidden configurations are $k$ intersecting edges, $k$ pairwise intersecting edges, $k$ crossing edges, $k$ pairwise crossing edges, $k$ edges that can be stabbed by an $i$-flat, etc. Some of our estimates are tight.

# 1 Introduction

In recent years, the study of graph drawings has become a rich separate discipline within computational geometry. Much of the research has been motivated by applications, including software engineering, CAD, database design, cartography, circuit schematics, automatic animation, visual interfaces, etc. (See [24].) It is quite remarkable that classical graph theory proved to be rather powerless to tackle many of the arising problems. Instead, one often had to develop new topological tools to deal with families of curves, i.e., graphs drawn in the plane or in some other surfaces. Perhaps the best known example is the Lipton–Tarjan

---

[1]Dept. of CSE, IIT, Kharagpur, India 721302. Research supported in part by DST-SR-OY-E-06-95 grant, India

[2]Mathematical Institute of the Hungarian Academy of Sciences and Courant Institute of Mathematical Sciences, New York University, New York, NY 10012. Research supported by NSF grant CCR-94-24398, PSC-CUNY Research Award 663472, and OTKA-4269.

Separator Theorem for planar graphs [13], which has many extensions, generalizations, and a broad spectrum of applications ranging from numerical analysis to complexity theory. In particular, it enables us to use the divide-and-conquer paradigm to construct various geometric representations of abstract graphs and networks. Another important example is the following result discovered independently by Ajtai-Chvátal-Newborn-Szemerédi and Leighton. It can be used to obtain e.g. sharp bounds for the area requirement of graph layouts. Let $\kappa(G)$ denote the *crossing number* of a graph $G$, i.e., the minimum number of crossing pairs of edges over all planar drawings of $G$.

**Theorem 1.1** *[1], [12] Let $G$ be a simple graph with $n$ vertices and $e(G)$ edges. If $e(G) \geq 4n$, then $\kappa(G) \geq \frac{e(G)^3}{100n^2}$.*

As Székely [21] pointed out, this result almost immediately implies the Szemerédi-Trotter theorem [22], [23] on the number of incidences between points and lines. His argument is so nice and short that we cannot resist adapting it to establish the following generalization of the Szemerédi-Trotter theorem, which was found by Clarkson-Edelsbrunner-Guibas-Sharir-Welzl and has numerous algorithmic consequences. (For some other applications of Székely's idea, see [17].)

**Theorem 1.2** *[6] The total number of sides of $n$ distinct cells determined by $m$ lines in general position in the plane is at most $O(m^{2/3}n^{2/3} + m)$.*

PROOF. Notice that it is sufficient to prove the assertion for a system of cells $\mathcal{C}$, no two of which share an edge. Pick a point $p_i$ in each cell $c_i \in \mathcal{C}$. For any pair $(s_i, s_j)$ of collinear edges belonging to $c_i \in \mathcal{C}$ and $c_j \in \mathcal{C}$, respectively, connect $p_i$ to $p_j$ by a polygonal chain of length three via the midpoints of the segments $s_i$ and $s_j$, provided that this polygon is not adjacent to any other member of $\mathcal{C}$. The collection of these polygonal chains can be regarded as the edge set of a graph $G'$ whose vertices are $p_1, \ldots, p_n$. If a line is adjacent to $k$ cells in $\mathcal{C}$, then it contributes to exactly $k - 1$ edges of $G'$. Hence, $X$, the total number of sides of all cells in $\mathcal{C}$, differs from the number of edges of $G'$ by at most $m$. Removing the multiple edges from $G'$, we obtain a simple graph whose number of edges satisfies $e(G) \geq e(G')/4 \geq (X - m)/4$. In view of the fact that any crossing between two edges of $G$ occurs at a crossing of some pair of lines of the arrangement, Theorem 1.1 implies that either $e(G) < 4n$ or

$$\binom{m}{2} \geq \kappa(G) \geq \frac{e(G)^3}{100n^2}.$$

Since $X \leq 4e(G) + m$, Theorem 1.2 follows.  ▢

Given a set of $n$ points in general position in the plane, join two of them by a line segment if there are exactly $k$ points on one side of the line connecting them. Let $G$ denote the resulting graph. Lovász [14] proved that no straight line can cross more than $2k$ edges of $G$. Now Theorem 1.1 implies that the number of edges (the number of so called *k-sets*) is at most $O(k^{1/2}n)$. Indeed, if the number of edges is $e$, then either we have $e < 4n$ or there exists an edge crossing at least $e^2/(50n^2)$ other edges. Thus, $e^2/(50n^2) \leq 2k$, as required. (See [19] for a slight improvement.) It was shown by Dey-Edelsbrunner [8] that a similar approach can be used to establish an $O(n^{8/3})$ upper bound on the number of *halving planes* in 3-space, which improved some earlier results of [5] and [4].

A graph drawn in the plane by possibly crossing straight-line segments is called a *geometric graph*. More precisely, a geometric graph $G$ consists of a set of points $V$ in general position in the plane and a set of segments $E$ whose endpoints belong to $V$. As was demonstrated above, for a number of applications it was necessary to solve some extremal problems for geometric graphs. The systematic study of these problems was initiated by P. Erdős, Y. Kupitz [11], and M. Perles. (For a recent survey, see [16].)

It seems plausible that to extend the incidence results to higher dimensions, to improve the upper bound for the number of times the unit distance can occur among $n$ points in 3-space, or to make further progress concerning the $k$-set problem, one has to find the right generalizations of Theorem 1.1 to systems of surfaces or surface patches in $d$-space. For simplicity, we will only discuss the case when these surface patches are flat (simplices).

**Definition 1.1** *A $d$-dimensional geometric $r$-hypergraph $H_r^d$ is a pair $(V, E)$, where $V$ is a set of points in general position in $\Re^d$, and $E$ is a set of closed $(r-1)$-dimensional simplices induced by some $r$-tuples of $V$. The sets $V$ and $E$ are called the* vertex set *and* edge set *of $H_r^d$, respectively.*

Akiyama and Alon [3] proved the following theorem. Let $V = V_1 \cup \ldots \cup V_d$ ($|V_1| = \ldots = |V_d| = n$) be a $dn$-element set in general position in $\Re^d$, and let $E$ consist of all $(d-1)$-dimensional simplices having exactly one vertex in each $V_i$. Then $E$ contains $n$ disjoint simplices. Combining this with a result of Erdős [9], we obtain a non-trivial upper bound for the number of edges of a $d$-dimensional geometric $d$-hypergraph of $n$ vertices that contains no $k$ pairwise disjoint edges.

If we want to exclude *crossings* rather than disjoint edges, or want to generalize Theorem 1.1 to geometric hypergraphs, we face the following problem. Even if we restrict our attention to systems of triangles induced by 3-dimensional point sets in general position, it is not completely clear how a "crossing" should be defined, let alone the notion of "crossing number". If two segments cross, they do not share an endpoint. Should this remain true for triangles? We have to clarify the terminology.

**Definition 1.2** *Two simplices are said to have a* non-trivial intersection, *if their relative interiors have a point in common. If, in addition, the two simplices are vertex disjoint, then they are said to* cross.

*More generally, $k$ simplices are said to have a* non-trivial intersection, *if their relative interiors have a point in common. If, in addition, all simplices are vertex disjoint, then they are said to* cross.

Consider $k$ simplices. It is important to note that the fact that *every pair* of them has a non-trivial intersection does not imply that *all* of them do. To emphasize that this stronger condition is satisfied, we often say that the simplices have a *non-trivial intersection in the strong sense*, or simply that they *strongly intersect*. Similarly, a set of *pairwise* crossing simplices is not necessarily *crossing*. If want to emphasize that they *all* cross, we will say that they *cross in the strong sense*, or shortly that they *strongly cross*.

As we pick more and more distinct $(r-1)$-dimensional simplices induced by a set of $n$ points in $\Re^d$, the number of crossings between them will usually increase. The aim of this paper is to generalize the planar results to obtain some information about the growth rate of this process. In the inverse formulation, one can ask for the maximum number of edges that a $d$-dimensional geometric $r$-hypergraph $H_r^d$ of $n$ vertices can have without containing some fixed crossing pattern. Throughout this paper, let $f_r^d(\mathcal{F}, n)$ denote this maximum, where $\mathcal{F}$ is the family of *forbidden* configurations, i.e., forbidden geometric subhypergraphs. Most of our bounds will be asymptotic: $d$ and $r$ are thought to be fixed, while $n$ tends to infinity.

In the next two sections, we estimate $f_r^d(\mathcal{F}, n)$ for various families $\mathcal{F}$. In Section 4, we generalize Theorem 1.1. Finally, we discuss some related questions and give a few applications of our results.

## 2 Full Dimensional Simplices

Let $\mathcal{I}_k^r$ (resp. $\mathcal{SI}_k^r$) denote the class of all geometric hypergraphs consisting of $k$ $(r-1)$-dimensional simplices, any two of which have a non-trivial intersection (resp. all of which are strongly intersecting). Similarly, let $\mathcal{C}_k^r$ (resp. $\mathcal{SC}_k^r$) denote the class of all geometric hypergraphs consisting of $k$ pairwise crossing (resp. strongly crossing) $(r-1)$-simplices.

**Theorem 2.1** *For any fixed $k > 1$, one can select at most $O(n^{\lceil d/2 \rceil})$ $d$-dimensional simplices induced by $n$ points in $d$-space with the property that no $k$ of them share a common interior point. This bound cannot be improved. That is,*

$$f_{d+1}^d(\mathcal{I}_k^{d+1}, n) = \Theta(n^{\lceil d/2 \rceil}), \qquad f_{d+1}^d(\mathcal{SI}_k^{d+1}, n) = \Theta(n^{\lceil d/2 \rceil}).$$

**Theorem 2.2** *Let $E$ be any set of $d$-dimensional simplices induced by an $n$-element point set $V \subseteq \Re^d$. If $E$ has no two crossing elements, then $|E| = O(n^d)$, and this bound is asymptotically tight. In notation,*

$$f_{d+1}^d(\mathcal{C}_2^{d+1}, n) = \Theta(n^d).$$

PROOF. To prove the lower bound, consider a geometric hypergraph consisting of all $d$-dimensional simplices induced by $V$ that contain a given vertex $v \in V$.

Next we establish the upper bound. If $E$ has no two simplices having a non-trivial intersection, then $|E| \leq O(n^{\lceil d/2 \rceil})$, by the previous theorem. Otherwise, choose two $d$-simplices $\Delta_1, \Delta_2 \in E$ whose intersection is non-trivial. It is easy to show (see e.g. [7], [8]) that there exist an $\ell_1$-face $\Delta_1'$ of $\Delta_1$ and an $\ell_2$-face $\Delta_2'$ of $\Delta_2$ with $\ell_1 + \ell_2 = d$ such that $\Delta_1'$ and $\Delta_2'$ are crossing.

Assume first that there is an edge $e \in E$ which is vertex disjoint from $\Delta_1'$ and contains $\Delta_2'$ as a face. Then every edge $f \in E$ that contains $\Delta_1'$ as a face must share at least one vertex with $e$. The number of such simplices $f$ is at most $(d+1)\binom{n}{d-\ell_1-1}$. Let us remove all of them from $E$.

In the second case, every edge $e \in E$ that contains $\Delta_2'$ as a face shares a vertex with $\Delta_1'$. Obviously, the number of such simplices $e$ is at most $(\ell_1+1)\binom{n}{d-\ell_2-1}$. Remove all of them from $E$.

We continue this procedure until there remain no non-trivial intersections in $E$. At this point, $E$ has at most $O(n^{\lceil d/2 \rceil})$ elements, and the total number of simplices that have been removed is at most

$$\binom{n}{\ell_1+1}(d+1)\binom{n}{d-\ell_1-1} + \binom{n}{\ell_2+1}(\ell_1+1)\binom{n}{d-\ell_2-1} = O(n^d).$$

◻

# 3  $(d-1)$-simplices in $d$-space

**Theorem 3.1** *Let $E$ be a family of $(d-1)$-dimensional simplices induced by an $n$-element point set $V \subseteq \Re^d$ such that $E$ has no $k$ members with pairwise non-trivial intersections $(d, k > 1)$. Then, for $k = 2$ and $3$, we have $|E| = O(n^{d-1})$. Otherwise, $|E| = O(n^{d-1} \log^{2k-6} n)$. In notation,*

$$f_d^d(\mathcal{I}_k^d, n) = \begin{cases} O(n^{d-1}) & \text{if } k = 2, 3; \\ O(n^{d-1} \log^{2k-6} n) & \text{otherwise.} \end{cases}$$

*This result is asymptotically tight if $d, k \leq 3$.*

PROOF. For $d = 2$, the assertion is true, by the results of [18] and [2]. Assume that $d \geq 3$. For any $(d-3)$-simplex $\Delta$ induced by $V$, let $E_\Delta$ denote the family of all members of $E$ that contain $\Delta$ as a face. Pick any point $p_\Delta$ in the relative interior of $\Delta$, and let $F_\Delta$ denote the 3-dimensional flat orthogonal to $\Delta$ and passing through $p_\Delta$.

Every $e \in E_\Delta$ meets $F_\Delta$ in a polygon, whose two sides incident to $p_\Delta$ are the intersections of $F_\Delta$ with the two $(d-2)$-faces of $e$ containing $\Delta$. Thus, the total number of sides incident to $p_\Delta$ that occur in some $e \cap F_\Delta$ ($e \in E_\Delta$) is at most $n - d + 2 < n$. Take a small 2-dimensional sphere $S^2 \subseteq F_\Delta$ centered at $p_\Delta$. The intersections of $S^2$ with the elements of $E_\Delta$ form the edge set of a graph with at most $n$ vertices. It follows from the properties of $E$ that this graph has no $k$ pairwise crossing edges, so, by the planar results, its number of edges, $|E_\Delta|$, satisfies

$$|E_\Delta| = \begin{cases} O(n) & \text{if } k = 2, 3; \\ O(n \log^{2k-6} n) & \text{otherwise.} \end{cases}$$

Summing over all $(d-3)$-simplices $\Delta$ induced by $V$, we obtain $\binom{d}{2}|E| = \sum_\Delta |E_\Delta|$, and hence the upper bound.

To show that the result is tight for $d = 3, k = 2$, consider a nested sequence of $n/2$ pyramids based on the same 2-dimensional convex $n/2$-gon. These pyramids have a total of $n^2/4$ triangular faces, no two of which have a non-trivial intersection. $\square$

It is an outstanding open problem to decide whether the order of magnitude of the above bound can be improved e.g. for $d = 4, k = 2$. However, modifying the procedure described in the proof Theorem 2.2, one can show that the following related result is asymptotically tight. The details are left to the reader.

**Theorem 3.2** *Let $E$ be a family of $(d-1)$-dimensional simplices induced by an $n$-element point set $V \subseteq \Re^d$. If $E$ has no two crossing members, then $|E| = O(n^{d-1})$, and this bound cannot be improved. In notation,*

$$f_d^d(C_2^d, n) = \Theta(n^{d-1}).$$

The results in the next section enable us to establish the following generalization of Theorem 3.2.

**Theorem 3.3** *Let $E$ be a family of $(d-1)$-dimensional simplices induced by an $n$-element point set $V \subseteq \Re^d$, where $d, k > 1$. If $E$ has no $k$ pairwise crossing members, then $|E| = O(n^{d-(1/d)^{k-2}})$. In notation,*

$$f_d^d(C_k^d, n) = O(n^{d-(1/d)^{k-2}}).$$

# 4  $k$-tuples of Strongly Crossing Simplices

Given any $d$-dimensional geometric $r$-hypergraph $H_r^d = (V, E)$ with $n$ vertices, let $x_k(H_r^d)$ denote the number of *strongly crossing* $k$-tuples of edges. Using our notations, $|E| > f_r^d(SC_k^r, n)$ obviously implies that $x_k(H_r^d) > 0$. Define

$$x_{k,r}^d(n, e) = \min_{H_r^d} x_k(H_r^d),$$

where the minimum is taken over all $H_r^d = (V, E)$ with $|V| = n$ vertices and $|E| = e$ edges (simplices).

The following theorem provides us with a recipe how to give a lower bound on the *number* of crossing $k$-tuples of edges, if we know how many edges are necessary to guarantee the existence of *one* such $k$-tuple. This result generalizes Theorem 1.1. (See also [7, 8, 15, 20], for related problems and results.)

**Theorem 4.1** *Assume that $f_r^d(SC_k^r, n) < c_1\binom{n}{r}/n^\delta$ and $e \geq (c_1 + 1)\binom{n}{r}/n^\delta$ for suitable constants $c_1$ and $0 \leq \delta \leq 1$. Then there exists $c_2 > 0$ such that*

$$x_{k,r}^d(n, e) > c_2\binom{n}{kr}e^\gamma \Big/ \binom{n}{r}^\gamma,$$

*where $\gamma = 1 + \frac{(k-1)r}{\delta}$.*

PROOF. By induction on $n$. Let $H = H_r^d = (V, E)$ be a $d$-dimensional geometric $r$-hypergraph with $n$ vertices and $e$ edges. Suppose further that $H$ has the smallest possible number of crossing $k$-tuples, i.e., $x_k(H) = x_{k,r}^d(n, e) = x(n, e)$. We can also assume that $n > kr$, for otherwise the assertion is trivial. First we consider the range $(c_1 + 1)\binom{n}{r}/n^\delta \leq e \leq (c_1 + r + 1)\binom{n}{r}/n^\delta$. It follows from the assumptions that if $H$ has more than $c_1\binom{n}{r}/n^\delta$ edges, then any additional edge will participate in a new $k$-tuple of strongly crossing edges. Hence, the number of crossing $k$-tuples is at least

$$e - c_1\frac{\binom{n}{r}}{n^\delta} \geq \frac{\binom{n}{r}}{n^\delta} > c\binom{n}{kr}\frac{e^\gamma}{\binom{n}{r}^\gamma},$$

as long as $c$ is sufficiently small. Next we assume that $e > (c_1 + r + 1)\binom{n}{r}/n^\delta$. For any point $p \in V$, let $H_p$ denote the geometric hypergraph obtained from $H$ by removing $p$ together with all edges ($(r-1)$-simplices) that contain $p$ as a vertex. The number of edges of $H_p$ is denoted by $e_p$. If we sum over all $p$ the number of crossing $k$-tuples of edges in $H_p$, then every crossing $k$-tuple of $H$ will be counted exactly $n - kr$ times. Therefore,

$$(n - kr) \cdot x(n, e) = (n - kr) \cdot x_k(H) = \sum_{p \in V} x_k(H_p) \geq \sum_{p \in V} x(n - 1, e_p).$$

Note that $\binom{n-1}{r-1} \leq r\binom{n}{r}/n^\delta$, because $\delta \leq 1$. Since $p$ can be a vertex of at most $\binom{n-1}{r-1}$ edges of $H$, we have

$$e_p > (c_1 + r + 1)\frac{\binom{n}{r}}{n^\delta} - \binom{n-1}{r-1} \geq (c_1 + 1)\frac{\binom{n}{r}}{n^\delta} \geq (c_1 + 1)\frac{\binom{n-1}{r}}{(n-1)^\delta}.$$

Thus, we can apply the induction hypothesis to every $H_p$ to obtain

$$(n - kr) \cdot x(n, e) > c\frac{\binom{n-1}{kr}}{\binom{n-1}{r}^\gamma}\sum_{p \in V} e_p^\gamma.$$

Obviously, $\sum_{p \in V} e_p = (n - r)e$. Using the fact that $\gamma > 1$, we have $\sum_{p \in V} e_p^\gamma \geq n(\frac{(n-r)e}{n})^\gamma$. This finally yields

$$x_{k,r}^d(n, e) = x(n, e) > c \cdot \frac{n\binom{n-1}{kr}}{n - kr} \cdot \frac{(\frac{n-r}{n})^\gamma}{\binom{n-1}{r}^\gamma} \cdot e^\gamma = c\binom{n}{kr}\frac{e^\gamma}{\binom{n}{r}^\gamma}.$$

$\square$

Now we are ready to apply the last – somewhat technical – result.

**Theorem 4.2** *Let $x_{2,d}^d(n,e)$ denote the minimum number of crossing pairs in any $e$-element set of $(d-1)$-simplices induced by $n$ points in $\Re^d$ in general position. Then, for every $n$ and $e \geq cn^{d-1}$, we have*

$$c_1 \frac{e^{d+1}}{n^{d(d-1)}} < x_{2,d}^d(n,e) < c_2 \frac{e^{2+1/\lfloor d/2 \rfloor}}{n^{d/\lfloor d/2 \rfloor}}.$$

PROOF. By Theorem 3.2, we can apply Theorem 4.1 with $k = 2, r = d$, and $\delta = 1$, and the lower bound immediately follows. The proof of the upper bound uses points on the moment curve. $\qquad\qquad\square$

## 5 Related Problems

**Lower dimensional simplices.** So far, most of our results concerned full-dimensional simplices or $(d-1)$-simplices in $\Re^d$. Now we apply a theorem of Vrećica and Živaljeviċ [25] to deduce a result on geometric hypergraphs with lower dimensional edges.

Consider a *complete $r$-partite geometric hypergraph $K_r^d$*, whose vertex set is the union of $r$ disjoint sets $V_1, V_2, \ldots, V_r$ of size $\ell$ each, and whose edge set consists of all $(r-1)$-dimensional simplices that have a vertex in each $V_i$. Generalizing a result of [26], Vrećica and Živaljević [25] have shown that if $\ell \geq 2p - 1$ for some prime $p \leq \frac{d}{d-r+1}$, then $K_r^d$ contains $p$ strongly crossing edges. Note that for $r < \lceil \frac{d}{2} \rceil + 1$, we have $p \leq 1$, which is impossible.

**Proposition 5.1** *Let $f_r^d(SC_k^r, n)$ denote the maximum number of edges that a $d$-dimensional geometric $r$-hypergraph of $n$ vertices can have without containing $k$ strongly crossing edges. Suppose that there is a prime $p$ such that $k \leq p \leq d/(d-r+1)$. Then*

$$\Omega(n^{r-1}) \leq f_r^d(SC_k^r, n) \leq O(n^{r-\delta}),$$

*where $\delta = 1/(2p-1)^{r-1}$.*

We conjecture that the lower bound in Proposition 5.1 is asymptotically tight.

**Crossing many simplices by a flat.** In [15], we posed the following question. What is the largest number $g = g_{k,r}^d(n,e)$ such that for any $d$-dimensional geometric $r$-hypergraph with $n$ vertices and $e$ edges, one can find a $k$-flat crossing at least $g$ edges.

Applying the results of the previous sections, one can obtain the following two bounds.

**Proposition 5.2** $g_{k,d}^d(n,e) = \Omega(e^d/n^{d(d-1)})$ *for $k \geq \lfloor d/2 \rfloor$.*

**Proposition 5.3** $g_{k,r}^d(n,e) = \Omega(e^{3^{(r-1)}r}/n^{r(3^{(r-1)}r-1)})$ *for* $\lfloor d/2 \rfloor \leq k \leq r$, *and* $\lceil d/2 \rceil + 1 \leq r$.

**Ramsey-type questions.** Let us color with two colors all $(r-1)$-dimensional simplices induced by $n$ points in general position in $\Re^d$. Is it true that one of the color classes necessarily contains certain special subconfigurations, provided that $n$ is large enough? If the answer is in the affirmative, then we can ask for the smallest $n$ for which this will occur. A variety of questions of this type are discussed in [10], in the planar case. Some of the results can be generalized to higher dimensions.

**Theorem 5.1** *Let us color with two colors all $(d-1)$-dimensional simplices induced by $(d+1)n-1$ points in general position in $\Re^d$. Then one can always find $n$ disjoint simplices of the same color. This result cannot be improved.*

**Acknowledgement.** We are grateful to Herbert Edelsbrunner, whose ideas and valuable suggestions are reflected in several parts of this paper.

# References

[1] M. Ajtai, V. Chvátal, M. M. Newborn, and E. Szemerédi. Crossing-free subgraphs. *Ann. Discrete Math.* **12** (1982), 9–12.

[2] P. K. Agarwal, B. Aronov, J. Pach, R. Pollack, and M. Sharir. Quasi-planar graphs have a linear number of edges. In: *Graph Drawing '95, Lecture Notes in Computer Science* **1027**, Springer-Verlag, Berlin, 1996, 1–7. Also in: *Combinatorica* (to appear).

[3] J. Akiyama and N. Alon. Disjoint simplices and geometric hypergraphs. *Combinatorial Mathematics* (G. S. Bloom et al., eds.), Annals of the New York Academy of Sciences **555** (1989), 1–3.

[4] B. Aronov, B. Chazelle, H. Edelsbrunner, L. Guibas, M. Sharir, and R. Wenger. Points and triangles in the plane and halving planes in space. *Discrete and Computational Geometry* **6** (1991), 435–442.

[5] I. Bárány, Z. Füredi, and L. Lovász. On the number of halving planes. *Combinatorica* **10** (1990), 175–183.

[6] K. Clarkson, H. Edelsbrunner, L. Guibas, M. Sharir, and E. Welzl. Combinatorial complexity bounds for arrangements of curves and spheres. *Discrete Comput. Geom.* **5** (1990), 99–160.

[7] T. K. Dey. On counting triangulations in d dimensions. *Computational geometry: Theory and Applications.* **3** (1993), 315–325.

[8] T. K. Dey and H. Edelsbrunner. Counting triangle crossings and halving planes. *9th Sympos. Comput. Geom.* (1993), 270–273. Also: *Discrete and Computational Geometry* **12** (1994), 281–289.

[9] P. Erdős. On extremal problems on graphs and generalized graphs. *Israel J. Math.* **2** (1964), 183–190.

[10] G. Károlyi, J. Pach, and G. Tóth. Ramsey-type results for geometric graphs. *12th Sympos. Comput. Geom.* (1996). Also in: *Discrete and Computational Geometry* (to appear).

[11] Y. Kupitz. *Extremal Problems in Combinatorial Geometry, Aarhus University Lecture Notes Series* **53**, Aarhus University, Denmark, 1979.

[12] F. T. Leighton. *Complexity Issues in VLSI, Foundations of Computing Series*, MIT Press, Cambridge, Mass., 1983.

[13] R. Lipton and R. Tarjan. A separator theorem for planar graphs. *SIAM J. Applied Mathematics* **36** (1979), 177–189.

[14] L. Lovász. On the number of halving lines. *Annales Universitatis Scientarium Budapest, Eötvös, Sectio Mathematica* **14** (1971), 107–108.

[15] J. Pach. Notes on geometric graph theory. *DIMACS Ser. Discr. Math. and Theoret. Comput. Sc.* **6** (1991), 273–285.

[16] J. Pach and P. K. Agarwal. *Combinatorial Geometry*, Wiley, New York, 1995.

[17] J. Pach and M. Sharir. On the number of incidences between points and curves (to appear).

[18] J. Pach, F. Shahrokhi, and M. Szegedy. Applications of crossing numbers. *10th ACM Sympos. Comput. Geom.* (1994), 198–202.

[19] J. Pach, W. Steiger, and M. Szemerédi. An upper bound on the number of planar k-sets. *Discrete and Computational Geometry* **7** (1992), 109–123.

[20] J. Pach and J. Törőcsik. Some geometric applications of Dilworth's theorem. *9th Sympos. Comput. Geom.* (1993), 264–269. Also in: *Discrete and Computational Geometry* **12** (1994), 1–7.

[21] L. A. Székely, Crossing numbers and hard Erdős problems in discrete geometry. *Combinatorics, Probability, and Computing*, (to appear).

[22] E. Szemerédi and W. T. Trotter. Extremal problems in discrete geometry. *Combinatorica* **3** (1983), 381–392.

[23] E. Szemerédi and W. T. Trotter. A combinatorial distinction between the Euclidean and projective planes. *European J. Combinatorics* **4** (1983), 385–394.

[24] R. Tamassia and I. Tollis (eds.). *Graph Drawing, Lecture Notes in Computer Science* **894**, Springer-Verlag, Berlin, 1995.

[25] S. Vrećica and R. Živaljević. New cases of the colored Tverberg's theorem. In: *Jerusalem Combinatorics '93, Contemp. Math.* **178**, Amer. Math. Soc., Providence, 1994, 325–334.

[26] R. Živaljević and S. Vrećica. The colored Tverberg's problem and complexes of injective functions. *J. Combin. Theory Ser. A* **61** (1992), 309–318.

# Computing Fair and Bottleneck Matchings in Geometric Graphs

Alon Efrat[1], Matthew J. Katz[2]

[1] School of Mathematical Sciences, Tel-Aviv University, Tel-Aviv 69982, Israel
alone@math.tau.ac.il
[2] Department of Computer Science, Utrecht University, P.O.Box 80.089, 3508 TB
Utrecht, The Netherlands    matya@cs.ruu.nl

**Abstract.** Let $A$ and $B$ be two sets of $n$ points in the plane, and let $M$ be a (one-to-one) matching between $A$ and $B$. Let $\min(M)$, $\max(M)$, and $\Sigma(M)$ denote the length of the shortest edge, the length of the longest edge, and the sum of the lengths of the edges of $M$ respectively. The *uniform matching* problem (also called the *balanced assignment* problem, or the *fair matching* problem) is to find $M_U^*$, a matching that minimizes $\max(M) - \min(M)$. A *minimum deviation matching* $M_D^*$ is a matching that minimizes $(1/n)\Sigma(M) - \min(M)$. We present algorithms for computing $M_U^*$ and $M_D^*$ in roughly $O(n^{10/3})$ time. These algorithms are more efficient than the previous $O(n^4)$-time algorithms of Martello and Toth [19] and Gupta and Punnen [11], who studied these problems for general bipartite graphs.

We also consider the (non-bipartite version of the) bottleneck matching problem in higher dimensions. We extend the planar results of Chang et al. [4] and Su and Chang [22], and show that given a set $A$ of $2n$ points in $d$-space, it is possible to compute a bottleneck matching of $A$ in roughly $O(n^{3/2})$ time, for $d \le 6$, and in subquadratic time, for $d > 6$.

## 1   Introduction

Let $G = (A \cup B, E)$ be a bipartite graph. A *Matching* $M$ in $G$ is a subset of $E$ such that each $v \in A \cup B$ belongs to at most one edge of $M$. $M$ is *maximum* if $|M| \ge |M'|$, for any other matching $M'$ in $G$. $M$ is *perfect* if each $v \in A \cup B$ belongs to an edge of $M$. (Obviously, if a perfect matching exists then $|A| = |B|$.) The problem of computing a perfect matching in $G$ has been studied extensively (see [15, 20]).

When weights are associated with the edges of $G$, it is often desirable to compute a perfect matching that is optimal with respect to some criterion. A *minimum weight matching* minimizes the sum of the weights of the edges of the matching, a *bottleneck matching* minimizes the maximum weight of an edge of the matching, a *most uniform matching* minimizes the difference between the maximum weight and the minimum weight, and a *minimum deviation matching* minimizes the difference between the average weight and the minimum weight (alternatively, minimizes the difference between the maximum weight and the average weight). Much work has been done on the problems of finding efficient algorithms for computing these matchings; see e.g. [9, 11, 17, 18, 19].

In this paper we consider the last two of these four problems from a geometric point of view. That is, we assume that the sets of vertices $A$ and $B$ are sets of points in the plane, and that $G$ is the complete bipartite graph over $A,B$. The weight associated with an edge $(a, b)$ is the Euclidean distance between $a$ and $b$. The geometric versions of the first two problems already have solutions that are more efficient than the corresponding general solutions; see for example [3, 7, 23].

Martello and Toth [19] consider the problem of computing a most uniform matching (or a *balanced* assignment, as they call it) in general bipartite graphs, and present an $O(n^4)$-time solution. Here we present an $O(n^{10/3} \log n)$-time solution for this problem in the geometric setting. Our solution is based on a technique for batched range searching, where the ranges are congruent annuli [16], and it also borrows ideas from [7].

The problem of computing a most uniform Euclidean matching $M_U^*$ is of relevance to the field of pattern matching, as we show in Section 3.

The problem of computing a minimum deviation matching for an arbitrary graph was considered by Gupta and Punnen [11] who gave an $O(n^4)$-time solution. For this problem we present an $O(n^{10/3+\varepsilon})$-time solution[3] in the geometric setting. This solution is based on the algorithm of [3] for computing a minimum-weight Euclidean matching.

We also study the (non-bipartite version of the) Euclidean bottleneck matching problem in higher dimensions, that is, given a set $A$ of $2n$ points in $\mathbb{R}^d$, compute a bottleneck matching in $G$, the (complete) Euclidean graph over $A$. (The weight associated with an edge $(a, b)$ is the Euclidean distance between $a$ and $b$.) The best known algorithm for computing a bottleneck matching in a general graph is due to Gabow and Tarjan [9]; it runs in time $O((n \log n)^{1/2} m)$. Chang et al. [4] have shown for $d = 2$ that a linear-size subgraph of $G$, called the *17 relative neighborhood graph* of $A$ and denoted 17RNG($A$) (see definition in Section 5), already contains a bottleneck matching of $G$. Thus, by applying the algorithm of Gabow and Tarjan to 17RNG($A$) a bottleneck matching can be computed in $O(n^{3/2} \log^{1/2} n)$ time, after computing $17RNG(A)$, which can be done within the same time bound. See also [22]. We extend this result and show that for any dimension $d$ there exists a constant $k = k(d)$ such that $k$RNG($A$) contains a bottleneck matching of $G$. Since $k$RNG($A$) (or related linear-size supergraphs of it) can be computed efficiently, we obtain subquadratic algorithms for computing a bottleneck matching in any fixed dimension. In particular, for $d \leq 6$ a bottleneck matching of $G$ can be computed in roughly $O(n^{3/2})$ time.

## 2 Computing a Most Uniform Matching

We assume some familiarity of the reader with the notion of augmenting paths in the context of computing a maximum (cardinality) matching in a bipartite graph. See [18] for further details.

Let $A$ and $B$ be two sets of $n$ points in the plane. Let $G[r, r']$ denote the bipartite graph whose set of vertices is $A \cup B$, and there is an edge between $a \in A$ and $b \in B$ if $r \leq \|a - b\| \leq r'$, where $\|a - b\|$ is the Euclidean distance between $a$ and $b$. Let $d^{(1)} \ldots d^{(n^2)}$ denote the $n^2$ distances between points of $A$

---

[3] Throughout the paper, $\varepsilon$ stands for a positive constant which can be chosen arbitrarily small with an appropriate choice of other constants of the algorithms.

and points of $B$, in increasing order. We refer to them as *critical distances*, and we assume, for simplicity of exposition, that they are distinct. We will maintain a maximum matching in $G[r, r']$, where $r, r'$ are critical distances that vary. We start with $G[d^{(i)}, d^{(j)}]$ for $i = j = 1$ and with the empty matching. The top level of the algorithm consists of the following loop. If the current (maximal) matching in $G[d^{(i)}, d^{(j)}]$ is of cardinality less than $n$, increase $j$ by one, else increase $i$ by one; in either case compute a maximum matching in the new graph, and repeat. Increasing $j$ adds a single edge to the graph, and we must check whether the size of the maximum matching increases. Increasing $i$ deletes a single edge from the graph, and, if this edge was in the current maximum matching, we must check whether the size of the maximum matching remains as before or decreases by one. Both these conditions can be checked by trying to compute an alternating path for the current matching (in the latter case, after deleting the edge corresponding to the distance $d^{(i)}$ from the current matching). If such a path exists then the answer is positive and we update the current maximum matching, otherwise, the answer is negative. If a perfect matching was found, then we compare the appropriate difference, i.e., either $d^{(j+1)} - d^{(i)}$ or $d^{(j)} - d^{(i+1)}$, with the difference corresponding to the best matching found so far. Clearly, the most uniform matching will be discovered in this way, and the number of times we need to compute an augmenting path is $O(n^2)$.

We next describe how to compute an augmenting path in $G[r, r']$ (if such a path exists) in time $O(n^{4/3} \log n)$. Let $S \subseteq A$ be the set of all vertices in $A$ that are currently unmatched (exposed). If we can find an augmenting path in $G[r, r']$ from some vertex $a \in S$ to an exposed vertex $b \in B$, then we can augment the current matching. Otherwise, the current matching is also a maximum matching of the new graph.

We need a data structure $\mathcal{D}_{r,r'}(P)$ over a set of points $P$, supporting the following operations.

- **neighbor**$_{r,r'}(P, q)$: For a query point $q$, return a point $p \in P$ whose distance from $q$ is between $r$ and $r'$. If no such $p$ exists, then **neighbor**$_{r,r'}(P, q) = \emptyset$.

- **delete**$_{r,r'}(P, p)$: Delete the point $p$ from $P$.

Using these operations, we can compute an augmenting path, if such exists, with the following simple procedure, which is a variant of a procedure that appeared in [7]. The underlying data structure is described immediately afterwards.

---

$L_1 \leftarrow S$ ; $\quad L_2 \leftarrow \emptyset$ ; $\quad \mathcal{D} \leftarrow \mathcal{D}_{r,r'}(B)$ ;
Repeat forever
  <u>For</u> each $a \in L_1$ <u>Do</u>   /* Find all $b$'s that are neighbors of $a$ in $G[r, r']$ */
    <u>While</u> **neighbor**$_{r,r'}(\mathcal{D}, a) \neq \emptyset$
      $b \leftarrow$ **neighbor**$_{r,r'}(\mathcal{D}, a)$ ;
      <u>If</u> $b$ is exposed, then stop. An augmenting path was found.
      Add $b$ to $L_2$ and **delete**$_{r,r'}(\mathcal{D}, b)$ ;   /* prevent re-finding $b$ */
  <u>If</u> $L_2 = \emptyset$, then stop. No augmenting path exists.
  $L_1 \leftarrow$ all vertices connected to some $b \in L_2$ by a matching edge ; $\quad L_2 \leftarrow \emptyset$ ;

Note that this procedure performs $O(n)$ operations of **neighbor**$_{r,r'}$ and **delete**$_{r,r'}$, and finds an augmenting path (if such exists). The procedure actually computes a forest of trees whose roots are the exposed vertices of $A$ (i.e., the vertices in $S$). When an exposed vertex $b$ of $B$ is reported, the path leading to $b$ from the root of the tree to which $b$ belongs is an augmenting path. Below we describe the underlying data structure $\mathcal{D}_{r,r'}(B)$ that enables us to perform the operations **neighbor**$_{r,r'}$ and **delete**$_{r,r'}$ in amortized time $O(n^{1/3}\log n)$. The data structure can be constructed in $O(n^{4/3}\log n)$ time. Thus, an augmenting path can be computed in total time $O(n^{4/3}\log n)$, and, since we repeat this $O(n^2)$ times, we obtain an $O(n^{10/3}\log n)$-time algorithm for computing a most uniform matching.

**The data structure:** The data structure is based on the following theorem.

**Theorem 1.** *(Katz [16]) Let $\mathcal{M}$ be a set of $m$ congruent annuli and $A$ a set of $n$ points in the plane. One can compute the set of pairs of the form $(c, a)$, where $c \in \mathcal{M}$, $a \in A$, and $a$ lies inside $c$, as a collection $\{\mathcal{M}_u \times A_u\}_u$ of complete edge-disjoint bipartite graphs, in $O((m^{2/3}n^{2/3}+m+n)\log m)$ time and space. The number of graphs obtained is $O(m^{2/3}n^{2/3} + m + n)$, and we have $\sum_u |A_u|, \sum_u |\mathcal{M}_u| = O((m^{2/3}n^{2/3} + m + n)\log m)$. Each such point-annulus containment pair appears in exactly one of these graphs.*

For each $b \in B$ draw the annulus of radii $r$ and $r'$ that is centered at $b$. Let $\mathcal{M}$ be the set of these annuli. Clearly, $r \leq \|q - b\| \leq r'$ for some point $q$ iff $q$ lies inside the annulus associated with $b$. We apply Theorem 1 to the sets $A$ and $\mathcal{M}$ and obtain in time $O(n^{4/3}\log n)$ a collection of $O(n^{4/3})$ bipartite graphs $H_u$ such that $\sum_u |A_u|, \sum_u |\mathcal{M}_u| = O(n^{4/3}\log n)$. We now create a few auxiliary linked lists so that we don't report a point in $B$ more than once. First, we convert the vertex sets of the graphs $H_u$ into doubly linked lists. Next, for each $a \in A$ we create a doubly linked list of pointers to the occurrences of $a$ in the lists $A_u$, and have each such occurrence of $a$ point back at the pointer to it. For each point $b \in B$ we create a linked list of pointers to its occurrences in the lists $\mathcal{M}_u$. All this can be done in $O(n^{4/3}\log n)$ time and space by traversing the vertex sets of the graphs $H_u$.

Now in order to report all the neighbors of a point $a \in L_1$ that have not been reported yet for some other point, that is, in order to execute the <u>While</u> loop in the procedure above, we proceed as follows. Traverse the list associated with $a$. For each occurrence of $a$ in a list $A_u$ report all the points $b$ in $\mathcal{M}_u$. For each such reported $b$, remove all occurrences of $b$ in the lists $\mathcal{M}_u$. For each vertex $a'$ in a list $A_u$ containing $a$, remove from the linked list associated with $a'$ the pointer to this occurrence. Clearly, the overall number of basic operations that are performed during the construction of the forest is only $O(n^{4/3}\log n)$.

*Remark.* Note that if the underlying norm is $L_\infty$, then, even if the input points are in $d$-space for some $d > 2$, it is easy to find a most uniform matching in time $O(n^3 \text{polylog } n)$. This is done by constructing a two level orthogonal range tree, where the first level enables us to find the points of $B$ lying at distance at least $r$ from a query point, as a polylogarithmic number of disjoint subsets of $B$, and the second level enables us to find out of the points found in the first level those lying at distance at most $r'$ from the query point. Details are standard and hence omitted from this version. Summarizing, we have:

**Theorem 2.** *Let $A$ and $B$ be two sets of $n$ points in $\mathbb{R}^d$. It is possible to compute $M_U^*$ in time $O(n^{10/3} \log n)$ when $d = 2$ and the underlying norm is $L_2$, or in time $O(n^3 polylog\, n)$ when $d \geq 2$ and the underlying norm is $L_\infty$.*

## 3 Applications to Pattern Matching

Note that an obvious variant of the algorithm of the preceding section (with the same running time) finds a matching $M^*$ for which $\max(M) - \min(M)$ is minimum among all matchings $M$ between $A$ and $B$ of cardinality $k$, where $1 \leq k \leq n$ is some pre-determined parameter. This variant has the following interesting application.

Assume that the set $B' \subseteq B$ is a translated copy of the set $A' \subseteq A$, but every point in $B'$ has been independently slightly perturbed. In addition, $A - A'$ and $B - B'$ consist of spurious points; that is, they were created as a result of noise. Given $A$ and $B$, the problem is therefore to find $A'$, $B'$ and the translation by which $B'$ was translated with respect to $A'$. Assume furthermore that we have some priori knowledge that the noisy points are spread randomly and that $|A'| \geq k$. One approach for solving this problem is to find a translation that minimizes the *Hausdorff distance* between large enough subsets of $A$ and $B$; see [5, 6] for further details. This approach has the drawback that the computed matching is not necessarily one-to-one, and hence might not be appropriate, since we are certain in this case that a one-to-one matching between $A'$ and $B'$ exists.

On the other hand, it is reasonable to assume that the most uniform matching of size $k$ will fit points of $A'$ to their images in $B'$, while most of the points of $A - A'$ and $B - B'$ will remain unmatched. (If we don't have an exact estimation of the size of $A'$, we can compute such an estimation by a binary search, since we expect a significant increase in the differences corresponding to the most uniform matchings that are computed when moving from sizes below $|A'|$ to sizes above $|A'|$.) Hence our algorithm is most likely to identify the sets $A'$ and $B'$ and to give a useful initial translation to the algorithms of Efrat and Itai [7], Heffernan and Schirra [13], or Heffernan [14] that compute the exact translation (once $A'$ and $B'$ are known). We are not aware of any other way to solve this problem in less than $O(n^{3.5})$ time.

## 4 Computing a Minimum Deviation Matching

In this section we show how to find $M_D^*$, the matching that minimizes $(1/n)\Sigma(M) - \min(M)$. We continue to use the same notation as in Section 2. In addition, we define $M^i$ to be the perfect matching between $A$ and $B$ whose sum of distances is minimum among all matchings $M$ for which $\min(M) \geq d^{(i)}$. Let $d^{(i)}$ be the shortest edge of $M_D^*$. An easy but crucial observation is that $M_D^* = M^i$.

In [3] an $O(n^{2+\epsilon})$-time algorithm for finding $M^1$ was proposed. This algorithm is a variant of Vaidya's algorithm for solving the same problem. We use a variant of the algorithm of [3] for finding $M_D^*$. First we find $M^1$, using the algorithm of [3] mentioned above. Next we perform $n^2 - 1$ phases of the algorithm, where in the $j$-th phase we use $M^{j-1}$ to find $M^j$, and compute $(1/n)M^j - d^{(j)}$. The previous observation shows that the smallest value encountered is $M_D^*$. It will be shown that the time needed for a phase (actually for obtaining $M^j$ from $M^{j-1}$) is $O(n^{4/3+\epsilon})$, so the total running time is $O(n^{10/3+\epsilon})$.

Let $G$ be a general weighted bipartite graph on $A \cup B$, where each edge $(a_i, b_j)$ is associated with a weight $d(a_i, b_j)$. The bipartite weighted matching problem can then be formulated as a linear program of minimizing $\sum_{i,j} d(a_i, b_j)x_{ij}$, where

$$\sum_{j=1}^n x_{ij} = 1, \quad i = 1, \ldots, n,$$
$$\sum_{i=1}^n x_{ij} = 1, \quad j = 1, \ldots, n,$$
$$x_{ij} \geq 0, \quad i, j = 1, \ldots, n,$$

The edge $(a_i, b_j)$ is an edge of $M$ if and only if $x_{ij} = 1$. The dual linear program is to maximize $\sum_i \alpha_i + \sum_j \beta_j$ subject to $\alpha_i + \beta_j \leq d(a_i, b_j)$ (for $i, j = 1, \ldots, n$), where $\alpha_i$ (resp. $\beta_j$) is the dual variable associated with the point $a_i$ (resp. $b_j$). A necessary and sufficient condition for the optimality of the solution is that the orthogonal conditions below hold:

$$x_{ij} > 0 \Rightarrow \alpha_i + \beta_j = d(a_i, b_j) \quad i, j = 1 \ldots n \tag{1}$$
$$\alpha_i \neq 0 \quad \Rightarrow \sum_j x_{ij} = 1 \quad i = 1 \ldots n \tag{2}$$
$$\beta_j \neq 0 \quad \Rightarrow \sum_i x_{ij} = 1 \quad j = 1 \ldots n \tag{3}$$

Assume that in the last phase we have computed $M^{k-1}$. That is, we have found values $x_{i,j}, \alpha_i, \beta_j$ (for $1 \leq i, j \leq n$) such that all orthogonal conditions (1), (2), and (3) hold. We now seek $M^k$. If the edge $(a_{i'}, b_{j'})$ whose length is $d^{(k-1)}$ is not in $M^{k-1}$ (and then all its edges are of length at least $d^{(k)}$), then surely $M^k = M^{k-1}$, and this phase is done. Otherwise we want to delete this edge, and make sure that no matching in the future will use it. This will be achieved by setting $d(a_{i'}, b_{j'}) \equiv \infty$, $x_{i',j'} = 0$ (that is, delete $(a_{i'}, b_{j'})$ from the matching), $\alpha_{i'} = 0$, and $\beta_{j'}$ is not changed. Hence all conditions remain valid, except for (3), which is violated. Note that at this stage, for each pair $a_i, b_j$ whose (Euclidean) distance $\|a_i - b_j\|$ is less than $d^{(k)}$, we have $d(a_i, b_j) = \infty$. For all other edges $d(a_i, b_j) = \|a_i - b_j\|$.

The Hungarian method computes a matching in $n$ stages, each of which augments the matching by one edge and updates the dual variables. Let $X$ be the current matching, i.e., $X = M^{k-1} - \{(a_{i'}, b_{j'})\}$. An edge $(a_i, b_j)$ is called admissible if $d(a_i, b_j) = \alpha_i + \beta_j$. Due to the orthogonal conditions, all the edges of $X$ are admissible. As in the previous section, we use an augmenting path to increase the cardinality of the matching back to $n$.

We search for an augmenting path consisting only of admissible edges, as follows. From the unique exposed vertex $b_{j'} \in B$, we grow (in an implicit manner) an 'augmenting tree' whose paths are augmenting paths starting at the root $b_{j'}$. More precisely, each point of $A \cup B$ in the augmenting tree is reachable from $b_{j'}$ by an augmenting path that consists only of admissible edges. For a point $w$ of $A$ (resp. $B$), the path leading to $w$ ends at an edge not in $X$ (resp. in

$X$). Let $S \subseteq B$ and $T \subseteq A$ denote the set of points of $B$ and of $A$ that lie in the augmenting tree. At the beginning of a phase, $S = \{b_{j'}\}$ and $T = \emptyset$. Let $\delta = \min_{a_i \in A - T, b_j \in S} \{d(a_i, b_j) - \alpha_i - \beta_j\}$. At each step, the algorithm takes one of the following actions, depending on whether $\delta = 0$ or $\delta > 0$:

**Case 1:** $\delta = 0$. Let $(a_i, b_j)$, for $a_i \in A - T$ and $b_j \in S$, be an admissible edge ($\delta = 0$ implies that such an edge must exist). If $a_i$ is the exposed vertex $a_{i'}$, then an augmenting path has been found. Otherwise ($a_i$ is matched), let $b_k$ be the vertex matched (in $X$) to $a_i$. We add the edges $(a_i, b_j)$ and $(a_i, b_k)$ to the augmenting tree, the point $a_i$ to $T$, and the point $b_k$ to $S$.

**Case 2:** $\delta > 0$. The algorithm updates the dual variables, as follows. For each vertex $a_i \in T$, it sets $\alpha_i = \alpha_i - \delta$ and, for each $b_j \in S$, it sets $\beta_j = \beta_j + \delta$. Note that every edge of the augmenting tree remains admissible.

The algorithm repeats these steps until it reaches an exposed vertex of $A$, thereby obtaining an augmenting path. If an augmenting path $\Pi$ is found, we delete the edges of $\Pi \cap X$ from the current matching $X$ and add the other edges of $\Pi$ to $X$ (thereby increasing the size of the current matching by 1) to obtain the new matching $M^k$. Note that (1), (2), and (3) hold and hence the optimality is provided. This completes the description of the algorithm. Further details of the algorithm and the proof of its correctness can be found in [18, 23].

Vaidya [23] suggested the following approach to expedite the running time of each step. Maintain a variable $\Delta$ and associate a weight with each point in $A \cup B$. In the beginning of each phase, $\Delta = 0$ and $w(a_i) = \alpha_i$, $w(b_j) = \beta_j$ for each $1 \leq i, j \leq n$. During each step, the weights and $\Delta$ are updated, but the values of the dual variables remain unchanged. This is done as follows. If Case 1 occurs, then we set $w(a_i) = \alpha_i + \Delta$ and $w(b_k) = \beta_k - \Delta$, and do not change the value of $\Delta$. (Note that $a_i \notin T$ and $b_k \notin S$, so the values of $\alpha_i$ and $\beta_k$ are the same as at the beginning of the phase.) If Case 2 occurs, then we set $\Delta = \Delta + \delta$, and do not change the weights. Notice that, for each $b_j \in S$, the current value of $\beta_j$ is equal to $w(b_j) + \Delta$, and, similarly, for each $a_i \in T$, the current value of $\alpha_i$ is equal to $w(a_i) - \Delta$. Also, the current values of the dual variables for other points are equal to their values at the beginning of the phase. At the end of each phase, the values of the dual variables can be computed from $\Delta$ and from the weights of the corresponding points. The weight of each point changes only once during a phase, namely, when it is added either to $S$ or to $T$. Moreover, at any time during a phase,
$$\delta = \min_{a_i \in A - T, b_j \in S} \{d(a_i, b_j) - w(a_i) - w(b_j)\} - \Delta.$$

For a parameter $r > 0$, let $\mathbf{F}_r = \{d_1(x), \ldots, d_n(x)\}$ denote a family of functions where $d_i(x) = \|x - a_i\| + w(a_i)$ if $\|x - a_i\| \geq r$, and $\infty$ otherwise (for $i = 1, \ldots, n$). Using obvious abuse of notation, we say that a point $a_i \in A - T$ is the closest in $A - T$ to a point $q \in \mathbb{R}^2$ if $d_i(q) \leq d_j(q)$, for $a_j \in A - T$. Our problem is thus to maintain the closest pair (using the distance functions defined above) between $A - T$ and $S$. To bound the running time of each operation, we need to argue about the complexity of the lower envelope of $\mathbf{F}_r$. The best bound we are aware of is $O(n^{2+\varepsilon})$, which results from [12, 21]. Hence we can maintain $\delta$ using the algorithm of Agarwal et al. [3, Thm. 6.8] that maintain (under deletions and insertions) the closest pair between $S$ and $A - T$, when the distance is measured using the distance functions $\mathbf{F}_{d(k)}$, in time $O(n^{1/3+\epsilon})$ for each operation. This data structure can be built in time $O(n^{4/3+\epsilon})$. Since each step requires at most two update operations (inserting a point into $S$ and deleting a point from $A - T$), each phase can be performed in time $O(n^{4/3+\epsilon})$. Moreover, at the beginning of the $k$-th phase, we can build the data structure for $\mathbf{F}_{d(k)}$ in time $O(n^{4/3+\epsilon})$. Thus, we have:

**Theorem 3.** *Given sets $A, B \subseteq \mathbb{R}^2$ of $n$ points, we can find $M_D^*$ in time $O(n^{10/3+\epsilon})$, for every $\epsilon > 0$.*

## 5 Bottleneck Matching in Higher Dimensions

Let $A$ be a set of $n$ points in $\mathbb{R}^d$, where $n$ is even and $d \geq 3$. In this section, we show how to compute a bottleneck matching $M_B^*$ in the (complete) Euclidean graph over $A$. As in the preceding sections, we assume that the $\binom{n}{2}$ distances between pairs of points in $A$ are distinct. Let $B_r(p)$ denote the ball of radius $r$ centered at point $p$. For two points $p, q \in \mathbb{R}^d$, let $lune(p, q)$ denote the region $B_{\|p-q\|}(p) \cap B_{\|p-q\|}(q)$, where $\|p - q\|$ is the distance between $p$ and $q$. The $k$ *relative neighborhood graph* of $A$, denoted $k\mathrm{RNG}(A)$, is a graph $(A, E)$, where a pair of points $(p, q) \in E$ if and only if the number of points of $A$ other than $p$ and $q$ that lie in $lune(p, q)$ is less than $k$. In the plane, Chang et al. [4] have proven that $17\mathrm{RNG}(A)$ contains a bottleneck matching. Thus, in order to compute $M_B^*$ in this case, it is enough to consider $17\mathrm{RNG}(A)$ whose size is only $O(n)$. We then apply to it the general $O((n \log n)^{1/2}m)$ algorithm of Gabow and Tarjan [9]. Chang et al. compute the $17\mathrm{RNG}$ in time $O(n^2)$, but it can be computed in time $O(n \text{ polylog } n)$. Su and Chang [22] describe how to construct in $O(n \log n)$ time another linear-size graph, called the $17$-Gabriel graph, that contains the $17\mathrm{RNG}$. Thus in both cases $M_B^*$ can be computed in total time $O(n^{1.5} \log^{0.5} n)$.

We show below that for any fixed dimension $d$, there exists a constant $k = k(d)$ such that $k\mathrm{RNG}(A)$ contains a bottleneck matching. Since the size of a $k\mathrm{RNG}$ for a constant $k$ remains linear also in higher dimensions (if the distances between pairs of points are distinct [4]), and since it can be computed in subquadratic time, $M_B^*$ can also be computed in total time that is subquadratic even in higher dimensions. Moreover, we show that if the underlying norm is $L_\infty$, then $M_B^*$ can be computed in time $O(n^{1.5} \log^{0.5} n)$ in any fixed dimension.

The proof of Chang et al. [4] to the claim that $k\mathrm{RNG}(A)$ contains an optimal matching applies also in higher dimensions, provided that the following lemma, which is proven for the planar case in [4], remains correct for an appropriate constant $k = k(d)$. Below we show that indeed this is the case.

Let $M$ be a bottleneck matching, and let $(p, q)$ be an edge of $M$. Let $S$ be the subset of $A$ consisting of the points that are matched with points lying in the interior of $lune(p, q)$, and assume that $S \cap lune(p, q) = \emptyset$, since, if $u \in S$ lies inside $lune(p, q)$, then we could replace the edges $(p, q)$ and $(u, v)$ of $M$, where $v$ is the point inside $lune(p, q)$ matched to $u$, by the edges $(p, u)$ and $(q, v)$ and obtain a matching $M'$ which is no worse than $M$, but the sum of lengths of its edges is

---

[4] This size grows in degenerate situations; see [1]

smaller than the sum of lengths of the edges of $M$. Hence, we assume that all the points in (the interior of) $lune(p, q)$ are matched with points lying outside of $lune(p, q)$. For a point $u \in S$, let $n_u$ be its nearest point on the boundary of $lune(p, q)$ and put $r = \|p - q\|$.

**Lemma 4.** *Assume the underlying norm is either $L_2$ or $L_\infty$. If for every $u, v \in S$*
(i) $\|u - p\| > r$ *and* $\|u - q\| > r$,
(ii) $\|u - v\| > r$, *and*
(iii) $\|u - v\| > \|u - n_u\|$ *and* $\|u - v\| > \|v - n_v\|$,
*then $|S|$ is a constant depending on the dimension $d$.*

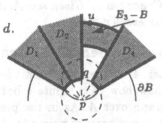

**Proof:** This proof is for the $L_2$ case. The proof for the (easier) $L_\infty$ case appears in the full version of this paper. For clarity, we restrict ourselves in the proof to $\mathbb{R}^3$, however, the same proof with obvious modifications also holds for $d > 3$. We divide $\mathbb{R}^3 - lune(p, q)$ into three regions. Let $C_p$ (resp. $C_q$) be the cone whose origin is $p$ (resp. $q$) and its boundary contains the boundary of the disk $(\partial B_r(p) \cap \partial B_r(q))$. The first region $R_1$ is $C_p - lune(p, q)$, the second region $R_2$ is $C_q - lune(p, q)$, and the third region $R_3$ is $\mathbb{R}^3 - C_p \cup C_q$. We will show that the number of points in each region is less than some constant. The difficult case is to show this for the regions $R_1, R_2$, since the boundary of these regions contains a 2-dimensional face of $lune(p, q)$, while the boundary of $R_3$ contains only a 1-dimensional face of $lune(p, q)$. We will show this for $R_1$ (the proof for $R_2$ is completely analogous). Assume that $p$ is the origin of our coordinate system and that the segment $pq$ lies on the $z$-axis. We partition $R_1$ into a constant number of pyramid-like regions as follows. Consider the planes containing the $y$-axis that form angles of $30 + \alpha$, $30 + 2\alpha$, etc. with the $xy$ plane, and the planes containing the $x$-axis forming these angles with the $xy$ plane, for some sufficiently small but fixed angle $\alpha$. These planes together with the boundary of $C_p$ partition the first region $R_1$ into a constant number of pyramid-like regions (see the figure above for an illustration in the plane). We will prove that each of these pyramids $D_i$ contains only a constant number of points. Let $B = B_c(p)$, where $c$ is an appropriate constant. Let $u$ be the point in $D_i \cap S$ that is furthest from $p$, and let $E_i = D_i \cap B_{\|p-u\|}(p)$. If $u \in B$ then so are all the other points in $D_i$, and clearly the number of these points is bounded by some constant, due to assumption (ii) above. Otherwise, $u$ lies outside of $B$, but then $B_{\|u-n_u\|}(u) \supseteq E_i - B$, so there can be no other point in $D_i$ outside of $B$, since such a point would necessarily lie inside $B_{\|u-n_u\|}(u)$ violating assumption (iii) above. $\qquad \square$

Lemma 4 together with the remarks preceding it lead to the conclusion that for some appropriate constant $k$ a bottleneck matching is contained in $k\text{RNG}(A)$. However, instead of computing $k\text{RNG}(A)$, we prefer to compute a graph $G = (A, E)$, such that $|E| = O(n)$, and $k\text{RNG}(A) \subseteq G$. The containment follows from arguments similar to those given in [4]. We first deal with the $L_\infty$ case.

Using a standard orthogonal multi-level range tree, we can count (and find) all points lying in a query range in time $O(\log^{d-1} n)$ (using fractional cascading). Let $u \in A$, and let $R^+$ denote the region of $\mathbb{R}^d$ consisting of all points $x$ such that all the components of the vector $x - u$ are non-negative (e.g., in the plane, $R^+$ is the north-eastern quadrant of $u$). The idea is to find the $k$ nearest neighbors of $u$ in $A \cap R^+$. This process is repeated for all $u \in A$, and for all the $2^d$ regions of $\mathbb{R}^d$ symmetric to $R^+$.

To find these neighbors, we check how many points of $A$ lie in a cube $B$ of (yet unknown) radius, whose most-negative corner coincides with $u$. We seek the one that contains exactly $k$ points. To determine the size of the cube, we use the technique of Frederickson and Johnson [8] to perform a binary search in the (implicitly defined) matrix $M$ containing all differences of the form $u.x_i - v.x_i$, where $u, v \in A$, and $u.x_i$ is the $i$'th coordinate of $u$. Hence we can find $G$ in time $O(n \log^d n)$. We now apply the algorithm of Gabow and Tarjan to the graph $G$ to obtain a bottleneck matching of $A$ in time $O(n^{1.5} \log^{0.5} n)$.

As to the $L_2$ case, Agarwal and Matoušek [1] show how to compute the $k$RNG, where $k$ is some constant. They construct first a super graph of the $k$RNG which is also of linear size, so we may take this supergraph as input to the algorithm of Gabow and Tarjan. The construction time of this supergraph is only $O(n^{4/3} \log^2 n)$ for $d = 3$, and $O(n^{2-2/(\lceil d/2 \rceil + 1) + \epsilon})$ for $d \geq 4$ (See [2]). Thus we can compute $M_B^*$ in $O(n^{1.5} \log^{0.5} n)$ for $2 \leq d \leq 4$, and $O(n^{2-2/(\lceil d/2 \rceil + 1) + \epsilon})$ for $d \geq 5$. Hence we have:

**Theorem 5.** *Let $A$ be a set of $2n$ points in d-space. A bottleneck matching of $A$ can be computed in $O(n^{1.5} \log^{0.5} n)$ time for $2 \leq d \leq 4$, in $O(n^{1.5+\epsilon})$ time for $5 \leq d \leq 6$, and in $O(n^{2-2/(\lceil d/2 \rceil + 1) + \epsilon})$ for $d > 6$. Under the $L_\infty$ norm, a bottleneck matching of $A$ can be computed in $O(n^{1.5} \log^{0.5} n)$ time for any fixed $d \geq 2$.*

*Remark.* The following observation is due to Pankaj Agarwal, it implies that finding a faster algorithm for computing a bottleneck matching in d-space, for $d > 6$, is probably a non-trivial problem. Assume that $A$ consists of the two subsets $A_1$ and $A_2$, where $|A_1| = |A_2| = n/2$, and $n/2$ is odd. Assume furthermore that the diameter of $A_1$ and the diameter of $A_2$ are very small compared to the diameter of $A$. In this case, a bottleneck matching of $A$ must match the closest mixed pair $a_1, a_2$ where $a_1 \in A_1$ and $a_2 \in A_2$. But this is actually the famous *bichromatic closest pair problem in d-space* (see [2]), which for several years now has no faster solution that the one we gave.

## Acknowledgment

We would like to thank Pankaj Agarwal, Rafi Hassin, Alon Itai and Micha Sharir for helpful discussions on the contents of this paper.

## References

[1] P.K. Agarwal and J. Matoušek, Relative neighborhood graphs in three dimensions, *Comp. Geom. Theory and App.* 2 (1992), 1–14.

[2] P.K. Agarwal, J. Matoušek and S. Suri, Farthest neighbors, maximum spanning trees and related problems in higher dimensions, *Comp. Geom. Theory and App.* 1 (1992), 189–201.

[3] P.K. Agarwal, A. Efrat and M. Sharir, Vertical decomposition of shallow levels in 3-dimensional arrangements and its applications, *Proceedings 11 Annual Symposium on Computational Geometry*, 1995, 39–50.

[4] M.S. Chang, C.Y. Tang, and R.C.T. Lee, Solving the Euclidean bottleneck matching problem by $k$-relative neighborhood graphs, *Algorithmica* 8 (1992), 177–194.

[5] L.P. Chew, D. Dor, A. Efrat and K. Kedem, Geometric pattern matching in $d$-dimensional space, *ESA'95*, LNCS 979, 264–279.

[6] L.P. Chew and K. Kedem, Improvements on geometric pattern matching problems, *SWAT'92*, LNCS 621, 318–325.

[7] A. Efrat and A. Itai, Improvements on bottleneck matching and related problems using geometry, *Proceedings 12 Annual Symposium on Computational Geometry*, 1996, 301–310.

[8] G.N. Frederickson and D.B. Johnson, Generalized selection and ranking sorted matrices, *SIAM J. Computing* 13 (1984), 14–30.

[9] H.N. Gabow and R.E. Tarjan, Algorithms for two bottleneck optimization problems, *J. Algorithms* 9 (1988), 411–417.

[10] Z. Galil and B. Schieber, On finding most uniform spanning trees, *Disc. App. Mathematics* 20 (1988), 173–175.

[11] S.K. Gupta and A.P. Punnen, Minimum deviation problems, *Oper. Res. Lett.* 7 (1988), 201–204.

[12] D. Halperin and M. Sharir, New bounds for lower envelopes in three dimensions, with applications to visibility of terrains, *Discand CompGeom* 12 (1994), 313–326.

[13] P.J. Heffernan and S. Schirra, Approximate decision algorithms for point set congruence, *SIAM J. Computing* 8 (1992), 93–101.

[14] P.J. Heffernan, Generalized approximate algorithms for point set congruence, *WADS'93*, LNCS 709, 373–384.

[15] J. Hopcroft and R.M. Karp, An $n^{5/2}$ algorithm for maximum matchings in bipartite graphs, *SIAM J. Computing* 2 (1973), 225–231.

[16] M. Katz, Improved algorithms in geometric optimization via expanders, *Proc. 3rd Israel Symp. on Theory of Computing and Systems*, 1995, 78–87. (See also, M.J. Katz and M. Sharir, An expander-based approach to geometric optimization, *SIAM J. Computing*, to appear.)

[17] H. Kuhn, The Hungarian method for the assignment problem, *Naval Research Logistics Quarterly* 2 (1955), 83–97.

[18] E. Lawler, Combinatorial Optimization: Networks and Matroids, *Holt, Rinehart and Winston*, New-York, 1976.

[19] S. Martello and P. Toth, Linear assignment problems, Annals of Disc. Mathematics 31 (1987), 259–282.

[20] S. Micali and V.V. Vazirani, An $O(\sqrt{|V|} \cdot |E|)$ algorithm for finding maximum matching in general graphs, *Proc. 21 Annual ACM Symp. on Theory of Comp.*, 1980, 17–27.

[21] M. Sharir, Almost tight upper bounds for lower envelopes in higher dimensions, *Discand CompGeom* 12 (1994), 327–345.

[22] T.H. Su and R.C. Chang, The $k$-Gabriel graphs and their applications, *1st Annual SIGAL International Symp. Algorithms*, LNCS 450, 1990, 66–75.

[23] P.M. Vaidya, Geometry helps in matching, *SIAM J. Computing* 18 (1989), 1201–1225.

# Computing the Maximum Overlap of Two Convex Polygons Under Translations*

Mark de Berg,[1] Olivier Devillers,[2] Marc van Kreveld,[1] Otfried Schwarzkopf,[3]
Monique Teillaud[2]

[1] Dept. of Computer Science, Utrecht University, P.O. Box 80.089, 3508 TB Utrecht,
the Netherlands.
[2] INRIA, BP93, 06902 Sophia-Antipolis cedex.
[3] Dept. of Computer Science, Pohang University of Science and Technology,
Pohang 790-784, South Korea.

**Abstract.** Let $P$ be a convex polygon in the plane with $n$ vertices and let $Q$ be a convex polygon with $m$ vertices. We prove that the maximum number of combinatorially distinct placements of $Q$ with respect to $P$ under translations is $O(n^2 + m^2 + \min(nm^2 + n^2m))$, and we give an example showing that this bound is tight in the worst case. Second, we present an $O((n+m)\log(n+m))$ algorithm for determining a translation of $Q$ that maximizes the area of overlap of $P$ and $Q$.
We also prove that the placement of $Q$ that makes the centroids of $Q$ and $P$ coincide realizes an overlap of at least 9/25 of the maximum possible overlap. As an upper bound, we show an example where the overlap in this placement is 4/9 of the maximum possible overlap.

## 1 Introduction

Matching plays an important role in areas such as computer vision. Typically one is given two 'shapes'—point sets or polygons, for instance—and one wants to determine how much these shapes resemble each other. More precisely, one wants to find a rigid motion of one shape that maximizes the resemblance with the other shape. There are several ways to measure resemblance. For example, for point sets or polygonal chains one can use the Hausdorff distance [2, 1, 9, 15, 16]; for polygonal chains one can also use the Fréchet distance [3].

The resemblance of two convex polygons can also be measured by looking at the Hausdorff or Fréchet distance between their boundaries. For an application in computer vision, however, it seems more appropriate to look at the area of the symmetric difference of the two polygons, since this distance measure is less sensitive to noise in the image: noise may add thin features to the boundary but is unlikely to add large areas.

* This work was supported by ESPRIT Basic Research Action No. 7141 (project ALCOM II: *Algorithms and Complexity*). M.d.B. and O.S. were supported by the Netherlands' Organisation for Scientific Research (NWO). O.S. also acknowledges partial support by Pohang University of Science and Technology Grant P96005, 1996.

Notice that minimizing the area of the symmetric difference of two polygons is equivalent to maximizing the area of overlap of the polygons. An algorithm with $O(n(n + m))$ time complexity is known for finding the maximum overlap area for two convex polygons, one of which is allowed to rotate with one point on its boundary sliding on the other polygon's boundary [20]. Mount et al. [18] studied the behavior of the area of overlap for two simple polygons under translations of one polygon. They pose the case of two convex polygons as an open problem.

We consider the matching problem for convex polygons in the plane, and the rigid motions that we allow are translations. In other words, we are given two convex polygons $P$ and $Q$ in the plane, and our goal is to find a translation of $Q$ that maximizes the area of overlap with $P$. Our results are as follows. Let $n$ and $m$ denote the number of vertices of $P$ and $Q$, respectively. We start by studying a combinatorial question: how many combinatorially distinct placements of $Q$ with respect to $P$ are there? Here we define two placements to be combinatorially equivalent if the same pairs of edges (one from $P$ and one from $Q$) intersect— see Section 2 for a more precise definition. We show that the number of distinct placements is $O(n^2 + m^2 + \min(nm^2 + n^2m))$, and we give an example showing that this bound is tight in the worst case. To our surprise, this result appears to be new: previous work on bounding the number of placements of a polygon in a polygonal environment is usually motivated by motion planning problem and, hence, only deals with the case where the polygon is not allowed to intersect the environment at all—see Latombe's book [17] or Halperin's thesis [14]. Our main result is presented in Section 3, where we give an $O((n + m)\log(n + m))$ time algorithm for computing a placement of $Q$ that maximizes the area of overlap with $P$. Our algorithm is based on the fact that the area-of-overlap function is unimodal. To round off our exposition, we show that one can, in a sense, approximate the maximum possible overlap of two convex polygons by simply superposing them such that their centroids coincide. We show that that placement realizes an overlap that is at least 9/25 of the maximum possible overlap, and we give an upper bound example where the factor is 4/9.

Our work can also be seen as a generalization of the problem of placing a copy of one polygon inside another polygon. Chazelle [6] studied several variants of this problem. One of his results is that, given two convex polygons $P$ and $Q$, one can decide in linear time whether $Q$ can be translated such that it is contained in $P$. Other papers compute the largest copy of a polygon that can be placed inside another one [5, 10, 11, 19].

## 2 The Number of Distinct Placements

Let $P$ be a simple polygon with $n$ vertices in the plane and let $Q$ be a simple polygon with $m$ vertices. The position and orientation of $P$ are fixed, but $Q$ is free to translate. In this section we bound the number of distinct placements of $Q$ with respect to $P$. We first define formally when we call two placements distinct.

We denote the boundary of $P$ by $\partial P$, and the boundary of $Q$ by $\partial Q$. We consider boundary edges to be relatively open sets, that is, their endpoints are not included. Let $r_Q$ be a reference point on $Q$, say the lexicographically smallest vertex. For a point $r$ in the plane, $Q(r)$ denotes $Q$ with its reference point placed at $r$. Similarly, for an edge $e$ or a vertex $v$ of $Q$, $e(r)$ and $v(r)$ denote the edge $e$ and vertex $v$ when $Q$ is placed at $r$. We call $Q(r)$ a *placement* of $Q$. The space of all possible placements of $Q$—in our case this is a 2-dimensional space–is called the *configuration space* [17].

**Definition 1.** The *intersection set* of $P$ and a placement $Q(r)$, denoted $I(r)$, is the set consisting of all pairs $(f, g)$ such that $f$ is the interior of $P$, an edge of $P$, or a vertex of $P$, $g$ is the interior of $Q(r)$, an edge of $Q(r)$, or a vertex of $Q(r)$, and $f$ and $g$ intersect. Two placements $Q(r)$ and $Q(r')$ are *combinatorially distinct* if and only if $I(r) \neq I(r')$.

The configuration space can be partitioned into regions according to the intersection sets of the corresponding placements: two points are in the same region if and only if the corresponding placements are combinatorially equivalent. Hence, the number of combinatorially distinct placements is bounded by the number of regions in the configuration space.

Let's have a closer look at this space. Fix an edge $e$ of $P$ and an edge $e'$ of $Q$, and consider the locus of all points $r$ such that $e$ intersects $e'(r)$. This region is a parallelogram, denoted $\pi(e, e')$, spanned by a translated copy of $e$ and a translated copy of $e'$. Observe that for points $r$ in the interior of the edges of $\pi(e, e')$, a vertex of $e$ lies on $e'(r)$ or a vertex of $e'(r)$ lies on $e$; for a point $r$ that is a vertex of $\pi(e, e')$, a vertex of $e$ coincides with a vertex of $e'(r)$. Let

$$\Pi = \{\pi(e, e') : \ e \text{ is an edge of } P, \ e' \text{ is an edge of } Q\}.$$

The arrangement $\mathcal{A}(\Pi)$ induced by $\Pi$ is the partitioning of configuration space we mentioned above: there is a one-to-one correspondence between the combinatorially distinct placements and the faces, arcs,[4] and nodes of $\mathcal{A}(\Pi)$. So a bound on the complexity of $\mathcal{A}(\Pi)$ immediately implies a bound on the number of distinct placements.

We proceed to bound the complexity of $\mathcal{A}(\Pi)$. Because $\mathcal{A}(\Pi)$ is a planar subdivision defined by $nm$ parallelograms, its complexity is bounded by $O(n^2 m^2)$. For simple polygons, this bound is tight in the worst case. In fact, the complexity of the free space can already be $\Theta(n^2 m^2)$, as mentioned above.

For convex polygons, the situation is different: We show in the following theorem that the maximum number of combinatorially distinct placements is $\Theta(n^2 + m^2 + \min(nm^2, n^2 m))$. While this is significantly less than $\Theta(n^2 m^2)$, it is much larger than the complexity of the free space for two convex polygons. The proof can be found in the full version of this paper.

---

[4] To avoid confusion between the edges of the polygons and the edges of $\mathcal{A}(\Pi)$, we call the latter arcs. Similarly, we call the vertices of $\mathcal{A}(\Pi)$ nodes.

**Theorem 2.** *The maximum number of combinatorially distinct placements of two convex polygons with $n$ and $m$ vertices, respectively, is:*

$$\Theta(n^2 + m^2 + \min(nm^2, n^2m)).$$

## 3 Computing the Maximum Overlap

We now get to the main problem studied in this paper: given two convex polygons $P$ and $Q$, find a placement of $Q$ that maximizes the overlap with $P$. First, we need to introduce some notation. The *overlap function* $\omega(r) : \mathbb{R}^2 \to \mathbb{R}$ of $P$ and $Q$ is defined as

$$\omega(r) := \text{ the area of } P \cap Q(r).$$

Our problem is thus to find a placement $Q(r)$ that maximizes $\omega(r)$. We call such a placement a *goal placement*.

We first look at a restricted version of the problem, where $Q$ is only allowed to translated into a fixed direction. Without loss of generality, we assume this direction to be horizontal. Thus, for a given value $y^*$, we define the *(horizontal) overlap function at $y^*$*, denoted by $\omega_{y^*}(t)$, as

$$\omega_{y^*}(t) := \omega((t, y^*)).$$

**Theorem 3.** *Let $P$ and $Q$ be two convex polygons then the function $r \mapsto \sqrt{\omega(r)}$ is downwards concave, that is the volume below its graph is convex.*

*Proof.* Since a function $\mathbb{R}^2 \to \mathbb{R}$ is downwards concave if any cross-section along a line is downwards concave, it suffices to prove the latter fact. Without loss of generality, we can restrict ourself to the case of horizontal lines, and will prove that the monovariate function $t \mapsto \sqrt{\omega_{y^*}(t)}$ is downwards concave.

Imagine moving $Q$ from left to right over the plane, starting with $Q((-\infty, y^*))$ and ending at $Q((+\infty, y^*))$. Define $Q(t) := Q((t, y^*))$, and $A(t) := P \cap Q(t)$. Thus $A(t)$ is the intersection of $P$ and $Q$ at time $t$. We define a three-dimensional polytope $\mathcal{P}_{PQ}$ by viewing time as the third dimension, and taking the union of all polygons $A(t)$:

$$\mathcal{P}_{PQ} := \{(x, y, t) : (x, y) \in A(t)\}.$$

Since $\mathcal{P}_{PQ}$ can be written as the intersection of two convex polytopes, it is a convex polytope itself:

$$\mathcal{P}_{PQ} = \{(x, y, t) : (x, y) \in P\} \cap \{(x, y, t) : (x, y) \in Q(t)\}$$

Following Avis et al. [4], we can now apply the Brunn-Minkowski theorem [13], which states that the square root of the function that describes the area of intersection of $\mathcal{P}_{PQ}$ and a horizontal plane $h$ is downwards concave, as we sweep $h$ through $\mathcal{P}_{PQ}$. Since the cross-section of $\mathcal{P}_{PQ}$ with the horizontal plane $t = t^*$ is exactly the intersection $A(t^*)$, the theorem follows. □

A non-negative function $\chi : D \to \mathbb{R}$ is called *unimodal* if there is an interval $D = [a_0 : a_1]$ and points $b_0, b_1 \in D$ with $b_0 \leqslant b_1$ such that $\chi$ is zero outside $D$, strictly increasing from $a_0$ to $b_0$, constant from $b_0$ to $b_1$, and strictly decreasing from $b_1$ to $a_1$. Our algorithm is based on the unimodality of the monovariate overlap function.

**Corollary 4.** *Let $P$ and $Q$ be two convex polygons, and let $y^* \in \mathbb{R}$. Then the horizontal overlap function $\omega_{y^*}(t)$ is unimodal.*

*Proof.* From the downwards concavity of $\sqrt{\omega_{y^*}(t)}$ immediately follows that it is a unimodal function, which in turn implies that $\omega_{y^*}(t)$ is unimodal as well. □

Theorem 4 can be used to compute the maximum overlap of $P$ and $Q$ for the case where $Q$ is confined to translate along a fixed line. This algorithm will be an important ingredient of the general algorithm.

**Lemma 5.** *[Avis et al.] For a line $\ell$ we can compute $\max_{r \in \ell} \omega(r)$ in $O(n + m)$ time.*

*Proof.* Using Chazelle's algorithm [7] the convex polytope $\mathcal{P}_{PQ}$ can be computed in linear time, and then Avis et al.'s algorithm [4] can be used to compute the horizontal section of $\mathcal{P}_{PQ}$ of maximal area in linear time. □

We now turn our attention to the general case, where arbitrary translations are allowed. Our algorithm consists of two stages.

*The first stage.* Here we locate a horizontal strip that contains the reference point of a goal placement. This will be done by a binary search that uses the algorithm from Lemma 5 as a subroutine. Consider a placement where $Q$ is completely below $P$, and imagine moving $Q$ upward until it is entirely above $P$. Let $Y = y_1, y_2, \ldots, y_{nm}$ be the sorted sequence of $y$-values where a vertex of $Q$ and a vertex of $P$ align horizontally. In other words, $Y$ contains the values $y_i$ such that there are vertices $v$ of $P$ and $w$ of $Q((x, y_i))$ with the same $y$-coordinate. We shall do a binary search on $Y$ to locate a horizontal strip $[-\infty : \infty] \times [y_i : y_{i+1}]$ that contains a goal placement. We do not compute the set $Y$ explicitly, however, because $Y$ can contain $nm$ elements and we do not want to spend that much time.

Let's look more closely at the set $Y$. Let $A = \{a_1, \ldots, a_n\}$ be the set of $y$-coordinates of the vertices of $P$, sorted in increasing order, and let $B = \{b_1, \ldots, b_n\}$ be the set of $y$-coordinates of the vertices of $Q((0,0))$, sorted in decreasing order. The sets $A$ and $B$ can be computed in linear time. The elements of the set $Y$ are exactly the entries of the matrix

$$\mathcal{M} = (c_{ij}), \text{ where } c_{ij} = a_i - b_j.$$

Because the sets $A$ and $B$ are sorted, every row and every column of $\mathcal{M}$ is sorted. Furthermore, an entry $c_{ij}$ can be evaluated in constant time. Hence, for any parameter $k$ with $1 \leqslant k \leqslant nm$, we can compute the $k$-th largest entry of

$\mathcal{M}$ in $O(m \log(2n/m)) \leqslant O(n+m)$ time with an algorithm by Frederickson and Johnson [12].

The binary search now proceeds as follows. In a generic step we have two values, $k_{\min}$ and $k_{\max}$ such that there is a goal placement in the horizontal strip $[-\infty : \infty] \times [y_{k_{\min}} : y_{k_{\max}}]$. Initially $k_{\min} = 0$ and $k_{\max} = nm$. We first compute the values $y_k$ and $y_{k+1}$, where $k = \lfloor (k_{\min}+k_{\max})/2 \rfloor$, with the algorithm of Frederickson and Johnson. Then we compute $\max_t \omega_{y_k}(t)$ and $\max_t \omega_{y_{k+1}}(t)$ using Lemma 5. There are three cases to consider, depending on the computed values:

If $\max_t \omega_{y_k}(t) < \max_t \omega_{y_{k+1}}(t)$ then we set $k_{\min} := k$.

If $\max_t \omega_{y_k}(t) > \max_t \omega_{y_{k+1}}(t)$ then we set $k_{\max} := k + 1$.

If $\max_t \omega_{y_k}(t) = \max_t \omega_{y_{k+1}}(t)$ then we set $k_{\min} := k$ and $k_{\max} := k + 1$ and we have found the strip.

The binary search continues until $k_{\max} - k_{\min} = 1$. The correctness of the algorithm is based on the following lemma.

**Lemma 6.** *Let $\ell_1$ and $\ell_2$ be two lines, and let $r_1$ and $r_2$ be points on $\ell_1$ and $\ell_2$, respectively, such that $\omega(r_1) = \max_{r \in \ell_1} \omega(r)$ and $\omega(r_2) = \max_{r \in \ell_2} \omega(r)$. If $\omega(r_1) \geqslant \omega(r_2) > 0$ and $r_1$ does not lie on $\ell_2$ then the open half-plane bounded by $\ell_2$ and containing $r_1$ contains a goal placement.*

The running time of this binary search algorithm is $O((n+m)\log(n+m))$, and it results in a horizontal strip that contains a goal placement. For any placement $Q(r)$ in the interior of this strip, the vertical order of the vertices of $P$ with respect to those of $Q(r)$ is fixed. This means that the complexity of the part of $\mathcal{A}(\Pi)$ within $R$ is significantly less than the total complexity of $\mathcal{A}(\Pi)$:

**Lemma 7.** *After the first stage of the algorithm we have located a horizontal strip $\sigma = [-\infty : \infty] \times [y : y']$ containing a goal placement such that the part of $\mathcal{A}(\Pi)$ inside $\sigma$ is formed by $O(n+m)$ segments.*

*The second stage.* We enter the second stage with a horizontal strip $\sigma = [-\infty : \infty] \times [y : y']$ that contains a goal position. Let $S(\sigma)$ be the set of segments defining $\mathcal{A}(\Pi)$ inside $\sigma$. The number of segments in $S(\sigma)$ is $O(n+m)$ by Lemma 7, and they can be computed in linear time.

We construct a $(1/4)$-cutting $\Xi(S(\sigma))$. This cutting consists of $O(1)$ triangles, each intersected by $|S(\sigma)|/4$ segments, and can be constructed in linear time [8]. The idea is to find a triangle in $\Xi(S(\sigma))$ that contains a goal placement, and to proceed recursively inside that triangle. (Actually, we will recurse in two triangles.) To decide in which triangle to recurse we proceed as follows.

Let $L = \{\ell_1, \ldots, \ell_a\}$ be the set of lines through the edges of the cutting $\Xi(S(\sigma))$. On each line $\ell_i$ we compute the maximum overlap $\xi_i = \max_{r \in \ell_i} \omega(r)$ in $O(n+m)$ time using the algorithm of Lemma 5. In this extended abstract we will assume that all the maxima are distinct. Let $i^*$ be such that $\xi_{i^*} = \max_i \xi_i$. By Lemma 6 we know for each line $\ell_i$ with $i \neq i^*$ to which side we can restrict our attention. This implies that we can restrict our attention to at most two triangles (separated by the line $\ell_{i*}$). The number of segments on which we must recurse is thus at most $|S(\sigma)|/2$. After $O(\log(n+m))$ recursive calls we are left with two triangular regions that are not intersected by any of the segments of $\mathcal{A}(\Pi)$. Inside each of these regions, the overlap function is a second-degree polynomial, and can be computed in linear time. Once we have the polynomial we can compute its maximum in constant time, giving us the desired goal placement. The total running time for the second stage is $O((n+m)\log(n+m))$.

This completes the proof of our main result, which is summarized in the following theorem.

**Theorem 8.** *Let $P$ be a convex polygon in the plane with $n$ vertices, and let $Q$ be a convex polygon with $m$ vertices. Then a placement of $Q$ that maximizes the area of $P \cap Q$ can be computed in $O((n+m)\log(n+m))$ time.*

## 4   Bounds on the Overlap for a Particular Translation

We prove in this section that we can approximate the area of the largest possible overlap by simply looking at the placement where the centroids of $P$ and $Q$ coincide. We prove that the overlap in that placement is at least 9/25 of the maximum possible overlap. We also give an upper bound example where the ratio is 4/9.

Let's first define some notations. The centroids of $P$ and $Q$ are denoted by $c_P$ and $c_Q$. In this section we will choose the origin 0 so that the overlap function is maximal at the origin, that is the reference position $Q(0)$ for $Q$ is a maximal overlap position. The maximal overlap area is thus denoted $\omega(0)$. In the sequel, we will use the polar coordinates $(r, \theta)$ with respect to that origin, and the horizontal direction; The point with polar coordinates $(1, \theta)$ will be denoted as $e_\theta$.

We denote by $\Omega$ the three-dimensional object bounded above by the graph of $\omega$ and below by the horizontal plane $z = 0$.

$$\Omega = \{(x, y, z) \in \mathbb{R}^3 : 0 \leqslant z \leqslant \omega(x, y)\}.$$

**Lemma 9.** *The translation $r$ that superimposes the centroids of $P$ and $Q(r)$ is given by the projection of the centroid of $\Omega$ onto $\mathbb{R}^2$.*

$p(c_\Omega)$ can now be evaluated in polar coordinates:

$$p(c_\Omega) == \int_{\theta=0}^{2\pi} e_\theta \rho(\theta) A(\theta) d\theta \Bigg/ \int_{\theta=0}^{2\pi} e_\theta A(\theta) d\theta , \tag{1}$$

where $A(\theta) = \int_{r=0}^{\infty} \omega(re_\theta) r\, dr$ is defined as the area of the intersection of $\Omega$ with a vertical half plane with polar coordinates $(r, \theta)$, $r \geqslant 0$. The polar coordinates of the horizontal projection of the centroid of that cross section are $(\rho(\theta), \theta)$ where $\rho(\theta) = \frac{1}{A(\theta)} \int_{r=0}^{\infty} \omega(re_\theta) r^2\, dr$

**Lemma 10.** *The value of $\omega$ at the projection centroid of a cross section of $\Omega$ is greater than 9/25 times the maximum of $\omega$. That is $\omega(\rho(\theta)e_\theta) \geqslant (9/25)\omega(0)$, for any $\theta$.*

*Proof.* Since $\omega(re_\theta)$ is strictly decreasing from its maximum to 0 (Theorem 4), there is a unique value $r_\theta$ such that $\omega(r_\theta e_\theta) = \frac{9}{25}\omega(0)$. Now we consider the function $\omega_\theta'(r) = \omega(0)\left(1 - \frac{2r}{5r_\theta}\right)^2$ for $r \in [0, \frac{5}{2}r_\theta]$.

Using the downwards concavity of the function $\sqrt{\omega}$ (Theorem 3), the relative position of $\omega_\theta'$ and $\omega$ are the following (see Figure 1):

$$\omega(0) = \omega_\theta'(0)$$
$$\omega(re_\theta) \geqslant \omega_\theta'(r) \, , \, r \in [0, r_\theta]$$
$$\omega(re_\theta) \leqslant \omega_\theta'(r) \, , \, r \in [r_\theta, \frac{5}{2}r_\theta]$$
$$\omega(re_\theta) = 0 \, , \, r \in [\frac{5}{2}r_\theta, \infty]$$

Since the weighted barycenter of function $\omega_\theta'$ is

$$\int_{r=0}^{\frac{5}{2}r_\theta} \omega_\theta'(r)r^2 dr \Big/ \int_{r=0}^{\frac{5}{2}r_\theta} \omega_\theta'(r)r\,dr = r_\theta$$

from inequalities above, we deduce that $\rho(\theta) \leqslant r_\theta$ and thus by Theorem 4: $\omega(\rho(\theta)e_\theta) \geqslant \omega(r_\theta) = \frac{9}{25}\omega(0)$. $\qquad\square$

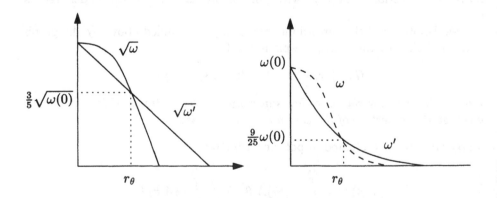

Fig. 1.: Relative position of $\omega$ and $\omega'$

**Lemma 11.** *The curve $(r_\theta e_\theta)_{0 \leqslant \theta < 2\pi}$ is convex.*

*Proof.* It follows directly from the downwards concavity of function $\sqrt{\omega}$ (Theorem 3). The curve $\theta \mapsto r_\theta e_\theta$ is the intersection of the 3D surface defined by $z = \sqrt{\omega}$ and the horizontal plane $z = \frac{3}{5}\sqrt{\omega(0)}$. □

**Theorem 12.** *The translation which matches the centroids of two convex polygons realizes an overlap area of at least 9/25 of the maximal overlap area.*

*Proof.* The overlap at the placement where the centroids of $P$ and $Q$ coincide is $\omega(c_Q - c_P)$, which is $\omega(p(c_\Omega))$ by Lemma 9. $p(c_\Omega)$ is the centroid of points $\rho(\theta)e_\theta$ weighted by the positive function $A(\theta)$ (Equation 1). The curve $\rho(\theta)e_\theta$ is inside the convex curve $r_\theta e_\theta$ (using Lemma 10), which is convex by Lemma 11. Thus $p(c_\Omega)$ is inside the convex curve $r_\theta e_\theta$ and thus, by the downwards concavity of $\sqrt{\omega}$ (Theorem 3), $\omega(p(c_\Omega)) \geqslant \omega(r_\theta e_\theta) = \frac{9}{25}\omega(0)$. □

We mention the generalization to higher dimensions.

**Theorem 13.** *The translation which matches the centroids of two d-dimensional convex polyhedra realizes an overlap volume of at least $(3/(d + 3))^d$ times the maximal overlap volume.*

Upper bounds for the factor between the overlap at the centroid position and the maximal overlap are shown in Figures 2 and 3. The factor is 4/9 in two dimensions and 9/32 in three dimensions.

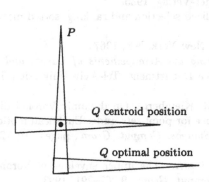

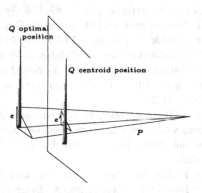

Fig. 2.: Example reaching $\frac{4}{9}$ upper bound.

Fig. 3.: Example reaching $\frac{9}{32}$ upper bound.

*Acknowledgment.* Thanks to H. Alt for helpful discussions about the centroid problem.

# References

1. Pankaj K. Agarwal, Micha Sharir, and Sivan Toledo. Applications of parametric searching in geometric optimization. In *Proc. 3rd ACM-SIAM Sympos. Discrete Algorithms*, pages 72–82, 1992.

2. H. Alt, B. Behrends, and J. Blömer. Approximate matching of polygonal shapes. In *Proc. 7th Annu. ACM Sympos. Comput. Geom.*, pages 186–193, 1991.

3. H. Alt and M. Godau. Measuring the resemblance of polygonal curves. In *Proc. 8th Annu. ACM Sympos. Comput. Geom.*, pages 102–109, 1992.

4. D. Avis, P. Bose, T. Shermer, J. Snoeyink, G. Toussaint, and B. Zhu. On the sectional area of convex polytopes. In *Proc. 12th Annu. ACM Sympos. Comput. Geom.*, pages C11–C12, 1996.

5. F. Avnaim and J.-D. Boissonnat. Polygon placement under translation and rotation. In *Proc. 5th Sympos. Theoret. Aspects Comput. Sci.*, volume 294 of *Lecture Notes in Computer Science*, pages 322–333. Springer-Verlag, 1988.

6. B. Chazelle. The polygon containment problem. In F. P. Preparata, editor, *Computational Geometry*, volume 1 of *Advances in Computing Research*, pages 1–33. JAI Press, London, England, 1983.

7. B. Chazelle. An optimal algorithm for intersecting three-dimensional convex polyhedra. *SIAM J. Comput.*, 21(4):671–696, 1992.

8. B. Chazelle. Cutting hyperplanes for divide-and-conquer. *Discrete Comput. Geom.*, 9(2):145–158, 1993.

9. L. P. Chew, M. T. Goodrich, D. P. Huttenlocher, K. Kedem, J. M. Kleinberg, and D. Kravets. Geometric pattern matching under Euclidean motion. In *Proc. 5th Canad. Conf. Comput. Geom.*, pages 151–156, Waterloo, Canada, 1993.

10. L. P. Chew and K. Kedem. Placing the largest similar copy of a convex polygon among polygonal obstacles. *Comput. Geom. Theory Appl.*, 3(2):59–89, 1993.

11. S. J. Fortune. A fast algorithm for polygon containment by translation. In *Proc. 12th Internat. Colloq. Automata Lang. Program.*, volume 194 of *Lecture Notes in Computer Science*, pages 189–198. Springer-Verlag, 1985.

12. G. Frederickson and D. Johnson. Generalized selection and ranking: sorted matrices. *SIAM J. Comput.*, 13:14–30, 1984.

13. B. Grünbaum. *Convex Polytopes*. Wiley, New York, NY, 1967.

14. D. Halperin. *Algorithmic Motion Planning via Arrangements of Curves and of Surfaces*. Ph.D. thesis, Computer Science Department, Tel-Aviv University, Tel Aviv, 1992.

15. D. P. Huttenlocher, K. Kedem, and J. M. Kleinberg. On dynamic Voronoi diagrams and the minimum Hausdorff distance for point sets under Euclidean motion in the plane. In *Proc. 8th Annu. ACM Sympos. Comput. Geom.*, pages 110–120, 1992.

16. D. P. Huttenlocher, K. Kedem, and M. Sharir. The upper envelope of Voronoi surfaces and its applications. *Discrete Comput. Geom.*, 9:267–291, 1993.

17. J.-C. Latombe. *Robot Motion Planning*. Kluwer Academic Publishers, Boston, 1991.

18. D.M. Mount, R. Silverman, and A. Wu. On the area of overlap of translated polygons. *SPIE Vision Geometry II*, 2060:254–264, 1993.

19. M. Sharir, R. Cole, K. Kedem, D. Leven, R. Pollack, and S. Sifrony. Geometric applications of Davenport-Schinzel sequences. In *Proc. 27th Annu. IEEE Sympos. Found. Comput. Sci.*, pages 77–86, 1986.

20. R. Venkatasubramanian. On the area of intersection of two closed 2D objects. *Information Sciences*, 82:25–44, 1995.

# OBDDs of a Monotone Function and of Its Prime Implicants

Kazuyoshi HAYASE[1]* and Hiroshi IMAI[2]

[1] NTT Multimedia Networks Laboratories, 1-2356 Take, Yokosuka 238-03, Japan
E-mail: `hayase@nttmhs.tnl.ntt.jp`
[2] Department of Information Science, University of Tokyo, Hongo, Tokyo 113, Japan
E-mail: `imai@is.s.u-tokyo.ac.jp`

**Abstract.** Coudert made a breakthrough in the two-level logic minimization problem with Ordered Binary Decision Diagrams (OBDDs, in short) recently [3]. This paper discusses relationship between the two OBDDs of a monotone function and of its prime implicant set to clarify the complexity of this practically efficient method. We show that there exists a monotone function which has an $O(n)$ size sum-of-products but cannot be represented by a polynomial size OBDD. In other words, we cannot obtain the OBDD of the prime implicant set of a monotone function in an output-size sensitive manner, once we have constructed the OBDD of that function as in [3], in the worst case. A positive result is also given for a meaningful class of matroid functions.

## 1 Introduction

Boolean function theory is very fundamental in VLSI design and other many areas such as artificial intelligence in computer science. From the viewpoint of combinatorics, Boolean function theory can be regarded as a model for general set systems. Many problems concerned with Boolean functions like two-level logic minimization are known to be intractable in complexity theory. Until Ordered Binary Decision Diagrams (OBDDs, in short) [2] had been proposed by Bryant, these hard problems of large size could not be solved reasonably in practice. With the power of OBDDs, Coudert has proposed an implicit method to do two-level logic minimization and has demonstrated that it is much more efficient (stated as "100 times faster" in the title) than other conventional logic minimizers based on the Quine-McCluskey method [3]. The Quine-McCluskey method has a doubly-exponential complexity because we compute the set of all the prime implicants which could become exponential size in the number of variables and then solve the set cover problem, which is known to be *NP*-complete, on the prime implicant set. Coudert's method also computes the OBDD of the prime implicant set from the OBDD of the original function. One of the reasons for its success is that the prime implicant set can be represented with a dramatically small OBDD.

---

* Part of this research was performed while the first author was at Department of Information Science, University of Tokyo.

Coudert poses several theoretical questions in this practically useful approach. Two important questions among them are

(1) What is the relation between the size of a sum-of-products and the size of the BDD of the function it represents[1]?

(2) What is the relation between the size of a BDD and the size of the CS (Combinational Set[2]) of the set of its prime implicants?

In this paper, we investigate the combinatorial complexity of OBDDs of a monotone Boolean function and of its prime implicant set. We give answers to Coudert's questions in the following way.

(i) There exists a monotone Boolean function such that it has a polynomial size sum-of-products and the OBDD of the function is exponential in size for any variable ordering.

(ii.a) There exists a monotone Boolean function such that the OBDD of the function is exponential in size and the OBDD of its prime implicant set is polynomial, both for any variable ordering.

(ii.b) There exits a meaningful class of monotone Boolean functions (related to matroids) such that the OBDD of the function and the OBDD of its prime implicant set are of almost same size.

(i) and (ii.a) are shown by investigating the characteristic function of the family of stable sets in a mesh graph $\chi(SS(M_k))$, which has been analyzed in [5]. We have shown that the OBDD of the maximal stable sets, which correspond to prime implicants of the function, can be constructed in an output-size sensitive manner if the variables are ordered in row-major manner. This is because the two OBDDs have exponential size in that variable ordering. On the other hand, the characteristic function of the family of stable sets of a graph $\chi(SS(G))$ has a polynomial size product-of-sums. It follows from these two facts that the two OBDDs of $\neg\chi(SS(M_k))$ and of its prime implicant set are different exponentially in size if we choose the row-major variable ordering. This observation is, however, not satisfactory to show a very bad case for Coudert's method because the OBDDs of a function may different exponentially in size if we change the variable ordering. The major result of the paper is to prove that the characteristic function of the stable sets in a mesh graph cannot be represented by a polynomial size OBDD. One should note that other conventional methods can minimize the polynomial size sum-of-products in polynomial time because this is composed of only positive literals.

This indicates that there still requires a new method of computing the prime implicants more directly. For some problems related to reliability computation,

[1] It is well-known that the parity function can be represented by a polynomial size OBDD but has an exponential lower bound on the size of sum-of-products. Hence, this question should be regarded as whether there is a function with a polynomial size sum-of-products and exponential size OBDDs (for any variable ordering).

[2] CS is a kind of OBDD which had been originally proposed in [9] and called 0-Sup-BDD (0-suppressed BDD).

such a method is proposed by utilizing the underlying structure [11]. Although it is left open whether there exists a Boolean function such that its OBDD is small and the OBDD of its prime implicant set is large, we thus answered main parts of Coudert's open questions.

(ii.b) shows that there is a good case for Coudert's approach, although still it is left open for what type of Boolean functions the OBDDs of them and their prime implicant sets can be constructed in an output-size sensitive manner. The top-down breadth-first construction approach succeeded in constructing many Boolean functions related graphs, codes, matroids in an output-size sensitive manner [4, 5, 7, 6, 10, 12, 13], but there still need more investigations in this direction.

## 2 Preliminaries

*Two-level logic minimization* of a Boolean function is an optimization problem to minimize the cost of representing a Boolean function with a sum-of-products, which have quite many applications in computer science. A Boolean function $f : \{0,1\}^n \rightarrow \{0,1\}$ is assumed to have its variable set as $X = \{x_1, x_2, \ldots, x_n\}$. We denote the value of $f$ at a *truth assignment* $a \in \{0,1\}^n$ by $[f, a]$. A *product* $p$ is a conjunction of some literals which are made from different variables each other. A *sum-of-products* is a disjunction of several products and its cost is measured by the total number of literals. A product $p$ is called an *implicant* of $f$ if $[p, a] \leq [f, a]$ at any $a \in \{0,1\}^n$. An implicant $p$ is called *prime* if there is no other implicant $q$ such that $\forall a \in \{0,1\}^n$, $[p, a] \leq [q, a]$. $PI(f)$ denotes the prime implicant set of a Boolean function $f$. The smallest sum-of-products must be composed of prime implicants by definition.

Monotone Boolean functions have desirable properties on prime implicants and two-level logic minimization. A Boolean function $f$ is *positive* if $[f, a] \leq [f, b]$ for any $a \leq b$. $f$ is *negative* if $\neg f$ (negation of $f$) is positive. A Boolean function which is positive or negative is called *monotone*. Although a Boolean function has multiple smallest sum-of-products in general, that of a monotone Boolean function is uniquely determined because it consists of all prime implicants.

We introduce how we can treat a family of subsets of a finite universe by means of a Boolean function. We identify the universe with the variable set $X$ $\{x_1, x_2, \ldots, x_n\}$. The *characteristic vector* $\xi(S)$ of a subset $S$ is an $n$-dimensional Boolean vector in $\{0,1\}^n$ such that $\xi_i(S) = 1$ if and only if $x_i$ is in $S$. The *characteristic function* $\chi(\mathcal{F})$ of the family $\mathcal{F}$ of subsets is a Boolean function such that $[\chi(\mathcal{F}), \xi(S)] = 1$ if and only if $S$ belongs to $\mathcal{F}$.

The most important technique in Coudert's method is that OBDDs are employed to represent not only the original Boolean function but also its prime implicant set where a product is regarded as a subset of literals. Although we must consider $2n$ positive and negative literals to treat sets of products in general, we can manage with $n$ literals in the case of monotone Boolean functions because a prime implicant of a positive function $f$ is a set of positive literals.

Thus in the rest of the paper, $\chi(PI(f))$ is a Boolean function of $n$ variables $\{x_1, x_2, \ldots, x_n\}$ for a monotone Boolean function $f$.

We now define the OBDD and introduce a canonical form QOBDD of the OBDD (see Fig. 1). An OBDD $D$ is a labelled DAG (Directed Acyclic Graph) with a root [2]. Each non-terminal node $v$ is labelled by a variable $label(v) \in X$ and called a *variable node*. Each terminal node $v$ is labelled by a constant $label(v) \in \{0,1\}$ and called a *constant node*. The size of an OBDD $D$ is measured by the number of its nodes. Each directed path follows a total order $\pi$ on the variable set $X$ which is called the *variable ordering*. We assume the variable ordering $\pi$ to be $x_1 < x_2 < \ldots < x_n$ unless otherwise specified. From each variable node $v$, there are two outgoing edges labelled by 0 and 1, which we call 0-edge and 1-edge, respectively. An OBDD represents a Boolean function $f$ in the way that we get to the constant node of value $[f, a]$ from the root by choosing 0-edge at any variable node if and only if the input value of the variable is 0.

We need to consider *subfunctions* of a Boolean function for investigating structure of the OBDD. $[f, a]$ denotes the subfunction of a Boolean function $f$ obtained by applying a *partial truth assignment* $a \in \{0,1\}^l$ $(0 \le l \le n)$ to $f$. A partial truth assignment $a$ of length $l$ assigns the value $a_i$ to the variable $x_i$ for each $1 \le i \le l$. A *supplementary truth assignment* $b$ of $a$ has length of $n - l$ and assigns $b_j$ to $x_{j+l}$ for each $1 \le j \le n - l$. The following equation also describes this definition, where $a \cdot b$ denotes the *concatenation* of $a$ and $b$.

$$\forall a \in \{0,1\}^l, \ \forall b \in \{0,1\}^{n-l} \quad [[f, a], b] = 1 \quad \text{if and only if} \quad [f, a \cdot b] = 1$$

To obtain the canonical form QOBDD of OBDDs, we share same structures (subgraphs) maximally with the restriction that any directed path from the root to a constant node contains exactly $n + 1$ nodes. The following proposition comes from uniqueness of the QOBDD. The width $W_l(D)$ of the QOBDD $D$ of a function $f$ in level $l$ is the number of the variable nodes which are labelled by the $l$-th variable $x_l$ of the variable ordering.

**Proposition 1.** *The width $W_l(D)$ is the number of the distinct subfunctions of $f$ which are obtained by applying all partial truth assignments $a \in \{0,1\}^{l-1}$ to $x_1, x_2, \ldots, x_{l-1}$.*

## 3 Stable and Maximal Stable Sets of a Mesh Graph

In this section, we give a negative answer to Coudert's questions by analyzing the OBDD of the characteristic function of the family of non-stable sets in a mesh graph $M_k$ which is illustrated in Fig. 2. We begin with definition of stable sets in a graph. Let $G$ be a simple graph with vertex set $X$ and edge set $E$. A subset $S$ of $X$ is called *stable* if any pair of distinct vertices in $S$ are not adjacent to each other. Let $SS(G)$ denote the family of stable sets in $G$. We consider the negation of $\chi(SS(G))$, which is the characteristic function of the family of non-stable sets in a graph. This is positive and its smallest sum-of-products is

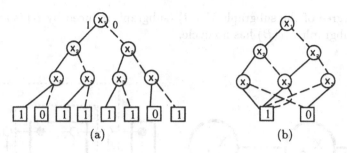

**Fig. 1.** (a) An OBDD of a function $f$ such that $[f, x] = 0$ if $x = (0, 0, 1)$, $(1, 1, 0)$ and 1 otherwise. (b) The QOBDD of $f$.

$\bigvee_{(x_i, x_j) \in E}(x_i \wedge x_j)$ because a subset is non-stable if and only if it has an adjacent pair of vertices. Our $k \times k$ mesh graph has $n = k^2$ vertices and $2k(k-1)$ edges, hence the prime implicant set is of only linear size in this case. Proposition 1. is also utilized to show the following.

**Lemma 2.** *The characteristic function of non-stable sets in $M_k$ has the smallest sum-of-products of $O(n)$ size, and the size of the QOBDD of the characteristic function of its prime implicant set is $O(n^2)$ for any variable ordering.*

The rest of the section is given to show an exponential lower bound on the size of the QOBDD of $\neg\chi(SS(M_k))$ for arbitrary variable ordering $\pi$. Before discussing the mesh graph, we count the number of independent sets in a *path graph*. A path graph $P_n$ is a simple path of $n$ vertices. The number of stable sets in $P_n$ can be expressed by a Fibonacci number sequence $F_n$ defined recursively by $F_0 = 1$, $F_1 = 2$, and $F_n = F_{n-1} + F_{n-2}$ for $n \geq 2$. This Fibonacci number sequence has an exponential lower bound for all $n \geq 1$.

**Lemma 3.** *The number $F_n$ of stable sets in $P_n$ grows exponentially like:*

$$2^{\frac{n}{2}} \leq F_n \leq (3\sqrt{5}/5)((1 + \sqrt{5})/2)^n \quad (for\ n \geq 1).$$

To show the bound for any variable ordering, we take any vertex subset $A$ of size $n/2$. Such a subset has many adjacent vertices outside of it as shown in [8]. In the following, $\Gamma(A)$ denotes the set of vertices adjacent to a vertex in $A$.

**Lemma 4.** *Let $A$ be a subset of $X$ of $M_k$ such that $|A| = n/2 = k^2/2$. Let $C$ be the subset $C := \Gamma(A) \setminus A$. Then $k/2 \leq |C|$.*

We can find a sufficiently large subset $B$ of $A$ called *path collection* which has the following properties. See also Fig. 3.

**Properties**

(i) Any vertex $u$ in $B$ has an adjacent vertex in $C$ ($\Gamma(u) \cap C \neq \emptyset$).
(ii) Any pair of two different vertices $u$ and $v$ in $B$ has no common adjacent vertices in $C$ ($\Gamma(u) \cap \Gamma(v) \cap C = \emptyset$).

(iii) The degree of the subgraph $M_k(B)$ (subgraph induced by $B$) is at most 2.
(iv) The subgraph $M_k(B)$ has no cycle.

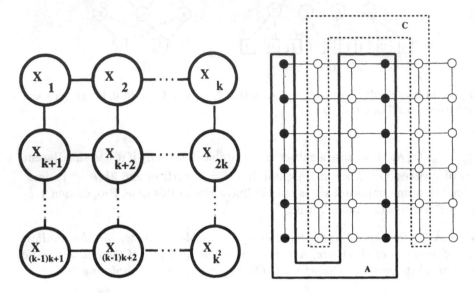

**Fig. 2.** The Mesh Graph $M_k$       **Fig. 3.** In this case, the path collection $B$ consists of black vertices.

**Lemma 5.** *There is a path collection $B$ which contains at least $3k/640$ vertices.*

*Proof.* Let F be the subset of vertices which are in $A$ and have adjacent vertices in $C$. We can find a path collection $B$ of size at least $3k/640$ in F by the following procedure.

(1) We can pick a vertex in F one after one by consuming at most 40 vertices in $C$ without violating the property (ii).
(2) The degree of the subgraph induced by the picked vertices is at most three. We can satisfy the property (iii) by removing at most half of picked vertices.
(3) We obtain a path collection $B$ by removing a vertex from each cycle. □

The connected components $B_1, B_2, \ldots, B_p$ of the subgraph $M_k(B)$ induce path graphs. Our lower bound comes from the number of the combinations of stable sets in $M_k(B_i)$'s.

**Lemma 6.** *The width of the QOBDD of $\neg\chi(SS(M_k))$ in level $\frac{n}{2}+1$ is at least the number of stable sets in a path collection $M_k(B)$, which is equal to the arithmetic product of the numbers of stable sets in all $M_k(B_i)$'s.*

*Proof.* We count the number of the subfunctions $[\neg\chi(SS(M_k)), a]$ where $a$ is a partial truth assignment of length $n/2$ (to the variables in $A$) such that:

– if $x_i \in A - B$, then we assign $x_i := 0$ and,
– the subset of the vertices assigned to be 1 is stable in $M_k$.

The number of such partial truth assignments is obviously equal to the arithmetic product of the numbers of stable sets in $M_k(B_i)$ because two vertices $u$ and $v$ of different components $B_i$ and $B_j$ are not adjacent each other. It is sufficient to show that there exists a supplementary truth assignment $b \in \{0, 1\}^{n-n/2}$ such that $[\neg\chi(SS(M_k)), a_1 \cdot b] \neq [\neg\chi(SS(M_k)), a_2 \cdot b]$ for any two different $a_1$ and $a_2$. We assume w.l.o.g. that there exists a vertex $x_i$ such that $x_i$ is assigned to be 1 in $a_1$ but to be 0 in $a_2$. We select one vertex $x_j$ in $\Gamma(x_i) \cap C$, whose existence is guaranteed by the property (i) of the path collection. Our supplementary truth assignment $b$ assigns $x_j$ to be 1 and the others to be 0. □

To sum up, we have the following answer to Coudert's questions.

**Theorem 7.** *There exists a monotone function which has an $O(n)$ size sum-of-products, and an exponential lower bound in OBDD size. Furthermore, its prime implicant set can be expressed by a polynomial size OBDD for arbitrary variable ordering.*

## 4 Matroid Function

In the previous section, it is shown that there exists a monotone Boolean function whose sum-of-products is polynomial in size (and hence the QOBDD of prime implicants of the function is of polynomial size for any variable ordering) while the QOBDD of the original function is of exponential size for any variable ordering with respect to the number of variables.

This section provides another extremal case such that these two QOBDDs necessarily have almost same structure and size. We prove this property for so-called matroid functions [1] which form a quite important and useful class of Boolean functions in combinatorics (see for example [14]).

A non-empty family $\mathcal{B}$ of subsets of $X$ is the set of bases of a matroid if and only if it satisfies the following condition:

– If $B_1, B_2 \in \mathcal{B}$ and $x_i \in B_1 - B_2$, $\exists x_j \in B_2 - B_1, (B_1 - \{x_i\}) \cup \{x_j\} \in \mathcal{B}$.

A set in $\mathcal{B}$ is called a *base*. A subset of some base is called *independent*. Denote by $\mathcal{I}$ the set of all independent sets of the matroid. A superset of some base is called *spanning*. Denote by $\mathcal{S}$ the set of all spanning sets of the matroid. A subset of $X$ which is not independent is called *dependent*.

A Boolean function is called a matroid function if it is the characteristic function $\chi(\mathcal{S})$ of the family $\mathcal{S}$ of all spanning sets of some matroid [1]. Since a superset of a spanning set is spanning, this function is positive. In this paper, we further consider two additional functions $\chi(\mathcal{I})$ and $\chi(\mathcal{B})$ related to the same matroid. $\chi(\mathcal{I})$ is the characteristic function of the family $\mathcal{I}$ of all independent sets of the matroid and $\chi(\mathcal{B})$ is that of the family $\mathcal{B}$ of all bases. Among these functions, the following proposition holds.

**Proposition 8.** $\chi(\mathcal{B}) = \chi(\mathcal{I}) \wedge \chi(\mathcal{S})$. $\mathcal{B}$ *corresponds to* $PI(\chi(\mathcal{I}))$ *and* $PI(\chi(\mathcal{S}))$.

An element $x$ is a *loop* of matroid if no base contains $x$ (hence, $x$ itself is dependent). An element $x$ is a *coloop* of matroid if any base contains $x$.

For an element $x \in X$ and a matroid $(X, \mathcal{B})$, define $\mathcal{B} \backslash e$ by $\mathcal{B} \backslash x = \{B \mid x \notin B \in \mathcal{B}\}$ if $x$ is not a coloop; otherwise, $\mathcal{B} \backslash x$ is defined by $\mathcal{B} \backslash x = \{B - \{x\} \mid B \in \mathcal{B}\}$. $(E - \{x\}, \mathcal{B} \backslash x)$ is a matroid, and is called a matroid obtained by deleting $x$.

For an element $x \in X$ and a matroid $(X, \mathcal{B})$, define $\mathcal{B}/x$ by $\mathcal{B}/x = \{B - \{x\} \mid x \in B \in \mathcal{B}\}$ if $x$ is not a loop; otherwise, $\mathcal{B}/x$ is defined simply to be $\mathcal{B}$. $(X - \{x\}, \mathcal{B}/x)$ is a matroid, and is called a matroid obtained by contracting $x$.

A matroid obtained by deleting elements in $S_1$ and contracting elements in $S_2$ for disjoint subsets $S_1$ and $S_2$ of $X$ is again a matroid, called a *minor* of the original matroid. The minor is invariant with respect to ordering of deletions and contractions.

With these definitions, let us consider the QOBDDs of $\chi(\mathcal{B})$, $\chi(\mathcal{I})$ and $\chi(\mathcal{S})$. Each node in these QOBDDs corresponds to a minor of the matroid as follows.

**Lemma 9.** *Consider a subfunction* $[\chi(\mathcal{B}), a]$ *(resp.,* $\chi(\mathcal{I})$, $\chi(\mathcal{S})$*) which is not equal to the constant function* 0. *This subfunction is equal to the characteristic function of bases (resp., independent sets, spanning sets) of a minor obtained by contracting elements* $x_i$ *with* $a_i = 1$ *and deleting elements* $x_i$ *with* $a_i = 0$ *from the original matroid.*

*Proof.* Using a deletion/contraction as one step, the lemma can be proved by induction. Note that the minor obtained by deleting $x$ and that by contracting $x$ are isomorphic if $x$ is coloop or loop in the matroid. □

The following theorem describes a transformation rule from the QOBDD of $\chi(\mathcal{B})$ into those of $\chi(\mathcal{S})$ and $\chi(\mathcal{I})$.

**Theorem 10.** *We can transform the QOBDD of* $\chi(\mathcal{B})$ *into that of* $\chi(\mathcal{S})$ *(resp.,* $\chi(\mathcal{I})$*) by processing each variable node as follows, and then removing redundant intermediate nodes having no entering edge recursively.*

*For a node labelled by* $x_i$, *if* $x_i$ *is a loop (resp., coloop) in the minor of the node, update its* 1-*edge (resp.,* 0-*edge) to be parallel with its* 0-*edge (resp.,* 1-*edge).*

We here only show an example for $K_3$ in Fig. 4. The structures of these QOBDDs differ only in nodes whose subfunctions are 0, and we obtain the following.

**Corollary 11.** *The sizes of QOBDDs of* $\chi(\mathcal{B})$, $\chi(\mathcal{S})$ *and* $\chi(\mathcal{I})$ *differ in size by at most the number of variables.*

For the uniform matroid $U_{n,k}$ of rank $k$ of $n$ elements, whose bases are all $k$-element subsets, the QOBDDs of $\chi(\mathcal{B})$, $\chi(\mathcal{S})$ and $\chi(\mathcal{I})$ are all $\Theta(kn)$ in size where $k$ is constant.

In the graphic case, that is, spanning trees, spanning sets and forests of a graph, it is known that the QOBDDs of spanning trees are of exponential size

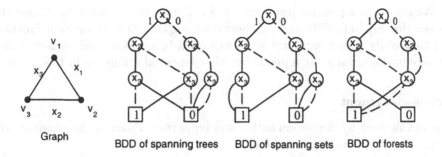

**Fig. 4.** QOBDD of spanning trees $\chi(\mathcal{B})$, spanning sets $\chi(\mathcal{S})$, and forests $\chi(\mathcal{I})$ of $K_3$

for a complete graph and a mesh graph for the *canonical edge orderings* [12]. In canonical edge orderings, variables (edges) are ordered in the increasing lexicographic order where edges are regarded as pairs of vertices. Furthermore, in this case, we can analyze relationship of the sizes of the QOBDD of $\neg\chi(\mathcal{I})$, the characteristic function of dependent sets, and that of its prime implicants.

**Theorem 12.** *For a complete graph and a mesh graph, the sizes of the QOBDDs of $\chi(PI(\neg\chi(\mathcal{I})))$, representing all simple circuits, are exponential in size for the canonical edge orderings.*

*Proof.* Due to space limitation, we briefly state an outline for the case of complete graphs. As in the proof of the previous section, we estimate the width of the QOBDD in a middle level. Following the counting technique in [12], we consider the elimination front and elimination partition for the canonical edge ordering. Our elimination front has half of the vertices and our elimination partition consists of just two subsets. The number of such partitions has an exponential lower bound.                                                                                    □

Note that the size of QOBDD of $\neg\chi(\mathcal{I})$ is equal to that of the QOBDD of $\chi(\mathcal{I})$ (just replacing 0-node and 1-node), and hence its size is exponential as mentioned above. Thus, unlike the stable set case, the QOBDDs concerning matroids basically have almost same sizes.

## 5   Conclusion

We have investigated the combinatorial complexity of QOBDDs of a monotone Boolean function and of its prime implicants. As a result, we found a monotone Boolean function $\neg\chi(SS(M_k))$ which has an $O(n)$ size sum-of-products but cannot be represented by a polynomial size QOBDD with any variable ordering. This indicates that computational complexity of the practically useful approach by Coudert has an exponential lower bound in the sizes of input and output. In other words, it is not always a good route to compute the QOBDD of $f$ in advance for the purpose of constructing the QOBDD of $\chi(PI(f))$.

We also gave a positive result for a class of matroid functions. In Hayase [4], the relationship in OBDD structure between the QOBDD of a monotone function and that of its prime implicant set is expressed in terms of their subfunctions. These expressions suggest a possibility of exponential difference in their sizes.

## Acknowledgment

Part of this work by the second author was supported in part by the Grant-in-Aid of the Ministry of Education, Japan.

## References

1. M. O. Ball and J. S. Provan. Disjoint products and efficient computation of reliability. *Operations Research*, 36:703–715, 1988.
2. R. Bryant. Graph-based algorithms for Boolean function manipulation. *IEEE Transactions on Computers*, C-35(8):677–691, 1986.
3. O. Coudert. Doing two-level logic minimization 100 times faster. In *Proc. ACM-SIAM Symposium on Discrete Algorithms*, pages 112–121, 1995.
4. K. Hayase. On the complexity of constructing OBDDs of a monotone function and of the set of its prime implicants. Master's thesis, University of Tokyo, 1996. (available at http://naomi.is.s.u-tokyo.ac.jp/theses.html).
5. K. Hayase, K. Sadakane, and S. Tani. Output-size sensitiveness of OBDD construction through maximal independent set problem. In *COCOON'95, Lecture Notes in Computer Science*, volume 959, pages 229–234, 1995.
6. H. Imai, S. Iwata, K. Sekine, and K. Yoshida. Combinatorial and geometric approaches to counting problems on linear matroids, graphic arrangements and partial orders. In *COCOON'96, Lecture Notes in Computer Science*, volume 1090, pages 68–80, 1996.
7. H. Imai, K. Sekine, and K. Yoshida. Binary decision diagrams and generating functions of sets related to graphs and codes. In *9th Karuizawa Workshop on Circuits and Systems*, pages 91–96, 1996.
8. R. J. Lipton, D. J. Rose, and R. E. Tarjan. Generalized nested dissection. *SIAM J. Numer. Anal.*, 16(2):346–358, 1979.
9. S. Minato. Zero-suppressed BDDs for set manipulation in combinatorial problems. In *Proc. 30th ACM/IEEE DAC*, pages 272–277, 1993.
10. J. Niwa, K. Sadakane, K. Hayase, and H. Imai. Parallel top-down construction of OBDDs of monotone functions. In *JSPP'96, Joint Symposium on Parallel Processing*, pages 161–168, 1996. (in Japanese).
11. K. Sekine and H. Imai. A unified approach via BDD to the network reliability and path numbers. Technical Report 95-09, Department of Information Science, University of Tokyo, 1995.
12. K. Sekine, H. Imai, and S. Tani. Computing the Tutte polynomial of a graph of moderate size. In *ISAAC'95, Lecture Notes in Computer Science*, volume 1004, pages 224–233, 1995.
13. S. Tani and H. Imai. A reordering operation for an ordered binary decision diagram and an extended framework for combinatorics of graphs. In *ISAAC'94, Lecture Notes in Computer Science*, volume 834, pages 575–583, 1994.
14. D. J. A. Welsh. *Matroid Theory*. Academic Press, 1976.

# Algorithms for Maximum Matching and Minimum Fill-in on Chordal Bipartite Graphs

Maw-Shang Chang *

Department of Computer Science and Information Engineering
National Chung Cheng University
Ming-Hsiun, Chiayi 621, Taiwan
Republic of China

**Abstract.** We define an ordering of vertices of a chordal bipartite graph. By using this ordering, we give a linear time algorithm for the maximum matching problem and an $O(n^4)$ time algorithm for the minimum fill-in problem on chordal bipartite graphs improving previous results.

## 1   Introduction

Standard graph theory terminology is used in the most part of the paper. Some of the notation and definitions will be explicitly stated for clarity. We will often denote $|V|$ by $n$ and $|E|$ by $m$ for a graph $G = (V, E)$. An edge is a *chord* of a cycle if it connects two vertices of the cycle but is not itself an edge within the cycle. A graph is *chordal* if every cycle of length greater than three has a chord. For an arbitrary graph $G = (V, E)$, a *minimum fill-in* for $G$ refers to a minimum cardinality set $E'$ from $\overline{E} = \{(u, v) \notin E : u, v \in V\}$ such that $G' = (V, E \cup E')$ is chordal. In other words, the problem of finding a minimum fill-in for $G$, referred to as the *minimum fill-in problem*, is equivalent to finding a chordal embedding of $G$ with minimum number of edges. The minimum fill-in problem is sometimes also called the *chordal graph completion problem* and has applications to the solution of sparse systems of linear equations by Gaussian elimination [11]. The problem of computing the minimum fill-in is NP-complete for general graphs [15]. In fact, Yannakakis [15] proved that the problem is NP-complete for cobipartite graphs which are subclasses of cocomparability graphs;

---

* Supported partly by the National Science Council of the Republic of China under grant NSC 83-0208-M-194-017. Email: mschang@cs.ccu.edu.tw

and Tarjan proved that the problem is NP-complete for bipartite graphs [14]. Corneil et al. gave a polynomial time algorithm for cographs [3]. A graph is *bipartite* if its vertices can be partitioned into two independent sets. The length of any cycle in a bipartite graph is greater than three and is even. Clearly, any chord of a cycle in a bipartite graph is odd. A graph is *chordal bipartite* if it is bipartite and every cycle of length greater than four has a chord. Note that chordal bipartite graphs are not necessarily chordal. A bipartite permutation graph is a graph which is both a bipartite graph and a permutation graph [13]. Chordal bipartite graphs properly contains bipartite permutation graphs. A chordal bipartite graph can be recognized in $O(n^2)$ time [12]. In [13], Spinrad et al. gave an $O(n^5)$ time algorithm for finding a minimum fill-in of a bipartite permutation graph by constructing a bipartite graph with $O(n^2)$ vertices and finding a maximum matching of this bipartite graph. Kloks [7] gave an $O(n^5)$ time algorithm for finding a minimum fill-in for the class of chordal bipartite graphs. For more algorithms, readers are referred to [1, 8, 9]. In this paper, we show that the algorithm proposed by Spinrad et al. for the class of bipartite permutation graphs works correctly on chordal bipartite graphs. Besides, the bipartite graph constructed by the algorithm is a chordal bipartite graph. By giving a linear-time algorithm for computing a maximum matching of a chordal bipartite graph, we show that the algorithm proposed by Spinrad et al. can be implemented in $O(n^4)$ time.

## 2 Preliminaries

Given a graph $G = (V, E)$ and a subset $X$ of $V$, let $G[X]$ denote the subgraph of $G$ induced by $X$. That is, $G[X] = (X, E_X)$, where

$E_X = \{(u, v) \in E : u \in X, \text{ and } v \in X\}$.

Given a graph $G = (V, E)$, we use the notation $N_G(v)$ ($N(v)$ if $G$ is understood) for the set of neighbors of a vertex $v$ in $G$. And we use the notation $N_G[v]$ ($N[v]$ if $G$ is understood) to denote $N_G(v) \cup \{v\}$. A vertex $v$ is *simple* if for all $x, y \in N[v]$,

$N[x] \subseteq N[y]$ or $N[y] \subseteq N[x]$. An ordering of vertices $V = \{v_1 < v_2 < \cdots < v_n\}$ is a *strong elimination ordering* if, for each $i \in \{1, 2, \ldots, n\}$, $v_i$ is a simple vertex of $G_i$, where $G_i = G[\{v_i, v_{i+1}, \ldots, v_n\}]$. Farber showed [4] that if $V = \{v_1 < v_2 < \cdots < v_n\}$ is a strong elimination ordering, then for $1 \leq i \leq j < k \leq n$ where $v_j, v_k \in N[v_i]$, we have that $N_{G_i}[v_j] \subseteq N_{G_i}[v_k]$. Let $C = (v_{c_1}, v_{c_2}, \ldots v_{c_k}, v_{c_1})$ be an even length cycle, i.e. $k$ is even. A chord $(v_{c_h}, v_{c_j})$ is an *odd chord* of $C$ if $|h - j|$ is odd. A graph is *strongly chordal* if it is chordal and every even cycle of length greater than four has an odd chord. Farber [4] showed that a graph is strongly chordal iff it admits a strong elimination ordering.

A graph is *split* if its vertices can be partitioned into an independent set and a clique. For a bipartite graph $G = (S, T, E)$, let $Split(G)$ be the split graph obtained from $G$ by making that $S$ induces a complete subgraph in $Split(G)$, i.e., $Split(G) = (S \cup T, E_s)$ where $E_s = E \cup \{(s_i, s_j) : s_i, s_j \in S, 1 \leq i < j \leq |S|\}$. The following lemma can be derived from a theorem in [2].

**Lemma 1.** *If $G = (S, T, E)$ is a chordal bipartite graph, then $Split(G)$ is a strongly chordal graph.*

For a bipartite graph $G = (S, T, E)$, an ordering of vertices, $S = \{s_1 < s_2 < \cdots < s_p\}$ and $T = \{t_1 < t_2 < \cdots < t_q\}$, is a *strong $T$-elimination ordering* if, for each $1 \leq i \leq q$ and $1 \leq j < k \leq p$ where $s_j, s_k \in N(t_i)$, we have that $N_{G_{t_i}}(s_j) \subseteq N_{G_{t_i}}(s_k)$ where $G_{t_i}$ is the subgraph of $G$ induced by $S \cup \{t_h \in T : h \geq i\}$. The following theorem is crucial to our algorithms.

**Theorem 2.** *A bipartite graphs is a chordal bipartite graph iff it admits a strong $T$-elimination ordering.*

*Proof.* Let $G = (S, T, E)$ be a bipartite graph that admits a strong T-elimination ordering where $S = \{s_1 < s_2 < \cdots < s_p\}$ and $T = \{t_1 < t_2 < \cdots < t_q\}$. Let $C = (s_{c_1}, t_{d_1}, s_{c_2}, t_{d_2}, \ldots s_{c_k}, t_{d_k}, s_{c_1})$ be a cycle in $G$ of length $2k \geq 6$. Without loss of generality, assume $d_1 < d_i$ for all $1 < i \leq k$ and $c_1 < c_2$. For simplicity, let $t = t_{d_1}$. By definition, $N_{G_t}(s_{c_1}) \subseteq N_{G_t}(s_{c_2})$. Thus $C$ has a chord $(s_{c_2}, t_{d_k})$ since $t_{d_k} \in N_{G_t}(s_{c_1})$. Hence, $G$ is a chordal bipartite graph.

Let $G = (S, T, E)$ be a chordal bipartite graph. By Lemma 1, $Split(G)$ is a strongly chordal graph. Hence $Split(G)$ admits a strong elimination ordering $(z_1, z_2, \ldots, z_n)$ where $n = |S| + |T|$. Let $j$ be the smallest index such that $z_j$ is a vertex of $S$ and $i$ be the largest index such that $z_i$ is a vertex of $T$. If $i < j$, then $j = i+1$, $T = \{z_1, z_2, \ldots, z_i\}$ and $S = \{z_j, z_{j+1}, \ldots, z_n\}$. It is straightforward to verify that this ordering is also a strong T-elimination ordering of $G$. Suppose $i > j$. Let $h$ be the smallest index such that $h > j$ and $z_h$ is a vertex of $T$. Then $z_{h-1}$ is a vertex of $S$. We can easily verify that $z_h$ and $z_{h-1}$ are simple vertices of $Split(G[\{z_h, z_{h-1}, z_{h+1}, z_{h+2}, \ldots, z_n\}])$ and $Split(G[\{z_{h-1}, z_{h+1}, z_{h+2}, \ldots, z_n\}])$, respectively. Thus $(z_1, z_2, \ldots, z_{h-2}, z_h, z_{h-1}, z_{h+1}, z_{h+2}, \ldots, z_n)$ is also a strong elimination ordering of $Split(G)$. By repeatedly applying the above operation, we can obtain a strong elimination ordering of $Split(G)$ such that $T = \{z_1, z_2, \ldots, z_i\}$ and $S = \{z_{i+1}, z_{i+2}, \ldots, z_n\}$ where $i = |T|$, which is also a strong T-elimination ordering of $G$. □

## 3 Maximum Matching

A matching $M$ of a graph $G = (V, E)$ is a subset of $E$ such that no two edges in $M$ are incident to a common vertex. The maximum matching problem involves finding a matching of maximum cardinality. In this section, we give a linear time algorithm for the maximum matching problem on chordal bipartite graphs given a strong T-elimination ordering. Given a strong T-elimination ordering of a chordal bipartite graph $G = (S, T, E)$, the algorithm scans the vertices of $T$ one by one in the strong T-elimination ordering to select a vertex of $S$ to match with the currently visited vertex $t$ by using a simple greedy rule that selects the minimum exposed neighbor of $t$. The details of the algorithm are as follows.

**Algorithm MM.** Finding a maximum matching of a chordal bipartite graph $G = (S, T, E)$.

**Input.** A chordal bipartite graph $G = (S, T, E)$ and a strong T-elimination ordering of $G$.

**Output.** A maximum matching $M$ of $G$.

**Method.**

> $M \leftarrow \emptyset$;
>
> for all $s \in S$, $exposed(s) = 1$;
>
> **for** $i = 1$ to $|T|$ **do**
>
> > $U \leftarrow \{s : s \in N_G(t_i), exposed(s) = 1\}$;
> >
> > **if** $U \neq \emptyset$ **then**
> >
> > > $s \leftarrow \min U$;
> > >
> > > $M \leftarrow M \cup \{(s, t_i)\}$;
> > >
> > > $exposed(s) \leftarrow 0$;
> >
> > **end if**
>
> **end for**
>
> output $M$ as a maximum matching of $G = (S, T, E)$;

**Theorem 3.** *Given a strong T-elimination ordering of a chordal bipartite graph $G = \{S, T, E\}$, a maximum matching of $G$ can be found in $O(n + m)$ time.*

*Proof.* Omitted.

## 4 Minimum Fill-in

Given a graph $G = (V, E)$ and a subset $X$ of $V$, $G[X]$ is called an *induced cycle of length $l$*, denoted by $C_l$, if $G[X]$ is a cycle and its length is $l$. For a graph $G = (V, E)$, a set $E'$ from $\overline{E} = \{(u, v) \notin E : u \in V \text{ and } v \in V\}$ is said to be $C_4$-destroying if for every induced cycle $(w, x, y, z, w)$ of length four in $G$, the edge $(w, y)$ or $(x, z)$ is in $E'$. The concept of $C_4$-destroying set was first defined by Spinrad et al. for bipartite graphs [13]. The following theorem was given by them.

**Theorem 4.** [13] *Let $G = (S, T, E)$ be a bipartite permutation graph and $E'$ a minimum cardinality $C_4$-destroying set. Then, the graph $G' = (S \cup T, E \cup E')$ is a chordal graph.*

Based upon the above theorem, Spinrad et al. [13] gave an $O(n^5)$ time algorithm for solving the minimum fill-in problem in bipartite permutation graphs. For the paper to be self-contained, we repeat their algorithm in the following.

If $(u, v)$ is in a minimum $C_4$-destroying set of a bipartite graph $G = (S, T, E)$, then both $u$ and $v$ are in either $S$ or $T$. Given a bipartite graph $G = (S, T, E)$, we construct graph $H(G) = (V_H, E_H)$, where

$$V_H = S_H \cup T_H, \ S_H = \{ss' : s, s' \in S, s < s'\}, \ T_H = \{tt' : t, t' \in S, t < t'\},$$

and

$$E_H = \{(ss', tt') : ss' \in S_H, tt' \in T_H, (s, t), (s, t'), (s', t), (s', t') \in E\}.$$

Vertices of $H(G)$ correspond to possible edges of a $C_4$-destroying set, and edges of $H(G)$ correspond to cycles of length four in $G$. Thus a set of edges is a $C_4$-destroying set in $G$ if and only if the corresponding set of vertices in $H(G)$ is a vertex cover of $H(G)$. By Theorem 4, we can find a minimum fill-in of a bipartite permutation graph $G$ by finding a minimum vertex cover of $H(G)$. Note that $H(G)$ is a bipartite graph. A minimum vertex cover of a bipartite graph can be constructed in $O(n + m)$ time from a maximum matching of the graph. A maximum matching in a bipartite graph can be found in $O(mn^{0.5})$ time [6]. Since $H(G)$ has $O(n^2)$ vertices and $O(n^4)$ edges, a minimum fill-in can be found in $O(n^5)$ time.

We will show Theorem 4 remains true for chordal bipartite graphs. Thus the algorithm for bipartite permutation graphs works correctly even in chordal bipartite graphs. Let $G = (S, T, E)$ be a chordal bipartite graph. By Theorem 2, $G$ admits a strong T-elimination ordering where $S = \{s_1 < s_2 < \cdots < s_p\}$ and $T = \{t_1 < t_2 < \cdots < t_q\}$. For two distinct vertices, $z_1, z_2 \in S \cup T$, let $mt(z_1, z_2) = z_1$ if (1) $z_1, z_2 \in T$ and $z_1 < z_2$, or (2) $z_1 \in T$ and $z_2 \in S$; let $mt(z_1, z_2) = z_2$ if (1) $z_1, z_2 \in T$ and $z_2 < z_1$, or (2) $z_2 \in T$ and $z_1 \in S$;

let $mt(z_1, z_2)$ be the vertex $t_i$ of $N(z_1) \cap N(z_2)$ with the smallest index $i$ if $z_1, z_2 \in S$ and $N(z_1) \cap N(z_2) \neq \emptyset$; and let $mt(z_1, z_2)$ be undefined for other cases. If $mt(z_1, z_2)$ is defined, then clearly it is a vertex of $T$. Let $E'$ be a minimum cardinality $C_4$-destroying set and $C$ be a cycle in graph $G' = (S \cup T, E \cup E')$ of length greater than three. We claim that $C$ has a chord. Hence $G'$ is chordal. It is straightforward to verify that, if $(z_1, z_2) \in E \cup E'$, than $mt(z_1, z_2)$ is defined. Let $C = (z_1, z_2, \ldots, z_k, z_1)$ where $k \geq 4$ be a cycle in $G'$. Assume that $(z_1, z_2)$ is an edge in $C$ with the smallest index $i$ such that $t_i = mt(z_1, z_2)$. Consider the following cases.

**Case 1**, one of $z_1$ and $z_2$ is a vertex of $T$. Without loss of generality, assume that $z_1 \in T$ and $z_1 < z_2$ if $z_2$ is also a vertex of $T$. For clarity, let $z_1 = t_i$. Consider the following subcases.

**Case 1.1**, both $z_2$ and $z_k$ are vertices of $S$. Without loss of generality, assume $z_2 < z_k$. By strong T-elimination ordering, $N_{G_{t_i}}(z_2) \subseteq N_{G_{t_i}}(z_k)$. Suppose $z_3$ is a vertex of $T$. By definition, $z_3 > t_i = z_1$. Therefore $z_3$ is a vertex of $G_{t_i}$ and is in $N_{G_{t_i}}(z_2)$. Obviously, $z_3 \in N_{G_{t_i}}(z_k)$. Hence $(z_1, z_2, z_3, z_k, z_1)$ is a cycle in $G$ of length four. Since $E'$ destroys all cycles in $G$ of length four, either $(z_1, z_3) \in E'$ or $(z_2, z_k) \in E'$. Thus $C$ has a chord. Next, suppose $z_3$ is a vertex of $S$, i.e. $(z_2, z_3) \in E'$. Since $E'$ is a minimum cardinality $C_4$-destroying set, there exist two vertices $t$ and $t'$ of $T$ such that $(z_2, t, z_3, t', z_2)$ is a cycle in $G$ of length four and $(t, t') \notin E'$. By the definition of $z_1$, we have $t \geq z_1$ and $t' \geq z_1$. Thus $t, t' \in N_{G_{t_i}}(z_k)$. Then, $(z_2, t, z_k, t', z_2)$ is also a cycle in $G$ of length four. Apparently, $(z_2, z_k) \in E'$ since $(t, t') \notin E'$. Thus $C$ has a chord in this case.

**Case 1.2**, both $z_2$ and $z_k$ are vertices of $T$. Without loss of generality, assume $z_2 < z_k$. Since $E'$ is a minimum cardinality $C_4$-destroying set, there exist two vertices $s$ and $s'$ of $S$ such that $(z_1, s, z_2, s', z_1)$ is a cycle in $G$ of length four and $(s, s') \notin E'$. Similarly, there exist two vertices $s^*$ and $s"$ of $S$ such that $(z_1, s", z_k, s^*, z_1)$ is a cycle in $G$ of length four and $(s", s^*) \notin E'$. Without loss of generality, assume $s < s'$ and $s" < s^*$. Apparently, $s, s', s", s^* \in N(t_i)$. We claim that either $z_2, z_k \in N_{G_{t_i}}(s) \cap N_{G_{t_i}}(s')$ or $z_2, z_k \in N_{G_{t_i}}(s'') \cap N_{G_{t_i}}(s^*)$. In either case, we have that $(z_2, z_k) \in E'$, That is, $C$ has a chord. The claim can be easily seen by the following arguments. By strong T-elimination ordering, it is straightforward to verify that $z_2 \in N_{G_{t_i}}(s'') \cap N_{G_{t_i}}(s^*)$ if $s \leq s''$; and $z_k \in N_{G_{t_i}}(s) \cap N_{G_{t_i}}(s')$ otherwise.

**Case 1.3**, exactly one of $z_2$ and $z_k$ is a vertex of $T$. Without loss of generality, assume that $z_2$ is a vertex of $S$ and $z_k$ is a vertex of $T$. Clearly, $z_1 < z_k$. Since $(z_1, z_k) \in E'$, there exist two vertices $s$ and $s'$ such that $(s, s') \notin E'$ and $(z_1, s, z_k, s', z_1)$ is a cycle in $G$ of length four. Without loss of generality, assume $s < s'$. If $(z_2, z_k) \in E$, then $C$ has a chord. Suppose $(z_2, z_k) \notin E$. If $s \leq z_2$, then $N_{G_{t_i}}(s) \subset N_{G_{t_i}}(z_2)$ by strong T-elimination ordering. Since $s$ is adjacent to vertex $z_k$ which is not adjacent to $z_2$, we have that $z_2 < s < s'$. Suppose $z_3$ is a vertex of $S$. That is $(z_2, z_3) \in E'$. Then, there exist two vertices $t$ and $t'$ of $T$ such that $(t, t') \notin E'$ and $(t, z_2, t', z_3, t)$ is a cycle in $G$ of length four. That is, $t, t' \in N_{G_{t_i}}(z_2) \subseteq N_{G_{t_i}}(s) \subseteq N_{G_{t_i}}(s')$. In other words, we have a cycle, $(t, s, t', s', t)$ of $G$ of length four which is not destroyed by $E'$, a contradiction. Thus $z_3$ is a vertex of $T$. Then, $z_3 \in N_{G_{t_i}}(s) \cap N_{G_{t_i}}(s')$. Hence $(z_1, s, z_3, s', z_1)$ is a cycle of length four. Since $E'$ destroys all cycles in $G$ of length four and $(s, s') \notin E'$, clearly $(z_1, z_3) \in E'$. Hence $C$ has a chord.

**Case 2**, both $z_1$ and $z_2$ are vertices of $S$. Without loss of generality, assume that $z_1 < z_2$. For clarity, let $t_i = mt(z_1, z_2)$. Consider the following subcases.

**Case 2.1**, $z_k$ is a vertex of $T$. By the definition of $z_1$, we have that $z_k > t_i$. By strong T-elimination ordering, $z_k \in N_{G_{t_i}}(z_1) \subseteq N_{G_{t_i}}(z_2)$. Thus $(z_2, z_k) \in E$, i.e., $C$ has a chord.

**Case 2.2**, $z_k$ is a vertex of $S$. Since $(z_1, z_2) \in E'$, there exist two vertices $t$ and $t'$ of $T$ such that $(t, t') \notin E'$ and $(t, z_1, t', z_2, t)$ is a cycle in $G$ of length four. By the definition of $z_1$, $t, t' \geq t_i$. Similarly, there exist two vertices $t''$ and $t^*$ of $T$ such that $(t'', z_1, t^*, z_k, t'')$ is a cycle in $G$ of length four, $(t'', t^*) \notin E'$, and

$t", t^* \geq t_i$. Suppose $z_k < z_1$. Then, $t", t^* \in N_{G_{t_i}}(z_k) \subseteq N_{G_{t_i}}(z_1) \subseteq N_{G_{t_i}}(z_2)$, and hence $E'$ includes $(z_2, z_k)$ to destroy cycle $(t", z_k, t^*, z_2, t")$ in $G$ of length four. Therefore $C$ has a chord. Similarly, if $z_1 < z_k$, then $E'$ includes $(z_2, z_k)$ to destroy cycle $(t, z_k, t', z_2, t)$ in $G$ of length four and therefore $C$ has a chord.

We have proved that $C$ has a chord in all above cases. Hence, $G'$ is chordal. Thus we have the following theorem.

**Theorem 5.** *Let* $G = (S, T, E)$ *be a chordal bipartite graph, and let* $E'$ *be a minimum cardinality* $C_4$-*destroying set. Then, the graph* $G' = (S \cup T, E \cup E')$ *is a chordal graph.*

Immediately following from the above theorem, we can compute the minimum fill-in for chordal bipartite graphs in polynomial time by using the algorithm proposed by Spinrad et al. [13]. In other words, the minimum fill-in problem can be solved in $O(n^5)$ time on chordal bipartite graphs. But we would like to improve the time complexity of this algorithm. We observe that graph $H(G)$ is a special bipartite graph. In fact, $H(G)$ is a chordal bipartite graph. Given a strong T-elimination ordering of a chordal bipartite graph $G$, a maximum matching of $G$ can be found in $O(m+n)$ time. This leads to the conclusion that the minimum fill-in problem can be solved in $O(n^4)$ time on chordal bipartite graphs.

Now, we show that $H(G)$ is a chordal bipartite graph by giving a strong T-elimination ordering of vertices in $H(G)$. Let vertices of both $T_H$ and $S_H$ in lexicographic ordering. That is, for $tt'$ and $t^*t"$ in $T_H$ where $t < t'$ and $t^* < t"$ (resp. $ss'$ and $s^*s"$ in $S_H$ where $s < s'$ and $s^* < s"$), we have that $tt' < t^*t"$ (resp. $ss' < s^*s"$) if $t < t^*$ or $t = t^*$ and $t' < t"$ (resp. $s < s^*$ or $s = s^*$ and $s' < s"$). Let $H(G)_{tt'}$ denote $S_H \cup \{t^*t" \in T_H : tt' \leq t^*t"\}$. We can show that lexicographic ordering is a strong T-elimination ordering by showing that if $ss', s^*s" \in S_H$, $ss' < s^*s"$, and $ss', s^*s" \in N_{H(G)}(tt')$, then $N_{H(G)_{tt'}}(ss') \subseteq N_{H(G)_{tt'}}(s^*s")$. Clearly $s \leq s^* < s"$. Suppose $t^*t" \in N_{H(G)_{tt'}}(ss')$. Then, $t \leq t^* < t"$ and hence $t^*, t" \in N_{G_t}(s)$. By strong T-elimination ordering of $G$, $t^*, t" \in N_{G_t}(s) \subseteq$

$N_{G_t}(s^*) \subseteq N_{G_t}(s'')$. Therefore $N_{H(G)_{tt'}}(ss') \subseteq N_{H(G)_{tt'}}(s^*s'')$. We have shown that $H(G)$ admits a strong T-elimination ordering. By Theorem 2, $H(G)$ is a chordal bipartite graph. Thus we have the following theorem.

**Theorem 6.** *The minimum fill-in of a chordal bipartite graph can be computed in $O(n^4)$ time.*

# References

1. H. Bodlaender, T. Kloks, D. Kratsch, and H. Müller, Treewidth and minimum fill-in on d-trapezoid graphs, Technical Report UU-CS-1995-34, Utrecht University.
2. A. Brandstädt, Classes of bipartite graphs related to chordal graphs, *Discrete Appl. Math.* **32** (1991) 51-60.
3. D. G. Corneil, Y. Perl, and L. K. Stewart, Cographs: recognition, applications, and algorithms, *Congressus Numerantium,* **43** (1984) 249-258.
4. M. Farber, Characterization of strongly chordal graphs, *Discrete Math.* **43** (1983) 173-189.
5. F. Gavril, Algorithms for maximum k-colorings and k-coverings of transitive graphs, *Networks* **17** (1987) 465-470.
6. J. E. Hopcroft and R. M. Karp, A $n^{5/2}$ algorithm for maximum matching in bipartite graphs, *SIAM J. Comput.* **2** (1973) 225-231.
7. T. Kloks, Minimum fill-in for chordal bipartite graphs, Technical report RUU-CS-93-11, Department of CS, Utrecht Universtity, Utrecht The Netherlands (1993).
8. T. Kloks, D. Kratsch, and C. K. Wong, Minimum fill-in on circle and circular-arc graphs, To appear in the proceedings of ICALP'96.
9. A. Parra, Triangulating multitolerance graphs, Technical Report, 392/1994, Technische Universität Berlin.
10. D. J. Rose, Triangulated graphs and the elimination process, *J. Math. Anal. Appl.* **32** (1970) 597-609.
11. D. J. Rose, A graph-theoretic study of the numerical solution of sparse positive definite systems of linear equations, in: *Graph Theory and Computing*, R. Read, (ed.), Academic Press, New York, (1973) 183-217.
12. J. P. Spinrad, Doubly lexical ordering of dense 0-1 matrices, *Information Processing Letters* **45** (1993) 229-235.
13. J. P. Spinrad, A. Brandstädt, and L. Stewart, Bipartite permutation graphs, *Discrete Appl. Math.* **18** (1987) 279-292.
14. R. E. Tarjan, Decomposition by clique seperators, *Discrete Math.* **55** (1985) 221-232.
15. M. Yannakakis, Computing the minimum fill-in is NP-complete, *SIAM J. Algebraic Discrete Methods* **2** (1981) 77-79.

# Graph Searching on Chordal Graphs

Sheng-Lung Peng[*1], Ming-Tat Ko[2], Chin-Wen Ho[3], Tsan-sheng Hsu[*2], and Chuan-Yi Tang[1]

[1] National Tsing Hua University, Taiwan
[2] Academia Sinica, Taiwan
[3] National Central University, Taiwan

**Abstract.** Two variations of the graph searching problem, edge searching and node searching, are studied on several classes of chordal graphs, which include split graphs, interval graphs and $k$-starlike graphs.

## 1 Introduction

The graph searching problem was first proposed by Parsons [16]. A graph represents a system of tunnels. Initially, all edges of the graph are contaminated by a gas. We would like to obtain a state of the graph in which all edges are simultaneously cleared by a sequence of moves using searchers. The objective is to achieve the desired state by using the least number of searchers.

In this paper, two versions of graph searching problem, the edge searching problem [13] and the node searching problem [10], are discussed. These two problems are different in the way to clear a contaminated edge. In *node searching*, the allowable moves are (1) placing a searcher on a vertex and (2) removing a searcher from a vertex. In node searching, a contaminated edge is *cleared* if both its two endpoints contain searchers. In *edge searching*, besides the allowable moves in the node-searching, one more allowable move, (3) moving a searcher along an edge, is needed. In edge searching, there are two ways in which a contaminated edge becomes clear: (i) if one endpoint of a contaminated edge contains a searcher, it can be cleared by moving a second searcher from that endpoint to the other endpoint along the edge, (ii) if one endpoint of a contaminated edge contains a searcher and all the other edges incident to it are cleared, the searcher can be moved along the edge to the other endpoint to clear the edge.

A vertex is *guarded* if it contains a searcher. A path that contains a guarded vertex is also called *guarded*. A cleared edge may be *recontaminated* once there is an unguarded path connecting the edge with a contaminated one.

An *edge-search strategy* (resp., *node-search strategy*) of a graph $G$ is a sequence of allowable moves in edge searching (resp., node searching) that clears the initially contaminated graph. A search strategy is called *optimal* if it clears the graph and the maximum number of searchers on the graph at any time step is as small as possible. In edge searching (resp., node searching), this number is

---

* Part of this research was supported by NSC85-2213-E-001-003.

called the *edge-search number* (resp., *node-search number*) of $G$, and is denoted by $es(G)$ (resp., $ns(G)$).

For edge searching, it was shown in [12] and [4] that there is always an optimal edge-search strategy for searching $G$ in which no tunnel is searched twice. In other words, recontamination does not help in edge-searching a graph. It was shown that solving the edge searching problem is NP-hard on general graphs and can be solved in linear time for trees [13]. This problem remains NP-hard for planar graphs with maximum vertex degree three [15].

For node searching, it was also shown in [10] and [4] that recontamination does not help in node-searching a graph. It is interesting to note that the path-width problem [17], the interval thickness problem [9] and several others [14] [11][7] are all equivalent to the node searching problem. These problems have been extensively studied on many special classes of graphs. The node searching problem is NP-hard on planar graphs with vertex degree at most three [10], starlike and chordal graphs [6], bipartite graphs [8], and cobipartite graphs (complement of bipartite graphs) [1]. There are linear-time algorithms on trees [18][14] and cographs [3], $O(pn)$-time algorithm on permutation graphs [2] (where $p$ represents the node-search number of the input graph), and $O(n^{2k+1})$-time algorithm on $k$-starlike graphs [6], where $m$ and $n$ are the numbers of edges and vertices in the input graph respectively.

In this paper, we use a seemly uniform high-level approach to solve the graph searching problem on several special classes of chordal graphs. In particular, for the edge searching problem, we give an $O(mn^2)$-time algorithm on split graphs, an $O(m+n)$-time algorithm on interval graphs, and an $O(mn^k)$-time algorithm on $k$-starlike graphs for any fixed $k$ and $k \geq 2$. All of the above problems were not previously known to be solved in polynomial time. In addition, we also give an $O(mn^k)$-time algorithm for solving the node searching problem on $k$-starlike graphs for any fixed $k$, which greatly improves a previous $O(n^{2k+1})$-time algorithm in [6]. We also show that the edge searching problem is NP-hard on chordal graphs.

## 2 Preliminaries

Let $G = (V, E)$ be a finite, simple and undirected graph, where $V$ and $E$ denote the vertex set and the edge set of the graph $G$ respectively. Let $n = |V|$ and $m = |E|$. Let $N(v) = \{w \mid (v, w) \in E\}$ denote the neighborhood of the vertex $v$ and $\deg(v) = |N(v)|$. The *close neighborhood* of $v$ is denoted as $N[v] = N(v) \cup \{v\}$. For vertex sets $X$ and $Y$, let $N(X) = \{w \notin X \mid \exists v \in X, (v, w) \in E\}$ and let $N_Y(X) = N(X) \cap Y$. For convenience, we use $N(v, w)$ to denote $N(\{v, w\})$ and $X^{<c}$ (resp., $X^{>c}$) to denote the set of vertices in $X$ with degree smaller (resp., greater) than $c$.

A graph $G$ is called a *split graph* if its vertex set $V$ can be partitioned into two subsets $C$ and $I$ such that $C$ induces a maximal clique and $I$ is an independent set of $G$. Throughout this paper, we use $G = (C \cup I, E)$ to denote a split graph in which $C$ induces a maximal clique and $I$ is an independent set.

A graph $G$ is a *chordal graph* if $G$ is the intersection graph of a family of subtrees of a tree. Specially, there exists a tree model, called *clique tree*, in which the tree nodes one-to-one correspond to the maximal cliques of $G$ [5]. A chordal graph is called an *interval* graph if one of its clique trees is a path. A chordal graph $G$ is called *starlike* if one of its clique trees is a star. Let $\{X_0, X_1, \ldots, X_r\}$ be the set of all maximal cliques of a starlike graph. Throughout this paper, $X_0$ denotes the clique corresponding to the central node of the star clique tree of $G$. $X_0$ is called the *central clique* and the other maximal cliques $X_i$'s($i \neq 0$) are called *peripheral cliques*. A peripheral clique $X$ of a starlike graph is called an $i$-p-clique if $|X \setminus X_0| = i$. We call the vertices in $X_0$ (resp., $V \setminus X_0$) the *central* (resp., *peripheral* ) *vertices* of $G$. For an integer $k$, we call a starlike graph $G$ $k$-*starlike* if $\max_{i=1}^{r} |X_i \setminus X_0| = k$. Note that split graphs are 1-starlike graphs.

During an edge-searching of any graph, there are three types of vertices: clear, contaminated, and partially clear [12]. A vertex is called *clear* if all of its incident edges are cleared, *contaminated* if all of its incident edges are contaminated, and *partially clear* if there are both contaminated and clear edges incident to it. A partial clear vertex is called a *spot* vertex if it has only one contaminated incident edge and is called a *stain* vertex otherwise. In [12], LaPaugh defined the *clearing move* composed of several allowable moves. A clearing move clears an edge and reaches a state that satisfies the following two conditions, (I) no clear or contaminated vertex contains a searcher; and (II) no vertex contains more than one searcher.

The clearing moves are of the following two types.

1. $m^+(x, y)$: Vertex $x$ has at least two incident contaminated edges. If vertex $x$ contains no searcher, we first place a searcher on $x$. Then move a second searcher from $x$ to $y$ along the edge $(x, y)$. Remove searchers from $y$ if vertex $y$ violates conditions (I) or (II).

2. $m^-(x, y)$: Vertex $x$ has only one incident contaminated edges. If vertex $x$ contains no searcher, we first place a searcher on $x$. Then move the searcher at $x$ to $y$ along the edge $(x, y)$. Remove searchers from $y$ if vertex $y$ violates conditions (I) or (II).

Note that a clearing move clears exactly one edge and it does not cause any recontamination. In other words, any edge-search strategy composed of clearing moves has $|E|$ clearing moves. LaPaugh proved the following important theorem that implies recontamination can not help in the optimal edge-search strategy.

**Theorem 2.1** *[12]There is an optimal edge-search strategy composed of clearing moves.*

By Theorem 2.1, in the following we only consider edge-search strategies composed of clearing moves.

In an edge-search strategy composed of clearing moves, the state at time step $t$ is defined as $S_t = (Y_t, Z_t)$ where $Y_t \subseteq V$ is the set of guarded vertices and $Z_t \subseteq E$ is the set of remaining contaminated edges. An edge-search strategy of $G$ can be represented as a sequence of states $S_i$'s and clearing moves $m_i$'s:

$$S_0, m_1, S_1, ..., m_{|E|}, S_{|E|},$$

where $S_0 = (Y_0 = \emptyset, Z_0 = E)$, $S_{|E|} = (Y_{|E|} = \emptyset, Z_{|E|} = \emptyset)$, and state $S_i$ is obtained from $S_{i-1}$ by applying clearing move $m_i$ for $i = 1, \ldots, |E|$. A vertex $u$ is clear in a state $S_t$ if all its incident edges are not in $Z_t$. We say a vertex $u$ is *cleared* at $S_t$ if it is the first state $u$ becomes clear.

With respect to a state, any searcher not guarding a vertex is called a *free* searcher. If the number of searchers to clear a graph is given, the *deadlock state* is the state with no free searcher and every guarded vertex is a stain vertex. Note that there is no deadlock state in any edge-search strategy.

The contaminated edge $(x, y)$ incident to a spot vertex $x$ can be cleared by $m^-(x, y)$ in which no free searcher is used. The following lemmas can be obtained straightforward.

**Lemma 2.2** *There exists an optimal edge-search strategy* $S_0, m_1, S_1, ..., m_{|E|}, S_{|E|}$ *such that for every $S_i$ containing spot vertices, $m_{i+1} = m^-(x, y)$, where $x$ is a spot vertex and $(x, y)$ is the contaminated edge incident to $x$.*

**Lemma 2.3** *Let $G = (V, E)$ be a starlike graph with maximal cliques $\{X_0, X_1, \ldots, X_r\}$ and $es(G) = |X_0|$. Then there exists an optimal edge-search strategy such that the last cleared edge incident to the first cleared vertex in $X_0$ is in the central clique.*

In Lemma 2.2, if there are many spot vertices in a state, their incident contaminated edges can be cleared in an arbitrary order. In the rest of this paper, all the edge-search strategies considered satisfy Lemmas 2.2 and 2.3.

## 3 Edge searching on split graphs

The node searching problem on split graphs was first solved implicitly in [6]. Let $G = (C \cup I, E)$ be a split graph. It was proved that $ns(G) = |C|$ or $|C| + 1$ [6]. Since $ns(G) - 1 \leq es(G) \leq ns(G) + 1$ [10], we obtain $es(G) = |C| - 1, |C|, |C| + 1$ or $|C| + 2$. Since $es(K_n) = n$ for $n \geq 4$, $es(G) = |C| - 1$ may happen only if $|C| < 4$. In the case $|C| = 2$, if $G$ is a path then $es(G) = 1$ otherwise $es(G) = 2$. In the case $|C| = 3$, it can be proved that (1) $es(G) = 2$ if and only if $\exists w \in C$, $\deg(w) = 2$ and $\forall u, v \in C \cup I$, $|N(u) \cap N(v)| \leq 1$; (2) $es(G) = 4$ if and only if $\forall u, v \in C$, $|N_I(u) \cap N_I(v)| \geq 2$; (3) $es(G) = 3$, otherwise.

In the following, we consider the split graph with $|C| \geq 4$. The following two lemmas state the necessary and sufficient conditions for $G$ to be of $es(G) = |C|$ and $es(G) = |C| + 1$ respectively. Due to the space limitation, the proof of Lemma 3.5 is omitted.

**Lemma 3.4** *Let $G = (C \cup I, E)$ be a split graph with $|C| \geq 4$. Then $es(G) = |C|$ if and only if $\exists u, v \in C$, such that (1) $|C \setminus N(N_I^{\geq 2}(u, v) \setminus (A \cup W))| \geq 2$ and (2) $|C \setminus N(N_I(u) \cup N_I^{\geq 2}(v) \setminus (B \cup W))| \geq 1$, where (i) $A \subseteq N_I^{\geq 2}(u, v)$ and $|A| \leq 1$, (ii) $W \subseteq N_I^{\geq 2}(v) \setminus N_I^{\geq 2}(u)$ and $|W| \leq 1$, and (iii) $B \subseteq N_I^{=2}(u)$ and $|B| \leq 1$.*

**Proof.** Suppose that $es(G) = |C|$. Consider an optimal edge-search strategy, $S = S_0, m_1, S_1, \ldots, m_{|E|}, S_{|E|}$, in which $u$ and $v$ are the first and the second cleared vertex in $C$ respectively, and $y$ and $x$ are the last and the second last guarded vertex in $C$ respectively. Let $S_{t_3}$ (resp., $S_{t_4}$) be the first state in which $u$ (resp., $v$) is cleared. Let $S_{t_1}$ (resp., $S_{t_2}$) be the first state in which $x$ (resp., $y$) is guarded.

At $S_{t_1}$, since there are $|C| - 1$ searchers on the central clique, at most one peripheral vertex is guarded. If such a peripheral vertex $z$ exists, $z$ has to be a spot vertex and by strategy assumption, $z$ is cleared at $S_{t_1+1}$. Otherwise, $z$ is a stain vertex and then there is no free searcher between $S_{t_1}$ and $S_{t_3}$. In order that $u$ can be cleared, edge $(u, x)$ has to be cleared just before $S_{t_1}$ and the scenario must be $m^+(u, x), S_{t_1}, m^-(u, y), S_{t_2}(= S_{t_3})$. Then every vertex $w$ in $C \setminus \{u\}$ is a stain vertex, since at least $(w, x)$ and $(w, y)$ are contaminated. Thus, $S_{t_2}$ is a deadlock state which leads to a contradiction. Between $S_{t_1}$ and $S_{t_4}$, since at most one free searcher can be used, no peripheral vertices at $S_{t_1}$ of degree greater than 2 except $z$ are cleared. Hence only peripheral vertices of degree no greater than 2 are cleared between $S_{t_1+1}$ and $S_{t_2}$. Since $S$ satisfies Lemma 2.3, $S_{t_3}$ is obtained from $S_{t_2}$ by applying $m^-(u, y)$ or $t_2 = t_3$. In either case, we obtain that all the vertices in $N_I(u)$ are already clear in $S_{t_3-1}$. Otherwise, there is a guarded vertex $z' \in N_I(u)$ with $(z', u)$ being clear at $S_{t_3}$. If $deg(z') = 2$, by Lemma 2.2, we may change $S$ to another optimal edge-search strategy such that $z$ is cleared earlier than $u$. If $deg(z') \geq 3$, then $(z', u)$ is cleared just before $S_{t_3-1}$. In $S_{t_3-2}$, $u$ and $z'$ both are stain vertices. To continue the strategy, $v$ must be a spot vertex in $S_{t_3}$, which implies $v$ has been a spot vertex at $S_{t_3-2}$. It contradicts to that the edge-search strategy satisfying Lemma 2.2.

Let $W = \{w | w \in N_I^{\geq 2}(v) \setminus N_I^{\geq 2}(u)$ and $w$ is contaminated in $S_{t_1}\}$, $A = \{a | a \in N_I^{\geq 2}(u, v) \setminus W$ and $a$ is unclear in $S_{t_1-1}\}$ and $B = \{b | b \in N_I^{=2}(u)$ and $b$ is unclear in $S_{t_2-1}\}$. By the definition of $W$, $A$ and $B$ and the fact that vertices in $N_I^{\geq 2}(u)$ are clear in $S_{t_1+1}$, vertices in $N_I^{\geq 2}(v) \setminus W$ are clear in $S_{t_1+1}$, vertices in $N_I^{\geq 2}(u, v) \setminus (W \cup A)$ are clear in $S_{t_1-1}$ and vertices in $N_I(u) \setminus B$ are clear in $S_{t_2-1}$. Thus, vertices in $N_I(u) \cup N_I^{\geq 2}(v) \setminus (B \cup W)$ are clear in $S_{t_2-1}$. Therefore, $x$ and $y$ is not in $N(N_I^{\geq 2}(u, v) \setminus (W \cup A))$ and $y$ is not in $N(N_I(u) \cup N_I^{\geq 2}(v) \setminus (B \cup W))$. Namely, condition (1) and (2) hold. To complete the proof, it remains to prove that $|W| \leq 1$, $|B| \leq 1$ and $|A| \leq 1$.

Since $v$ is cleared at $S_{t_4}$, the edges between $v$ and vertices in $W$ are not in $Z_{t_4}$. Thus, those edges are cleared between $S_{t_1}$ and $S_{t_4}$ by at most two searchers, a free searcher and the searcher on $v$, that implies $|W| \leq 2$. Suppose $W = \{w_1, w_2\}$. Let $e = (v, w_2)$ be the last cleared edge incident to $v$. Then the clearing scenario before $S_{t_4}$ has to be $S_{t_4-2}$, $m^+(v, w_1)$, $S_{t_4-1}$, $m^-(v, w_2)$, $S_{t_4}$. In $S_{t_4}$, $w_1$ and $w_2$ are stain vertices. Since $S$ satisfies Lemma 2.2, every vertex in $C \setminus \{u, v\}$ is a stain vertex. Thus, $S_{t_4}$ is a deadlock state which is a contradiction. Hence, we conclude that $|W| \leq 1$.

Since $B \subset N_I(u)$, it has to be clear in $S_{t_2}$. From $S_{t_2-1}$ to $S_{t_2}$, there is at most one vertex becomes clear. Thus, $|B| \leq 1$.

By definition of $A$, vertices in $A$ are not clear at $S_{t_1-1}$, but are clear at $S_{t_1+1}$.

Since there is at most one peripheral vertex of degree greater than 2 becomes clear from $S_{t_1-1}$ to $S_{t_1+1}$, we conclude that $|A| \leq 1$.

Conversely, assume that conditions (1) and (2) hold. Let $C_1$ denote $C \setminus N(N_I^{>2}(u, v) \setminus (A \cup W))$ and $C_2$ denote $C \setminus N(N_I(u) \cup N_I^{>2}(v) \setminus (B \cup W))$. Notice that $C_1 \supseteq C_2$ and by the conditions we have two central vertices $x$ and $y$ such that $x, y \in C_1$ and $y \in C_2$. Without loss of generality, assume that $A = \{a\}$, $W = \{w\}$ and $B = \{b\}$ and $(a, x), (b, y) \in E$. By clearing the peripheral vertices of $G$ and central vertices $u$ and $v$ in the following order, $N_I^{>2}(u, v) \setminus \{a, w\}, a, N_I^{<3}(u) \setminus \{b\}, b, u, I^{<3} \setminus N_I^{<3}(u), v, w, I^{>2} \setminus N_I(u, v)$, with necessary searchers guarding the central vertices, it is easy to obtain an edge-search strategy using $|C|$ searchers. **Q.E.D.**

**Lemma 3.5** Let $G = (C \cup I, E)$ be a split graph with $|C| \geq 4$ and $es(G) > |C|$. Then $es(G) = |C| + 1$ if and only if $\exists u, v \in C$, such that $|N_I^{>2}(u) \cap N_I^{>2}(v)| \leq 2$.

By checking the conditions in Lemma 3.4 and Lemma 3.5, we have an algorithm to determine $es(G)$ of a split graph $G$. The time complexity depends on checking the conditions of Lemma 3.4 and Lemma 3.5. The condition of Lemma 3.5 can be checked in $O(mn^2)$ time. Thus the bottleneck is to check the conditions of Lemma 3.4. However, by the following observation and careful implementation, it also can be done in $O(mn^2)$ time. For each pair $u, v \in C$, we want to find whether there exist sets $A, B$ and $W$ satisfy the conditions of Lemma 3.4 or not. If the sets $A, B$ and $W$ do exist, then there exist two central vertices $x, y$, called *indicator pair* for $u, v$, such that $x, y \in C \setminus N(N_I^{>2}(u, v) \setminus (A \cup W))$ and $y \in C \setminus N(N_I(u) \cup N_I^{>2}(v) \setminus (B \cup W))$. Instead of checking the existence of sets $A, B$ and $W$ directly, our algorithm is to check the existence of an indicator pair $x, y$ for $u, v$. By a careful investigation, we can see that a central vertex $c$ is a candidate of $x$ if (1) $|N_I(c) \cap N_I^{>2}(u)| \leq 1$ and (2) $|N_I(c) \cap N_I^{>2}(u, v)| \leq 2$ and a candidate of $y$ if (1) $|N_I(c) \cap N_I^{>2}(u)| = 0$, (2) $|N_I(c) \cap (N_I^{>2}(v) \setminus N_I^{>2}(u))| \leq 1$ and (3) $|N_I(c) \cap N_I^{=2}(u)| \leq 1$. Let $c_1$ be a candidate of $x$ and $c_2$ be a candidate of $y$. In case that $|N_I(c_1) \cap N_I^{>2}(u, v)| \leq 1$ or $|N_I(c_2) \cap (N_I^{>2}(v) \setminus N_I^{>2}(u))| = 0$, $c_1, c_2$ can serve as an indicator pair for $u$ and $v$. In case that $|N_I(c_1) \cap N_I^{>2}(u, v)| = 2$ and $|N_I(c_2) \cap (N_I^{>2}(v) \setminus N_I^{>2}(u))| = 1$, $c_1, c_2$ can serve as an indicator pair for $u$ and $v$ only if $N_I(c_2) \cap (N_I^{>2}(v) \setminus N_I^{>2}(u)) \subset N_I(c_1) \cap N_I^{>2}(u, v)$. By the above observation and careful implementation, we may have an $O(m)$-time algorithm to check the existence of indicator pairs for $u, v$. For $O(n^2)$ pairs of $u, v$, totally it takes $O(mn^2)$ time to check the conditions in Lemma 3.4.

**Theorem 3.6** There is an $O(mn^2)$-time algorithm to determine the edge-search number of a split graph.

## 4 Edge searching on interval graphs

For any interval graph $G$, $ns(G)$ is trivially equal to its maximum clique size. Since $es(K_n) = n$ for a clique of $n \geq 4$ vertices, if $G$ is an interval graph with the maximum clique size $\omega \geq 4$, then $es(G) = \omega$ or $\omega + 1$. For the case $\omega < 4$, it is easy to verify $\omega - 1 \leq es(G) \leq \omega$. If $\omega = 2$ then $es(G) = 1$ if and only if $G$ is a path. If $\omega = 3$ then $es(G) = 2$ if and only if $\forall u, v, |N(u) \cap N(v)| \leq 1$.

Thus we consider the case that $\omega \geq 4$. The maximal cliques of an interval graph $G$ can be linear ordered such that each vertex in $G$ occurs in consecutive maximal cliques. With such an order of maximal cliques, the correspondence between vertex and the interval of maximal cliques containing it gives an interval model of $G$, which is used in our algorithm for determining $es(G)$. Our algorithm clears the vertices (intervals) of $G$ according to their right endpoints in nondecreasing order and in increasing order of the left endpoints if right endpoints are equal. Let $u$ be the next interval to be cleared and $W$ be the set of remaining unclear vertices at that state. The only situation that the algorithm needs $\omega + 1$ searchers to clear $u$ occurs when $|N[u] \cap W| = \omega$, all vertices in $N(u) \cap W$ are guarded and $u$ contains no searcher. Let $K = N[u] \cap W$. In this case, we can find another two vertices $u_1$ and $u_2$ such that $u_1$ and $u_2$ are cleared earlier and later than $u$ respectively, $N[u_1] \cap K = N[u_2] \cap K = K - \{u\}$, and $u_1, u_2, u$ are mutually nonadjacent. Now, the set $K \cup \{u_1, u_2\}$ induces a split graph that does not satisfy the conditions of Lemma 3.4. Hence, $\omega + 1$ searchers is necessary. The algorithm takes $O(n)$ time to sort the intervals and $O(m + n)$ time to clear the intervals.

**Theorem 4.7** *The edge-search number of an interval graph can be computed in linear time.*

# 5 Node searching on k-starlike graphs

According to [9], a node-search strategy can be represented by a sequence of vertex sets $(Y_i) = (Y_0, ..., Y_s)$, where $Y_i$ is the set of guarded vertices set at step $i$. Since recontamination does not occur, thus if $v$ is guarded at step $i$ and cleared at step $j$ then $v \in Y_t$ for $i \leq t \leq j$, and we call such a property the *consecutive property*. Moreover, every edge is cleared at some step $i$, i.e. for every edge $(u, v) \in E$, there exists some $i$ such that $u, v \in Y_i$. On the other hand, every sequence of vertex sets $(Y_i)$ satisfying the above two properties must correspond to a node-search strategy[9].

Let $(Y_i)$ be a node-search strategy. For any maximal clique $W$ of $G$, there exists a step $t$ such that $W \subseteq Y_t$ [6], i.e. $W$ is cleared at step $t$. Assume that the clearing order of maximal cliques is given by a permutation $\pi : \{0, ..., r\} \rightarrow \{0, ..., r\}$, i.e. $X_i$ is cleared before $X_j$ if $\pi(i) < \pi(j)$. Then, it is possible to define a node-search strategy $(Y_i)$ corresponding to such a clearing order as: $Y_0 = X_{\pi^{-1}(0)}$, $Y_r = X_{\pi^{-1}(r)}$, $Y_i = (Y_{i-1} \cap X_0) \cup X_{\pi^{-1}(i)}$ for $0 < i \leq \pi(0)$ and $Y_i = (Y_{i+1} \cap X_0) \cup X_{\pi^{-1}(i)}$ for $\pi(0) \leq i < r$. Such a node-search strategy is called *normalized*. In [6], Gustedt implicitly proved that every starlike graph $G$ has an optimal node-search strategy that is normalized. Since each node-search strategy $(Y_i)$ uses $\max_i |Y_i|$ searchers, we can conclude that if $G$ is a $k$-starlike graph, then $|X_0| \leq ns(G) \leq |X_0| + k$. From now on, we only consider that $(Y_i)$ is normalized.

Assume that $ns(G) = |X_0| + s$ for some integer $s$, $0 \leq s \leq k - 1$. Let $(Y_i)$ be an optimal node-search strategy, $t_0$ be the first step when all vertices in $X_0$ are guarded (i.e. $X_0 \subseteq Y_{t_0}$ and $X_0 \not\subseteq Y_{t_0-1}$), and $f_l$ be the $l$-th cleared vertex in $X_0$ for $1 \leq l \leq k - s$. We claim that all the $i$-p-cliques containing $f_l$ with $i \geq s + l$ must be cleared before step $t_0$. Otherwise, there exists an $i$-p-clique $X$ such that $f_l \in X, i \geq s + l$, and $X \subset Y_{t_l}$ where $t_l \geq t_0$. According to the consecutive property, we know that at step $t_l$, at least $|X_0| - l + 1$ vertices in $X_0$ are guarded, and we need another $i \geq s + l$ searchers to clear the peripheral vertices in $X$ which contradicts the assumption $ns(G) = |X_0| + s$.

Let $Q_i$ be the set of all $i$-p-cliques, $F_l = \{f_1, f_2, \ldots, f_l\}$, and $Q_i^l = \{X \in Q_i | X \cap F_l \neq \emptyset\}$. According to the above discussion, those cliques in $Q_k^{k-s} \cup Q_{k-1}^{k-s-1} \cup \ldots \cup Q_{s+1}^1$ must be cleared before step $t_0$. Now, for $1 \leq i \leq k - s$, consider the step $t_i' < t_0$ such that $|Y_{t_i'} \cap X_0| \leq |X_0| - i$ and $|Y_{t_i'+1} \cap X_0| > |X_0| - i$. Since between step $t_i' + 1$ and step $t_0$, at most $s + i - 1$ searchers are available to clear peripheral vertices, hence cliques in $Q_k^{k-s} \cup Q_{k-1}^{k-s-1} \cup \ldots \cup Q_{s+i}^i$ must be cleared before step $t_i' + 1$. Let $N_i^{k-s} = (Q_k^{k-s} \cup Q_{k-1}^{k-s-1} \cup \ldots \cup Q_{s+i}^i) \cap X_0$. We have $|N_i^{k-s}| \leq |X_0| - i$, or equivalently $|X_0 \setminus N_i^{k-s}| \geq i$.

Conversely, if there exist vertices $f_1, \ldots f_{k-s} \in X_0$, such that $|X_0 \setminus N_i^{k-s}| \geq i$, for $1 \leq i \leq k - s$, consider a clearing order as follows. Clear the cliques in $Q_{s+i}^i$ from $i = k - s$ down to 1, clear the cliques in $Q_i$ with $i \leq s$, and then clear the cliques in $Q_i \setminus Q_{s+i}^i$ from $i = 1$ to $k - s$. It is easy to argue that the resulting node-search strategy corresponding to this clearing order uses at most $|X_0| + s$ searchers. Hence, we have the following lemma.

**Lemma 5.8** *Let $G$ be a $k$-starlike graph with maximal cliques $\{X_0, X_1, \ldots, X_r\}$ and $s$ be an integer, $0 \leq s \leq k - 1$. If $ns(G) \geq |X_0| + s$ then $ns(G) = |X_0| + s$ if and only if $\exists f_1, \ldots, f_{k-s} \in X_0$, such that $|X_0 \setminus N_i^{k-s}| \geq i$, for $1 \leq i \leq k - s$.*

By Lemma 5.8, we have an algorithm to determine $ns(G)$ of a $k$-starlike graph $G$. The algorithm iteratively executes the following test from $s = 0$ to $k - 1$: Is there any sequence of vertices $f_1, f_2, \ldots, f_{k-s}$ in $X_0$ such that $|X_0 \setminus N_i^{k-s}| \geq i$, for $1 \leq i \leq k - s$? If the test condition holds, then the algorithm terminates with the output $ns(G) = |X_0| + s$. Otherwise, the algorithm concludes that $ns(G) > |X_0| + s$ and executes the next iteration if $s < k - 1$, and terminates with output $ns(G) = |X_0| + k$ if $s = k - 1$.

In each iteration, when a sequence of vertices $f_1, f_2, \ldots, f_{k-s}$ is given, a standard graph traversal algorithm can be used to compute $N_i^{k-s}$, for $1 \leq i \leq k - s$ in $O(m)$ time. Since the algorithm tries all possible $(k-s)$-sequences of vertices in $X_0$ to test the condition, hence $O((k - s)! \cdot C_{k-s}^{|X_0|} \cdot m) = O((k - s)! \cdot n^{k-s} \cdot m)$ time is needed in the iteration. If $k$ is a fixed constant, then totally the time bound of the algorithm is $O(n^k m)$.

**Theorem 5.9** *There is an $O(mn^k)$-time and $O(m)$-space algorithm to determine the node-search number of a $k$-starlike graph.*

# 6 Edge searching on k-starlike graphs

As in Sections 3 and 5, we derive necessary and sufficient conditions for a $k$-starlike graph to have a given edge-search number by discussing the first $k$ cleared vertices of the central clique in an optimal edge-search strategy. The argument to derive the conditions that determine whether an $i$-p-clique, $i \geq 2$, should be cleared earlier than the first cleared central vertex is similar to that for the node search of $k$-starlike graphs, but a little more complicated. For the 1-p-cliques, the argument is the same as that for the edge search of the split graphs. Putting them together, we obtain the following lemmas whose proofs are omitted due to the space limitation.

Let $Q_i$, $F_l$, $Q_i^l$, and $N_i^k$ be the sets defined as in the Section 5. Let $P_1(v) = \{w \in X | w \in N(v) \setminus X_0 \text{ and } X \in Q_1\}$. Analogous to Lemmas 3.4, 3.5 and 5.8, we have the following lemmas.

**Lemma 6.10** *Let $G$ be a $k$-starlike graph with $k \geq 2$. Then $es(G) = |X_0|$ if and only if $\exists f_1, f_2, \ldots, f_k \in X_0$ such that (1) $|X_0 \setminus N_i^k| \geq i$, for $i = 2, \ldots, k$, (2) $|X_0 \setminus (N_2^k \cup N(P_1^{>2}(f_1, f_2) \setminus (A \cup W))| \geq 2$, and (3) $|X_0 \setminus (N_2^k \cup N(P_1(f_1) \cup P_1^{>2}(f_2) \setminus (B \cup W))| \geq 1$, where (i) $A \subseteq P_1^{>2}(f_1, f_2)$ and $|A| \leq 1$, (ii) $W \subseteq P_1^{>2}(f_2) \setminus P_1^{>2}(f_1)$ and $|W| \leq 1$, and (iii) $B \subseteq P_1^{=2}(f_1)$ and $|B| \leq 1$.*

**Lemma 6.11** *Let $G$ be a $k$-starlike graph with $k \geq 2$. If $es(G) \geq |X_0| + 1$ then $es(G) = |X_0| + 1$ if and only if $\exists f_1, f_2, \ldots, f_{k-1} \in X_0$, such that the following conditions hold. (1) $|X_0 \setminus N_i^{k-1}| \geq i$, for $i = 1, \ldots, k-1$, and (2) $\exists v \in X_0 \setminus N_1^{k-1}$ such that $|P_1^{>2}(f_1) \cap P_1^{>2}(v)| \leq 2$.*

**Lemma 6.12** *Let $G$ be a $k$-starlike graph with $k \geq 3$ and $s$ be an integer with $2 \leq s \leq k - 1$. If $es(G) \geq |X_0| + s$ then $es(G) = |X_0| + s$ if and only if $\exists f_1, \ldots, f_{k-s} \in X_0$, such that $|X_0 \setminus N_i^{k-s}| \geq i$, for $1 \leq i \leq k - s$.*

It is not hard to see that if $G$ is a $k$-starlike graph with $k \geq 2$, then $|X_0| \leq es(G) \leq |X_0| + k$. Thus, we have the following algorithm to determine the $es(G)$ of a $k$-starlike graph $G$. The algorithm determines if $es(G) = |X_0|$ by checking the conditions in Lemma 6.10. If it is not true then determines if $es(G) = |X_0| + 1$ by checking the conditions in Lemma 6.11. If it also fails, then we check the condition in Lemma 6.12 until we can find an $s$ ($\leq k - 1$) such that $|X_0 \setminus N_i^{k-s}| \geq i$ holds, for all $1 \leq i \leq k - s$. If such an $s$ is found, then $es(G) = |X_0| + s$; otherwise, $es(G) = |X_0| + k$.

The time complexity depends on checking the conditions of Lemmas 6.10, 6.11 and 6.12 for each sequence $f_1, f_2, \ldots, f_{k-s}$, for $0 \leq s \leq k - 1$. By the same technique used in Sections 3 and 5, we can check them in $O(mn^k)$-time.

**Theorem 6.13** *There is an $O(mn^k)$-time algorithm to determine the edge-search number of a $k$-starlike graph for $k \geq 2$.*

A $k$-starlike graph is called a *pure* $k$-starlike graph if all its peripheral cliques are $k$-p-cliques. By Lemmas 5.8 and 6.12, we conclude that for any pure $k$-starlike graph $G$ with $k \geq 3$, $es(G) = ns(G)$. Gustedt [6] proved the pathwidth problem on chordal graph is NP-hard by reducing the vertex separator problem on an arbitrary graph $G = (V, E)$ with $|V| = n$ and $|E| = m$ to the pathwidth problem on a pure $n$-starlike graph $H$ with $|V(H)| = mn$. Using the above observation and his result, we have the following theorem.

**Theorem 6.14** *The edge searching problem is NP-hard on chordal graphs.*

# References

1. S. ARNBORG, D.G. CORNEIL, and A. PROSKUROWSKI, *Complexity of finding embeddings in a k-tree*, SIAM J. Alg. Disc. Meth., 8(1987), pp. 277-284.

2. H.L. BODLAENDER, T. KLOKS, and D. KRATSCH, *Treewidth and pathwidth of permutation graphs*, 20th ICALP, LNCS 700(1993), pp. 114-125.

3. H.L. BODLAENDER and R.H. MOHRING, *The pathwidth and treewidth of cographs*, SIAM J. Disc. Math., 6(1993), pp. 181-188.

4. D. BIENSTOCK and P. SEYMOUR, *Monotonicity in graph searching*, J. Algorithms, 12(1991), pp. 239-245.

5. F. GAVRIL, *The intersection graphs of subtrees in trees are exactly the chordal graphs*, J. Comb. Theory Ser. B, 16(1974), pp.47- 56.

6. J. GUSTEDT, *On the pathwidth of chordal graphs*, Discr. Appl. Math. 45(1993), pp. 233-248.

7. N.G. KINNERSLEY, *The vertex separation number of a graph equals its path-width*, Inform. Process. Letter, 42(1992), pp. 345-350.

8. T. KLOKS, *Treewidth*, Ph.D. Thesis, Utrecht University, The Netherlands, 1993.

9. L.M. KIROUSIS and C.H. PAPADIMITRIOU, *Interval graph and searching*, Disc. Math. 55(1985), pp. 181-184.

10. L.M. KIROUSIS and C.H. PAPADIMITRIOU, *Searching and pebbling*, Theoretical Comp. Scie. 47(1986), pp. 205-218.

11. A. KORNAI and Z. TUZA, *Narrowness, pathwidth, and their application in natural language processing*, Disc. Appl. Math., 36(1992), pp. 87-92.

12. A.S. LAPAUGH, *Recontamination does not help to search a graph*, J. of the Assoc. Comput. Mach., 40(1993), pp. 224-245.

13. N. MEGIDDO, S.L. HAKIMI, M.R. GAREY, D.S. JOHNSON, and C.H. PAPADIMITRIOU, *The complexity of searching a graph*, J. of the Assoc. Comput. Mach., 35(1988), pp. 18-44.

14. R.H. MOHRING, *Graph problems related to gate matrix layout and PLA folding*, in: G. TINNHOFER et al., eds., Computational Graph Theory (Springer, Wien, 1990), pp. 17-32.

15. B. MONIEN and I.H. SUDBOROUGH, *Min cut is NP- complete for edge weighted trees*, Theoretical Comp. Scie. 58(1988), pp. 209-229.

16. T.D. PARSONS, *Pursuit-evasion in a graph*, in Y. ALAVI and D.R. LICK, eds., Theory and applications of graphs, Springer-Verlag, New York, 1976, pp. 426-441.

17. N. ROBERTSON and P.D. SEYMOUR, *Graph minors I. Excluding a forest*, J. Comb. Theory Ser. B, 35(1983), pp. 39-61.

18. P. SCHEFFLER, *A linear algorithm for the pathwidth of trees*, in: R. BODENDIEK and R. HENN, eds., Topics in Combinatorics and Graph Theory (Physica-Verlag, Heidelberg, 1990), pp. 613-620.

# An Algorithm for Enumerating all Directed Spanning Trees in a Directed Graph

Takeaki UNO *

**Abstract:** A directed spanning tree in a directed graph $G = (V, A)$ is a spanning tree such that no two arcs share their tails. In this paper, we propose an algorithm for listing all directed spanning trees of $G$. Its time and space complexities are $O(|A| + ND(|V|, |A|))$ and $O(|A| + DS(|V|, |A|))$, where $D(|V|, |A|)$ and $DS(|V|, |A|)$ are the time and space complexities of the data structure for updating the minimum spanning tree in an undirected graph with $|V|$ vertices and $|A|$ edges. Here $N$ denotes the number of directed spanning trees in $G$.

**Keywords** directed spanning tree, listing, enumerating algorithm

## 1  Introduction

Let $G = (V, A)$ be a directed graph with vertex set $V$ and arc set $A$. An arc is specified by both of its endpoints. One of them is called its *head* and the other is called its *tail*. A directed spanning tree of $G$ is a spanning tree in which no two arcs share their tails. Each vertex is the tail of exactly one arc of the directed spanning tree except for a special vertex $r$. We call $r$ the *root* of the spanning tree.

Directed spanning trees have been studied in many fields. For instance, many problems on road and telephone networks have been formulated as some optimization problems of directed spanning trees. Some of them have complicated objective functions, and we can hardly solve them in efficient time. For those problems, one of the most simple approaches is to use enumerating. The branch-and-bound method, which is one of the most popular approaches, can be also considered as a kind of enumerating. In this method, the time complexity of the enumerating algorithm greatly influences to its speed. Therefore, improvements of enumerating algorithms largely enhance the efficiency for those problems.

In this paper, we consider the problem of enumerating all directed spanning trees with the root $r$ of the given graph $G$. This problem has been studied for nearly 20 years. In 1978, H. N. Gabow and E. W. Myers proposed an algorithm which runs in $O(|A| + N|A|)$ time and $O(|A|)$ space [1]. Here $N$ denotes the number of directed spanning trees in $G$.

---
*Department of Systems Science, Tokyo Institute of Technology, 2-12-1 Oh-okayama, Meguro-ku, Tokyo 152, Japan. uno@is.titech.ac.jp

In 1992, H. N. Kapoor and H. Ramesh improved the time complexity to $O(|A| + N|V|)$ [2]. Since a directed spanning tree requires $O(|V|)$ time for outputting, it is the optimal algorithm in the sense of both time and space complexities if we have to output all of them explicitly. On the other hand, in undirected graphs, *compact output methods* have been studied to shorten the size of outputs [2, 3]. They output a spanning tree by differences from the previous outputted tree. By outputting all spanning trees in a special order, we can reduce the total size of outputs to $O(N)$. A. Shioura, A. Tamura and T. Uno proposed an optimal algorithm with the compact output method for enumerating all undirected spanning trees [3]. In their algorithm, they utilizes the reverse search. Our method in this note is also based on this reverse search technique. It can be also considered a kind of the back tracking method, which is used in [2].

Here we propose an algorithm for enumerating all directed spanning trees in $O(|A| + ND(|V|, |A|))$ time. Our algorithm constructs and preserves an undirected graph with $|V|$ vertices and at most $|A|$ weighted edges. This graph is modified after each directed spanning tree is encountered. To update the data structure for finding new directed spanning trees, we delete, add or change the weight of an edge, and find the minimum spanning tree of the graph in each iteration of the algorithm. Since only few edges are changed in each iteration, we construct the minimum spanning tree of the changed graph from the previous one by some *updating spanning tree algorithm.*

Therefore the time complexity of our algorithm depends on the time complexity of the updating algorithm, which is denoted by $D(|V|, |A|)$. The space complexity also depends on the updating algorithm. We denote its space complexity by $DS(|V|, |A|)$. The update operations are adding, deleting and changing the weight of an edge. Some data structures are proposed for these updating operations of the minimum spanning tree. G. N. Frederickson proposed an $O(|A|^{1/2})$ time data structure [5], and D. Eppstein, Z. Galil, G. F. Italiano and A. Nissenzweig improved the time complexity into $O(|V|^{1/2} \log(|A|/|V|))$ [4]. Their time complexities are smaller than $O(|V|)$. There use only $O(|A|)$ space, thus we do not lose the optimality of space complexity by our improvement.

## 2 The Reverse Search and the Parent-Child Relationship

We use a technique called reverse search for enumeration problems. The reverse search requires a parent-child relationship on those objects to be enumerated. The relationship has to satisfy: i) each object except for a specified object $r_0$ has a parent object and ii) each object is not a proper ancestor of itself.

Let us consider the graph representation of this parent-child relationship. An object corresponds to a node $v$ of the graph and an edge $(v, u)$ is included in the graph if and only if $u$ corresponds to its parent. By the condition of the relationship, the graph contains no cycle and forms a rooted spanning tree. This tree is called an *enumeration tree*. In Figure

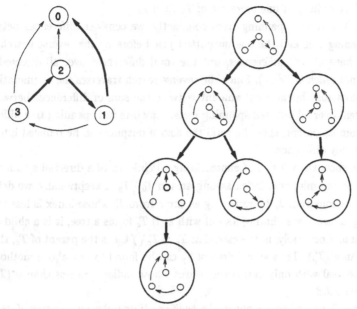

Fig1: The enumeration tree. The emphasized line of the graph is $T_0$.

1, we show an example of an enumeration tree. The reverse search traverses all nodes of the enumeration tree by some depth-first-search scheme from the specified object $r_0$. Even if the size of the enumeration tree is huge, only a small memory space is required, since we can traverse it by only finding the parent and all children of the current object. Hence an important point in speeding up reverse search is "How to enumerate all children efficiently fast."

To enumerate all directed spanning trees, we define a parent-child relationship among them as below. Let $T_0$ be a depth-first-search tree of the given graph $G$. A depth-first-search tree is a directed spanning tree which arises from the traversal route of a depth-first-search. We assume that any search traverses an arc from its head to tail. Using this specified directed tree $T_0$, we introduce the parent-child relationship. In the relationship, $T_0$ corresponds to the specified object $r_0$. Let the indices of the vertices of $G$ be as the order of the traversal of the depth first search. Similarly, we define the indices of arcs of $G$ by indices of their tails. The parent of a directed spanning tree $T_c \neq T_0$ is defined by the directed spanning tree $T_p = T_c \setminus f \cup e$ where $e$ is the minimum index arc $e$ in $T_0 \setminus T_c$ and $f$ is the arc of $T_c$ sharing its tail with $e$. Since $f$ shares its tail only one arc of $T_c$, $T_p$ is uniquely defined. From the definition of the index, there always exists a path from the root to the head of $e$. Hence $e$ is not contained in any cycle and $T_p$ forms a directed spanning tree. In this parent-child relationship, $T_c$ is not a proper ancestor of itself, as $T_p$ contains just one more arc of $T_0$ than $T_c$.

To output directed spanning trees compactly, we consider differences between a directed spanning tree and the one outputted just before by the reverse search. Some of them may have $O(|V|)$ differences, but the total differences over all directed spanning trees does not reach $O(|V|N)$. Since the reverse search traverses the enumeration tree by the depth-first-search, the total amount is twice the sum of differences between a child and its parent over all directed spanning trees. Any directed spanning tree differs by only two arcs from its parent, thus the total the size of outputs can be reduced into $O(N)$ by outputting only differences.

Next we show the method of enumerating all children of a directed spanning tree $T_p$. Let $v^*(T_p)$ be the minimum index among arcs in $T_0 \setminus T_p$. Exceptionally, we define $v^*(T_0)$ by $\infty$. Let us construct $T_c$ by removing an arc $e$ from $T_p$ whose index is less than $v^*(T_p)$ and adding an arc $f \neq e$ sharing its tail with $e$. If $T_c$ forms a tree, it is a child of $T_p$ from the definition. Conversely, in the case that $T_p = T_c \setminus f \cup e$ is the parent of $T_c$, the index of $e$ is less than $v^*(T_p)$. Thus all children of $T_p$ can be found by the above method. To find children, we deal with only vertices and arcs whose indices are less than $v^*(T_p)$. Hence we call them *valid*.

For a tree $T$, we call an arc not in $T$ a *back-arc*, if its tail is an ancestor of its head, and otherwise a *non-back-arc*. In the above method, if $f$ is a back-arc of $T_p$, then $T_c = T_p \setminus e \cup f$ contains a cycle and is not a directed spanning tree. If $f$ is a non-back-arc of $T_p$, there always exists a path from $r$ to the head of $f$ in $T_c$ and $T_c$ contains no cycle. Hence each child of $T_p$ is obtained by adding a valid non-back-arc $f$ and removing an arc $e$ sharing its tail with $f$. Valid non-back-arcs have a one-to-one correspondence with the children and the number of children is same as the number of valid non-back-arcs. By maintaining the set of all valid non-back-arcs, finding all children can be accomplished sufficiently easily.

We now describe the framework of our algorithm.

**ALGORITHM:** ENUM_DIRECTED_SPANNING_TREES($G$)

**Step 1:** Find $T_0$ by a depth-first-search.

**Step 2:** Assign indices for each vertex.

**Step 3:** Classify arcs not in $T_0$ into the back-arc set and the non-back-arc set.
Sort them in order of their indices.

**Step 4:** Call ENUM_DIRECTED_SPANNING_TREES_ITER($T_0$).


**ALGORITHM:** ENUM_DIRECTED_SPANNING_TREES_ITER($T_p$)

**Step 1:** For each valid non-back-arc $f$ of $T_p$, do the following.

**Step 2:** Construct $T_c$ by adding $f$ and removing an arc $e$.
Output it by the difference from the previous outputted one.

**Step 3:** List all valid non-back-arcs of $T_c$ in order of their indices.

**Step 4:** Call ENUM_DIRECTED_SPANNING_TREES($T_c$) recursively.

Step 1 to 3 of the algorithm ENUM_DIRECTED_SPANNING_TREES($G$) runs in $O(|A|)$ time. The construction of a child of $T_p$ in Step 2 of ENUM_DIRECTED_SPANNING_TREES_ITER may be done in $O(1)$ by swapping two arcs. Step 3 lists all valid non-back-arcs of $T_c$, and the number of them is equal to the number of children of $T_c$ in the enumeration tree. Thus total time spent until Step 3 is the time to find one valid non-back-arc of $T_c$ per one outputted directed tree.

The earlier algorithm [2] takes $O(|V|)$ time for finding one non-back-arc in the worst case and it is the bottle neck of the time complexity while other parts of the algorithm take only $O(1)$ time. Our improvement on this part reduces its time complexity to the time necessary to update the minimum spanning tree of an undirected graph. In the next section, we show a data structure which is the key to the improvement.

# 3   An Improved Data Structure

To enumerate all valid non-back-arcs sufficiently fast in **Step 3**, we show the following two conditions. They are also stated in [2]. We denote the tail of $e$ by $v$ and the nearest common ancestor of $v$ and the head of $f$ by $w$.

**Lemma 1** *Let $T_c = T_p \setminus e \cup f$ be a child of $T_p$. A valid non-back-arc of $T_p$ is a valid non-back-arc of $T_c$ if its index is less than $v^*(T_c)$.*

*Proof:* Since $T_0$ is a depth-first-search tree, the index of the tail of $f$ is larger than $v^*(T_c)$. Thus for all valid vertices of $T_c$, all its descendants in $T_c$ are also descendants $T_p$. Therefore any valid vertex is not an ancestor of a vertex $v$ in $T_c$ if it is not an ancestor in $T_p$. ∎

**Lemma 2** *Let $P$ be the vertex set of the interior points of the path from the head of $e$ to $w$ on $T_p$, where interior points of the path are vertices of the path which are not its endpoints. A back-arc of $T_p$ is a non-back-arc of $T_c$ if and only if its head is a descendant of $v$ and its tail is in $P$.*

*Proof:* A descendant of $v$ is also a descendant of all vertices of $P$ but not in $T_c$. Thus if the head of a back-arc is a descendant of $v$ and its tail is in $P$, it is a non-back-arc of $T_c$. Conversely, only descendants of $v$ have ancestors which are not ancestors in $T_c$ and only vertices of $P$ have descendants which are not descendants in $T_c$. Thus any back-arc of $T_p$ has the head in descendants of $v$ and a tail in $P$ if it is a non-back-arc of $T_c$. ∎

From these lemmas, the valid non-back-arc set of $T_c$ is composed by those back-arcs of $T_p$ and valid non-back-arcs of $T_p$ whose indices are less than $v^*(T_c)$. By listing both of them in order of their indices, we can obtain the non-back-arc set sorted by their indices. The former can be listed easily if we have the set of valid non-back-arcs of $T_p$ sorted by

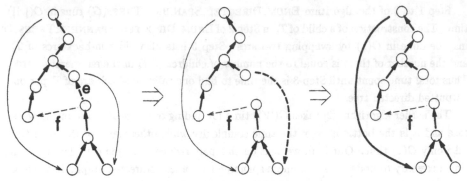

Fig2: The way of $B(T)$ changes.

their indices. But the latter is hard to list with only simple data structures. We use the following data structure shown as below.

For a directed spanning tree $T_p$, let $B(T_p)$ be an undirected graph with vertex set $V$ and edge set composed by arcs of $T_p$, valid back-arcs of $T_p$ and some of other rest back-arcs. $B(T_0)$ is an undirected graph composed by $T_0$ and all of its back-arcs. The weight of edges in the graph $B(T_p)$ is defined as 0 for arcs of $T_p$ and ($|V|+1-$ index ) for others. The minimum spanning tree of $B(T_p)$ is $T_p$.

By removing a valid arc $e$ of $T_p$ from the undirected graph $B(T_p)$, the minimum spanning tree of $B(T_p)$ is split into two. One of them contains all descendants of $v$ and the other contains the rest. The minimum spanning tree of the graph $B(T_p) \setminus e$ is obtained by adding the minimum weight cut edge $b$, which has endpoints in both of those two trees. From above lemmas, $b$ is a valid non-back-arc of $T_c$ if and only if the tail of $b$ is in $P$. In the case that $b$ does not exist, no back-arc of $T_p$ is a non-back-arc of $T_c$. Since $b$ is a back-arc of $T_p$, the tail of $b$ is in the path from $v$ to $r$ and its head is a descendant of $v$. From the definition of the weight, no back-arc connects a descendant of $v$ and an interior point of the path from the tail of $b$ to $v$. We show an example of the way of $B(T)$ changes in Figure 2. Emphasized lines are a part of the minimum spanning tree. Doted lines are eliminated and a new edge will be inserted.

Therefore we can find a new valid non-back-arc of $T_c$ by updating the minimum spanning tree. By removing $b$ and applying the method repeatedly, we can enumerate all new valid non-back-arcs of $T_c$. The algorithm is showed as follows. It outputs all new non-back-arcs and updates the graph from $B(T_p)$ to $B(T_c)$.

**ALGORITHM:** ENUM_NON-BACK-ARCS$(T_p, B(T_p), e, f)$
**Step 1:** Remove $e$ from $B(T_p)$.
**Step 2:** Update the minimum spanning tree by adding some edge $b$.
      If $b$ does not exist, then add $f$ to $B(T_p)$, and stop.

**Step 3:** If $b$ is not a back-arc of $T_c = T_p \setminus e \cup f$, then remove $b$ and add $f$ to $B(T_p)$. Stop.

**Step 4:** Output $b$. Remove $b$ from $B(T_p)$. Go to **Step 2**

**Lemma 3** *The algorithm* ENUM_NON-BACK-ARCS *outputs all back-arcs of $T_p$ which are valid non-back-arcs of $T_c$ in the order of their indices. It takes $O(D(|V|, |A|))$ time and $O(D(|V|, |A|))$ time for one output. Its space complexity is $O(|A| + DS(|V|, |A|))$.*

*Proof:* Since the algorithm finds the maximum index one among those back-arcs, back-arcs are outputted in the order of their indices. Therefore merging operation is done in linear time of the number of non-back-arcs. To identify both the endpoints of the path $P$, we have to find the nearest common ancestor of the head of $e$ and the head of $f$. It can be found in $O(\log |V|)$ time by D. D. Slater and R. E. Tarjan's dynamic tree data structure [6]. The updating operation, which is adding an edge, deleting an edge and changing the root, of it is also done in $O(\log |V|)$ time. The time complexity of the rest of the algorithm is $O(D(|V|, |A|))$ for one iteration. Because of the number of iterations, the time complexity is $O(D(|V|, |A|))$ per one output. ∎

By using this algorithm, we obtain the following theorem.

**Theorem 3.1** *Algorithm* ENUM_DIRECTED_SPANNING_TREES *enumerates all directed spanning trees in $O(|A| + ND(|V|, |A|))$ time. It uses $O(|A| + DS(|V|, |A|))$ space.*

*Proof:* By above lemmas, the time complexity of algorithm is $O(|A|)$ for preprocessing, $O(\log |V| + D(|V|, |A|))$ time for one directed spanning tree and $O(D(|V|, |A|))$ time for one new non-back-arc. Thus the time complexity satisfies the condition. The algorithm changes the data structure for updating the minimum spanning tree if a recursive call occurs. When the recursive call ends, we have to restore it. The restoring operation is done by adding the deleted edge, deleting the added edge and changing the modified weight. The total amount of these changes may increase by recursive calls, but the total size of each of the added edges, deleted edges and weight changed edges does not exceed $|A|$. Therefore they are stored in $O(|A|)$ space.

The set of valid non-back-arcs and back-arcs are treated similarly. After a recursive call, we can restore them by adding and removing the removed or added arcs. The total space for storing them is $O(|A|)$. Hence the space complexity is $O(|A| + DS(|V|, |A|))$. ∎

# Acknowledgment

We greatly thank to Associate Professor Akihisa Tamura of University of Electro-Communications for his kindly advise. We also owe a special debt of gratitude of Research Assistant Yoshiko T. Ikebe of Science University of Tokyo.

# References

[1] H. N. Gabow, E. W. Myers, "Finding All Spanning Trees of Directed and Undirected Graphs," SIAM J. Comp., 7, 280-287, 1978.

[2] H. N. Kapoor and H. Ramesh, "Algorithms for Generating All Spanning Trees of Undirected, Directed and Weighted Graphs," Lecture Notes in Computer Science, Springer-Verlag, 461-472, 1992.

[3] A. Shioura, A. Tamura and T. Uno, "An Optimal Algorithm for Scanning All Spanning Trees of Undirected Graphs, " SIAM J. Comp., to be appeared.

[4] D. Eppstein, Z. Galil, G. F. Italiano and A. Nissenzweig, "Sparsification - A Technique for Speeding up Dynamic Graph Algorithms, " FOCS 33, 60-69, 1992.

[5] G. N. Fredrickson, "Data Structure for On-line Updating of Minimum Spanning Trees, with Applications, " SIAM J. Comp., 14, No 4, 781-798, 1985.

[6] D. D. Sleator and R. E. Tarjan, "A Data Structure for Dynamic Trees," J. Comp. Sys. Sci. 26, 362-391, 1983.

[7] R. E. Tarjan, "Depth-First Search and Linear Graph Algorithm," SIAM J. Comp. 1, 146-169, 1972.

# Vertex Ranking of Asteroidal Triple-Free Graphs

Ton Kloks[1]*, Haiko Müller[2]** and C. K. Wong[3]***

[1] Department of Applied Mathematics
University of Twente, University for technical and social sciences
P.O.Box 217
7500 AE Enschede, The Netherlands
[2] Friedrich-Schiller-Universität Jena
Fakultät für Mathematik und Informatik
07740 Jena, Germany
[3] Department of Computer Science and Engineering
The Chinese University of Hong Kong
Shatin, Hong Kong

**Abstract.** We present an efficient algorithm for computing the vertex ranking number of an asteroidal triple-free graph. Its running time is bounded by a polynomial in the number of vertices and the number of minimal separators of the input graph.

## 1 Introduction

A vertex ranking of a graph is a vertex coloring by a linear ordered set of colors such that for every path in the graph with end vertices of the same color there is a vertex on this path with a higher color. The vertex ranking problem asks for a vertex ranking with a minimum number of colors.

The vertex ranking problem is mainly considered because of its great importance for the parallel Cholesky factorization of matrices [4, 10, 19]. Let $Ax = b$ be a large system of linear equations, where $A$ is a symmetric, positive definite matrix. Solving such a system via Cholesky factorization involves computing a triangular matrix $L$, such that $A = LL^\mathsf{T}$, and then solving the systems $Ly = b$ and $L^\mathsf{T}x = y$. If $A$ is sparse, then one often applies some preprocessing on the graph $G$ whose edges correspond with the nonzero entries of $A$. For example, one splits $A$ into smaller matrices (which can be handled in parallel) by removing some elements on the main diagonal of $A$ together with their rows and columns. Clearly these rows and columns correspond with a separator $S$ of $G$. The same procedure is applied recursively to the connected components of $G - S$. This leads to the problem of finding a minimum-height elimination tree of a graph,

---

* This research was done while this author was with the department of computer science and engineering of the Chinese university of Hong Kong, Shatin, Hong Kong, as a research fellow. Email: A.J.J.Kloks@math.utwente.nl

** Email: hm@minet.uni-jena.de

*** On leave from IBM T.J. Watson Research Center, P.O.Box 218, Yorktown Heights, NY 10598, U.S.A. Email: wongck@cs.cuhk.hk

which is equivalent to the vertex ranking problem [23, 8]. Yet other applications of the vertex ranking problem lie in the field of VLSI-layout [17, 22].

Asteroidal triples in graphs were introduced in [18], where the interval graphs are characterized as those chordal graphs without asteroidal triples. In the meantime asteroidal triples (or ATs, for short) turned out to be an important concept. Their absence in a graph forces several useful structures [6], however, a nice characterization of AT-free graphs as intersection graphs (for example such as for interval graphs) is not known.

The AT-free graphs do not form a class of perfect graphs. Nevertheless efficient algorithms are known for some domination problems restricted to this class [7]. The complements of comparability graphs are an important subclass of AT-free graphs, which shows that the class of AT-free graphs is a relatively large class of graphs. While in general the number of minimal separators of an AT-free graph cannot be bounded by any polynomial of the number of vertices, the complements of comparability graphs of partial orders of (interval) dimension bounded by $d$ have at most $n^d$ and $(2n)^d$ minimal separators, respectively, where $n$ denotes the number of vertices [15]. Especially, interval graphs on $n$ vertices have at most $n$ minimal separators, permutation graphs on $n$ vertices have at most $n^2$ minimal separators, and trapezoid graphs on $n$ vertices have at most $4n^2$ minimal separators.

The dimension is one of the most carefully studied parameters of a partial order [24]. Yannakakis [25] showed that determining whether a partial order has dimension at most $d$ is NP-complete for any fixed $d \geq 3$. Furthermore, while many problems have been shown to be efficiently solvable on partial orders of dimension 2, no NP-complete partial order problem was known to be solvable by a polynomial time algorithm for partially ordered sets of some fixed dimension greater than 2. This changed when a polynomial time algorithm was found for the treewidth of a cocomparability graph of fixed dimension [16], where the intersection model is not given in advance. In this paper we show that, for the vertex ranking problem, a similar result can be obtained.

If the intersection model of a cocomparability graph is part of the input, it was already shown in [8] that the vertex ranking problem can be solved. However, if only the graph is given as input, this does not yield a polynomial time algorithm, since we cannot find the representation efficiently. Thus, until now, it was unclear whether the problem was easy because the class of graphs is well behaved, or because having the representation (which is the solution to an NP-complete problem) is such a powerful tool that it gave the solution.

Much work has been done in finding optimal rankings of trees. There is now a linear time algorithm finding an optimal vertex ranking of any tree [20]. For the closely related edge ranking problem on trees an $O(n^3 \cdot \log n)$ algorithm has been presented in [23]. Recently an $O(n^2 \cdot \log \Delta)$ algorithm has been given in [26], where $\Delta$ is the maximum degree of the input tree. Efficient vertex ranking algorithms were known for very few other classes of graphs. The vertex ranking problem is trivial on split graphs and it is solvable in linear time on cographs [21]. Recently a $O(n^4)$ algorithm for vertex ranking of interval graphs was presented in [1]. The approach presented in [8] can be used to design a $O(n^3)$ algorithm computing an optimal vertex ranking for interval graphs. In [8] also an $O(n^6)$ algorithm was presented computing the vertex ranking of permutation graphs.

The decision problem 'Given a graph $G$ and a positive integer $k$, has $G$ a vertex ranking with at most $k$ colors' is NP-complete, even when restricted to cobipartite or bipartite graphs [3]. In view of this it is interesting to notice that for each *constant* $t$, the class of graphs with vertex ranking number at most $t$ is recognizable in linear time [3]. This follows from the fact that for each $t$, the class of graphs with vertex ranking number at most $t$ is minor closed and from the recent results of Robertson and Seymour.

In [12], among other things, an $O(\sqrt{n})$ bound is given for the vertex ranking number of a planar graph and the authors describe a polynomial time algorithm which finds a ranking using only $O(\sqrt{n})$ colors. For graphs in general there is an approximation algorithm known with factor $O(\log^2 n)$ [4, 13]. In [4] it is also shown that one plus the *pathwidth* of a graph is a lower bound for the vertex ranking number of the graph (hence a planar graph has pathwidth $O(\sqrt{n})$, (which was also shown in [13] using different methods).

For definitions and properties of classes of well-structured graphs not given here we refer to [5, 9, 11].

In this paper we show that the vertex ranking problem can be solved efficiently for AT-free graphs with a polynomial number of minimal separators.

## 2 Preliminaries

### 2.1 Preliminaries on AT-free graphs

An independent set of three vertices is called an *asteroidal triple* if between each pair in the triple there exists a path that avoids the neighborhood of the third. A graph is *AT-free* if it does not contain an asteroidal triple. Recently the structure of AT-free graphs has been studied extensively (see [6]). The class of AT-free graphs contains various well-known graph classes, as e.g., interval, permutation, trapezoid and cocomparability graphs. A good reference for more information on all these subclasses is [11].

### 2.2 Preliminaries on separators

All graphs in this paper are simple and undirected. $G[W]$ will be the subgraph of $G$ *induced* by the vertices of $W$.

Our main tool will be the set of minimal separators.

**Definition 1.** A vertex set $S$ is an $a, b$-*separator* if the removal of $S$ separates $a$ and $b$ in distinct connected components. If no proper subset of $S$ is an $a, b$-separator then $S$ is a *minimal $a, b$-separator*. $S$ is a *minimal separator* if there exist non adjacent vertices $a$ and $b$ such that $S$ is a minimal $a, b$-separator. An *inclusion minimal separator* is a separator such that no other separator is properly contained in it.

**Definition 2.** Let $G = (V, E)$ be a graph and let $S$ be a separator of $G$. Then $C \subset V$ is *S-full* in $G$ if every vertex in $S$ has a neighbor in $C$. We say that $G[C]$ is a *full component* of $G[V \setminus S]$ if $C$ is $S$-full in $G$ and $G[C]$ is a connected component of $G[V \setminus S]$.

Obviously a vertex set $S$ is a minimal $a, b$-separator of $G$ iff $a$ and $b$ are vertices in different full components of $G[V \setminus S]$. Furthermore, $S$ is an inclusion minimal separator of $G$ iff $G[V \setminus S]$ is disconnected and all connected components of $G[V \setminus S]$ are $S$-full in $G$.

**Lemma 3.** *Let $G = (V, E)$ be a graph with separator $S_1$. Let $a$ and $b$ be non-adjacent vertices in a connected component $G[C]$ of $G[V \setminus S]$. Let $S \subset C$ be a minimal $a, b$-separator in $G[C]$. Then there exists a minimal $a, b$-separator $S_2$ of $G$ with the following properties:*

1. *$S \subseteq S_2 \subseteq S \cup S_1$*
2. *The connected component $G[A]$ of $G[C \setminus S]$ containing vertex $a$ is a full component of $G[V \setminus S_2]$.*

*Proof.* Let $G[A]$ be the connected component of $G[C \setminus S]$ and let $G[B]$ be the connected component of $G[C \setminus S]$ containing $b$. Clearly, $S^* := S_1 \cup S$ separates $a$ and $b$ in $G$ and the connected components containing $a$ and $b$ are $G[A]$ and $G[B]$ respectively.

Remove all vertices from $S^*$ which do not have a neighbor in $A$. Let the resulting set be $S_2$. Notice that no vertex of $S$ is removed in the process, hence $S \subseteq S_2 \subseteq S \cup S_1$. Since only vertices of $S^*$ are removed which do not have a neighbor in $G[A]$, $G[A]$ is a full component of $G[V \setminus S_2]$. (Notice that when we remove vertices of $S^*$, the component of $G[V \setminus S^*]$ containing the vertex $b$ may grow, but not the component containing $a$.) □

In [14] the following result is shown.

**Theorem 4.** *For a positive number $R$, let $\mathcal{G}(R)$ be the class of graphs with at most $R$ minimal separators. There exists an algorithm running in time $O(n^5 R)$, which, for all $R$, given a graph $G$ with $n$ vertices, either detects that $G \notin \mathcal{G}(R)$ or lists all minimal separators in $G$.*

In general the number of minimal separators of an AT-free graph cannot be bounded by any polynomial of the number of vertices. To prove this we consider the complement of the cocktail party graph $(V, E \setminus M)$ where $(V, E)$ is the complete bipartite graph $K_{n,n}$, and $M \subseteq E$ is a perfect matching of $(V, E)$. This complement is AT-free and it has exactly $2^n - 2$ (inclusion) minimal separators.

## 2.3 Preliminaries on rankings

**Definition 5.** Let $G = (V, E)$ be a graph and let $t$ be some integer. A *(vertex) t-ranking* is a coloring $c : V \to \{1, \ldots, t\}$ such that for every pair of vertices $x$

and $y$ with $c(x) = c(y)$ and for every path between $x$ and $y$ there is a vertex $z$ on this path with $c(z) > c(x)$. The *vertex ranking number* of $G$, $\chi_r(G)$, is the smallest value $t$ for which the graph admits a $t$-ranking.

By definition a vertex ranking is a proper coloring. Hence $\chi_r(G) \geq \chi(G)$ for every graph $G$. We call a $\chi_r(G)$-ranking of $G$ an optimal ranking. Clearly, $\chi_r(K_n) = n$, where $K_n$ is a complete graph on $n$ vertices. Furthermore, the vertex ranking number of a disconnected graph is equal to the maximum vertex ranking number of its components.

**Lemma 6.** *Let $G = (V, E)$ be connected, and let $c$ be a $t$-ranking of $G$. Then there is at most one vertex $x$ with $c(x) = t$.*

*Proof.* Assume there are two vertices with color $t$. Since $G$ is connected, there is a path between these two vertices. By definition this path must contain a vertex with color at least $t + 1$. This is a contradiction. □

*Remark.* Notice that if $c$ is a $t$-ranking of a graph $G$ and $H$ is a subgraph of $G$, then the restriction $c'$ of $c$ to the vertices of $H$ is a $t$-ranking for $H$.

This observation together with Lemma 6 leads to the following lemma which appeared in [12].

**Lemma 7.** *A coloring $c : V \to \{1, \ldots, t\}$ is a $t$-ranking for a graph $G = (V, E)$ if and only if for each $1 \leq i \leq t$, each connected component of the subgraph $G[\{x \mid c(x) \leq i\}]$ of $G$ has at most one vertex $y$ with $c(y) = i$.*

The following theorem, presented in [8], is our main tool for designing efficient vertex ranking algorithms on special classes of graphs.

**Theorem 8.** *Let $G = (V, E)$ be a a graph and $S$ a nonempty collection of subsets of $V$ containing all inclusion minimal separators of $G$. Then*

$$\chi_r(G) = \min_{S \in \mathcal{S}}(|S| + \max_C \chi_r(G[C]))$$

*where $C$ ranges over the vertex sets of all connected components of $G[V \setminus S]$.*

## 3 Preliminaries on blocks

We introduce 1-blocks and 2-blocks, which are a basic concept for our algorithm.

**Definition 9.** *A 1-block is a pair $(S, C)$, where $S$ is a minimal separator of $G$ and $G[C]$ is a $S$-full component of $G[V \setminus S]$.*

We want to determine the vertex ranking number $\chi_r(G[C])$ for all 1-blocks $(S, C)$, by decomposing it into smaller 1-blocks and 2-blocks. The vertex ranking for the graph $G$ then follows from Theorem 8. We introduce 2-blocks.

**Definition 10.** A triple $(S_1, C, S_3)$ is called *2-block* of a graph $G = (V, E)$ if $S_1$ and $S_3$ are minimal separators of $G$ with $S_1 \not\subseteq S_3$ and $S_3 \not\subseteq S_1$, and $C \subset V$ induces a connected component $G[C]$ of $G[V \setminus (S_1 \cup S_3)]$ such that for $i = 1, 3$ there exist $S_i$-full components $G[C_i]$ of $G[V \setminus S_i]$ with $C \subseteq C_i$, $S_1 \setminus S_3 \subseteq C_3$ and $S_3 \setminus S_1 \subseteq C_1$.

## 4 Decomposing 1-blocks

Consider a 1-block $(S, C)$. If $C$ is a clique then $\chi_r(G[C]) = |C|$. Henceforth, assume this is not the case.

The following theorem gives the decomposition of 1-blocks.

**Theorem 11.** *Let $(S_1, C)$ be a 1-block of an AT-free graph $G = (V, E)$, let $S$ be an inclusion minimal separator of $G[C]$, and let $G[A]$ be a connected component of $G[C \setminus S]$. Then there is a minimal separator $S_2$ of $G$ such that $(S_2, A)$ is a 1-block of $G$ or $(S_1, A, S_2)$ is a 2-block of $G$.*

*Proof.* We choose a vertex $a \in A$ and $b \in C \setminus (A \cup S)$. Then $S_1 \cup S$ is an $a, b$-separator of $G$. Let $S_2$ be a minimal $a, b$-separator with $S_2 \subseteq S_1 \cup S$. Then we have $S \subseteq S_2$ and $A$ is contained in one full component of $G[V \setminus S_2]$.

If $S_2 = N(A) \setminus A$ then $(S_2, A)$ is a 1-block of $G$.

Assume $N(A) \setminus (A \cup S_2) \neq \emptyset$. We show that $(S_1, A, S_2)$ is a 2-block. Clearly, $S_2 \not\subseteq S_1$, since $S \not\subseteq S_1$. But also, $S_1 \not\subseteq S_2$, since there is a vertex of $A$ with a neighbor in $S_1 \setminus S_2$.

Clearly, $S_2 \setminus S_1 = S$ is contained in the full component $G[C]$ of $G[V \setminus S_1]$.

Consider a full component $G[C^*]$ other than $G[C]$ of $G[V \setminus S_1]$. Hence every vertex of $S_1 \setminus S_2$ has a neighbor in $C^*$. It follows that there is connected component $G[D]$ of $G[V \setminus S_2]$ containing all vertices of $S_1 \setminus S_2$. Since there is a vertex of $A$ with a neighbor in $S_1 \setminus S_2$, this component $D$ contains all vertices of $A$. It follows that $G[D]$ is a full component of $G[V \setminus S_2]$, since every vertex of $S_2$ has a neighbor in $A$. This proves that $(S_1, A, S_2)$ is a 2-block. $\qquad\square$

## 5 Decomposing 2-blocks

We need the following lemma.

**Lemma 12.** *Let $(S_1, C, S_3)$ be a 2-block of an AT-free graph $G = (V, E)$ and let $S$ be an inclusion minimal separator of $G[C]$. Then for all minimal separators $S_2$ of $G$ with $S \subseteq S_2 \subseteq S \cup S_1 \cup S_3$ holds $S_1 \subseteq S_2 \cup S_3$ or $S_3 \subseteq S_2 \cup S_1$ or the vertices of $S_1 \setminus S_2$ and $S_3 \setminus S_2$ are in different connected components of $G[V \setminus S_2]$.*

*Proof.* Suppose $S_1 \not\subseteq (S_2 \cup S_3)$ and $S_3 \not\subseteq (S_2 \cup S_1)$. Choose vertices $s_1$ and $s_3$ in $S_1 \setminus (S_2 \cup S_3)$ and in $S_3 \setminus (S_2 \cup S_1)$ respectively. Clearly, there also exists a vertex $s_2 \in S_2 \setminus (S_1 \cup S_3)$.

Assume $S_1 \setminus S_2$ and $S_3 \setminus S_2$ are contained in one connected component of $G[V \setminus S_2]$. Then there is another full component of $G[V \setminus S_2]$ without vertices of

$S_1 \cup S_3$. Choose a vertex $v_2$ in this component. Choose a vertex vertices $v_1$ in a full component of $G[V \setminus S_1]$ that contains no vertices of $S_2 \cup S_3$ and a vertex $v_3$ in a full component of $G[V \setminus S_3]$ that contains no vertex of $S_1 \cup S_2$. Notice that there is a path from $v_1$ to $v_3$, via $s_1$ and $s_3$, avoiding the neighborhood of $v_2$. Similar paths exist from $v_1$ to $v_2$ and from $v_3$ to $v_2$. It follows that $v_1$, $v_2$ and $v_3$ form an AT. $\qquad\square$

This section shows how 2-blocks are decomposed. et $(S_1, C, S_3)$ be a 2-block of an AT-free graph $G$. Clearly, if $C$ is a clique, then the vertex ranking number of $G[C]$ is $|C|$. Assume henceforth this is not the case.

**Theorem 13.** *Let $(S_1, C, S_3)$ be a 2-block of an AT-free graph $G = (V, E)$, let $S$ be an inclusion minimal separator of $G[C]$, and let $G[A]$ be a connected component of $G[C \setminus S]$. Then there is a minimal separator $S_2$ of $G$ such that $(S_2, A)$ is a 1-block of $G$ or one of $(S_1, A, S_2)$ and $(S_3, A, S_2)$ is a 2-block of $G$.*

*Proof.* We choose a vertex $a \in A$ and $b \in C \setminus (S \cup A)$. Then $S_1 \cup S \cup S_3$ is an $a, b$-separator of $G$. Let $S_2$ be a minimal $a, b$-separator with $S_2 \subseteq S_1 \cup S \cup S_3$. Then we have $S \subseteq S_2$ and $A$ is in one full component of $G[V \setminus S_2]$.

First assume that there exist vertices $s_1$ and $s_3$ in $S_1 \cap N(A) \setminus (S_2 \cup S_3)$ and in in $S_3 \cap N(A) \setminus (S_2 \cup S_1)$ respectively. Then by lemma 12 $s_1$ and $s_3$ would belong to different connected components of $G[V \setminus S_2]$ contradicting the fact that $s_1$ and $s_3$ are both in $N(A)$.

If $S_2 = N(A) \setminus A$ then $(S_2, A)$ is a 1-block of $G$.

Finally assume that $S_2 \subset N(A) \setminus A$. Without loss of generality we can now assume that that $N(A) \setminus (A \cup S_2) \subseteq S_1$. Then $(S_1, A, S_2)$ is a 2-block of $G$ which follows in the same manner as in the proof of Theorem 11. $\qquad\square$

## 6 The algorithm

Assume our input is an AT-free graph $G = (V, E)$ on $n$ vertices with $R$ minimal separators. Using the algorithm of [14] we first compute the set $\Delta$ of all minimal separators of $G$. This takes time $O(n^5 R)$. Next we compute a list $\mathcal{B}$ of all vertex sets $C \subseteq V$ such that $C = V$ or there is a separator $S \in \Delta$ such that $(S, C)$ is a 1-block of $G$ or there exist separators $S_1, S_2 \in \Delta$ such that $(S_1, C, S_2)$ is a 2-block of $G$. Creating list $\mathcal{B}$ is possible in time $O(n^2 R^2)$ since we have to run $R(R + 1)/2$ times a subroutine computing connected components. We sort the elements of $\mathcal{B}$ by the number of vertices in time $O(n^2 R^2)$. Now the set $\mathcal{B}$ contains at most $nR^2$ different vertex sets $C \subseteq V$. For each $C \in \mathcal{B}$ we compute the vertex ranking number $\chi_r(G[C])$ in the following way.

If $C$ is a clique of $G$ then $\chi_r(G[C]) = |C|$. Otherwise $C$ is representable as block of $G$ and decomposes into smaller blocks with ranking numbers computed before. By theorem 8 we have

$$\chi_r(G[C]) = \min_{S \in \Delta}(|S \cap C| + \max_{C'} \chi_r(G[C']))$$

where $G[C']$ ranges over all connected components of $G[C \setminus S]$. The last ranking number computed this way is $\chi_r(G)$ for $C = V$. By lemma 3 the set $\{S \cap C : S \in \Delta\}$ is non-empty and contains all minimal separators of $G[C]$. By Theorems 11 and 13 for the inclusion minimal separators of $G[C]$ the ranking numbers of the connected components $G[C']$ are computed before. For components $G[C']$ with $C' \notin B$ we assume ranking number $|C'|$. These components cannot realize the minimum. Per component $G[C]$ we need time $O(n^2 R)$ for this step. Hence the total running time of our algorithm is $O(n^5 R + n^3 R^3)$.

# 7   Conclusion

We have given a efficient algorithm computing the ranking number of AT-free graphs. This generalizes in a non trivial way the result of [8], where an intersection model is explicitly used. Notice that in this paper we do not assume any intersection model. As a matter of fact, no intersection model for AT-free graphs is known.

To us it is unknown whether similar results as obtained in this paper hold for graphs in general. To be more precise, we do not know if there is an algorithm computing the vertex ranking number for graphs in general, with a running time polynomial in the number of vertices and the number of minimal separators.

# References

1. Aspvall, B. and P. Heggernes, Finding minimum height elimination trees for interval graphs in polynomial time, BIT vol. 34, 1994, pp. 484–509.
2. Bodlaender, H., T. Kloks and D. Kratsch, Treewidth and pathwidth of permutation graphs, *Proceedings of the 20th International Colloquium on Automata, Languages and Programming*, Springer-Verlag, Lecture Notes in Computer Science vol. 700, 1993, pp. 114–125.
3. Bodlaender, H., J. S. Deogun, K. Jansen, T. Kloks, D. Kratsch, H. Müller and Z. Tuza, Rankings of graphs, *Proceedings of the 20th International Workshop on Graph-Theoretic Concepts in Computer Science WG'94*, Springer-Verlag, Lecture Notes in Computer Science vol. 903, 1995, pp. 292–304.
4. Bodlaender, H. L., J. R. Gilbert, H. Hafsteinsson and T. Kloks, Approximating treewidth, pathwidth and minimum elimination tree height, *Proc. 17th International Workshop on Graph-Theoretic Concepts in Computer Science WG'91*, Springer-Verlag, Lecture Notes in Computer Science vol. 570, 1992, pp. 1–12.
5. Brandstädt, A., Special graph classes — a survey, Schriftenreihe des Fachbereichs Mathematik, SM-DU-199, Universität Duisburg Gesamthochschule, 1991.
6. Corneil, D. G., S. Olariu, and L. Stewart, Asteroidal triple-free graphs, *Proceedings of the 19th International Workshop on Graph-Theoretic Concepts in Computer Science WG'93*, Springer-Verlag, Lecture Notes in Computer Science vol. 790, 1994, pp. 211–224.
7. Corneil, D. G., S. Olariu and L. Stewart, Linear time algorithms to compute dominating pairs in asteroidal triple-free graphs, *Proceedings of the 22nd International Colloquium on Automata, Languages and Programming*, Springer-Verlag, Lecture Notes in Computer Science vol. 944, 1995, pp. 292–302.

8. Deogun, J. S., T. Kloks, D. Kratsch and H. Müller, On vertex ranking for permutation and other graphs, *11th Annual Symposium on Theoretical Aspects of Computer Science*, Springer-Verlag, Lecture Notes in Computer Science vol. 775, 1994, pp. 747–758.

9. Duchet, P., Classical perfect graphs, in: *Topics on Perfect Graphs*, C. Berge and V. Chvátal, (eds.), Annals of Discrete Mathematics **21**, 1984, pp. 67–96.

10. Duff, I. S. and J. K. Reid, The multifrontal solution of indefinite sparse symmetric linear equations. *ACM Transactions on Mathematical Software* **9** (1983), 302–325.

11. Golumbic, M. C., *Algorithmic Graph Theory and Perfect Graphs*, Academic Press, New York, 1980.

12. Katchalski, M., W. McCuaig and S. Seager, Ordered colourings, *Discrete Mathematics* **142**, (1995), pp. 141–154.

13. Kloks, T., *Treewidth - Computations and Approximations*, Springer-Verlag, Lecture Notes in Computer Science vol. 842, 1994.

14. Kloks, T. and D. Kratsch, Finding all minimal separators of a graph, *Proceedings of the 11th Annual Symposium on Theoretical Aspects of Computer Science*, Springer-Verlag, Lecture Notes in Computer Science vol. 775, 1994, pp. 759–768.

15. Kloks, T., D. Kratsch and H. Müller, Measuring the vulnerability for classes of intersection graphs, to appear in *Discrete Applied Mathematics*.

16. Kloks, T., D. Kratsch and J. Spinrad, Treewidth and pathwidth of cocomparability graphs of bounded dimension, *Computing Science Notes* 93/46, Eindhoven University of Technology, Eindhoven, The Netherlands (1994).

17. Leiserson, C. E., Area efficient graph layouts for VLSI, *Proceedings of the 21st Annual IEEE Symposium on Foundations of Computer Science*, 1980, pp. 270–281.

18. Lekkerkerker, C. G., and J. Ch. Boland Representation of a finite graph by a set of intervals on the real line, *Fundamenta Mathematicae* **51**, (1962), pp. 45–64.

19. Liu, J. W. H., The role of elimination trees in sparse factorization. *SIAM Journal of Matrix Analysis and Applications* **11** (1990), 134–172.

20. Schäffer, A. A., Optimal node ranking of trees in linear time, *Information Processing Letters* **33**, (1989/1990), pp. 91–96.

21. Scheffler, P., Node ranking and Searching on Graphs (Abstract), in: U. Faigle and C. Hoede, (eds.), *3rd Twente Workshop on Graphs and Combinatorial Optimization*, Memorandum No.1132, Faculty of Applied Mathematics, University of Twente, The Netherlands, 1993.

22. Sen, A., H. Deng and S. Guha, On a graph partition problem with application to VLSI layout. *Information Processing Letters* **43** (1992), 87–94.

23. de la Torre, P., R. Greenlaw and A. A. Schäffer, Optimal ranking of trees in polynomial time, *Proceedings of the 4th Annual ACM-SIAM Symposium on Discrete Algorithms*, Austin, Texas, 1993, pp. 138–144.

24. Trotter, W. T., *Combinatorics and Partially Ordered Sets: Dimension Theory*, The John Hopkins University Press, Baltimore, Maryland, 1992.

25. Yannakakis, M., The complexity of the partial order dimension problem, *SIAM J. Alg. Discrete Methods* **3**, (1982), pp. 351–358.

26. Zhou, X., M. A. Kashem and T. Nishizeki, Generalized edge-rankings of trees, to appear in *Proceedings of the 22th International Workshop on Graph-Theoretic Concepts in Computer Science WG'96*, Springer-Verlag, Lecture Notes in Computer Science.

# Recursively Divisible Problems

Rolf Niedermeier

Wilhelm-Schickard-Institut für Informatik, Universität Tübingen, Sand 13,
D-72076 Tübingen, Fed. Rep. of Germany, niedermr@informatik.uni-tuebingen.de

**Abstract.** We introduce the concept of a $(p, d)$-*divisible* problem, where $p$ reflects the number of processors and $d$ the number of communication phases of a parallel algorithm solving the problem. We call problems that are *recursively* $(n^\epsilon, O(1))$-divisible in a work-optimal way with $0 < \epsilon < 1$ *ideally divisible* and give motivation drawn from parallel computing for the relevance of that concept. We show that several important problems are ideally divisible. For example, sorting is recursively $(n^{1/3}, 4)$-divisible in a work-optimal way. On the other hand, we also provide some results of lower bound type. For example, ideally divisible problems appear to be a proper subclass of the functional complexity class $FP$ of sequentially feasible problems. Finally, we also give some extensions and variations of the concept of $(p, d)$-divisibility.

## 1 Introduction

There have been several approaches to more realistic models of parallel computation in recent time, which can basically be separated into two groups. First, it is tried to make the classical PRAM (Parallel Random Access Machine) model and the corresponding classes more realistic by restricting the model of computation or by defining "more precise" complexity classes [1,5,9,10,13]. Second, there is the approach to completely abandon the world of PRAM's and to work with a great number of parameters characterizing the performance of a parallel machine [3], thus giving up conceptual simplicity. This paper is closer to the first line of research mentioned above in the sense that we stick to conceptual simplicity without burdening our model by a number of system parameters. We present an abstract framework for parallel algorithms that respects several demands drawn from practice on the one hand, but still admits a model simple enough in order to perform (structural) complexity theory on the other hand. So we adopt, in a sense, a very restricted view of parallelism and want to analyze how far one can get when one only allows a small (ideally constant) number of communication phases that interrupt long phases of local, sequential computations.

There are several issues that can make a parallel algorithm valuable or valueless from the viewpoint of existing and foreseeable parallel machines. Two main aspects are speedup and efficiency (in a broad sense), leading to demands like work-optimality, good use of locality, consideration of latency and contention effects, easy synchronization and structured communication, small number of blocked communications, modularization, reasonable processor numbers, scalability and adaptability to increasing numbers of processors, and so on. Those demands encourage the study of problems and classes dealing with the assumptions of significantly smaller numbers of processors than the input size is, a small number of local (i.e., sequential) computation phases interrupted by global communication phases and several other features, culminating in the concept of recursively $(p, d)$-divisible problems presented in Section 2. Note that the idea of communicating in phases also is important in the bulk-synchronous parallel model (BSP) [13]. The major point in the BSP model is that programs are written for $v$ virtual parallel processors to run on $p$ physical processors, where $v$ is rather larger than $p$ (e.g, $v = p \log p$). This "parallel slackness" is exploited by the compiler to schedule and pipeline computation and communication. It requires, however, quite a large administrative overhead. In our setting, by way of contrast, we design from the beginning $p$ programs for $p$ physical processors, thus avoiding the administrative overhead that arises with the "slackness principle" in the BSP. Moreover, we are mainly interested in algorithms with constantly bounded numbers of communication phases. Informally speaking, a problem is called $(p, d)$-divisible if it can be solved using $p$ processors, each having some part of size $O(n/p)$ of the input, where only $d$ phases of local computation and communication are sufficient. Recursive divisibility then means that the arising subproblems of size $O(n/p)$ again are $(p, d)$-divisible and so on. We call recursively $(n^\varepsilon, O(1))$-divisible problems ideally divisible if they can be solved in a work-optimal way and show that well-known problems like prefix sums and sorting are ideally divisible. On the other hand, we reveal the inherent limitations of the concept of ideal divisibility and, more generally, recursive $(p, d)$-divisibility. For example, unless $P = NC$ ideally divisible problems are a proper subclass of $FP$ and can be solved by $NC$-circuits of quasilinear size. Later we also provide extensions of our concept.

We assume familiarity with the basic concepts of parallel and sequential complexity theory. By $FP$ we denote the functional version of $P$. All circuits shall follow standard uniformity conditions, namely $ATIME(\log n)$-uniformity [12].

## 2 The Basic Concepts

In the following definition, the intuition behind parameter $w$ simply is that it shall represent the word size necessary to represent one of the $n$ input elements in binary code. Usually, $w = O(\log n)$. Note that in this paper the terms "problem" and "function" are used interchangeably. We focus attention on sequentially feasible problems, that is, problems in $FP$.

**Definition 1.** Let $f \in FP$ with $f : \{0, 1\}^{n \cdot w} \longrightarrow \{0, 1\}^{c \cdot n \cdot w}$ for some positive constant $c$. Function $f$ is $(p, d)$-divisible with natural numbers $n$, $w$, $p$, and $d$ and $p < n$ if it can be written as follows. Let $x \in \{0, 1\}^{n \cdot w}$. Then $f(x) = div(x, p, d)$, where for $d = 1$

$$div(x, p, 1) := s_1^1(z_1^1(x)) \circ s_2^1(z_2^1(x)) \circ \ldots \circ s_p^1(z_p^1(x))$$

and for $d > 1$ and $y := div(x, p, d - 1)$

$$div(x, p, d) := s_1^d(z_1^d(y)) \circ s_2^d(z_2^d(y)) \circ \ldots \circ s_p^d(z_p^d(y)).$$

Herein, "o" simply denotes concatenation of strings in $\{0,1\}^*$ and $s_i^k$ and $z_i^k$ with $1 \leq i \leq p$ and $1 \leq k \leq d$ are functions $z_i^k : \{0,1\}^{c_{1,i}^k \cdot n \cdot w} \longrightarrow \{0,1\}^{c_{2,i}^k \cdot n \cdot w/p}$ and $s_i^k : \{0,1\}^{c_{2,i}^k \cdot n \cdot w/p} \longrightarrow \{0,1\}^{c_{3,i}^k \cdot n \cdot w/p}$ where again $c_{1,i}^k$, $c_{2,i}^k$, and $c_{3,i}^k$ are some positive constants. In particular, we demand that each $s_i^k \in FP$ and that for each $k$, *all* $z_i^k$ with $1 \leq i \leq p$ can be simultaneously computed by an $ATIME(\log n)$-uniform $NC^0$-circuit of size $O(n \cdot w)$.

Definition 1 presented the notion of $(p,d)$-divisible problem in a highly abstract, but mathematical precise way. The task of functions $z_i^k$ is to do some kind of decomposition and functions $s_i^k$ represent local, sequential computations. Informally speaking, we call a functional problem concerning $n$ elements whose binary representation requires $w$ bits each $(p,d)$-divisible, if it can be solved by the following algorithmic scheme. Assume that the given input is partitioned in some way into $p$ sets of $O(n/p)$ input elements each. Repeat $d$ times:

1. Apply to all inputs with $O(n/p)$ elements of word size $w$ each a sequential algorithm that produces $O(n/p)$ output elements of word size $w$ each.
2. By means of an $NC^0$-circuit of size $O(n \cdot w)$ redistribute the output data from step 1, creating $O(p)$ new instances of problems (not necessarily the same type as the original one) of $O(n/p)$ elements each or the final output.

Roughly speaking, step one belongs to functions $s_i^k$ and step two belongs to functions $z_i^k$. For the sake of notational convenience, in the rest of the paper we, e.g., write $f : \{0,1\}^{n \cdot w} \longrightarrow \{0,1\}^{O(n \cdot w)}$, always standing for $f : \{0,1\}^{n \cdot w} \longrightarrow \{0,1\}^{c \cdot n \cdot w}$ for some arbitrary, but fixed positive constant $c$.

**Definition 2.** 1. We say that a problem is $(p,d)$-divisible *in a work-optimal way* if the time-processor-product of the parallel algorithm derived from the scheme given in Definition 1 is the same as the running time of the provably best sequential algorithm for that problem up to constant factors.

2. A $(p,d)$-divisible problem is *recursively $(p,d)$-divisible* if the size $O(n/p)$ instances produced above again are (recursively) $(p,d)$-divisible and so on, until we end up with instances of size $O(w)$.

3. Problems that are recursively $(p,d)$-divisible in a work-optimal way with $p = n^\epsilon$ for some $0 < \epsilon < 1$ and with $d = O(1)$ are *ideally divisible*.

It is important here to note that ideally divisible problems in general do *not* lead to algorithms that are work-optimal if the divisibility property is recursively applied a *non-constant* number of times. If a problem is (recursively) $(p,d)$-divisible in a work-optimal way, however, then there is a parallel algorithm using $d$ communication and $d$ local computation phases running in time $O(d \cdot s(n/p))$, where $s(n/p)$ denotes the maximum time to solve a size $O(n/p)$ problem instance sequentially.

Very recently, de la Torre and Kruskal [6] proposed a structural theory of recursively decomposable parallel processor networks. They study things like submachine locality basically on the network level. The seemingly tight connections to recursive divisibility, so to speak defined on the problem level, still remain to be investigated in detail. Heywood and Ranka [9] presented so-called Hierarchical PRAM's (HPRAM's) as "a framework in which to study the extent that general locality can be exploited in parallel computing." In a sense, Heywood and Ranka also study the issue of independent phases of local computations, but from a more technical, programmer's point of view than that of parallel complexity theory as we adopt.

**Proposition 3.** *The computation of the maximum of $n$ elements is recursively $(n^{i/(i+1)}, i+1)$-divisible in a work-optimal way for natural constant $i \geq 1$.*

*Proof.* (Sketch) The proof is an easy exercise. Start with $i = 1$. □

Proposition 3 exhibits the phenomenon that by using a larger number of processors, on the one hand, the running time gets reduced, but, on the other hand, the proportion of communication in the algorithm's overall complexity increases. This observation is generally neglected in the design of theoretically efficient algorithms and that is why they often might be of little use from the viewpoint of more realistic parallel computing. We end this section with some further justification for the relevance of the concepts we introduced.

- The requirement $d = O(1)$ means that we have only a constant number of communication (and, therefore, synchronization) phases.
- Demanding $p = n^\epsilon$ reflects massive parallelism (polynomial number of processors) as well as the fact that in practice the input size in general exceeds the number of processors by large [3].
- That $(p, d)$-divisibility has to hold recursively mirrors the scalability resp. adaptability of the algorithm.
- $NC^0$-circuits of linear size as laid down in Definition 1 represent the quest for efficient, data-independent, simple, and structured communications [8].
- Subproblems of size $O(n/p)$ guarantee efficient solutions having linearly bounded amounts of data to be transferred during each communication phase. In addition, this offers the opportunity for blocking of communications and serves to balance the distribution of data among the processors.

## 3 Some Ideally Divisible Problems

There are several important problems that are ideally divisible, perhaps the most prominent among them the sorting problem. The following theorem states that prefix sums is ideally divisible, possessing a parallel algorithm that runs in time $O((i + 2) n^{1/(i+1)})$ using $n^{i/(i+1)}$ processors and $(i + 2)$ communication phases for all $i \geq 1$. We omit the fairly straightforward proof.

**Theorem 4.** *For natural constant $i \geq 1$, prefix sums of $n$ elements is recursively $(n^{i/(i+1)}, i+2)$-divisible in a work-optimal way.*

**Corollary 5.** *For natural constant $i \geq 1$, the recognition of regular languages and constant width branching programs both are recursively $(n^{i/(i+1)}, i+2)$-divisible in a work-optimal way.*

*Proof.* (Hint) We reduce both problems to prefix sums computations. □

Barrington [2] showed that width 5 polynomial size branching programs recognize exactly those languages in $NC^1$. So one could be tempted to assume that Corollary 5 also might lead to recursive $(n^{i/(i+1)}, i+2)$-divisibility for all problems in $NC^1$ with linear size circuits. Here the difficulty arises that the branching program simulation of depth $d$ circuits needs branching program size $4^d$ [2]. So for $d = \log n$ we obtain quadratic size branching programs. Thus even assuming linear size $NC^1$-circuits, we end up with a polynomial blow-up in the size of the resulting branching programs. Thus it remains open whether all problems in (linear size) $NC^1$ are ideally divisible, because to apply Corollary 5, we need linear size branching programs.

**Theorem 6.** *The sorting of $n$ elements is recursively $(n^{1/3}, 4)$-divisible in a work-optimal way.*

*Proof.* (Sketch) Leighton's Columnsort [11] adapted to our framework works as follows. For the time being assume that we have $n^{1/3}$ processors, each holding $n^{2/3}$ data items of the in total $n$ items to be sorted. Now the algorithm roughly is as follows.

1. Each processor locally sorts its $n^{2/3}$ items.
2. Perform an all-to-all mapping where each processor gets $n^{1/3}$ items from each other one.
3. Again each processor locally sorts its $n^{2/3}$ (new) items.
4. Again perform an all-to-all mapping.
5. Locally sort the data items of each pair of neighboring processors.

Note that it suffices to know that an all-to-all mapping exchanges data between all processors in a very regular, fixed way. The correctness of the presented sorting algorithm in essence follows from [11]. We had to choose $p \leq n^{1/3}$ in order to be able to restrict ourselves to pairs of neighboring processors in the fifth, final step. From the above algorithm we get the $(n^{1/3}, 4)$-divisibility of sorting in the following way. The all-to-all mappings of steps 2 and 4 can clearly be done by some fixed data transports realized by linear size $NC^0$-circuits. The local sorts of steps 1 and 3 lead to the application of sequential sorting algorithms working on inputs of size $n^{2/3}$. Finally, the sorting of the items of neighboring processor pairs leads to two subsequent local phases of sequential sorting with inputs of size $2n^{2/3}$ each time. Altogether, four local sorting phases with input sizes $O(n^{2/3})$ suffice to get the whole size $n$ input sorted. So the sorting problem is solved reducing it to smaller sorting problems. Altogether this implies that sorting is recursively $(n^{1/3}, 4)$-divisible. The described algorithm is work-optimal. □

It is an interesting question whether parameter $n^{1/3}$ in Theorem 6 can be improved to some $p = n^\epsilon$ with $\epsilon > 1/3$. Cypher and Sanz [4] developed the so-called *Cubesort* algorithm that, for $p = n^\epsilon$, requires fewer than $5(1/(1 - \epsilon))^2$

rounds of local computation to sort. In particular, for $\epsilon = 1/3$ Cubesort needs 7 rounds, whereas Columnsort only needs 4 rounds as shown above. But for larger, constant $\epsilon$ Cubesort outperforms Columnsort and shows that sorting is recursively $(n^{i/(i+1)}, 5(i + 1)^2)$-divisible in a work-optimal way. By way of contrast, (recursive) application of Columnsort only yields that sorting is recursively $(n^{i/(i+1)}, (i+1)^\beta)$-divisible, where $\beta = 2/(\log 3 - 1) \approx 3.42$. It remains open, however, whether it is possible to deterministically sort in a work-optimal way with less than four communication phases using a polynomial number of processors.

# 4 Relations to Complexity Theory

In this section we present a circuit construction for recursively $(n^\epsilon, d)$-divisible problems. This construction in particular reveals the limitations of the concept of ideally divisible problems—in the natural case of a logarithmically bounded word length they can be computed by quasilinear size $NC$-circuits and form a proper subclass of $FP$.

**Definition 7.** $RD_w(p, d)$ is the class of functional problems $f : \{0, 1\}^{n \cdot w} \longrightarrow \{0, 1\}^{O(n \cdot w)} \in FP$ that are recursively $(p, d)$-divisible.

**Theorem 8.**
$RD_w(n^\epsilon, d) \subseteq NC\text{-}SIZE, DEPTH(n \log^{O(1)} n \cdot w^{O(1)}, \log^c n \cdot w^{O(1)})$ for $c = \log d / \log(1/(1 - \epsilon))$.

*Proof.* (Sketch) The basic idea of proof is to recursively apply the divisibility property in order to construct the desired circuit of bounded fan-in. If a problem is recursively $(n^\epsilon, d)$-divisible, then according to Definitions 1 and 2 we can solve it by working on subproblems with inputs of size $bn^{1-\epsilon}$ for some constant factor $b \geq 1$. These subproblems again can be replaced by smaller ones, now of size $b^2 n^{(1-\epsilon)^2}$ and so on. Iterating this process $x$ times leads to subproblems of size $b^x n^{(1-\epsilon)^x}$. For the time being we just assume that $b = 1$. Later on we will discuss the case $b > 1$. We iterate $x$ times until a constant number of elements, each requiring $w$ bits, is obtained, thus demanding

$$n^{(1-\epsilon)^x} = c, \qquad (*)$$

where $c$ is some positive constant. The point here is that problem instances comprising only a constant number of elements can be solved by bounded fan-in circuits of size and depth $w^{O(1)}$ both, because the considered problems are in $FP$. Resolving equation $(*)$ for $x$ yields

$$x = \log_{\frac{1}{1-\epsilon}}(\log n / \log c) = (\log \log n - \log \log c) / \log \frac{1}{1 - \epsilon} = O(\log \log n).$$

Because in this way we need $d^x$ layers of circuits requiring size and depth $w^{O(1)}$ and $d^x$ layers of linear size $NC^0$-circuits (cf. Definition 1 and the discussion there), the depth of the resulting circuit is in essence determined by

$$d^x = 2^{x \log d} = 2^{\log d (\log \log n - O(1)) / \log(1/(1-\epsilon))} \leq (\log n)^{\log d / \log(1/(1-\epsilon))}.$$

In total, we get the depth bounded by $d^x \cdot w^{O(1)}$.

Next we come to the general case $b > 1$. In the recursive construction, each subproblem of size $bn^{1-\epsilon}$ is replaced by $a(bn^{1-\epsilon})^{\epsilon}$ subproblems of size $b(bn^{1-\epsilon})^{1-\epsilon}$ and so on. This process continues $x$ times, where $x$ is as determined above. As can be easily verified, after $x$ iterations we end up with problems each having

$$b^{\sum_{i=0}^{x-1}(1-\epsilon)^i} n^{(1-\epsilon)^x} = b^{(1-(1-\epsilon)^x)/\epsilon} n^{(1-\epsilon)^x}$$

elements instead of $n^{(1-\epsilon)^x}$ as in $(*)$. Clearly, $b^{(1-(1-\epsilon)^x)/\epsilon} = O(1)$ for constant $0 < \epsilon < 1$. Thus $(*)$ is still valid. So we get the same depth estimation as for $b = 1$.

So far we have analyzed the depth of the circuit. It remains to also determine its size. For the moment neglecting the $NC^0$-communication layers, by our construction we end up with $d^x$ layers of subcircuits of size and depth $w^{O(1)}$ each. Let $s$ denote the number of subcircuits (or, equivalently, subproblems) per each such layer. Then $O(w^{O(1)} \cdot d^x \cdot s)$ obviously is an upper bound for the size of the constructed circuit. We now determine $s$. We started with $d$ layers, each layer consisting of $O(n^{\epsilon})$ subproblems of size $O(n^{1-\epsilon})$ each. Replace $O(n^{\epsilon})$ by $an^{\epsilon}$ and $O(n^{1-\epsilon})$ by $bn^{1-\epsilon}$ for some positive constants $a$ and $b$. Altogether, when this process terminates after $x$ recursive applications of divisibility, we have

$$s = an^{\epsilon} \cdot ab^{1-(1-\epsilon)^1} n^{(1-\epsilon)\epsilon} \cdot \ldots \cdot ab^{1-(1-\epsilon)^{x-1}} n^{(1-\epsilon)^{x-1}\epsilon}$$

subproblems of a constant number of elements per layer, each realizable by a size $w^{O(1)}$ circuit. It is easy to see that $s = a^x b^y n^z$, where $y \leq x - 1$ and $z = 1 - (1-\epsilon)^x$. So we have $s \leq a^x b^{x-1} n^z$, which for $x = O(\log \log n)$ means that $s = n(\log n)^{O(1)}$. (Note that $x$ was chosen such that $n^{(1-\epsilon)^x} = c$, so $n^z = O(n)$.) So far we still omitted the additional costs resulting from the layers of $NC^0$-communication. Because their size is linear and because we need $d^x = (\log n)^{O(1)}$ many of these layers, we altogether may conclude a size $n(\log n)^{O(1)} \cdot w^{O(1)}$ for the constructed circuit. To verify that the whole construction is $ATIME(\log n)$-uniform is straightforward. □

Applying Theorem 6 to the sorting problem demonstrates that there we can guarantee constants $a = 1$ and $b = 2$. Substituting this into the analysis done in the proof of Theorem 8 and making use of the recursive $(n^{1/3}, 4)$-divisibility as proved in Theorem 6, Theorem 8 now implies that sorting can be done by a depth $O((\log n)^{3.42})$ and size $O(n(\log n)^{5.13})$ $NC$-circuit that uses comparator gates with two $w$-bit-inputs. This reveals that ideally divisible problems do not necessarily lead to work-optimal parallel algorithms if the recursive implementation is iterated a non-constant number of times (compare the discussion following Definition 2). On the other hand, that also may indicate that even for non-work-optimal $NC$-algorithms, there may lie at their heart a work-optimal, practical parallel algorithm.

The following two direct corollaries of Theorem 8 exhibit the limitations of the concept of ideally divisible problems. Under the natural assumption of a logarithmically bounded word length $w$ we obtain that all ideally divisible problems possess quasilinear size $NC$-circuits.

**Corollary 9.** $RD_{\log n}(n^\epsilon, O(1)) \subseteq NC\text{-}SIZE, DEPTH(n \log^{O(1)} n, \log^{O(1)} n)$.

**Corollary 10.** $RD_{\log n}(n^\epsilon, O(1))$ *is strictly contained in FP.*

*Proof.* (Hint) The circuit construction of Theorem 8 can also be transformed into a (recursive) sequential algorithm.                                                □

Even if we allow a polylogarithmic instead of a constant number $d$ of communication phases, unless $P = NC$ the hardest problems in $FP$ still are resistant to efficient parallelization in the sense of recursive $(p, d)$-divisibility for sublinear, but still polynomial numbers $p$ of processors. For the following note that $(\log n)^{O(\log\log n)}$ is subpolynomial. Again we omit the proof.

**Theorem 11.** $RD_{\log n}(n^\epsilon, (\log n)^{O(1)}) \subseteq$
$$NC\text{-}SIZE, DEPTH(n(\log n)^{O(\log\log n)}, (\log n)^{O(\log\log n)}).$$

Subsequently we consider the levels of the so-called $NC$-hierarchy. It is not known whether this hierarchy is a proper one.

**Theorem 12.** *If* $NC^k \subseteq RD_{\log n}(n^\epsilon, d)$ *for* $d \le (\frac{1}{1-\epsilon})^{k-\delta'}$ *with arbitrary, but constant* $\delta' > \delta > 0$*, then* $NC = NC^{k-\delta}$.                        □

*Proof.* (Hint) Analogous to Theorem 8, but making use of the fact that the considered problems are in $NC^k$ instead of "only" $FP$.

## 5   Extensions and Variations

In the preceding section, assuming a logarithmic word length, we showed that recursively $(n^\epsilon, O(1))$-divisible problems possess quasilinear size $NC$-circuits (Theorem 8, Corollary 9) and are properly contained in $FP$ (Corollary 10). In particular, they can be solved with RAM's with unit cost measure in time $n \log^{O(1)} n$. In what follows we attack the problem to study and classify problems with sequential complexity $\Omega(n \log n)$. A central point here is to introduce the concept of *polynomial $(p, d)$-divisibility*, which is motivated by the matrix multiplication problem. Consider the problem of multiplying two $\sqrt{n} \times \sqrt{n}$-matrices. Having $p$ processors at hand, how to compute the product? The probably simplest and notwithstanding a fairly prospective approach is to partition each of the two matrices into $p$ submatrices and then to apply the standard matrix multiplication scheme already known from school. By this it is easy to show that multiplying two $\sqrt{n} \times \sqrt{n}$-matrices is recursively $(p, \sqrt{p})$-divisible for $p < n$. Thus for nonconstant $p$ we no longer have a constant number of communication phases, but need $\sqrt{p}$ many. But if we had $p\sqrt{p}$ processors, each dealing with inputs and outputs of size $O(n/p)$, then it is fairly easy to see that matrix multiplication could be done using only two local computation phases. This motivates the following definition of polynomial $(p, d)$-divisibility.

**Definition 13.** 1. Let $f \in FP$ with $f : \{0,1\}^{n \cdot w} \longrightarrow \{0,1\}^{O(n \cdot q \cdot w/p)}$, and let $n$, $w$, $p$, $q$, and $d$ be natural numbers with $p < n$ and $q = p^{O(1)}$. Function $f$ is *polynomially* $(p,d)$-*divisible* if it can be written as follows. Let $x \in \{0,1\}^{n \cdot w}$. Then $f(x) = div(x,q,d)$, where $div$ is the same as in Definition 1 with the exceptions that the functions $z_i^k$, $1 \leq i \leq q$, there now are mappings from $\{0,1\}^{c_i^k \cdot n \cdot q \cdot w/p}$ to $\{0,1\}^{c_{i,2}^k \cdot n \cdot w/p}$ and for all $k$, *all* $z_i^k$ have to be simultaneously computable by an $ATIME(\log n)$-uniform $NC^0$-circuit of size $O(n \cdot q \cdot w/p)$.

2. $RPD_w(p,d)$ is the class of problems $f : \{0,1\}^{n \cdot w} \longrightarrow \{0,1\}^{O(n \cdot q \cdot w/p)} \in FP$, $q = O(1)$, that are recursively polynomially $(p,d)$-divisible. Here "recursively" is defined as for $(p,d)$-divisible problems (see Definition 2).

Roughly speaking, the only difference comparing the above with Definition 1 of $(p,d)$-divisible problems is that we replace $p$ by $q = p^{O(1)}$, a polynomial increase of the number of processors. Under this new definition it is possible to show that matrix multiplication is recursively polynomially $(p,2)$-divisible for $p \leq \sqrt{n}$. Even if we allow recursive polynomial $(n^\epsilon, O(1))$-divisibility, not all problems in $FP$ do have this property unless $P = NC$. The proof of Theorem 14 is similar to that of Theorem 8 and is omitted, therefore.

**Theorem 14.**
$RPD_w(n^\epsilon, d) \subseteq NC\text{-}SIZE, DEPTH(n^{O(1)} \cdot w^{O(1)}, \log^c n \cdot w^{O(1)})$, where $c = \log d / \log(1/(1-\epsilon))$.

A comparison of Theorem 8 and Theorem 14 reveals that the difference between recursively $(n^\epsilon, d)$-divisible and recursively *polynomially* $(n^\epsilon, d)$-divisible problems is that, assuming $w = O(\log n)$, for the first ones we can construct quasilinear size circuits, whereas the latter ones require circuits of polynomial size. Using Theorem 14 we now may obtain results in analogy to that in the previous section. Perhaps the most important case is the statement concerning the relationship between $FP$ and $NC$, which we pick out at this place.

**Corollary 15.** *If* $FP \subseteq RPD_{\log n}(n^\epsilon, O(1))$, *then* $FP = NC$.

## 6   Conclusion

Several open questions arise from our work, for example: Can the result for sorting (cf. Theorem 6 and the subsequent discussion) be improved? Is it possible to show that the well-known list ranking problem is not recursively $(n^\epsilon, O(1))$-divisible? Does there exist a characterization of exactly those problems in $NC$ that are (polynomially) $(n^\epsilon, O(1))$-divisible?

This paper provides a link between "classical" (parallel) complexity theory (like $NC$-theory) and questions of parallel computing (like constant number of communication phases, blocked and structured communications, reuse of sequential algorithms). Whereas Vitter and Simons [14] demonstrated that there exist work-optimal PRAM algorithms for $P$-complete problems, we showed that there exist problems in *FP* that are inherently resistant to good divisibility properties. To an extent this (in contrast to Vitter and Simons) underpins the importance of the *P* versus *NC* debate [7] when considering a more restrictive framework for parallel computation (as we proposed) than the world of PRAM algorithms is.

# References

1. A. Aggarwal, A. K. Chandra, and M. Snir. Communication Complexity of PRAMs. *Theoretical Comput. Sci.*, 71:3–28, 1990.
2. D. A. Barrington. Bounded-width polynomial-size branching programs recognize exactly those languages in $NC^1$. *J. Comput. Syst. Sci.*, 38:150–164, 1989.
3. D. Culler, R. Karp, D. Patterson, A. Sahay, K. E. Schauser, E. Santos, R. Subramonian, and T. von Eicken. LogP: Towards a realistic model of parallel computation. In *4th ACM SIGPLAN Symposium on Principles and Practice of Parallel Programming*, pages 1–12, May 1993.
4. R. Cypher and J. L. Sanz. Cubesort: A parallel algorithm for sorting *N* data items with *S*-sorters. *Journal of Algorithms*, 13:211–234, 1992.
5. P. de la Torre and C. P. Kruskal. Towards a single model of efficient computation in real parallel machines. *Future Generation Computer Systems*, 8:395–408, 1992.
6. P. de la Torre and C. P. Kruskal. A structural theory of recursively decomposable parallel processor-networks. In *IEEE Symp. on Parallel and Distributed Processing*, 1995.
7. R. Greenlaw, H. J. Hoover, and W. L. Ruzzo. *Limits to Parallel Computation: P-Completeness Theory*. Oxford University Press, 1995.
8. T. Heywood and C. Leopold. Models of parallelism. In J. R. Davy and P. M. Dew, editors, *Abstract Machine Models for Highly Parallel Computers*, chapter 1, pages 1–16. Oxford University Press, 1995.
9. T. Heywood and S. Ranka. A practical hierarchical model of parallel computation: I and II. *Journal of Parallel and Distributed Computing*, 16:212–249, November 1992.
10. C. P. Kruskal, L. Rudolph, and M. Snir. A complexity theory of efficient parallel algorithms. *Theoretical Comput. Sci.*, 71:95–132, 1990.
11. T. Leighton. Tight bounds on the complexity of parallel sorting. *IEEE Transactions on Computers*, C-34(4):344–354, April 1985.
12. W. L. Ruzzo. On uniform circuit complexity. *J. Comput. Syst. Sci.*, 22:365–383, 1981.
13. L. G. Valiant. A bridging model for parallel computation. *Commun. ACM*, 33(8):103–111, 1990.
14. J. S. Vitter and R. A. Simons. New classes for parallel complexity: A study of unification and other complete problems for P. *IEEE Trans. Comp.*, C-35(5):403–418, 1986.

# $StUSPACE(\log n) \subseteq DSPACE(\log^2 n/ \log\log n)$

Eric Allender[*][1] and Klaus-Jörn Lange[**][2]

[1] Department of Computer Science
Rutgers University
P.O. Box 1179
Piscataway, NJ 08855-1179
USA
allender@cs.rutgers.edu
[2] Wilhelm-Schickard Institut für Informatik
Universität Tübingen
Sand 13
D-72076 Tübingen
Germany
lange@informatik.uni-tuebingen.de

## Abstract

We present a deterministic algorithm running in space $O\left(\log^2 n/ \log\log n\right)$ solving the connectivity problem on strongly unambiguous graphs. In addition, we present an $O(\log n)$ time-bounded algorithm for this problem running on a parallel pointer machine.

## 1  Introduction

One of the most central questions of complexity theory is to relate determinism and nondeterminism. Our inability to exhibit the precise relationship between these two notions motivates the investigation of intermediate notions such as symmetry or unambiguity. In this paper we concentrate on unambiguity in space-bounded computation, and present improved deterministic and parallel simulations.

Recently, surprising results have indicated that "symmetric" space bounded computation is weaker than nondeterminism. In particular, symmetric logspace has been shown to be contained in parity logspace [12], in $SC^2$ [18], and in $DSPACE(\log^{1.5} n)$ [19]. None of these upper bounds is known to hold in the nondeterministic case. If we consider these questions for space bounded unambiguous classes, we are confronted with the fact that there are several ways to define notions of unambiguity that apparently do not coincide [3]. In this paper we will concentrate on the notion of unambiguity (in the sense of unique existence of computation paths), and of strong unambiguity (in the sense of

[*] Supported in part by NSF grant CCR-9509603.
[**] Supported in part by NSF grant CCR-9509603.

uniqueness of computations between any pair of configurations). This yields the two classes $USPACE(\log n)$ and $StUSPACE(\log n)$.

By definition, $USPACE(\log n)$ is a subclass of parity logspace; this is not known to hold for $NSPACE(\log n)$ (although see [23]); there is no additional nontrivial containment known for $USPACE(\log n)$. However, $StUSPACE(\log n)$ is contained in $SC^2$, since strongly unambiguous logspace languages can be accepted by deterministic auxiliary pushdown automata in polynomial time [3, 5]. Still, it was unknown whether there are $o(\log^2 n)$ space algorithms for strongly unambiguous logspace languages. We answer this question affirmatively by showing that $StUSPACE(\log n)$ is contained in $DSPACE(\log^2 n/\log\log n)$.

Since $StUSPACE(\log n)$ is a subclass of $DAuxPDA\text{-}TIME(n^{O(1)})$ we know that there are logtime $CROW$-algorithms for the elements of $StUSPACE(\log n)$ [7]. To give better relative upper bounds on the complexity of $StUSPACE(\log n)$ it is interesting to consider intermediate classes between $DSPACE(\log n)$ and $DAuxPDA\text{-}TIME(n^{O(1)})$.

Trying to find these classes with sequential models could be difficult, since the usual restrictions of a pushdown store (e.g. the one-turn property, or using a counter instead of a push down) all collapse to logspace. Here, parallel machine models seem helpful, leading to two intermediate classes: the parallel pointer machine [6, 14] and the $OROW$-PRAM [20]. As a consequence of our main result we get $OROW$ algorithms for the elements of $StUSPACE(\log n)$ taking time $O(\log^2 n/\log\log n)$. We are also able to show that all sets in $StUSPACE(\log n)$ are accepted in logarithmic time on a parallel pointer machine. This latter containment is somewhat surprising, because there are characterizations in terms of parallel programs indicating that the class $PPM\text{-}TIME(\log n)$ is rather close to $DSPACE(\log n)$ [16].

## 2 Preliminaries

We assume the reader to be familiar with the basic notions of complexity theory (e.g. [10]). In addition, let $DTISP(f, g)$ be the set of all languages accepted by $O(g)$ space-bounded Turing machines in time $O(f)$. $NTISP(f, g)$ denotes the corresponding nondeterministic class.

We refer the reader to the survey article of Karp and Ramachandran [13] for coverage of the many varieties of parallel random access machines and their relationship to sequential classes. Let us remark here, that we deal in this paper only with algorithms and classes using PRAMs with a polynomial number of processors. The notion of a $CROW$-PRAM was introduced by Dymond and Ruzzo [7] and provides the tightest possible connections to deterministic machines [9].

$CROW$-PRAMs need only logarithmic time to recognize any given language in $DSPACE(\log n)$; There are two important ways to restrict $CROW$-PRAMs and still maintain this property. One way is to restrict the concurrent read access to the global memory, which leads to the $OROW$-PRAMs of Rossmanith [20]. The other way is to restrict the arithmetical capabilities of the instruction set leading to $rCROW$-PRAMs and to parallel pointer machines [6, 14].

# 3 Unambiguity

A concept intermediate in power between determinism and nondeterminism is *Unambiguity*. A nondeterministic machine is said to be unambiguous, if for every input there exists at most one accepting computation. This leads to the classes $UP$ and $USPACE(\log n)$; we have $P \subseteq UP \subseteq NP$ and $DSPACE(\log n) \subseteq USPACE(\log n) \subseteq NSPACE(\log n)$. The notion of ambiguity should be distinguished from that of *Uniqueness*, which uses the unique existence of an accepting path not as a restriction but as a tool. The resulting language classes $1NSPACE(\log n)$ and $1NP$ consists of languages defined by machines that accept their inputs if there is exactly one accepting path. Thus, the existence of two or more accepting computations is not forbidden, but simply leads to rejection. In the polynomial time case we have $Co\text{-}NP \subseteq 1NP$ [2]. In the logspace case inductive counting [11, 22] shows $1NSPACE(\log n) = NSPACE(\log n)$.

A more restrictive form of unambiguity is *Strong Unambiguity*. A nondeterministic machine is said to be strongly unambiguous, if for *every* pair of configurations there exits at most one computational path connecting these configurations. An ordinary unambiguous machine makes this restriction only for the initial and the accepting configurations of the machine.

While these two concepts coincide (yielding the class $UP$) in the case of time bounded computations, this is not known to be true in the space bounded case. There we end up with an additional class $StUSPACE(\log n)$ which is located between $DSPACE(\log n)$ and $USPACE(\log n)$. Correspondingly, $StUTISP(f, g)$ is the class of all languages accepted by strongly unambiguous Turing machines that are simultaneously $O(g)$ space- and $O(f)$ time-bounded. In fact, there are several more versions of unambiguity that are not known to coincide (depending for instance on whether or not there can be more than one accepting configuration, see [3]), but in this work we consider only the two classes $USPACE(\log n)$ and $StUSPACE(\log n)$. Our algorithms work for all unambiguous classes for which the unfoldings of the configuration graphs to trees are of polynomial size. This is the case for $StUSPACE(\log n)$ but not for $USPACE(\log n)$.

In the time-bounded setting, problems such as factoring and primality have efficient unambiguous algorithms but are not known to possess deterministic algorithms with a comparable running time [8]. Recently, Lange presented a problem that is complete for a subclass of $USPACE(\log n)$ [15]; this is the first explicit presentation of a problem in $USPACE(\log n)$ that is not known to be in $DSPACE(\log n)$. (Completeness is a tool that is not often available in studying unambiguous classes. None of $UP$, $USPACE(\log n)$, or $StUSPACE(\log n)$ is known or believed to have complete sets.)

No problem is known to be in $StUSPACE(\log n)$ that is not known to be in $DSPACE(\log n)$. Nonetheless, there is an important *class* of languages with this property: The class of unambiguous linear languages, $ULIN$, is contained in $StUSPACE(\log n)$ [3], and it is not known to be contained in $DSPACE(\log n)$. To date, however, no explicit example of a language in $ULIN$ has been exhibited for which a $DSPACE(\log n)$ algorithm is not known. (Note in this regard that the linear context free languages are $NSPACE(\log n)$–complete.)

Logspace classes are closely connected to graph accessibility problems. These connectivity problems in directed graphs, undirected graphs, and in trees are complete for $NSPACE(\log n)$, $SSPACE(\log n)$, resp. $DSPACE(\log n)$. For unambiguous classes the corresponding connectivity problems do not seem to be complete. Nevertheless it is useful to explain these notions in terms of directed graphs, since this will make the demonstration of our algorithms easier.

Let $G = (V, E)$ be a directed acyclic graph. If $G$ has $n$ nodes we assume $V$ to be $\{1, 2, \cdots, n\}$ and we will be interested in the existence of a path leading from 1 to $n$. For each pair of nodes $(x, y)$ let $d(x, y)$ be the length of the shortest path between $x$ and $y$. If $x$ and $y$ are not connected, $d(x, y)$ is infinite; $d(x, x)$ is 0. In the following we will work with complete binary graphs; that is, each node of $G$ is either a leaf with no outgoing edges, or an inner node with two outgoing edges. We assume the leaves to be accepting or rejecting. This is determined by a mapping $\phi : V \longrightarrow \{+, -, i\}$, which takes the value $+$ for accepting leaves, $-$ for rejecting leaves and $i$ otherwise. In particular, 1 is an inner node (unless the graph has only one vertex) and we assume $n$ to be the only accepting leaf, since we are only interested in paths from 1 to $n$. The two successors of an inner node $x$ will be denoted by $L(x)$ and $R(x)$.

Let $P_E := \{(a_0, a_1)(a_1, a_2) \cdots (a_{k-1}, a_k) | k \geq 1, (a_{i-1}, a_i) \in E\}$ be the set of all paths in $G$. We define a mapping $C : P_E \longrightarrow V \times V$ by mapping a path leading from node $x$ to node $y$ to the pair $(x, y)$, i.e.: we set

$$C((a_0, a_1)(a_1, a_2) \cdots (a_{k-1}, a_k)) := (a_0, a_k).$$

For $x \in V$ let $T(x) := \{y \in V | C^{-1}((x, y)) \neq \emptyset\} \cup \{x\}$ be the set of all nodes reachable from $x$.

$G$ is called *unambiguous* if there is at most one path from 1 to $n$, i.e.: if $|C^{-1}((1, n))| \leq 1$. $G$ is called *strongly unambiguous* or a *Mangrove* if for any pair $(x, y)$ of nodes there is at most one path leading from $x$ to $y$, i.e.: if $\forall_{x, y \in V} |C^{-1}((x, y))| \leq 1$. Although a mangrove does not need to be a tree, for each $x$ the subgraph of $G$ induced by $T(x)$ is indeed a tree and the same is true for the set of all nodes from which $x$ can be reached.

The following three examples show different version of unambiguous graphs. The first one is a very simple mangrove with a positive result since there is a path from node 1 to node 5. The second one is also a mangrove, but a negative one, since there is no path from 1 to 8. The third graph is unambiguous and positive, since there is exactly one path from 1 to 6, but it is not a mangrove, since there are two different paths from node 1 to node 5.

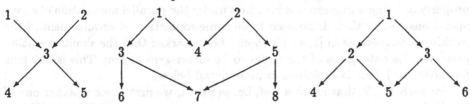

Typical examples of mangroves are butterfly graphs or the *Reinhardt Graphs* $R_n := (V_n^r, E_n^r)$ with $V_n^r := \{1, 2, \cdots, n\}^2$ and $E_n^r$ is $\{((i, j), (i, j + 1)) | 1 \leq i \leq$

$n, 1 \leq j < n\} \cup \{(\langle i,j \rangle, \langle i+j \bmod n, n \rangle)|1 \leq i \leq n, 1 \leq j < n\}$. (Observe that the Reinhardt Graph is not a complete binary graph.)

By definition a Turing machine is unambiguous if for each input word the reachability graph of its configurations is unambiguous. It is strongly unambiguous if for every input this graph forms a mangrove.

# 4 Contracting Mangroves

The class $StUSPACE(\log n)$ has been shown to be contained in $DAuxPDA\text{-}TIME(\text{pol})$ by unfolding the reachability graph of a mangrove to a tree of polynomial size [3]. This tree can be searched with the help of a push-down store in polynomial time. The $DAuxPDA\text{-}TIME(n^{O(1)})$-completeness of $DCFL$[21] together with the algorithm of Cook for deterministic context free languages [5] implies $StUSPACE(\log n) \subseteq SC^2$. Elements of Cook's algorithm were later used by Dymond and Ruzzo [7] and independently by Monien et al [17] to show $DCFL \subseteq CROW\text{-}TIME(\log n)$ (see also [9]). As a consequence, the elements of $StUSPACE(\log n)$ possess logtime algorithms running on a CROW–PRAM. But these algorithms for mangroves, generated by composing the constructions in [3] with those of [7, 9, 17], are rather complicated. Below, we give a very simple logtime algorithm that runs on a parallel pointer machine (which is a restricted CROW-PRAM).

**Theorem 1.**
$$StUSPACE(\log n) \subseteq PPM\text{-}TIME(\log n)$$

**Proof:** We first describe the parallel algorithm and then indicate how to run it on a parallel pointer machine. Let $A$ be a strongly unambiguous logspace machine and $G = (V, \phi, L, R)$ be its configuration mangrove induced by some input $w$ of size $n$. For each $x \in V$ the tree $T(x)$ is of polynomial size in $n$. What we would like to do on each $T(x)$ is pointer jumping. Unfortunately its pointers are directed towards the leaves instead of towards the root. The alternative is to perform the shunt operation (see e.g. [13]). This would need backpointers directed towards the root, which are usually provided by establishing an Euler tour through a tree. This is also not possible in a mangrove. The simple way out of this is to perform the shunt operation not by the parent node of a leaf but by its grandparent node, in parallel in each tree. Of course, these operations destroy any tree structure unless they are synchronized. But obviously the property of being a mangrove is invariant under the parallel application of shunt operations. Hence there is no need to do some sophisticated arrangement of the working processors as in [1, 4]. (It should be remarked that the shunt operation preserves the outdegrees of the nodes to be either zero or two. This is not true for indegrees.) This is explained in more detail below.

For each $x \in V$ that is not a leaf, i.e. $\phi(x) = i$, we first check whether one of its children is an accepting leaf, in this case we set $\phi(x) := +$. If both children are rejecting we set $\phi(x) := -$. Otherwise, one of the children has to be an inner node. If $\phi(L(x)) = i$ we check if the left grandson via the left subtree is a rejecting leaf. In that case we set $L(x) := R(L(x))$. If this was not the case we see whether the right grandson is a rejecting leaf which would lead to the assignment $L(x) := L(L(x))$. This is then repeated for the right son of $x$. After performing this parallel transformation $O(\log n)$ times, there is no inner node left; for each $x \in V$ we have either $\phi(x) = +$ or $\phi(x) = -$. The existence of an accepting path is then equivalent to $\phi(1) = +$.

Formally, the algorithm is expressed by the following statements:

**for** $j = 1, 2, \cdots, O(\log n)$ **do**
    **for all** $x \in V$ **do in parallel**
        **if** $\phi(x) = i$ **then**
            **if** $\phi(L(x)) = +$ **or** $\phi(R(x)) = +$ **then** $\phi(x) := +;$
            **if** $\phi(L(x)) = -$ **and** $\phi(R(x)) = -$ **then** $\phi(x) := -;$
            **if** $\phi(L(x)) = i$ **then**
                **if** $\phi(L(L(x))) = -$ **then** $L(x) := R(L(x))$
            **else**
                **if** $\phi(R(L(x))) = -$ **then** $L(x) := L(L(x));$
            **if** $\phi(R(x)) = i$ **then**
                **if** $\phi(L(R(x))) = -$ **then** $R(x) := R(R(x))$
            **else**
                **if** $\phi(R(R(x))) = -$ **then** $R(x) := L(R(x));$
**if** $\phi(1) = +$ **then accept**
**else reject**

We now shortly indicate how to put all this on a parallel pointer machine. Obviously, the algorithm itself is already suitable for a PPM. It remains only to show how to build the mangrove of configurations in logarithmic time, given the input word. As in the corresponding construction of Cook and Dymond [6] the initial PPM unit starts to build in logarithmic time a tree of logarithmic depth such that each leaf unit corresponds to a configuration of $A$ on $w$. Then the leaves are interconnected according to the successor relation of $A$. But instead of one pointer leading to a successor, each leaf unit will now have two pointers $L$ and $R$ representing the two subtrees hanging below an inner node of a mangrove. $\square$

Let us remark here that the number of PPM units, i.e. processors, is linear in the size of the mangrove. For the special case of $ULIN$, this leads to a quadratic number of processors. The best known sequential algorithm for solving membership questions in $ULIN$ needs quadratic time.

Our algorithm makes intensive use of concurrent reads and therefore does not pertain to owner read PRAMs. It works via $O(\log n)$ applications of the operations of searching and pruning trees of depth 2. Increasing this depth to $O(\log n)$ is the first step in reducing the number of iterations to $O(\log n / \log \log n)$. This yields an algorithm using $o(\log^2 n)$ space (or parallel time).

**Theorem 2.**

$$StUSPACE(\log n) \subseteq DSPACE(\log^2 n / \log \log n)$$

**PROOF**: Let $A$ be a strongly unambiguous logspace machine and let $w$ be an input of length $n$. This yields a reachability problem in a mangrove $G = (V, \phi, L, R)$ of polynomial size.

We will now construct a mangrove $G' = (V, \phi', L', R')$ of smaller size. Let $t \geq 1$ be an integer that determines the rate of contraction of $G'$ compared to $G$. We will fix the value of $t$ later. First we map each node $x$ to one of its successors which we call $f(x)$.

If in the following procedures an accepting leaf (i.e. a node $y$ with $\phi(y) = +$) is found, the search is stopped, and we set $\phi'(x) := +$ and $f(x) := x$. We search for each $x \in V$ the tree $T_t(x) := \{y \in V | d(x, y) \leq t\}$ of all nodes of distance to $x$ not greater than $t$.

Let $z$ be a pointer initialized to $x$. Consider the set $M$ of all nodes $y$ in $T_t(z)$ with $d(z, y) = t$ and $\phi(y) = i$, i.e. of all leaves of $T_t(z)$ that are not leaves in $T(z)$. If $M$ is empty, that is if $T(z) = T_t(z)$ then set $\phi'(x) := -$ and $f(x) := x$. If $M$ is nonempty we replace $z$ by the least common ancestor $z'$ of $M$ in $T_t(z)$. If $z = z'$ we stop the procedure for $x$ and set $f(x) := z$ and $\phi'(x) := i$. Otherwise we replace $z$ by $z'$ and continue the process.

This algorithm computing $f$ and $\phi'$ in a more formal way looks as follows, where $LCA(M)$ denotes the least common ancestor of a set $M$ of nodes in a tree:

```
z := x;
M := {y ∈ Tₜ(z) |d(z,y) = t, φ(y) = i} ;
while  + ∉ φ(Tₜ(z)) and  M ≠ ∅ and  LCA(M) ≠ z do
begin
        z := LCA(M);
        M := {y ∈ Tₜ(z) |d(z,y) = t, φ(y) = i}
end
if  + ∈ φ(Tₜ(z))
        then  φ'(x) := +; f(x) := x
        else if  M = ∅
                then  φ'(x) := −; f(x) := x
                else  φ'(x) := i; f(x) := z
```

After this process, for each $x \in V$ either $\phi'(x) = +$, i.e. an accepting leaf has been found in $T(x)$, or $\phi'(x) = -$, i.e. $T(x)$ has been totally searched and contains rejecting leaves, only, or $\phi'(x) = i$. In this case $x$ points to $f(x)$ which is a node such that both subtrees of $f(x)$ contain nodes of a distance larger than $t$ to $f(x)$. That is, both subtrees are of height not smaller than $t$, and, since both are complete binary trees, both of them are of a size not smaller than $2t + 1$.

Clearly, the computation of $f$ can be done in space $t + O(\log n)$. (Note that $M$ need not be stored explicitly.) We now construct $G'$ by setting $L'(x) := f(L(f(x)))$ and $R'(x) := f(R(f(x)))$ for each $x$ with $\phi'(x) = i$. We have for each $x \in V$ with $\phi'(x) = i$:

$$(P1) \quad |T(L(f(x)))| \geq 2t + 1 \text{ and } |T(R(f(x)))| \geq 2t + 1.$$

Furthermore, the construction implies

$$(P2) \ f(f(x)) = f(x) \text{ and } (P3) \ \phi'(x) = \phi'(f(x))$$

for every $x \in V$.

Now let $T'(x)$ be the tree below $x$ in $G'$. Although there may be nodes in $V$ that are not finished (i.e. $\phi'(x) = i$) and that are not contracted, (i.e. $f(x) = x$) the mangrove in total is smaller by a factor of $t$; we claim for each $x \in V$ with $\phi'(x) = i$:

$$|T'(x)| \leq |T(x)| / t$$

Proof of the claim: Let $L'(z)$ be a left leaf in the tree $T'(x)$. That is, $\phi'(z) = i$ and $\phi'(L'(z)) \neq i$. Then by (P3), we have that $\phi'(L(f(z))) = \phi'(f(L(f(z))))$, and by definition of $L'$ this is equal to $\phi'(L'(z)) \neq i$. Thus $L'(z) = f(L(f(z))) = L(f(z))$ by the construction of $f$. Hence by (P1), $|T(L'(z))| = |T(L(f(z)))| \geq 2t + 1$. Hence for each left leaf $L'(z)$ in $T'(x)$ there are at least $2t$ nodes that are removed from $T(x)$. And the same holds for right leaves. Hence $T(x)$ is not smaller than $2t$ times the number of leaves of $T'(x)$. Since at least half of the elements of a complete binary tree are leaves the result follows.

Repeating this process of shrinking $G$ ends in a mangrove of depth $O(1)$ after $O(\log_t n)$ phases. Setting $t := \log n$ we obtain $O(\log n / \log \log n)$ phases, each of which uses $O(\log n)$ space, to obtain a procedure to search a mangrove using $O(\log^2 n / \log \log n)$ space. □

We remark here, that the resulting algorithm is in general not polynomial time-bounded.

**Corollary 3.**

$$ULIN \subseteq DSPACE(\log^2 n / \log \log n)$$

All of $DSPACE(f)$ can be recognized by $OROW$-PRAMs in $O(f)$ steps. But generally these algorithms need $c^{O(f)}$ many processors, which in our case would be superpolynomial. But due to the recursive or iterative structure of our algorithm a polynomial number suffices, as we now show.

**Corollary 4.**

$$StUSPACE(\log n) \subseteq OROW\text{-}TIME \,(\log^2 n / \log \log n)$$

**Proof:** The PRAM works in $\log n / \log \log n$ phases, each taking $O(\log n)$ steps. In each phase the work of a logarithmically space bounded TM is simulated in logarithmic time by an $OROW$-PRAM with a polynomial number of processors [20]. The difference is here that we don't simulate an accepting machine, but one producing some output. This output is stored in global memory. In order to use this output as input of the next phase, we first have to distribute it to polynomially many processors in a tree-like way in $O(\log n)$ steps. In total this costs $O(\log^2 n / \log \log n)$ steps on an $OROW$-PRAM. The number of processors corresponds to the number of configurations of the simulated logspace machine and hence is polynomial. $\square$

We mention in passing that this result, applied to the unambiguous linear languages, leads to algorithms using $O(n^2)$ processors since this is the size of the corresponding mangrove.

We end this section by remarking that our main theorem generalizes in the following way: It is well-known that $NTISP(f, g)$ is a subset of $DSPACE(g \log f)$. If we restrict the nondeterminism to be strongly unambiguous we get the sharper bound $StUTISP(f, g) \subseteq DSPACE(g \cdot \log f / \log g)$.

# 5 Discussion and open questions

The most basic open question is, of course, to clarify the relationship between $StUSPACE(\log n)$ and $DSPACE(\log n)$. It might very well be that these two classes coincide. Indeed, there seem to be no drastic consequences implied by such a collapse. A first step in this direction would be to exhibit a logtime $OROW$-algorithm for $StUSPACE(\log n)$ (or even for $ULIN$).

Another open issue concerns the class $USPACE(\log n)$. Is it possible to place it in $SC^2$ or to give an $o(\log^2 n)$ space algorithm as it has been possible for symmetric logspace and the class $StUSPACE(\log n)$?

A third line of investigation is to consider these questions for polynomial time bounded auxiliary pushdown automata where results like our theorems 1 and 2 are probably harder to obtain. The complexity of $UCFL$, the class of unambiguous context free languages, is uncertain since we do not know whether they are complete for the strongly unambiguous auxiliary pushdown class, just as we don't know whether $ULIN$ is complete for $StUSPACE(\log n)$. Since $StUSPACE(\log n) \subseteq DAuxPDA\text{-}TIME(n^{O(1)})$ and since $DCFL$ is complete for the later class, we know $StUSPACE(\log n) \subseteq LOG(UCFL)$. It is not known whether $UCFL$ is also hard for $USPACE(\log n)$. On the other hand it might well be the case that $UCFL$ is contained in $SC^2$.

# References

1. R. J. Anderson and G. L. Miller. Deterministic parallel list ranking. In *VLSI Algorithms and Architectures, Proc. 3rd Aegean Workshop on Computing*, number 319 in LNCS, pages 81–90. Springer, 1988.
2. A. Blass and Y. Gurevich. On the unique satisfiability problem. *Inform. and Control*, 55:80–88, 1982.
3. G. Buntrock, B. Jenner, K.-J. Lange, and P. Rossmanith. Unambiguity and fewness for logarithmic space. In *Proc. of the 8th Conference on Fundamentals of Computation Theory*, number 529 in LNCS, pages 168–179, 1991.
4. R. Cole and U. Vishkin. Approximate parallel scheduling, part i: the basic technique with applications to optimal parallel list ranking in logarithmic time. *SIAM J. Comp.*, 17:128–142, 1988.

5. S. Cook. Deterministic CFL's are accepted simultaneously in polynomial time and log squared space. In *Proc. of the 11th Annual ACM Symp. on Theory of Computing*, pages 338–345, 1979.

6. S. Cook and P. Dymond. Parallel pointer machines. *Computational Complexity*, 3:19–30, 1993.

7. P. Dymond and W. Ruzzo. Parallel RAMs with owned global memory and deterministic context-free language recoginition. In *Proc. of 13th International Colloquium on Automata, Languages and Programming*, number 226 in LNCS, pages 95–104. Springer, 1986.

8. M. Fellows and N. Koblitz. Self-witnessing polynomial-time complexity and prime factorization. In *Proc. of the 7th IEEE Structure in Complexity Conference*, pages 107–110, 1992.

9. H. Fernau, K.-J. Lange, and K. Reinhardt. Advocating ownership. In *Proc. of the 17th FST&TCS*, 1996. to be published.

10. J. Hopcroft and J. Ullman. *Introduction to Automata Theory, Language, and Computation*. Addison-Wesley, Reading Mass., 1979.

11. N. Immerman. Nondeterministic space is closed under complementation. *SIAM J. Comp.*, 17:935–938, 1988.

12. M. Karchmer and A. Wigderson. On span programs. In *Proc. of the 8th IEEE Structure in Complexity Theory Conference*, pages 102–111, 1993.

13. R.M. Karp and V. Ramachandran. A Survey of Parallel Algorithms for Shared-Memory Machines. In J. van Leeuwen, editor, *Handbook of Theoretical Computer Science, Vol. A*, pages 869–941. Elsevier, Amsterdam, 1990.

14. T. Lam and W. Ruzzo. The power of parallel pointer manipulation. In *Proc. of the 1st ACM Symposium on Parallel Algorithms and Architectures (SPAA'89)*, pages 92–102, 1989.

15. K.-J. Lange. An unambiguous class possessing a complete set. Manuscript, 1996.

16. K.-J. Lange and R. Niedermeier. Data-independences of parallel random access machines. In *Proc. of 13th Conference on Foundations of Software Technology and Theoretical Computer Science*, number 761 in LNCS, pages 104–113. Springer, 1993.

17. B. Monien, W. Rytter, and H. Schäpers. Fast recognition of deterministic cfl's with a smaller number of processors. *Theoret. Comput. Sci.*, 116:421–429, 1993. Corrigendum, 123:427,1993.

18. N. Nisan. $RL \subseteq SC$. In *Proc. of the 24th Annual ACM Symposium on Theory of Computing*, pages 619–623, 1992.

19. N. Nisan, E. Szemeredi, and A. Wigderson. Undirected connectivity in $O(\log^{1.5} n)$ space. In *Proc. of 33th Annual IEEE Symposium on Foundations of Computer Science*, pages 24–29, 1992.

20. P. Rossmanith. The owner concept for PRAMs. In *Proc. of the 8th STACS*, number 480 in LNCS, pages 172–183. Springer, 1991.

21. I. Sudborough. On the tape complexity of deterministic context-free languages. *J. Assoc. Comp. Mach.*, 25:405–414, 1978.

22. R. Szelepcsényi. The method of forcing for nondeterministic automata. *Acta Informatica*, 26:279–284, 1988.

23. A. Wigderson. NL/poly $\subseteq \oplus$L/poly. In *Proc. of the 9th IEEE Structure in Complexity Theory Conference*, pages 59–62, 1994.

# Finding Edge-Disjoint Paths in Partial $k$-Trees
## —— An Extended Abstract ——

Xiao Zhou,* Syurei Tamura** and Takao Nishizeki***

Graduate School of Information Sciences
Tohoku University, Sendai 980-77, JAPAN

**Abstract.** For a given graph $G$ and $p$ pairs $(s_i, t_i)$, $1 \leq i \leq p$, of vertices in $G$, the edge-disjoint paths problem is to find $p$ pairwise edge-disjoint paths $P_i$, $1 \leq i \leq p$, connecting $s_i$ and $t_i$. Many combinatorial problems can be efficiently solved for partial $k$-trees (graphs of treewidth bounded by a fixed integer $k$), but it has not been known whether the edge-disjoint paths problem can be solved in polynomial time for partial $k$-trees unless $p = O(1)$. This paper gives two algorithms for the edge-disjoint paths problem on partial $k$-trees. The first one solves the problem for any partial $k$-tree $G$ and runs in polynomial time if $p = O(\log n)$ and in linear time if $p = O(1)$, where $n$ is the number of vertices in $G$. The second one solves the problem under some restriction on the location of terminal pairs even if $p \geq \log n$.
**Key words.** edge-disjoint paths, partial $k$-tree, bounded tree-width, polynomial-time algorithm, edge-coloring

## 1 Introduction

For a given graph $G$ and $p$ pairs $(s_i, t_i)$, $1 \leq i \leq p$, of vertices in $G$, the *edge-disjoint paths problem* is to find $p$ pairwise edge-disjoint paths $P_i$, $1 \leq i \leq p$, connecting terminals $s_i$ and $t_i$ in $G$. Figure 1 illustrates three edge-disjoint paths in a graph $G$. The edge-disjoint paths problem comes up naturally when analyzing connectivity questions or generalizing (integral) network flow problems. The *vertex-disjoint paths problem* is similarly defined. Both the edge-disjoint and vertex-disjoint paths problems are NP-complete even for planar graphs if $p$ is not bounded [MP93, Vyg95]. If $p = O(1)$, then the vertex-disjoint paths problem can be solved in polynomial time for any graph by Robertson and Seymour's algorithm based on their series of papers on graph minor theory [RS95]. The edge-disjoint paths problem on a graph $G$ can be reduced in polynomial time to the vertex-disjoint paths problem on a new graph $G'$ constructed from $G$ by replacing each vertex with a complete bipartite graph. Therefore, the edge-disjoint paths problem can also be solved in polynomial time for any graph by the algorithm if $p = O(1)$. However, the algorithm is not practically feasible due to the enormous constant factor [RS95]. On the other hand, practical algorithms for both problems with any $p$ have been given for various classes of planar graphs or plane grids in which there are restrictions on the location of terminals or on the degrees of vertices, and these algorithms have been applied to VLSI-routings [Fra82, KM86, MNS85, MNS86, MP86, NSS85, SAN90, Seb93, SIN90, SNS89, WW92].

---

 * Email: zhou@ecip.tohoku.ac.jp
 ** Email: tamura@nishizeki.ecei.tohoku.ac.jp
 *** Email: nishi@ecei.tohoku.ac.jp

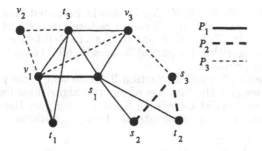

**Fig. 1.** Three edge-disjoint paths $P_1$, $P_2$ and $P_3$ in a partial 3-tree $G$.

The class of partial $k$-trees includes trees ($k = 1$), series-parallel graphs ($k = 2$) [TNS82], Halin graphs ($k = 3$), and $k$-terminal recursive graphs. Any partial $k$-tree can be decomposed into a tree-like structure $T$ of small "basis" graphs, each with at most $k + 1$ vertices. Many problems can be solved efficiently for partial $k$-trees with bounded $k$ by a dynamic programming algorithm based on the tree-decomposition [ACPS93, ALS91, BPT92, Cou90]. In particular, it is rather straightforward to design polynomial-time algorithms for vertex-type problems on partial $k$-trees. For example, the maximum independent vertex-set problem, minimum dominating vertex-set problem, vertex-coloring problem and vertex-disjoint paths problem with any $p$ can be solved all in linear time for partial $k$-trees with bounded $k$ [BPT92, Sch94, TP93]. (Indeed, Robertson and Seymour's algorithm for general graphs partly uses such an algorithm for partial $k$-trees.) However, this is not the case for edge-type problems such as the edge-coloring problem and the edge-disjoint paths problem with any $p$ [BPT92]. It needs sophisticated treatment tailored for individual edge-type problems to design efficient algorithms. (See, for example, sophisticated linear-time algorithms for edge-coloring partial $k$-trees or series-parallel multigraphs [Bod90, ZNN, ZN95, ZSN96].) This is partly due to the following facts: the number of vertices in a basis graph (a node of a tree-decomposition $T$) is bounded by $k + 1$ and hence the size of a DP table required to solve vertex-type problems can be easily bounded by a constant, say $2^{k+1}$ or $(k + 1)^{k+1}$; however, the number of edges incident to vertices in a basis graph is not always bounded and hence it is difficult to bound the size of a DP table for edge-type problems by a constant.

In this paper we give two algorithms to solve the edge-disjoint paths problem for partial $k$-trees $G$ with bounded $k$. Since the edge-disjoint paths problem with $p = O(1)$ can be expressed in the monadic second-order logic [Lag96], a linear-time algorithm can be automatically generated for partial $k$-trees by the general methods in [ACPS93, ALS91, BPT92, Cou90] although the constant factor of the complexity is huge. On the other hand, the first one of our two algorithms solves the edge-disjoint paths problem in polynomial time if $p = O(\log n)$. The computation time is bounded by $n^{O(k)}$, where $n$ is the number of vertices in $G$. Furthermore the algorithm runs in linear time (with a relatively small constant factor) if $p = O(1)$. The second one solves the edge-disjoint paths problem in polynomial time under some restriction on the location of terminal pairs even if $p \geq \log n$. In particular, the algorithm solves the problem for any $p$ in polynomial time if the graph $G^+$ obtained from $G$ by adding $p$ edges $(s_i, t_i)$, $1 \leq i \leq p$, is a partial $k$-tree.

Our idea is to formulate the edge-disjoint paths problem as a new type of an edge-coloring problem, and to bound the size of a DP table by $O((k + 4)^{(k+4)p})$ or $O(n^{(k+2)^{2k+8}})$, applying and extending techniques developed for the ordinary edge-coloring problem [Bod90, ZN95, ZNN]. It should be noted that the edge-disjoint paths problem on a partial $k$-tree $G$ can be reduced to the vertex-disjoint paths problem on a new graph $G'$, but $G'$ is not always a partial $k$-tree. Therefore, the vertex-disjoint paths algorithm for partial $k$-trees [Sch94] cannot be applied to the edge-disjoint paths problem.

The paper is organized as follows. In Section 2 we present some preliminary definitions. In Section 3 we give the first one of our two algorithms for the edge-disjoint paths problems on partial $k$-trees. In Section 4 we give the second. In Section 5 we conclude with some generalizations of our algorithms.

## 2 Terminology and Definitions

In this section we give some definitions. Let $G = (V, E)$ denote a graph with vertex set $V$ and edge set $E$. We often denote by $V(G)$ and $E(G)$ the vertex set and the edge set of $G$, respectively. We denote by $n$ the number of vertices in $G$. The paper deals with *simple undirected* graphs without multiple edges or self-loops. An edge joining vertices $u$ and $v$ is denoted by $(u, v)$. For $E' \subseteq E(G)$, $G[E']$ denotes the subgraph of $G$ induced by the edges in $E'$; $G[E']$ contains every vertex of $G$ to which at least one edge in $E'$ is incident, and hence $G[E']$ contains no isolated vertex.

The class of $k$-*trees* is defined recursively as follows:

(a) A complete graph with $k$ vertices is a $k$-tree.

(b) If $G = (V, E)$ is a $k$-tree and $k$ vertices $v_1, v_2, \cdots, v_k$ induce a complete subgraph of $G$, then $G' = (V \cup \{w\}, E \cup \{(v_i, w)|1 \le i \le k\})$ is a $k$-tree where $w$ is a new vertex not contained in $G$.

(c) All $k$-trees can be formed with rules (a) and (b).

A graph is a *partial $k$-tree* if it is a subgraph of a $k$-tree. Thus a partial $k$-tree $G = (V, E)$ is a simple graph, and $|E| < kn$. Figure 2 illustrates a process of generating a 3-tree. The graph in Figure 1 is indeed a subgraph of the 3-tree, and hence is a partial 3-tree. In this paper we assume that $k$ is a fixed constant.

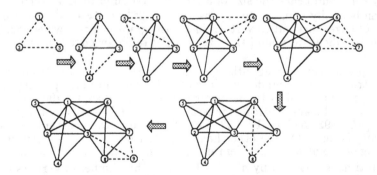

**Fig. 2.** A process of generating a 3-tree.

A *tree-decomposition* of a graph $G = (V, E)$ is a tree $T = (V_T, E_T)$ with $V_T$ a family of subsets of $V$ satisfying the following properties [RS86]:

- $\bigcup_{X_i \in V_T} X_i = V$;
- for every edge $e = (v, w) \in E$, there is a node $X_i \in V_T$ with $v, w \in X_i$; and
- if node $X_j$ lies on the path in $T$ from node $X_i$ to node $X_l$, then $X_i \cap X_l \subseteq X_j$.

The *width* of a tree-decomposition $T = (V_T, E_T)$ is $\max_{X_i \in V_T} |X_i| - 1$. The *treewidth* of graph $G$ is the minimum width of a tree-decomposition of $G$, taken over all possible tree-decompositions of $G$. It is known that every graph with treewidth $\leq k$ is a partial $k$-tree, and conversely, that every partial $k$-tree has a tree-decomposition with width $\leq k$. Bodlaender has given a linear-time sequential algorithm to find a tree-decomposition of $G$ with width $\leq k$ for fixed $k$ [Bod93].

Consider a tree-decomposition of a partial $k$-tree $G$ with treewidth $\leq k$. We transform it to a binary tree $T$ as follows [Bod90]: regard $T$ as a rooted tree by choosing an arbitrary node as the root $X_0$, and replace every internal node $X_i$ with $r$ children by $r + 1$ new nodes $X_{i_1}, X_{i_2}, \cdots, X_{i_{r+1}}$ such that $X_i = X_{i_1} = X_{i_2} = \cdots = X_{i_{r+1}}$, where $X_{i_1}$ has the same father as $X_i$, $X_{i_q}$ is the father of $X_{i_{q+1}}$ and the $q$th child $X_{j_q}$ of $X_i$ $(1 \leq q \leq r)$, and $X_{i_{r+1}}$ is a leaf of $T$. This transformation can be done in $O(n)$ time. $T$ is a tree-decomposition of $G = (V, E)$ with the following characteristics:

- the number of nodes in $T$ is $O(n)$;
- each internal node $X_i$ has exactly two children, say $X_j$ and $X_l$, and either $X_i = X_j$ or $X_i = X_l$; and
- for each edge $(v, w) \in E$ there is at least one leaf $X_l$ with $v, w \in X_l$.

Such a tree $T$ is called a *binary* tree-decomposition [Bod90]. Clearly $T$ has treewidth $\leq k$. For each edge $e = (v, w) \in E$, we choose an arbitrary leaf $X_i$ of $T$ such that $v, w \in X_i$ and denote it by $rep(e)$.

We next define an edge-set $E(X_i) \subseteq E$ for each node $X_i$ of $T$ as follows. If $X_i$ is a leaf of $T$, then let $E(X_i) = \{e \in E| rep(e) = X_i\}$. If $X_i$ is an internal node of $T$ having two children $X_l$ and $X_r$, then let $E(X_i) = E(X_l) \cup E(X_r)$. Note that the two edge-sets $E(X_l)$ and $E(X_r)$ are disjoint. Thus node $X_i$ of $T$ corresponds to a subgraph $G[E(X_i)]$ of $G$ induced by the edges in $E(X_i)$. The subgraph $G[E(X_i)]$ is denoted simply by $G[X_i]$ or $G_i$. Then $G_i$ is an edge-disjoint union of two subgraphs $G_l$ and $G_r$, which share common vertices only in $X_i$ because of the third property of a tree-decomposition.

## 3 The First Algorithm

In this section we give the first one of our two algorithms. Although all our algorithms only decide whether $G$ has $p$ edge-disjoint paths $P_i$, $1 \leq i \leq p$, connecting $s_i$ and $t_i$, they can be easily modified so that they actually find such $p$ edge-disjoint paths.

The main result of this section is the following theorem.

**Theorem 1.** *Let $G = (V, E)$ be a partial $k$-tree of $n$ vertices given by its tree-decomposition with treewidth $\leq k$. Let $(s_i, t_i)$, $1 \leq i \leq p$, be $p$ pairs of vertices in $G$. Then one can determine in time*

$$O(n\{(p + k^2)p^{k(k+1)/2} + p(k + 4)^{2(k+4)p+3}\})$$

*whether $G$ has $p$ pairwise edge-disjoint paths $P_i$, $1 \leq i \leq p$, connecting $s_i$ and $t_i$.*

Since $G$ is a partial $k$-tree with bounded $k$, $|E| \leq kn = O(n)$. Therefore one may assume without loss of generality that $p \leq kn = O(n)$. Thus the first term in the braces above, $(p + k^2)p^{k(k+1)/2}$, is bounded by a polynomial in $n$. The second term $p(k + 4)^{2(k+4)p+3}$ is also bounded by a polynomial if $p = O(\log n)$

since $p$ is in the single exponent over a constant $k + 4$. On the other hand, both of the terms are bounded by a constant if $p = O(1)$. Thus we have the following corollary.

**Corollary 2.** *If $p = O(\log n)$, then the edge-disjoint paths problem can be solved for partial $k$-trees in polynomial time. If $p = O(1)$, then the edge-disjoint paths problem can be solved for partial $k$-trees in linear time.*

In the remainder of this section we will give a proof of Theorem 1. Our idea is to formulate the edge-disjoint paths problem as a new type of an edge-coloring problem, and then to solve the coloring problem using dynamic programming with a table of size at most $(k + 4)^{(k+4)p}$. We employ techniques developed for the ordinary edge-coloring problem [Bod90].

Let $G = (V, E)$ be a graph, and let $(s_i, t_i)$, $1 \le i \le p$, be pairs of vertices in $V$ called *terminals*. Let $C = \{1, 2, \cdots, p\}$ be the set of colors. Any mapping $f : E \to C$ is called a *coloring* of graph $G$. For a color $c \in C$, we denote by $G(f, c)$ the subgraph of $G$ induced by the edges which are colored by $c$. We call $f$ a *correct coloring* of $G$ if, for each color $c \in C$, $G(f, c)$ has a connected component containing both terminals $s_c$ and $t_c$. Figure 3 depicts a correct coloring of the graph in Figure 1. Clearly the following lemma holds.

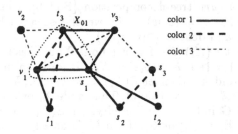

**Fig. 3.** A correct coloring of the partial 3-tree $G$ in Figure 1.

**Lemma 3.** *A graph $G$ has edge-disjoint paths connecting terminal pairs if and only if $G$ has a correct coloring.*

Thus the remaining problem is how to determine whether $G$ has a correct coloring.

Let $X_i$ be a node of a binary tree-decomposition $T$ of a partial $k$-tree $G$. We say that a coloring of graph $G_i = G[X_i]$ is *extensible* if it can be extended to a correct coloring of $G = G[X_{01}]$, where $X_{01}$ is the root of $T$. The number of vertices in node $X_i$ is bounded, but the number of edges of $G_i$ incident to vertices in $X_i$ is not bounded. Therefore, the number of distinct "patterns" of colors of these edges is not polynomially bounded, where the pattern runs over all colorings of $G_i$. However, the number of distinct "color vectors" of colorings is bounded by $(k + 4)^{(k+4)p}$, as we show below.

For a set $X$ we denote by $\mathcal{F}(X)$ the set of all families of pairwise disjoint subsets of $X$. If $x = |X| \ge 1$, then

$$|\mathcal{F}(X)| \le (x + 1)^{x+1}. \tag{1}$$

We call a $p$-tuple $\mathbf{C}(X_i) = (\mathcal{Y}_1, \mathcal{Y}_2, \cdots, \mathcal{Y}_p)$ of families $\mathcal{Y}_1, \mathcal{Y}_2, \cdots, \mathcal{Y}_p$ in $\mathcal{F}(X_i \cup \{s_c, t_c\})$ a *color vector* on node $X_i$. We say that a color vector $\mathbf{C}(X_i) = (\mathcal{Y}_1, \mathcal{Y}_2,$

$\cdots, \mathcal{Y}_p)$ on $X_i$ is *active* if $G_i$ has a coloring $f$ such that for each color $c \in C$

$$\mathcal{Y}_c = \{V(D) \cap (X_i \cup \{s_c, t_c\}) \mid D \text{ is a connected component of } G_i(f, c)\}.$$

Such a vector $\mathbf{C}(X_i)$ is called the *color vector of the coloring* $f$. Thus the coloring of $G = G[X_{01}]$ depicted in Figure 3 has a color vector $\mathbf{C}(X_{01}) = (\mathcal{Y}_{X_{01}1}, \mathcal{Y}_{X_{01}2}, \mathcal{Y}_{X_{01}3})$ such that

$$\begin{aligned}
\mathcal{Y}_{X_{01}1} &= \{\{v_1, s_1, t_1, t_3\}\}, \\
\mathcal{Y}_{X_{01}2} &= \{\{v_1, t_3\}, \{s_2, t_2\}\}, \text{ and} \\
\mathcal{Y}_{X_{01}3} &= \{\{v_1, s_3, t_3\}\}.
\end{aligned}$$

We now have the following lemma.

**Lemma 4.** *Let $X_i$ be any node of a tree-decomposition $T$ of a partial $k$-tree $G$. Let two colorings $f$ and $g$ of $G_i = G[X_i]$ have the same color vector. Then $f$ is extensible if and only if $g$ is extensible.*

Thus a color vector on $X_i$ characterizes an equivalence class of extensible colorings of $G_i$. Since $|X_i| \le k + 1$, $|X_i \cup \{s_c, t_c\}| \le k + 3$. Therefore by Eq. (1) we have $|\mathcal{F}(X_i \cup \{s_c, t_c\})| \le (k + 4)^{k+4}$. Hence there are at most

$$(k + 4)^{(k+4)p} \tag{2}$$

color vectors $\mathbf{C}(X_i) = (\mathcal{Y}_1, \mathcal{Y}_2, \cdots, \mathcal{Y}_p)$ on $X_i$.

The main step of our algorithm is to compute a table of all active color vectors on each node of $T$ from leaves to the root $X_{01}$ of $T$ by means of dynamic programming. From the table on $X_{01}$ one can easily check whether $G$ has $p$ edge-disjoint paths, as follows.

**Lemma 5.** *A partial $k$-tree $G$ has a correct coloring and hence has $p$ edge-disjoint paths if and only if the table on root $X_{01}$ has at least one active color vector $\mathbf{C}(X_{01}) = (\mathcal{Y}_1, \mathcal{Y}_2, \cdots, \mathcal{Y}_p)$ such that, for each $c \in C$, family $\mathcal{Y}_c \in \mathcal{F}(X_{01} \cup \{s_c, t_c\})$ contains a set $Y_{cj}$ with $s_c, t_c \in Y_{cj}$.*

We first compute the table of all active color vectors on each leaf $X_i$ of $T$ as follows:

(1) enumerate all colorings $f$ of $G_i$; and
(2) compute all active color vectors $\mathbf{C}(X_i) = (\mathcal{Y}_1, \mathcal{Y}_2, \cdots, \mathcal{Y}_p)$ on $X_i$ from the colorings $f$ of $G_i$.

Since $|C| = p$ and $|E(X_i)| \le k(k + 1)/2$ for leaf $X_i$, the number of colorings $f : E(X_i) \to C$ is at most $p^{k(k+1)/2}$. For each coloring $f$ of $G_i$, one can compute the color vector of $f$ in time $O(p + k^2)$. Therefore, steps (1) and (2) can be done for a leaf in time $O((p + k^2)p^{k(k+1)/2})$. Since $T$ has $O(n)$ leaves, the tables on all leaves can be computed in time $O(n(p + k^2)p^{k(k+1)/2})$, which corresponds to the first term in the braces of the complexity mentioned in Theorem 1.

We next compute all active color vectors on each internal node $X_i$ of $T$ from leaves to the root. Lemma 6 below shows how to compute all active color vectors on $X_i$ from all active color vectors on the left and right children $X_l$ and $X_r$ of $X_i$. We now introduce a notion of a family $\mathcal{U}(\mathcal{Y}_{lc}, \mathcal{Y}_{rc}; X)$ for families $\mathcal{Y}_{lc}, \mathcal{Y}_{rc}$ and a set $X$. Let $\mathbf{C}(X_l) = (\mathcal{Y}_{l1}, \mathcal{Y}_{l2}, \cdots, \mathcal{Y}_{lp})$ be an active color vector on $X_l$, and let $\mathbf{C}(X_r) = (\mathcal{Y}_{r1}, \mathcal{Y}_{r2}, \cdots, \mathcal{Y}_{rp})$ be an active color vector on $X_r$. Figure 4 illustrates

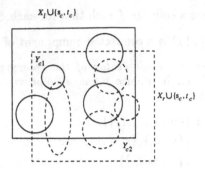

**Fig. 4.** Venn diagrams of $\mathcal{Y}_{lc}$ and $\mathcal{Y}_{rc}$ with two clusters $Y_{c1}$ and $Y_{c2}$.

Venn diagrams of $\mathcal{Y}_{lc}$ and $\mathcal{Y}_{rc}$, $c \in C$, where the sets in $\mathcal{Y}_{lc}$ are indicated by circles of solid lines and the sets in $\mathcal{Y}_{rc}$ by circles of dotted lines. All vertices shared by graphs $G_l$ and $G_r$ are contained in $X_i$, and $X_i \subseteq X_l \cup X_r$. Therefore each family $\mathcal{Y}_c$ in a color vector $\mathbf{C}(X_i)$ on $X_i$ corresponds to a "cluster" in Figure 4, formally defined as follows. For each color $c \in C$, let $G_{Bc} = (\mathcal{Y}_{lc} \cup \mathcal{Y}_{rc}, E_c)$ be a bipartite graph with partite sets $\mathcal{Y}_{lc}$ and $\mathcal{Y}_{rc}$, where a vertex $Y_{lc} \in \mathcal{Y}_{lc}$ and a vertex $Y_{rc} \in \mathcal{Y}_{rc}$ are joined by an edge in $E_c$ iff $Y_{lc} \cap Y_{rc} \neq \phi$. Let $D_{c1}, D_{c2}, \cdots, D_{cb}$ be the connected components of $G_{Bc}$, and for each $j$, $1 \leq j \leq b$, let

$$Y_{cj} = \bigcup_{Y \in V(D_{cj})} Y.$$

Then $Y_{cj}$ is a "cluster" mentioned above, and corresponds to the vertex set of a connected component of $G_i(f, c)$ for the coloring $f$ of $G_i$ extending the colorings of $G_l$ and $G_r$ having color vectors $\mathbf{C}(X_l)$ and $\mathbf{C}(X_r)$, respectively. For a set $X \subseteq V$ we define a family $\mathcal{U}(\mathcal{Y}_{lc}, \mathcal{Y}_{rc}; X)$ of vertex sets, as follows:

$$\mathcal{U}(\mathcal{Y}_{lc}, \mathcal{Y}_{rc}; X) = \{Y_{cj} \cap X \mid 1 \leq j \leq b\}.$$

We have the following lemma.

**Lemma 6.** *Let an internal node $X_i$ of $T$ have two children $X_l$ and $X_r$. Then a color vector $\mathbf{C}(X_i) = (\mathcal{Y}_1, \mathcal{Y}_2, \cdots, \mathcal{Y}_p)$ on $X_i$ is active if and only if there exist two active color vectors $\mathbf{C}(X_l) = (\mathcal{Y}_{l1}, \mathcal{Y}_{l2}, \cdots, \mathcal{Y}_{lp})$ on $X_l$ and $\mathbf{C}(X_r) = (\mathcal{Y}_{r1}, \mathcal{Y}_{r2}, \cdots, \mathcal{Y}_{rp})$ on $X_r$ such that $\mathcal{Y}_c = \mathcal{U}(\mathcal{Y}_{lc}, \mathcal{Y}_{rc}; X_i \cup \{s_c, t_c\})$ for each color $c \in C$.*

Since $|\mathcal{Y}_{lc}|, |\mathcal{Y}_{rc}| \leq k + 3$, the bipartite graph $G_{Bc} = (\mathcal{Y}_{lc} \cup \mathcal{Y}_{rc}, E_c)$ has at most $(k+3)^2$ edges, that is $|E_c| \leq (k+3)^2$. Clearly one can check in time $O(k+3)$ whether $Y_{lc} \cap Y_{rc} \neq \phi$. Therefore each bipartite graph $G_{Bc}$ can be constructed in time $O((k+3)^3)$, and hence all $p$ bipartite graphs can be constructed in time $O(p(k+3)^3)$. Thus one can compute an active color vector on $X_i$ from a pair of active color vectors on $X_l$ and on $X_r$ in time $O(p(k+3)^3)$. By Eq. (2) there are at most $(k+4)^{(k+4)p}$ active color vectors on $X_l$ and at most $(k+4)^{(k+4)p}$ active color vectors on $X_r$. Therefore there are at most $(k+4)^{2(k+4)p}$ pairs of active color vectors on $X_l$ and on $X_r$. Thus one can compute all active color vectors on $X_i$ in time $O(p(k+4)^{2(k+4)p+3})$. Since $T$ has $O(n)$ internal nodes, one can compute the tables for all internal nodes in time $O(np(k+4)^{2(k+4)p+3})$, which corresponds to the second term of the complexity in Theorem 1.

This completes a proof of Theorem 1.

# 4 The Second Algorithm

In this section we give the second one of our two algorithms. One may assume without loss of generality that any vertex $v$ in a graph $G$ is designated as at most one terminal and that every terminal has degree one, as follows. For each terminal, add $G$ a new vertex, join the terminal and the new vertex, and newly designate the new vertex as the terminal. Then the resulting graph is also a partial $k$-tree.

In the previous section all terminals can be located anywhere in a graph although $p = O(\log n)$. In this section we show that the edge-disjoint paths problem can be solved in polynomial time for partial $k$-trees under some restriction on the location of terminal pairs even if $p \geq \log n$. The main result of this section is the following theorem, whose proof is omitted in this extended abstract due to the space limitation but is included in our technical report [ZTN96].

**Theorem 7.** *Let $G$ be a partial $k$-tree given by its tree-decomposition $T$ with treewidth $\leq k$. Let $(s_i, t_i)$, $1 \leq i \leq p$, be $p$ pairs of terminals. Then one can determine in polynomial time whether $G$ has $p$ pairwise edge-disjoint paths $P_i$, $1 \leq i \leq p$, connecting $s_i$ and $t_i$ if there exist some nodes $X_1, X_2, \cdots, X_q$ of $T$ satisfying the following three conditions:*

(i) *each $X_j$, $1 \leq j \leq q$, is not a descendant of any other node $X_{j'}$, $1 \leq j' \leq q$,*
(ii) *each terminal pair $s_i, t_i, 1 \leq i \leq p$, are contained in graph $G_j = G[X_j]$ for some $j$, $1 \leq j \leq q$, and*
(iii) *each graph $G_j$, $1 \leq j \leq q$, contains $O(\log n)$ terminals.*

If the graph $G^+$ obtained from $G$ by adding $p$ edges $(s_i, t_i)$, $i \leq i \leq p$, is a partial $k$-tree, then $G$ has a binary tree-decomposition $T$ which contains leaves $X_1, X_2, \cdots, X_q$ such that each terminal pair $s_i, t_i$ are contained in graph $G_j$ for some $j$, $1 \leq j \leq q$, and each graph $G_j$, $1 \leq j \leq q$, contains at most $k(k + 1)/2 = O(1)$ terminal pairs. Therefore Theorem 7 immediately leads to the following corollary.

**Corollary 8.** *Let $(s_i, t_i)$, $1 \leq i \leq p$, be $p$ pairs of vertices in a graph $G$. If $G^+$ is a partial $k$-tree, then, for any $p$, one can determine in polynomial time whether $G$ has $p$ pairwise edge-disjoint paths $P_i$, $1 \leq i \leq p$, connecting $s_i$ and $t_i$.*

Middendorf and Pfeiffer announced in [MP93] that they were preparing a paper proving under the assumption $P \neq NP$ that the edge-disjoint paths problem for a partial $k$-tree $G$ is solvable in polynomial time if $G^+$ is a partial $k$-tree. However, they have not finished writing the paper [M96]. Corollary 8 implies the same claim without the assumption, and moreover provides a concrete algorithm of smaller complexity.

# 5 Conclusion

In this paper we gave two polynomial-time algorithms for the edge-disjoint paths problem on partial $k$-trees $G$ with bounded $k$. In the first algorithm all terminals can be located anywhere in $G$. If the number $p$ of terminal pairs is $O(\log n)$, then the computation time is bounded by $n^{O(k)}$. If $p = O(1)$, then the time is $O(n)$ where the hidden constant is singly exponential in $p$ and $k$. Thus the first algorithm is practically feasible when $p$ and $k$ are not large. On the other hand, the second algorithm solves the problem even if $p \geq \log n$ although there is a restriction on the location of terminal pairs. The computation time is bounded by a polynomial in $n$, the exponent of which is not a high tower of $k$ but is singly exponential in $k$. Thus the second algorithm looks to be practically feasible when $k$ is very small.

Our algorithms can be extended as follows:

(a) An extended algorithm finds the *maximum number* of edge-disjoint paths connecting terminal pairs in the same amount of time when $G$ does not have $p$ paths.

(b) An extended algorithm solves the edge-disjoint paths problem for a *multi-graph* whose underlying simple graph is a partial $k$-tree.

(c) Given a partial $k$-tree with nonnegative edge-lengths, an extended algorithm finds $p$ edge-disjoint paths with the *minimum total length* in the same amount of time.

(d) Given a partial $k$-tree with directed edges, an extended algorithm finds $p$ edge-disjoint *directed paths* in the same amount of time.

(e) Given a partial $k$-tree $G = (V, E)$ and subsets $N_i$, $1 \leq i \leq p$, of $V$ (called *nets*), an algorithm finds $p$ pairwise edge-disjoint *trees* $T_i$, $1 \leq i \leq p$, covering net $N_i$, that is, $N_i \subseteq V(T_i)$. The algorithm extended from the first one runs in polynomial time if $pr \log r = O(\log n)$ where $r = \max\{|N_1|, |N_2|, \cdots, |N_p|\}$. The algorithm extended from the second one runs in polynomial time if each $N_i$ is contained in some node.

(f) Our algorithms can be implemented to NC *parallel algorithms* which find $p$ edge-disjoint paths in $O(\log n)$ parallel time with a polynomial number of processors. The key idea is to apply the "tree contraction" to the bottom-up tree computation above in a sophisticated way like in [ZN95, ZNN].

The methodology we developed for the edge-disjoint paths problem can be applied to other edge-type problems, say the edge-set partition problem for various graph properties.

## Acknowledgment

We wish to thank Professors J. Lagergren and H. Suzuki for fruitful discussions.

## References

[ACPS93] S. Arnborg, B. Courcelle, A. Proskurowski, and D. Seese. An algebraic theory of graph reduction. *Journal of the Association for Computing Machinery*, Vol. 40, No. 5, pp. 1134–1164, 1993.

[ALS91] S. Arnborg, J. Lagergren and D. Seese. Easy problems for tree-decomposable graphs. *Journal of Algorithms*, Vol. 12, No. 2, pp. 308–340, 1991.

[Bod90] H. L. Bodlaender. Polynomial algorithms for graph isomorphism and chromatic index on partial $k$-trees. *Journal of Algorithms*, Vol. 11, No. 4, pp. 631–643, 1990.

[Bod93] H. L. Bodlaender. A linear time algorithm for finding tree-decompositions of small treewidth. In *Proc. of the 25th Ann. ACM Symp. on Theory of Computing*, pp. 226–234, San Diego, CA, 1993.

[BPT92] R. B. Borie, R. G. Parker, and C. A. Tovey. Automatic generation of linear-time algorithms from predicate calculus descriptions of problems on recursively constructed graph families. *Algorithmica*, Vol. 7, pp. 555–581, 1992.

[Cou90] B. Courcelle. The monadic second-order logic of graphs I: Recognizable sets of finite graphs. *Information and Computation*, Vol. 85, pp. 12–75, 1990.

[Fra82] A. Frank. Disjoint paths in rectilinear grids. *Combinatorica*, Vol. 2, No. 4, pp. 361–371, 1982.

[KM86] M. Kaufmann and K. Mehlhorn. Routing through a generalized switchbox. *J. Algorithms*, Vol. 7, pp. 510–531, 1986.

[Lag96] J. Lagergren. Private communication, May 2, 1996.

[MNS85] K. Matsumoto, T. Nishizeki, and N. Saito. An efficient algorithm for finding multicommodity flows in planar networks. *SIAM J. Comput.*, Vol. 14, No. 2, pp. 289–302, 1985.

[MNS86] K. Matshmoto, T. Nishizeki, and N. Saito. Planar multicommodity flows, maximum matchings and negative cycles. *SIAM J. Comput.*, Vol. 15, No. 2, pp. 495–510, 1986.

[MP86] K. Mehlhorn and F.P. Preparata. Routing through a rectangle. *J. ACM*, Vol. 33, No. 1, pp. 60–85, 1986.

[M96] M. Middendorf. Private communication, May 1996.

[MP93] M. Middendorf and F. Pfeiffer. On the complexity of disjoint paths problem. *Combinatorica*, Vol. 13, No. 1, pp. 97–107, 1993.

[NSS85] T. Nishizeki, N. Saito, and K. Suzuki. A linear-time routing algorithm for convex grids. *IEEE Trans. Comput.-Aided Design*, Vol. 4, No. 1, pp. 68–76, 1985.

[RS86] N. Robertson and P.D. Seymour. Graph minors. II. Algorithmic aspects of tree-width. *Journal of Algorithms*, Vol. 7, pp. 309–322, 1986.

[RS95] N. Robertson and P.D. Seymour. Graph minors. XIII. The disjoint paths problem. *J. of Combin. Theory, Series B*, Vol. 63, No. 1, pp. 65–110, 1995.

[SAN90] H. Suzuki, T. Akama, and T. Nishizeki. Finding Steiner forests in planar graphs. In *Proc. of the First Ann. ACM-SIAM Sympo. on Discrete Algorithms*, pp. 444–453, 1990.

[Sch94] P. Scheffler. A practial linear time algorithm for disjoint paths in graphs with bounded tree-width. Technical Report, 396, Dept. of Mathematics, Technische Universität Berlin, 1994.

[Seb93] A. Sebö. Integer plane multiflows with a fixed number of demands. *J. of Combin. Theory, Series B*, Vol. 59, pp. 163–171, 1993.

[SIN90] H. Suzuki, A. Ishiguro, and T. Nishizeki. Edge-disjoint paths in a grid bounded by two nested rectangles. *Discrete Applied Mathematics*, Vol. 27, pp. 157–178, 1990.

[SNS89] H. Suzuki, T. Nishizeki, and N. Saito. Algorithms for multicommodity flows in planar graphs. *Algorithmica*, Vol. 4, pp. 471–501, 1989.

[TNS82] K. Takamizawa, T. Nishizeki, and N. Saito. Linear-time computability of combinatorial problems on series-parallel graphs. *Journal of ACM*, Vol. 29, No. 3, pp. 623–641, 1982.

[TP93] J.A. Telle, and A. Proskurowski. Practical algorithms on partial $k$-trees with an application to domination like problems. *Proc. of Workshop on Algorithms and Data Structures, WADS'93*, Montreal, Lect. Notes on Computer Science, Springer-Verlag, 709, pp. 610–621, 1993.

[Vyg95] J. Vygen. NP-completeness of some edge-disjoint paths problems. *Discrete Appl. Math.*, Vol. 61, pp. 83–90, 1995.

[WW92] D. Wagner and K. Weihe. A linear-time algorithm for edge-disjoint paths in planar graphs. *Proc. of the First Europ. Symp. on Algorithms* (ESA'93), Lect. Notes in Computer Science, Springer-Verlag, 726, pp. 384–395, 1992.

[ZN95] X. Zhou and T. Nishizeki. Optimal parallel algorithms for edge-coloring partial $k$-trees with bounded degrees. *IEICE Trans. on Fundamentals of Electronics, Communication and Computer Sciences*, Vol. E78-A, pp. 463–469, 1995.

[ZNN] X. Zhou, S. Nakano, and T. Nishizeki. Edge-coloring partial $k$-trees. *Journal of Algorithms, to appear*.

[ZSN96] X. Zhou, H. Suzuki, and T. Nishizeki. A linear algorithm for edge-coloring series-parallel multigraphs. *Journal of Algorithms*, Vol. 20, pp. 174–201, 1996.

[ZTN96] X. Zhou, S. Tamura, and T. Nishizeki. Finding edge-disjoint paths in partial $k$-trees. *Tech. Rept. 96-1*, Dept. of Inf. Eng., Tohoku Univ., Sept. 1996.

# Optimal Augmentation for Bipartite Componentwise Biconnectivity in Linear Time (Extended Abstract)

Tsan-sheng Hsu*and Ming-Yang Kao**

**Abstract.** A graph is componentwise fully biconnected if every connected component either is an isolated vertex or is biconnected. We consider the problem of adding the smallest number of edges to make a bipartite graph componentwise fully biconnected while preserving its bipartiteness. This problem has important applications for protecting sensitive information in cross tabulated tables. This paper presents a linear-time algorithm for the problem.

## 1 Introduction

The problem of adding the minimum number of edges to make a given graph biconnected is called the *smallest biconnectivity augmentation problem*. This problem has been extensively studied for general graphs [3, 17, 18]. A simplified sequential algorithm which corrects an error in [18] and an efficient parallel algorithm are reported in [12]. Efficient polynomial-time algorithms for other vertex connectivity augmentation problems can be found in [8, 9, 20].

A graph is *componentwise fully biconnected* if every connected component either is biconnected or is an isolated vertex. This paper presents a linear-time algorithm for the problem of adding the smallest number of edges into a given bipartite graph to make it componentwise fully biconnected while maintaining its bipartiteness. Our problem arises naturally from research on data security of cross-tabulated tables [15]. To protect sensitive information in a cross tabulated table $T$, it is a common practice to suppress some of the cells in $T$. A fundamental issue concerning the effectiveness of this practice is how a table maker can suppress a small number of cells in addition to the sensitive ones so that the resulting $T$ does not leak significant information.

This protection problem can be reduced to various augmentation problems for a bipartite graph $G$ constructed from $T$ depending on different levels of security requirements [5, 6, 10, 15, 13, 14, 16]. The rows and the columns of $T$ form the two vertex sets of $G$, respectively; $G$ has an edge between row $i$ and column $j$ if and only if the value of the $(i, j)$-th cell in $T$ is suppressed [6]. We say that $T$ is *level-1 protected* if the value of each suppressed cell cannot be uniquely determined and no nontrivial information (e.g., partial sums of suppressed cells) about each row or column is revealed. It has been shown in [6] that $G$ has a cut

---

* Institute of Information Science, Academia Sinica, Nankang 11529, Taipei, Taiwan, ROC, E-mail: *tshsu@iis.sinica.edu.tw*. Research supported in part by NSC Grants 84-2213-E-001-005 and 85-2213-E-001-003.

** Department of Computer Science, Duke University, Durham, NC 27708, USA, E-mail: *kao@cs.duke.edu*. Research supported in part by NSF Grant CCR-9101385.

edge between row $i$ and column $j$ if and only if the value of the $(i, j)$-th cell in $\mathcal{T}$ can be deduced from the published data of $\mathcal{T}$. In [6], a linear-time algorithm is also given to augment a graph with the smallest number of edges such that the resulting graph contains no cut edge.

In addition to the above, a row vertex in $G$ is a cut vertex if and only if nontrivial information about that row in $\mathcal{T}$ is revealed [15]. The same holds for columns. As a consequence, the problem of suppressing the minimum number of additional cells to level-1 protect $\mathcal{T}$ is equivalent to that of adding the minimum number of edges to $G$ such that $G$ has no cut edge and cut vertex. Note that the newly added edge must preserve the bipartiteness of $G$. Note also that a graph with no cut vertex contains no cut edge. Thus we can solve the level-1 data protection problem by solving the smallest bipartite componentwise biconnectivity augmentation problem. There is no polynomial-time algorithm known previous for this problem. In this paper, we present a linear-time algorithm for this problem.

## 2 Definitions

Throughout this paper, let $G = (A, B, E)$ denote a bipartite graph, where $|A| \geq 2$ and $|B| \geq 2$ are two sets of vertices and $E$ is the set of edges.

A *trivial* connected component is an isolated vertex. A *cut vertex* (or *cut edge*) is one whose removal increases the number of connected components. A connected graph is *biconnected* if it has at least three vertices and no cut vertex. A *legal edge* of $G$ is an edge in $A \times B$ but not in $E$. A *biconnector* of $G$ is a set $L$ of legal edges such that $G \cup L$ is componentwise fully biconnected. A biconnector is *optimal* if it is one with the smallest number of edges. Note that if $|A| < 2$ or $|B| < 2$, then the optimal biconnector for $G$ is trivial. In light of this, the *optimal biconnector problem* is the following: given $G = (A, B, E)$ with $|A| \geq 2$ and $|B| \geq 2$, find an optimal biconnector of $G$.

A *block* in a graph is either the set of a single vertex that is not in any biconnected component or the set of vertices in a biconnected component. A block with exactly one vertex is a *singular block*. Let $nc(G)$ denote the number of connected components in $G$. A *strict* cut vertex is a cut vertex $c$ such that (1) $c$ is not an endpoint of a cut edge, or (2) $nc(G - \{c\}) - nc(G) \geq 2$.

A block is a *leaf-block* if it either (1) is a singular block which contains one endpoint of a cut edge or (2) contains exactly one strict cut vertex and no endpoint of any cut edge. A vertex is *demanding* if (1) it is the only vertex in a leaf-block or (2) it is neither a cut vertex nor an endpoint of a cut edge.

We use a well-known data structure $\Lambda(G)$ [7, 11, 19], called the *bc-forest* of $G$, to organize the non-strict cut blocks, cut edges, and strict cut vertices in $G$. The bc-forest is a tree if $G$ is connected and is a forest if $G$ is disconnected. Our augmentation algorithm works with this forest instead of $G$ directly. The vertices in $\Lambda(G)$ corresponding to blocks are called the *b-vertices*, and those corresponding to strict cut vertices and cut edges are called the *c-vertices*. Note that leaf-blocks correspond to b-vertices that are leaves in $\Lambda(G)$.

# 3 A lower bound on the size of optimal biconnectors

We classify the vertices and leaf-blocks of $G$ with the following definitions. A vertex is of *type-A* if it is in $A$. A vertex is of *type-B* if it is in $B$. A leaf-block is of *type-A* if all of its demanding vertices are in $A$. A leaf-block is of *type-B* if all of its demanding vertices are in $B$. A block is of *type-AB* if it has at least one demanding vertex in $A$ and one demanding vertex in $B$.

Let $\Psi'$ be a set of leaf-blocks in $G$. A *legal pair* of $\Psi'$ is two distinct elements in $\Psi'$ that are paired according to the following rules. Type-$A$ may pair with type-$B$ or type-$AB$. Type-$B$ may pair with type-$A$ or type-$AB$. Type-$AB$ may pair with all three types.

A *corresponding* legal edge of $G$ for a legal pair is a legal edge whose two endpoints are demanding vertices in the blocks. A *legal matching* of $\Psi'$ is a set of legal pairs of $\Psi'$ such that each element in $\Psi'$ is contained in at most one legal pair. A legal matching of $\Psi'$ with the largest cardinality can be obtained by iteratively applying any rule below whenever applicable:

- If all unpaired elements are type-$AB$, we pair two type-$AB$ elements.
- If there are one unpaired type-$A$ element and one unpaired type-$B$ element, we pair a type-$A$ element and a type-$B$ element.
- If there is no unpaired type-$B$ element and there are one unpaired type-$A$ element and one unpaired type-$AB$ element, we pair a type-$A$ element with a type-$AB$ element.
- If there is no unpaired type-$A$ element and there are one unpaired type-$B$ element and one unpaired type-$AB$ element, we pair a type-$B$ element with a type-$AB$ element.

$\Psi(G)$ denotes the set of leaf-blocks of $G$. For $\Psi' \subseteq \Psi(G)$, $\mathcal{M}(\Psi')$ denotes the maximum cardinality of a legal matching of $\Psi'$. For a maximum legal matching of $\Psi'$, $\mathcal{R}(\Psi')$ denotes the number of elements in $\Psi'$ that is not in the given maximum legal matching. Note that $\mathcal{R}(\Psi')$ is the same for any maximum legal matching of $\Psi'$.

For all vertices $u \in G$, $\mathcal{D}(u, G)$ denotes the number of connected components in $X - \{u\}$ where $X$ is the connected component of $G$ containing $u$. $\mathcal{C}(G)$ denotes the number of connected components in $G$ that are not fully biconnected. $\mathcal{B}(G)$ denotes the number of edges in an optimal biconnector of $G$. The next notation denotes our target size for an optimal biconnector of $G$: $\alpha(G) = \max_{u \in G}\{\mathcal{D}(u, G) + \mathcal{C}(G) - 2, \mathcal{M}(\Psi(G)) + \mathcal{R}(\Psi(G))\}$, which is shown in Theorem 1 and whose proof is omitted.

**Theorem 1.** $\mathcal{B}(G) \geq \alpha(G)$. □

# 4 Determining the optimal biconnector for a special case

In this section, we consider the case when the graph is connected with with at least two vertices in $A$ and two vertices in $B$. During the discussion, we also assume that $|\Psi(G)| > 3$, since otherwise it is obvious to prove $\alpha(G) = \mathcal{M}(\Psi(G)) +$

$\mathcal{R}(\Psi(G))$. Note that $\alpha(G) = \max_{u \in G}\{\mathcal{D}(u, G) - 1, \mathcal{M}(\Psi(G)) + \mathcal{R}(\Psi(G))\}$ for this case. We need the following definitions. (1) A cut vertex $u$ is *massive* if $\mathcal{D}(u, G) - 1 > \mathcal{M}(\Psi(G)) + \mathcal{R}(\Psi(G))$. (2) A cut vertex $u$ is *critical* if $\mathcal{D}(u, G) - 1 = \mathcal{M}(\Psi(G)) + \mathcal{R}(\Psi(G))$. (3) A graph with no massive c-vertex is *balanced*.

Theorem 2 is the main result of this section. If Theorem 2 is proved, then it follows from Theorem 1 that $\mathcal{B}(G) = \alpha(G)$.

**Theorem 2.** *There is a legal edge $e$ such that $\alpha(G \cup \{e\}) = \alpha(G) - 1$.*
*Proof.* Assume that $\Lambda(G)$ has more than three leaves. The following is well-known [3, 12, 18]. (1) There is at most one massive vertex. (2) If there is a massive vertex, then there is no critical vertex. (3) There are at most two critical vertices. (4) If there are two critical vertices, then $|\Psi(G)| = 2 \cdot \mathcal{M}(\Psi(G))$ and $\mathcal{R}(\Psi(G)) = 0$. Thus At least one of the following four cases holds for $G$.
*Case* I. $\mathcal{M}(\Psi(G)) = 0$.
*Case* II. $\mathcal{M}(\Psi(G)) > 0$ and there are two critical c-vertices.
*Case* III. $\mathcal{M}(\Psi(G)) > 0$, $G$ is balanced and there is at most one critical c-vertex.
*Case* IV. $\mathcal{M}(\Psi(G)) > 0$ and there is one massive c-vertex.

We will prove Theorem 2 through a sequence of lemmas in §4.1–§4.4 according to the above four cases. □

## 4.1 Case I of Theorem 2

This case assumes $\mathcal{M}(\Psi(G)) = 0$. We now construct an optimal biconnector as follows. Let $k = |\Psi(G)|$. By Theorem 1 and the fact that $\alpha(G) = |\Psi(G)|$, it suffices to construct a biconnector $L$ of $k$ edges for $G$. Let $Y_1, \ldots, Y_k$ be the leaf-blocks in $G$. Because $\mathcal{M}(\Psi(G)) = 0$, by the way a legal matching is defined these blocks are all type-$A$ or all type-$B$. Furthermore, each block is also singular. Assume without loss of generality that they are type-$A$. Then, there is a cut edge $(x_i, y_i)$ for each $Y_i = \{y_i\}$. Because we assume there are at least two type-$B$ vertices in $G$ and $\mathcal{M}(\Psi(G)) = 0$, there are at least two distinct vertices $u$ and $v$ among $x_1, \ldots, x_k$. We construct an optimal biconnector $L$ as follows:

- Let $\Psi_u$ (respectively, $\overline{\Psi}_u$) be the set of all $Y_i$ with $x_i = u$ (respectively, $x_i \neq u$).
- For each $Y_i \in \Psi_u$ (respectively, $\overline{\Psi}_u$), let $e_i = (y_i, v)$ (respectively, $(y_i, u)$).
- Let $L_u$ (respectively, $\overline{L}_u$) be the set of all $e_i$ associated with the blocks in $\Psi_u$ (respectively, $\overline{\Psi}_u$).
- Let $L = L_u \cup \overline{L}_u$.

Thus $|L| = k$. The part showing that $L$ is a biconnector is omitted.

## 4.2 Case II of Theorem 2

This case assumes that $\mathcal{M}(\Psi(G)) > 0$ and there are exactly two critical vertices. Let $P = u_1, \ldots, u_k$ be a path in $G$. $P$ is *branchless* if for all $i$ with $1 < i < k$ the degree of $u_i$ in $G$ is two. $P$ is *branching* if it is not branchless. A subtree rooted at a child of the root of a rooted tree is a *branch* of the given tree.

In this case, $\Lambda(G)$ has a very special shape [18] as described in the following. The graph $\Lambda(G)$ is a tree with $k$ leaves such that there are exactly two vertices $u$ and $v$ each of degree $1 + \frac{k}{2}$. Furthermore, (1) the value $k$ is even; (2) there is a unique branchless path in $T$ that connects $u$ and $v$; (3) for each leaf, there is a unique branchless path in $T$ that connects the leaf to either $u$ or $v$ and does not go through the other. We call $\Lambda(G)$ a *double star* centered at $u_1$ and $u_2$. The two vertices $u_1$ and $u_2$ are critical vertices. We say that a leaf is *clung* to $u_i$ if it is connected to $u_i$ through a path as described in the lemma. Note that $\mathcal{R}(\Psi(G)) = 0$ in this case.

We construct an optimal biconnector of cardinality $\mathcal{M}(\Psi(G))$ such that each pair in it consists of a leaf-block clung to $u_1$ and one clung to $u_2$ as follows. Let $b_1, \ldots, b_{r_1}, b_{r_1+1}, \ldots, b_{r_2}, b_{r_2+1}, \ldots, b_{r_3}$ be the leaves clung to $u_1$, where $b_i$, $1 \le i \le r_1$ are type-$A$, $b_i$, $r_1 < i \le r_2$ are type-$AB$, and $b_i$, $r_2 < i \le r_3$ are type-$B$. Let $w_1, \ldots, w_{s_1}, w_{s_1+1}, \ldots, w_{s_2}, w_{s_2+1}, \ldots, w_{s_3}$ be the leaves clung to $u_2$, where $w_i$, $1 \le i \le s_1$ are type-$B$, $w_i$, $s_1 < i \le s_2$ are type-$AB$, and $w_i$, $s_2 < i \le s_3$ are type-$B$. Since $\Lambda(G)$ is a double star, $r_3 = s_3 = \ell$, where $2 \cdot \ell$ is the number of leaves in $\Psi(G)$. We construct a legal matching $Q = \{(b_i, w_i) \mid 1 \le i \le \ell\}$. We omit the proof of $Q$ being a legal matching.

## 4.3 Case III of Theorem 2

This case assumes that $\mathcal{M}(\Psi(G)) > 0$, there is no massive c-vertex, and there is at most one critical c-vertex. Let $c_1$ be a c-vertex in $\Lambda(G)$ of the largest degree among all c-vertices. The next lemma identifies the conditions under which adding an edge decreases the number of leaves in $\Lambda(G)$ by two.

**Definition 3 [12].** Let $P$ be the path in $\Lambda(G)$ between two leaf-blocks $B_1$ and $B_2$ that are a legal pair. The pair $B_1$ and $B_2$ (or the path $P$) satisfy the *leaf-connecting condition* if $P$ has (1) a b-vertex of degree at least four or (2) two vertices each of degree at least three.

**Lemma 4 [12].** *Let $G'$ be the resulting graph obtained from $G$ after adding a corresponding legal edge between a legal pair that satisfy the leaf-connecting condition. Then $\Psi(G') = \Psi(G) - 2$ and $\mathcal{M}(\Psi(G')) + \mathcal{R}(\Psi(G')) = \mathcal{M}(\Psi(G)) + \mathcal{R}(\Psi(G)) - 1$.* $\square$

Thus we use Lemmas 5 and 6 to find a legal pair. Then adding an edge according to the legal pair can decrease $\alpha(G)$ by one. We keep on doing this until $\alpha(G)$ is 0. Thus we have added exactly $\alpha(G)$ edges to biconnect $G$.

**Lemma 5.** *Assume that $\Lambda(G)$ has more than three leaves and is rooted at a vertex $r$ of degree more than two. Let $h$ be a vertex in $\Lambda(G)$ other than $r$ with degree at least three. If $\mathcal{M}(\Psi(G)) > 0$, then there are two leaves $w_1$ and $w_2$ with the following properties: (1) $w_1$ and $w_2$ are a legal pair; (2) the path in $\Lambda(G)$ between $w_1$ and $w_2$ passes through $h$ and $z$, where $z$ is a vertex of degree at least three and $z \ne h$.*

*Proof.* Let $T_h$ be the subtree of $\Lambda(G)$ rooted at $h$. There are three cases.

*Case* 1. $T_h$ has a type-$AB$ leaf. Let this leaf be $w_1$. Let $w_2$ be a leaf in a branch of $r$ not containing $h$. Thus $w_1$ and $w_2$ are the pair we want, i.e., $z$ is $r$.

*Case* 2. $T_h$ has both type-$A$ and type-$B$ leaves, but no type-$AB$ leaves. Let $w_2$ be a leaf in a branch of $r$ not containing $h$. If $w_2$ is type-$AB$ or type-$A$ (respectively, type-$B$), then let $w_1$ be a type-$B$ (respectively, type-$A$) leaf in $T_h$. Thus $w_1$ and $w_2$ are the pair we want, i.e., $z$ is $r$.

*Case* 3. $T_h$ has type-$A$ (respectively, type-$B$) leaves only. Let $w_1$ be a leaf in $T$. Without loss of generality, $w_1$ is type-$A$. Since $\mathcal{M}(\Psi(G)) > 0$, some leaf $w_2$ can match with $w_1$. It is either the case that $w_2$ is in a branch of $r$ not containing $h$ or the case that there is a vertex $x \neq r$ in the path from $h$ to $r$ such that $w_2$ is a descendent of $x$. In the former case, let $z = r$. In the latter case, $x$ is of degree at least three. Thus let $z = x$. $\qquad\square$

**Lemma 6.** *Some legal pair in $\Lambda(G)$ satisfy the leaf-connecting condition. Moreover, if $c_1$ is critical, the path in $\Lambda(G)$ between this legal pair also contains $c_1$.*

*Proof.* There are two cases.

*Case* 1. The degree of $c_1$ is two. Since $|\Psi(G)| > 3$, there is no critical vertex. We root $\Lambda(G)$ at a b-vertex $r$ of the largest degree. Thus it is either the case that the degree of $r$ is at least four or the case that the degree of $r$ is three and there is another vertex $r'$ of degree at least three. In the former case, using a simple argument we can prove that there must exist two distinct leaves $w_1$ and $w_2$ such that they are in different branches of $r$ and they are a legal pair. This pair satisfy Condition (1) of the leaf-connecting condition. In the latter case, we use Lemma 5 by setting $h = r'$ to find a legal pair that satisfy Condition (2) of the leaf-connecting condition.

*Case* 2. The degree of $c_1$ is greater than two. Since $c_1$ is not massive, there must exist another vertex $r$ of degree at least three. We root $\Lambda(G)$ at $r$ and then use Lemma 5 by setting $h = c_1$ to find a legal pair that satisfy Condition (2) of the leaf-connecting condition. $\qquad\square$

### 4.4 Case IV of Theorem 2

This case assumes that $\mathcal{M}(\Psi(G)) > 0$ and there is exactly one massive c-vertex. Let $r$ be the massive cut vertex of $G$ with $\mathcal{D}(r, G) - 1 > \mathcal{M}(\Psi(G)) + \mathcal{R}(\Psi(G))$. For technical convenience, consider $\Lambda(G)$ as a tree rooted at $r$. We define a branch of a tree $T$ to be a *chain* if it contains exactly one leaf in $T$.

Let $Y_1, Y_2, T_1$ and $T_2$ be two leaves and two subtrees of $\Lambda(G)$ such that (1) $T_1$ is a chain; (2) $Y_1$ is in $T_1$ and $Y_2$ is in $T_2$; (3) $Y_1$ and $Y_2$ form a legal pair. Let $e$ be a legal edge between a demanding vertex in $Y_1$ and one in $Y_2$. Let $G' = G \cup \{e\}$. Let $P$ be the tree path of $\Lambda(G)$ between $Y_1$ and $Y_2$. It is well-known [3, 7, 18] that the blocks of $G$ on $P$ are merged into a new block $Y'$ in $G'$ and $r$ remains a cut vertex in $G'$. Because $T_1$ has only one leaf, it is absorbed into $T_2$ in $\Lambda(G')$. Thus, $\mathcal{D}(r, G') = \mathcal{D}(r, G) - 1$ and $\mathcal{D}(v, G') \leq \mathcal{D}(v, G)$ for all remaining cut vertices $v$ of $G$ in $G'$. Because $\mathcal{D}(r, G)$ is greater than $\mathcal{D}(u, G)$ for all other cut vertices

$u$ in $G$, $\mathcal{D}(r', G)$ is at least $\mathcal{D}(v, G')$ for all other cut vertices $v$ in $G'$. On the other hand, if $T_2$ has exactly one leaf, then $Y'$ is a leaf in $\Lambda(G')$; otherwise, it is an inner vertex. In either case, $\mathcal{M}(\Psi(G')) + \mathcal{R}(\Psi(G')) \leq \mathcal{M}(\Psi(G)) + \mathcal{R}(\Psi(G))$. Thus, $\alpha(G') = \alpha(G) - 1$. We repeat this process until the massive cut vertex becomes critical, which is Case III of Theorem 2.

## 5 A tight bound for the general case

Let $\mathcal{C}_1(G)$ be the number of connected components in $G$ that are not biconnected and have two or more vertices each. Let $\mathcal{C}_2(G)$, $\mathcal{C}_3(G)$ and $\mathcal{C}_4(G)$ be the numbers of isolated edges, isolated vertices, and biconnected components, respectively.

**Theorem 7.**

1. *If $\mathcal{C}_1(G) = 1$ and $\mathcal{C}_2(G) = 0$, then $\mathcal{B}(G) = \alpha(G)$.*
2. *If $\mathcal{C}_1(G) + \mathcal{C}_2(G) \geq 2$, then $\mathcal{B}(G) = \alpha(G)$.*
3. *If $\mathcal{C}_1(G) = 0$, $\mathcal{C}_2(G)=1$, and $\mathcal{C}_4(G) \geq 1$, then $\mathcal{B}(G) = 2$.*
4. *If $\mathcal{C}_1(G) = 0$, $\mathcal{C}_2(G)=1$, and $\mathcal{C}_4(G)=0$, then $\mathcal{B}(G) = 3$.*
5. *If $\mathcal{C}_1(G) = 0$ and $\mathcal{C}_2(G)=0$, then $\mathcal{B}(G) = 0$.* □

## 6 Computing an optimal biconnector in linear time

Since $\Lambda(G)$ can be computed in linear time [1, 2, 4], so can $\alpha(G)$. To find an optimal biconnector, we can iteratively add one legal edge at a time to reduce $\alpha(G)$ by one. With a naive implementation, this process may take quadratic time to find an optimal biconnector. However, with the data structure presented below, we can find an optimal biconnector in linear time. It is obvious to compute in linear time an optimal biconnector for Cases 3 through 5 of Theorem 7. For Cases 1 and 2, if $\mathcal{C}_1(G) > 1$, these two cases can be reduced in linear time to the case that $\mathcal{C}_1(G) = 1$ and $\mathcal{C}_i(G) = 0$ for $2 \leq i \leq 4$. It is also easy to compute an optimal biconnector in linear time for Cases I, II and IV of Theorem 2. For the rest of the section, we assume Case III of Theorem 2.

Given two blocks in $\Lambda(G)$, their *corresponding path* is the tree path in $\Lambda(G)$ that contains those two blocks. The linear time algorithm in [18] uses the fact that $\Lambda(G)$ is rooted at an internal b-vertex. Each time an edge $e$ is added, it is added between leaf-blocks whose corresponding path $P$ passes through the root. Thus $\Lambda(G \cup e)$ can be computed from $\Lambda(G)$ by local operations in $O(|P|)$ time. Note that if $P$ passes the root, then the new root of $\Lambda(G \cup e)$ is the new block created by merging all blocks in $P$.

Unfortunately, the key step of our algorithm as given in the proof of Lemma 6 (which uses Lemma 5) cannot guarantee that $P$ passes through the root because we have to find specified leaf-blocks satisfying the matching rules in addition to satisfying the leaf-connecting condition. We will prove in the following sections that we can satisfy the requirement of $P$ passing the root, if we reroot $\Lambda(G)$ while finding the legal edge to be added. In order to have a linear-time implementation, the total amount of time used for rerooting and finding two endpoints of the legal edge to be added must be also linear. Below, we describe a data structure to achieve all of the above goals.

## 6.1 Data structure

A vertex $u$ in $\Lambda(G)$ is *pure-A* if it is a type-$A$ leaf or all leaves in the subtree rooted at $u$ are type-$A$. We similarly define a *pure-B* vertex. A vertex $u$ in $\Lambda(G)$ is *hybrid* if it is neither type-$A$ nor type-$B$.

We use the following data structure to represent $\Lambda(G)$ as a tree rooted at a given non-leaf b-vertex. The siblings of a vertex are doubly linked from left to right. In this list, hybrid vertices whose branches contain type-$AB$ leaves appear first. Hybrid vertices whose branches contain no type-$AB$ leaves appear next. These vertices are followed by the pure-$A$ vertices and then the pure-$B$ vertices. Each vertex has two values: (1) a flag indicating whether it is pure-$A$, pure-$B$, or hybrid and (2) the number of leaves in the subtree rooted at it. Each hybrid vertex also keeps the number of type-$AB$ leaves in the subtree rooted at it. Each vertex maintains five pointers: (1) a parent pointer, (2) a child pointer to the leftmost pure-$A$ child, (3) a child pointer to the leftmost pure-$B$ child, (4) a pointer to the leftmost hybrid child who has a branch containing a type-$AB$ leaf and (5) a pointer to the leftmost hybrid child whose branches contain no type-$AB$ leaf. Each pointer is null if no such vertex exists. In addition to the above pointers, each vertex also has a pointer to the leftmost child and one to the rightmost child.

Given $\Lambda(G)$, we can construct the above data structure and root $\Lambda(G)$ at a given non-leaf b-vertex all in linear time. This construction and rooting process is called *ordering*. The resulting $\Lambda(G)$ together the data structure is called an *ordered tree*.

Using the ordered tree data structure, we can *walk down* the tree from a vertex $v$ toward one of its child pointers. If there is a leaf with a certain type in a subtree rooted at a vertex $v$. Using one of the 5 child pointers, we can walk down the tree to find such a leaf with the given type. We can *walk up* the tree from the vertex $v$ through its parent pointer. We can *walk down* the tree from a vertex $v$ toward one of its dependents $u$ by first find the path $P$ between $u$ and $v$ by walking up through the parent pointer until $v$ is encountered. Then we follow $P$ to walk down the tree. All of the walking operations mentioned above take time linear in the distance (number of vertices and edges) traversed.

**Lemma 8.** *Let $e$ be a legal edge in $G$ whose endpoints are in leaf-blocks $w_1$ and $w_2$. If the path $P$ between $w_1$ and $w_2$ in the ordered tree $\Lambda(G)$ passes through the root of the tree, then we can order $\Lambda(G \cup \{e\})$ in $O(|P|)$ time.* $\square$

It can be proved that Let $T$ be an ordered tree rooted at $r$. Let $r'$ be another vertex in $T$. Let $P$ be the path in $T$ from $r$ to $r'$. Given $P$, we can order $T$ to be rooted at $r'$ in $O(|P|)$ time. We can also prove that The vertices $w_1$, $w_2$, and $z$ of Lemma 5 can be found in $O(|P|)$ time in an ordered $\Lambda(G)$, where $P$ is the tree path between $w_1$ and $w_2$.

## 6.2 Choosing a legal pair in Lemma 6

By using the ordered tree data structure, it can be shown that The path between the legal pair found in Lemmas 6 passes through the root of $\Lambda(G)$ after rerooting if necessary. Without the rerooting processes, the algorithm runs in linear time. We now analyze the time spent in performing rerooting.

We define a *rerooting operation* to be the process of moving the root (if needed) from its current root before applying the proof of Lemma 6 to find a legal pair to its new root after finding a legal pair. We define a *rerooting step* to be the rerooting of a tree from its current root $r$ to a child $w$ of $r$. We say the above rerooting step *begins* from $r$ and *stops* at $w$. Given $w$, a rerooting step can be done in constant time. The following lemma bounds the total number of rerooting steps used in our algorithm.

**Lemma 9.** *During the entire execution of our algorithm for finding all legal pairs, rerooting steps are applied $O(q)$ times, where $q$ is the number of vertices in the initial input bc-forest.*

*Proof.* Assume that we need to perform a rerooting from $r$ to $r'$. Let the path between $r$ and $r'$ be $r = w_1, w_2, \ldots, w_{k-1}, w_k = r'$. Note that $r$ and $r'$ are both b-vertices. Thus $w_{k-1}$ is a c-vertex. We decompose a rooting operation into a sequence of rooting steps $w_1$ to $w_2$, $w_2$ to $w_3, \ldots, w_{k-1}$ to $w_k$. Those rerooting steps are classified into two categories. The first category consists of those rerooting steps from $w_s$ to $w_{s+1}$, $1 \leq s \leq k - 3$. The second category consists of the rerooting steps from $w_{k-2}$ to $w_{k-1}$ and from $w_{k-1}$ to $w_k$. We collect all rerooting operations performed during the entire execution of finding all legal pairs. Those rerooting operations are decomposed into rerooting steps. All rerooting steps are classified into the above two categories. We will analyze the number of rerooting steps in each category.

For a rooting step from $w$ to $w'$ in the first category, the following observations are useful. Let $T$ be the rooted tree before the rerooting step and let $T'$ be the rooted tree after the rerooting step. Either the vertex $w$ is a pure vertex in $T'$ or the subtree of $T'$ rooted at $w$ is a chain. Before the rerooting step the vertex $w'$ is neither pure-$A$ nor pure-$B$ in $T$. The subtree rooted at $r'$ in $T$ is also not a chain. After the rerooting step, a pure-$A$ (respectively, pure-$B$) vertex in $T$ remains pure-$A$ (respectively, pure-$B$) in $T'$. Given any vertex $v$, no rerooting step in the first category stops at $v$ twice. Thus the total number of rerooting steps in the first category is $O(q)$, where $q$ is the number of vertices in the bc-forest of the given graph.

For each legal pair found, we apply the rerooting operation once. For each rerooting operation, there are two rerooting steps in the second category. Note that we found only $O(q)$ legal pairs. Thus the total number of rerooting steps in the second category is $O(q)$. This proves the lemma. □

Following from the above discussion, we have the following theorem.

**Theorem 10.** *Given a bipartite graph with $n$ vertices and $m$ edges, an optimal biconnector can be computed in $O(n + m)$ time.* □

**Acknowledgments** The authors wish to thank Dan Gusfield for helpful discussions. The second author is grateful to Dan for his constant encouragements.

# References

1. A. Aho, J. Hopcroft, and J. Ullman. *The Design and Analysis of Computer Algorithms*. Addison-Wesley, Reading, Mass., 1974.
2. T. H. Cormen, C. E. Leiserson, and R. L. Rivest. *Introduction to Algorithms*. MIT Press, Cambridge, MA, 1990.
3. K. P. Eswaran and R. E. Tarjan. Augmentation problems. *SIAM J. Comput.*, 5(4):653–665, 1976.
4. S. Even. *Graph Algorithms*. Computer Science Press, Rockville, MD, 1979.
5. D. Gusfield. Optimal mixed graph augmentation. *SIAM Journal on Computing*, 16:599–612, 1987.
6. D. Gusfield. A graph theoretic approach to statistical data security. *SIAM Journal on Computing*, 17:552–571, 1988.
7. F. Harary. *Graph Theory*. Addison-Wesley, Reading, MA, 1969.
8. T.-s. Hsu. On four-connecting a triconnected graph (extended abstract). In *Proc. 33rd Annual IEEE Symp. on Foundations of Comp. Sci.*, pages 70–79, October 1992.
9. T.-s. Hsu. Undirected vertex-connectivity structure and smallest four-vertex-connectivity augmentation (extended abstract). In *Proc. 6th International Symp. on Algorithms and Computation*, volume LNCS #1004, pages 274–283. Springer-Verlag, 1995.
10. T.-s. Hsu and M. Y. Kao. Optimal bi-level augmentation for selectively enhancing graph connectivity with applications. In *Proc. 2nd International Symp. on Computing and Combinatorics*, volume LNCS #1090, pages 169–178. Springer-Verlag, 1996.
11. T.-s. Hsu and V. Ramachandran. A linear time algorithm for triconnectivity augmentation. In *Proc. 32th Annual IEEE Symp. on Foundations of Comp. Sci.*, pages 548–559, 1991.
12. T.-s. Hsu and V. Ramachandran. On finding a smallest augmentation to biconnect a graph. *SIAM J. Comput.*, 22(5):889–912, 1993.
13. M. Y. Kao. Linear-time optimal augmentation for componentwise bipartite-completeness of graphs. *Information Processing Letters*, pages 59–63, 1995.
14. M. Y. Kao. Total protection of analytic invariant information in cross tabulated tables. *SIAM Journal on Computing*, 1995. To appear.
15. M. Y. Kao. Data security equals graph connectivity. *SIAM Journal on Discrete Mathematics*, 9:87–100, 1996.
16. F. M. Malvestuto, M. Moscarini, and M. Rafanelli. Suppressing marginal cells to protect sensitive information in a two-dimensional statistical table. In *Proceedings of ACM Symposium on Principles of Database Systems*, pages 252–258, 1991.
17. J. Plesník. Minimum block containing a given graph. *ARCHIV DER MATHEMATIK*, XXVII:668–672, 1976.
18. A. Rosenthal and A. Goldner. Smallest augmentations to biconnect a graph. *SIAM J. Comput.*, 6(1):55–66, March 1977.
19. W. T. Tutte. *Connectivity in Graphs*. University of Toronto Press, 1966.
20. T. Watanabe and A. Nakamura. A minimum 3-connectivity augmentation of a graph. *J. Comp. System Sci.*, 46:91–128, 1993.

# Towards More Precise Parallel Biconnectivity Approximation

Ka Wong Chong[†] and Tak Wah Lam[†]

Department of Computer Science
University of Hong Kong, Hong Kong
Email: {kwchong, twlam}@cs.hku.hk

**Abstract.** The problem of finding a minimum biconnected spanning subgraph of an undirected graph is NP-hard. This paper, improving a chain of work [1, 10, 3, 4], gives the first NC algorithm for approximating a minimum biconnected subgraph with an approximation factor approaching $\frac{3}{2}$. This result matches the performance of the currently best sequential approximation algorithm [7].

## 1   Introduction

Let $G = (V, E)$ be any biconnected undirected graph. $G$ may contain edges whose removal does not affect the biconnected property. This paper is concerned with the problem of finding a biconnected spanning subgraph of $G$ that contains the smallest number of edges, such a subgraph is also called a minimum biconnected subgraph of $G$.

The problem of finding a minimum biconnected subgraph arises in designing communication networks that can tolerate a site failure. It is known to be NP-hard [6]. Nevertheless, simple approximation algorithms for the problem have been known for a while. For instance, a biconnected subgraph can be constructed (sequentially in polynomial time) as follows: Find a depth-first-search tree $T$ of $G$ and, for each vertex, pick the highest back edge. $T$ and these back edges together form a biconnected subgraph with at most $2n - 2$ edges, where $n$ is the number of vertices of $G$. As any biconnected subgraph has at least $n$ edges, this simple algorithm attains an approximation factor of $(2n - 2)/n \leq 2$.

The algorithm above may look simple as it is, but to achieve a factor better than 2 is not trivial. In the sequential context, Khuller and Vishkin [11] are the first to give a polynomial-time algorithm for finding a biconnected subgraph with an approximation factor of $\frac{5}{3}$. Later, Garg *et al.* [7] improved the factor to $\frac{3}{2}$.

In the parallel context, the work of Cheriyan and Thurimella [1] implies an NC algorithm for finding a biconnected subgraph with an approximation factor of 2. On the other hand, Kelsen and Ramachandran [10] gave a parallel algorithm for finding a *minimal* biconnected subgraph, which also approximates

---

[†] This research was supported by Hong Kong RGC Grant 338/065/0027.

to the minimum biconnected subgraph within a factor of 2. The sequential algorithms in [11, 7] do not admit efficient parallel implementation due to their sequential nature. They are all based on augmenting the depth-first-search tree, and at present deterministic NC algorithm for finding depth-first-search trees is not known [8]. Recently, Chong and Lam [3, 4] designed an NC approximation algorithm which does not compute a depth-first-search tree but still achieves a factor of $\frac{5}{3} + \epsilon$ for any constant $\epsilon > 0$. Since then, it has been an interesting open problem whether there is any NC algorithm that can approach a factor of $\frac{3}{2}$.

Interestingly, the factor of $\frac{3}{2}$ seems to be a barrier to parallel or sequential approximation algorithms for finding a minimum biconnected subgraph. This can be observed when we look at a related NP-hard problem—finding a minimum 2-edge connected subgraph (a graph is said to be 2-edge connected if it is still connected after removing any one of the edges). Obviously, 2-edge connectivity is a weaker requirement than biconnectivity. Yet in the literature the best approximation algorithms for 2-edge connectivity [11, 3] are also unable to give an approximation factor better than $\frac{3}{2}$.

The major contribution of this paper is a new NC algorithm that improves the approximation factor of the minimum biconnected subgraph problem to $\frac{3}{2} + \epsilon$ for any $\epsilon > 0$. Our work also implies a new sequential approach which is different from the traditional depth-first-search tree based approach, but which can still achieve an approximation factor of $\frac{3}{2}$.

From an algorithmic viewpoint, our improvement is rooted at two interesting findings: (i) Given a biconnected graph $G$, if $G$ contains a perfect matching $M$ then there always exist $n - 1$ other edges in $G$ which, together with edges of $M$, form a biconnected subgraph of $G$. (ii) If $G$ does not have a perfect matching, we can easily transform $G$ into another graph $G^*$ that has a perfect matching and the solution to $G^*$ naturally defines a solution to $G$. Moreover, the smaller the size of the maximum matching, the better the factor of our algorithm will attain. (In particular, if the maximum matching has no more than $x$ edges, our algorithm achieves a factor of at most $\frac{2n-x-1}{2n-2x-1} + \epsilon$.)

The remainder of this paper is organized as follows: Section 2 shows that the problem of approximating a minimum biconnected subgraph of a biconnected graph can be reduced to that of another graph containing a perfect matching. Section 3 gives the framework of the new approximation algorithm. Section 4 describes the way we merge biconnected components together. Section 5 discusses how a perfect matching of a graph can guide us picking no more than $n - 1$ additional edges to form a small biconnected subgraph. Section 6 shows the details of the parallel algorithm and its time complexity.

## 2  An approach based on perfect matching

Motivated by the work of [3], we have a new approach of using a sufficiently large matching of a biconnected graph $G$ to find a small biconnected subgraph of it. A characteristic of this approach is that the matching adopted initially will serve as a core to guide us picking other edges of $G$ to add to it, forming a biconnected subgraph eventually. For instance, if we are given a perfect matching $M$ of $G$, we are able to find at most $n-1$ other edges of $G$, which, together with the edges of $M$, form a biconnected subgraph of $G$ with less than $\frac{3}{2}n$ edges. However, $G$ may not have a perfect matching, i.e. the maximum matching $M$ of $G$ has less than $n/2$ edges. In this case, we need to add more than $n$ edges and the resultant biconnected subgraph may contain more than $\frac{3}{2}n$ edges. Yet, as shown in [3], a minimum biconnected subgraph of $G$ cannot contain too few edges when the maximum matching of $G$ is small.

**Lemma 1.** [3] If a maximum matching of $G$ has $x$ edges, a minimum biconnected subgraph of $G$ contains at least $\max(n, 2n - 2x - 1)$ edges.

Below, we show that, given any matching of $G$ which has $y$ edges (and $n-2y$ unmatched vertices), we can find a biconnected subgraph of $G$ with at most $2n - y - 1$ edges. An interesting idea behind our new approach is that we do not need a different algorithm to handle a graph $G$ of which the maximum matching has less than $n/2$ edges. We transform $G$ into a bigger graph $G^*$ that contains a perfect matching. Then we find a small biconnected subgraph $H^*$ of $G^*$, which can be easily transformed into a biconnected subgraph of $G$. $G^*$ is constructed as follows:

> Let $M$ be any matching of $G$, comprising $y < n/2$ edges. For each unmatched vertex $u$ in $G$, let $v$ be an arbitrary vertex adjacent to $u$. We add a new vertex $w$ and two edges $(u, w)$ and $(w, v)$. Also, we consider $(u, w)$ as a matched edge. Let $G^*$ denote the resultant graph.

Note that every new vertex is of degree two and both of its two edges must appear in any biconnected subgraph of $G^*$. $G^*$ contains $n + (n - 2y) = 2n - 2y$ vertices and has a perfect matching. As $G$ is biconnected, $G^*$ is biconnected, too. Using the approximation algorithm for graphs with perfect matching (to be described later), we can find a biconnected subgraph $H^*$ of $G^*$ which has at most $\frac{3}{2}n^* - 1$ edges, where $n^* = 2n - 2y$.

Next, we show how to obtain a biconnected subgraph $H$ of $G$ from $H^*$. For each unmatched vertex $u$ in $G$, we find the corresponding vertex $w$ in $H^*$ (which has been added to match with $u$ in $G^*$), and the two edges incident to $w$, $(u, w)$ and $(w, v)$. We replace the vertex $w$ and its two edges with the edge $(u, v)$, which is in $G$. The resultant graph $H$ is a biconnected subgraph of $G$ and contains at most $\frac{3}{2}n^* - 1 - (n - 2y) = 2n - y - 1$ edges.

In summary, if we are given a maximum matching $M$ of $G$ with $x < n/2$ edges, we know that any minimum biconnected subgraph of $G$ contains at least $2n - 2x - 1$ edges and the above algorithm achieves an approximation factor of at most $\frac{2n - x - 1}{2n - 2x - 1} \leq \frac{3}{2}$. However, there is no known deterministic NC algorithm for finding a maximum matching in $G$ [8], we instead use an approximate maximum matching which can be found in NC. For instance, the parallel algorithm of Fischer *et al.* [5] finds a matching of size at least $(1 - 1/c)x$ for any constant $c > 1$. Thus, the approximation factor becomes $\frac{3}{2} + \epsilon$ for any constant $\epsilon > 0$.

## 3   The framework

It follows from the previous section that we can restrict our attention to the case in which the input graph $G$ has a perfect matching $M$ and this matching is known.

We use an iterative process to find a biconnected subgraph of $G$. In each iteration, we start off with a connected subgraph $H$ of $G$ which contains all the edges of $M$ and at most $n - 1$ unmatched edges (i.e. other edges of $G$). If $H$ is biconnected then we can report $H$ immediately. Otherwise, let $D_1, D_2, \cdots, D_k$ be the biconnected components of $H$. Intuitively, we want to pick some edges in $G - H$ to add on $H$ so that the biconnected components of $H$ will be merged with others to form bigger biconnected components. However, this may increase the total number of unmatched edges too much. Here, we have a non-trivial observation that for some so-called "fully-matched" biconnected component of $H$, say, $D_i$, we can add an edge chosen from $G - H$ to merge $D_i$ with other biconnected component(s), and more importantly, if $D_i$ already contains too many unmatched edges then one of its unmatched edges must be redundant in the sense that it is a chord of some cycle in the combined biconnected component. Obviously, we can remove this redundant edge without affecting the biconnectivity. Note that such a "merge-and-remove" procedure, in the worst case, can reduce the number of biconnected components of $H$ by one.

Interestingly, we can prove that at least half of the biconnected components of $H$ are "fully-matched" (see Lemma 4) and each can be merged in parallel with other biconnected components in accordance with the merge-and-remove procedure. In other words, we can obtain a connected subgraph $H'$ from $H$, which has the number of biconnected components reduced by at least one quarter. $H'$ is also a connected subgraph of $G$ which contains all edges of $M$ and at most $n - 1$ unmatched edges. We proceed to the next iteration with $H'$.

It is clear that the above iterative process requires only $O(\log n)$ iterations to produce a biconnected subgraph of $G$. To generate the input for the first iteration, we can simply find a connected spanning subgraph $H_0$ comprising a spanning forest of $G - M$ and all the edges of $M$. Note that $H_0$ contains at most $n - 1$ unmatched edges and does satisfy the requirement of $H$ in the above iterative process. In the next section, we give the definition of a "fully-matched" biconnected component and prove the required properties of it. Then, we show an NC algorithm for selecting an edge from $G - H$ to merge a biconnected component of $H$ with others.

The following definition of biconnected components helps us to understand the merging process of our algorithm.

**Biconnected Components:** Define an equivalence relation $R$ as follows: For any edges $e_1, e_2 \in E$, $e_1 R e_2$ if and only if $e_1 = e_2$ or there is a simple (vertex-disjoint) cycle in $H$ containing $e_1$ and $e_2$. Partition the edges of $H$ into equivalence classes $E_1, E_2, \cdots, E_k$. For $1 \le i \le k$, let $V_i$ denote the set of vertices involved in $E_i$. Each subgraph $D_i = (V_i, E_i)$ forms a biconnected component of $H$. It is possible that some vertices belong to more than one biconnected component; actually, they are the cut-vertices in $H$. Also, a biconnected component may contain only one edge.

**Fact 1:** Let $n$ be the number of vertices in $H$ and $n_i$ the number of vertices in $D_i$. Then $\sum_{i=1}^{k}(n_i - 1) = n - 1$.

## 4 Merging

Let $H$ be a connected spanning subgraph of $G$. Assume $H$ is not biconnected. Let $D_1, D_2, \cdots, D_k$ be the biconnected components of $H$. In this section, we show how to merge a biconnected component with others by picking an edge in $G - H$.

**External path:** Consider a $D_i$ of $H$. In $G$, there is an "external path" $L$ connecting two distinct vertices of $D_i$, where $L$ contains more than one edge and all the vertices in $L$ except its two end-points are not in $D_i$.

To merge $D_i$ with other biconnected components, we find an edge $e_i$ in $G - H$ such that adding $e_i$ on $H$, there is an external path connecting two distinct vertices of $D_i$. The external path $L$ in $H \cup \{e_i\}$ forms a cycle involving edges of $D_i$ and edges of other biconnected components ($L$ contains at least one edge other than $e_i$). Suppose $L$ involves some edges of a biconnected component $D_j$. According to definition of biconnected components, the edges of $D_i$ and $D_j$ are in the same biconnected component of $H \cup \{e_i\}$. In this case, $D_i$ is said to be merged with other biconnected components and the number of biconnected components in $H \cup \{e_i\}$ is less than that of $H$.

Next we show how to find such an edge for $D_i$. In particular, for any $v \in D_i$ which is a cut-vertex in $H$, we can find an edge $e_i$ such that the external path in $H \cup \{e_i\}$ connecting $v$ to another vertex of $D_i$.

**Definition 2.** Let $a$ be a vertex in $D_i$. Denote $U(a)$ the set of vertices of $H$ such that $w \in U(a)$ if $w \ne a$ and there is a path in $H$ connecting $w$ to $a$ without going through another vertex in $D_i$. Let $\overline{U(a)}$ be the set of vertices $H - U(a) - \{a\}$. (Thus $H = U(a) \cup \overline{U(a)} \cup \{a\}$.)

Note that $a$ is a cut-vertex if and only if $U(a) \ne \phi$. Since $H$ is not biconnected, such an $a \in D_i$ must exist. If $D_i$ contains an unmatched vertex $u$, $u$ must be a cut-vertex in $H$.

**Lemma 3.** For any $u \in D_i$ which is a cut-vertex in $H$, we can find an edge $e_i \in G - H$ such that there is an external path in $H \cup \{e_i\}$ connecting $u$ to another vertex in $D_i$.

*Proof.* Let $w$ be a vertex in $U(u)$ and $v$ another vertex in $D_i$. In $G$, there is a path $P = \langle w = z_1, z_2, \cdots, z_l = v \rangle$ connecting $w$ to $v$ and $P$ does not contain $u$. Each $z_i$ is either in $U(u)$ or $\overline{U(u)}$. As $w \in U(u)$ and $v \in \overline{U(u)}$, we can find a $z_j$ such that $z_j \in U(u)$ and $z_{j+1} \in \overline{U(u)}$. Also, $(z_j, z_{j+1})$ should not exist in $H$.

Note that some edges of $P$, other than $(z_j, z_{j+1})$, may not appear in $H$. Nevertheless, $H$ must contain a path $P'$ connecting $z_j$ to $u$, and then to some other vertex in $D_i$, and finally ending at $z_{j+1}$. $P'$, together with the edge $(z_j, z_{j+1})$, must contain an external path connecting $u$ to another vertex in $D_i$. $\square$

For a particular vertex $u \in D_i$, such $e_i$ can be found by choosing any edge in $G - H$ that has its end-points in $U(u)$ and $\overline{U(u)}$.

## 5  Perfect Matching

Let $G$ be a biconnected graph with $n$ vertices and $m$ edges, where $n$ is even. Suppose $G$ contains a perfect matching $M$. We will give a parallel algorithm that finds a biconnected subgraph of $G$ involving all the edges of $M$ and no more than $n - 1$ unmatched edges of $G - M$.

Our algorithm runs in an iterative manner. Initially, we form a connected spanning subgraph $H_0$ of $G$ as follows: Let $F$ be a spanning forest of $G - M$. Then $H_0 = F \cup M$. Obviously, $H_0$ contains at most $n - 1$ unmatched edges. Moreover, for each biconnected component $D_i$ of $H_0$, $D_i$ contains at most $n_i - 1$ unmatched edges (i.e. tree edges of $F$), where $n_i$ is the number of vertices of $D_i$.

In general, at the beginning of each iteration, let $H$ be the connected spanning subgraph of $G$. Let $D_1, D_2, \cdots, D_k$ be the biconnected components of $H$. We want to maintain the following invariants: $H$ includes all the edges of $M$ and for every biconnected component $D_i$ of $H$, $D_i$ contains at most $n_i - 1$ unmatched edges. This implies the total number of unmatched edges in $H$ is no more than $\sum_{i=1}^{k}(n_i - 1) = n - 1$ (see Fact 1).

A biconnected component $D_i$ is said to be *fully-matched* if $D_i$ contains $n_i/2$ matched edges of $M$ if $n_i$ is even, or $(n_i - 1)/2$ if $n_i$ is odd.

**Lemma 4.** The number of full-matched $D_i$'s is strictly greater than $k/2$.

*Proof.* Let $T$ be a spanning tree of $H$ including all edges in $M$. Each $D_i$ contains exactly $n_i - 1$ tree edges of $T$. Denote $c(D_i)$ the number of matched tree edges minus the number of unmatched tree edges in $D_i$.

For each fully-matched $D_i$, $D_i$ contains $n_i/2$ matched tree edges if $n_i$ is an even, and $(n_i - 1)/2$ otherwise. Thus $0 \leq c(D_i) \leq 1$. On the other hand, if $D_i$ is not fully-matched, it has no more than $n_i/2 - 1$ matched tree edges if $n_i$ is an even, and $(n_i - 1)/2 - 1$ if $n_i$ is odd. In this case, $c(D_i) \leq -1$. Since $c(T) = \sum_{i=1}^{k} c(D_i) = 1$, the number of biconnected components that is fully-matched should be greater than the opposite. $\square$

We are going to merge those $D_i$'s which are fully-matched with other biconnected components. A biconnected component $D_i$ is said to be *excessive* if $D_i$ is fully-matched and contains exactly $n_i - 1$ unmatched edges. The following propositions ($P(n)$ and $Q(m)$), which will be proved in Lemmas 7 and 8, help us to show that after merging an excessive biconnected component $D_i$, there is an unmatched edge in $D_i$ which is a chord in the resultant biconnected component.

$P(n)$: Let $n \geq 4$ be an even integer, and let $\mathcal{G}$ be a biconnected graph with $n$ vertices. If $\mathcal{G}$ is excessive, one of the unmatched edges of $\mathcal{G}$ is a chord.

$Q(m)$: Let $m \geq 3$ be an odd integer, and let $\mathcal{G}$ be a biconnected graph with $m$ vertices. If $\mathcal{G}$ is excessive then, for any external path $L$ connecting the unmatched vertex $u$ and any other vertex of $\mathcal{G}$, one of the unmatched edges of $\mathcal{G}$ becomes a chord in $\mathcal{G} \cup L$.

Consider a biconnected component $D_i$ in $H$ which is excessive. If $n_i$ is even, as $P(n)$ is true, we can find a chord in $D_i$ which is unmatched. On the other hand, if $n_i$ is odd, by Lemma 3, we can find an edge $e_i \in G - H$ such that there is an external path connecting the unmatched vertex of $D_i$. Because $Q(m)$ is true, after adding $e_i$ to $H$, we can find an unmatched edge in $D_i$ which is a chord.

We merge those fully-matched biconnected components and for each excessive one, we identify the chord in it. After removing all these chords, let $H'$ be the resultant graph. Obviously, $H'$ contains all the edges of $M$. Let $D'_1, D'_2, \cdots, D'_{k'}$ be the biconnected components of $H'$ and $n'_i$ the number of vertices of $D'_i$.

**Lemma 5.** Every biconnected component $D'_i$ of $H'$ contains at most $n'_i - 1$ unmatched edges.

*Proof.* Consider a biconnected component $D'_i$ of $H'$. Let $X = \{D_{i_1}, D_{i_2}, \cdots, D_{i_j}\}$ be the set of biconnected components of $H$ that are merged together to form $D'_i$. Let $t$ be the total number of unmatched edges in $X$. Suppose there are $p$ ($\leq j$) fully-matched biconnected components in $X$ and $q$ ($\leq p$) of them are excessive. Then $t \leq \sum_{k=1}^{j}(n_{i_k} - 1) - (p - q)$. In forming $D'_i$, we have added no more than $p$ unmatched edges chosen from $G - H$ and also, removed $q$ unmatched edges from those excessive biconnected components in $X$. As a result, $D'_i$ contains no more than $t + p - q \leq \sum_{k=1}^{j}(n_{i_k} - 1)$ edges. It is easy to show that $\sum_{k=1}^{j}(n_{i_k} - 1) = n'_i - 1$. Thus, $D'_i$ contains at most $n'_i - 1$ unmatched edges. □

By the above lemma, $H'$ satisfies the invariants for the next iteration. By Lemma 4, at least half of the biconnected components of $H$ would have been merged with others. Then the number of biconnected components in $H'$ is at most $3/4$ of that of $H$. Therefore, the algorithm requires $O(\log n)$ iterations in order to form a single biconnected component, i.e. a biconnected subgraph of $G$.

Before we prove $P(n)$ and $Q(m)$, we first establish a relationship between $P(n)$ and $Q(m)$, which helps us to prove the propositions.

**Lemma 6.** Let $n \geq 4$ be an even integer. If $P(n)$ is true then $Q(n-1)$ is true.

*Proof.* Suppose $\mathcal{G}$ is a biconnected graph with $n-1$ edges and $\mathcal{G}$ is excessive. Let $u$ be the unmatched vertex in $\mathcal{G}$. We first consider the special case where the external path $L_0$ connecting $u$ to another vertex $v$ in $\mathcal{G}$ contains only one intermediate vertex $w$. That is, $L_0$ is composed of two edges, $(u, w)$ and $(w, v)$. Let $\mathcal{G}' = \mathcal{G} \cup L_0$ and consider $(u, w)$ as a matched edge. Note that $\mathcal{G}'$, being an $n$-vertex biconnected graph, is also excessive. By our assumption that $P(n)$ is true, one of the unmatched edges of $\mathcal{G}'$, say, $e$, is a chord of $\mathcal{G}'$. As $w$ is of degree two, $w$ cannot be an endpoint of a chord and the two edges incident to it cannot be a chord. Therefore, $e$ must be found in $\mathcal{G}$.

If we replace $L_0$ with any other external path $L$, $e$ is still a chord in $\mathcal{G} \cup L$. $\square$

**Lemma 7.** $P(n)$ is true for all $n \geq 4$.

*Proof.* We prove the lemma by induction. First of all, we find a spanning tree $\mathcal{T}$ of $\mathcal{G}$ satisfying two properties: (i) $\mathcal{T}$ includes all edges of $\mathcal{M}$; (ii) there exists an edge $e$ in $\mathcal{G} - \mathcal{T}$ such that $\mathcal{T} \cup \{e\}$ forms a cycle involving two matched edges. The existence of such $\mathcal{T}$ is guaranteed by the biconnectivity of $\mathcal{G}$.

**Base case, $n = 4$:** Then $\mathcal{T}$ has only an unmatched edge and $\mathcal{T} \cup \{e\}$ is biconnected. The remaining unmatched edge must be a chord in $\mathcal{G}$.

**Inductive case, $n \geq 4$:** As $\mathcal{G}$ is excessive, $\mathcal{G}$ contains a perfect matching $\mathcal{M}$ and $n-1$ unmatched edges. We give a sequential algorithm to find an unmatched edge which is a chord in $\mathcal{G}$: With respect to $\mathcal{T}$, we want to add $e$ as well as $\frac{n}{2} - 2$ (or less) other edges of $\mathcal{G} - \mathcal{T}$ to $\mathcal{T}$ such that every matched edge is involved in a cycle in the resultant graph $\mathcal{H}$. Note that $\mathcal{H}$ contains $n/2$ matched edges and at most $n-2$ unmatched edges. If $\mathcal{H}$ is biconnected then the unmatched edge in $\mathcal{G} - \mathcal{H}$ must be a chord. Suppose $\mathcal{H}$ is not biconnected. Let $\mathcal{D}_1, \mathcal{D}_2, \cdots, \mathcal{D}_k$ be the biconnected components of $\mathcal{H}$, and let $n_i$ be the number of vertices in $\mathcal{D}_i$. Due to the construction of $\mathcal{H}$, each $D_i$ has at most $n_i - 1$ unmatched edges. Moreover, if $\mathcal{D}_i$ contains a matched edge, it also contains a cycle and hence $n_i \geq 3$.

Since $\mathcal{H}$ contains a perfect matching, Lemma 4 implies that its number of fully-matched biconnected components is greater than $k/2$, or equivalently, at least 2 (as $k \geq 2$). If one of the fully-matched biconnected components, say, $\mathcal{D}_i$, is excessive, we can find the required chord as follows:

1. $n_i$ is even. $\mathcal{D}_i$ has at most $n-2$ vertices. By the inductive hypothesis, $P(n-2)$ is true and there is a chord in $\mathcal{D}_i$, and which is also a chord in $\mathcal{G}$.
2. $n_i$ is odd. By the inductive hypothesis, $P(n-2), P(n-4), \cdots, P(4)$ are all true. Then, by Lemma 6, $Q(m)$ is true for all odd integer $m \leq n - 3$. Recall that $\mathcal{H}$ contains another fully-matched biconnected component, which, by definition, contains some matched edge and at least three vertices. Together with the fact $\sum_{1 \leq j \leq k}(n_j - 1) = n - 1$, we can deduce that $n_i \leq n - 3$. Therefore, $Q(n_i)$ is true and, with respect to $\mathcal{D}_i$, adding an external path would cause an unmatched edge in $\mathcal{D}_i$ to become a chord. As $\mathcal{G}$ is

biconnected, we can form such a path whose intermediate vertices are chosen from $\mathcal{G} - \mathcal{D}_i$. The chord reported in $\mathcal{D}_i$ is also a chord in $\mathcal{G}$.

It remains to consider the case when none of the fully-matched biconnected components in $\mathcal{H}$ is excessive. That means, there are at least two biconnected components $\mathcal{D}_i$ and $\mathcal{D}_j$ such that they have at most $n_i - 2$ and $n_j - 2$ unmatched edges respectively. We can add an edge (chosen from $\mathcal{G} - \mathcal{H}$) to $\mathcal{H}$ to merge one of these two biconnected components with others. Let $\mathcal{D}'_i$ be the resultant biconnected component and $\mathcal{H}'$ be the resultant graph. $\mathcal{D}'_i$ has at most $n'_i - 1$ edges (see Lemma 5), where $n'_i$ denotes its number of vertices, and $\mathcal{H}'$ still has at most $n - 2$ unmatched edges but fewer biconnected components than $\mathcal{H}$. We can repeat the argument above to find a chord in $\mathcal{H}$ or to reduce the number of biconnected component. Thus, we will eventually find chord in $\mathcal{G}$. $\square$

**Lemma 8.** $Q(m)$ is true for all $m \geq 3$.

*Proof.* It follows from Lemmas 6 and 7.

## 6 The algorithm

Below is the algorithm for finding a biconnected subgraph of $G$. It achieves a factor of $\frac{3}{2}$ if a maximum matching $M$ of $G$ is given. The factor becomes $\frac{3}{2} + \epsilon$ for any $\epsilon > 0$ when we use the approximate maximum matching instead.

**Algorithm Biconnected Subgraph**
**Input:** A biconnected graph $G$ and its maximum matching $M$
**Output:** A biconnected subgraph $H$ with an approximation factor of $\frac{3}{2}$

1. For each unmatched vertex $u$, add a vertex $w$ and two edges $(u, w)$ and $(w, v)$. Include $(u, w)$ into the matching. Let $G^*$ be the resultant graph and $M^*$ the perfect matching.
2. Find a spanning forest $F$ in $G^* - M^*$. Let $H^* = F \cup M^*$.
3. **while** the subgraph is not biconnected **do**
   (a) Identify the biconnected components $D_1, D_2, \cdots, D_k$ of $H^*$.
   (b) For each $D_i$ which is fully-matched, find an edge $e_i \in G^* - H^*$ such that adding $e_i$ on $H^*$ will cause $D_i$ to merge with other biconnected components. In particular, if $n_i$ is odd, the external path must connect the unmatched vertex of $D_i$. If $D_i$ is excessive, identify the chord $e_i$.
   (c) Remove the chords found in the previous step.
4. For each unmatched vertex $u \in G$, let $w$ be the corresponding added vertex and $(u, w)$ and $(w, v)$ the edges. Replace $w$ and its two edges with $(u, v)$. Let $H$ be the resultant biconnected graph.

**Time Complexity:** The graph $G^*$ in Step 1 can be formed in $O(\log n)$ time. The number of vertices and edges in $G^*$ are $2n - 2x \leq 2n$ and $m + 2(n - 2x) \leq 2m$

respectively. In Step 2, the spanning forest $F$ can be found in $O(\log n \log \log n)$ time using a linear number of processors [2]. In Step 3(b), the edge for each fully-matched biconnected component to merge can be found in $O(\log n \log \log n)$ time using $n^2$ processors. Also, the chords in each excessive $D_i$ can be found in $O(\log n \log \log n)$ time using $n_i^2$ processors. Thus Step 3 requires at most $n^2$ processors. As mentioned before, the total number of merging phases is $O(\log n)$. Thus, Step 3 takes $O(\log^2 n \log \log n)$ time using $n^2$ processors. Finally, forming $H$ in Step 4 can be done in $O(\log n)$ time. Therefore, the running time of the algorithm, except that of finding the approximate maximum matching, is $O(\log^2 n \log \log n)$ using $n^2$ processors. The running time and the processor requirement of our algorithm is dominated by that of finding the approximate maximum matching, which takes $O(\log^3 n)$ time using $n^{2+1/\epsilon}$ EREW processors.

**Remark:** If we use randomized algorithms [9] for finding a maximum matching, we can obtain an RNC algorithm achieving an approximation factor of $\frac{3}{2}$.

# References

1. J. Cheriyan and R. Thurimella, Algorithms for Parallel $k$-Vertex Connectivity and Sparse Certificates, *STOC*, 1991, pp. 391-401.
2. K.W. Chong and T.W. Lam, Finding Connected Components in $O(\log n \, \mathrm{loglog} \, n)$ time on the EREW PRAM, *Journal of Algorithms*, 1995, pp. 378-402.
3. K.W. Chong and T.W. Lam, Approximating Biconnectivity in Parallel, *SPAA*, 1995, pp. 224-233.
4. K.W. Chong and T.W. Lam, Improving Biconnectivity Approximation via Local Optimization, *SODA*, 1996, pp. 26-35.
5. T. Fischer, A.V. Goldberg, D.J. Haglin, and S. Plotkin, Approximating Matchings in Parallel, *Information Processing Letter*, 46(1993), pp. 115-118.
6. M.R. Garey and D.B. Johnson, *Computers and Intractability: A guide to the theory of NP-completeness*, Freeman, San Francisco, 1978.
7. N. Garg, V.S. Santosh, and A. Singla, Improved Approximation Algorithms for Biconnected Subgraphs via Better Lower Bounding Techniques, *SODA*, 1993, pp. 103-111.
8. R.M. Karp and V. Ramachandran, Parallel Algorithms for Shared-Memory Machines, *Handbook of Theoretical Computer Science*, vol A, J. van Leeuwen Ed., MIT Press, Massachusetts, 1990, pp. 869-941.
9. V. Vazirani, Parallel graph matching, *Synthesis of Parallel Algorithms*, J.H. Reif Ed., Morgan Kaufmann, San Mateo, CA, 1991.
10. P. Kelsen and V. Ramachandran, On Finding Minimal 2-Connected Subgraphs, *Journal of Algorithms*, 1995, pp. 1-49.
11. S. Khuller and U. Vishkin, Biconnectivity approximations and graph carvings, *Journal of ACM*, Vol. 41 No. 2, 1994, pp. 214-235.

# The Complexity of Probabilistic versus Deterministic Finite Automata

Andris Ambainis

Institute of Mathematics and Computer Science
University of Latvia
Raina bulv. 29, Riga, Latvia
e-mail: ambainis@cclu.lv

**Abstract.** We show that there exists probabilistic finite automata with an isolated cutpoint and $n$ states such that the smallest equivalent deterministic finite automaton contains $\Omega(2^{n\frac{\log\log n}{\log n}})$ states.

**Keywords:** Automata theory, the complexity of finite automata, probabilistic finite automata.

## 1 Introduction

Rabin[Ra63] proved that arbitrary language which is accepted by a probabilistic finite automaton with an isolated cutpoint is also accepted by a deterministic finite automaton. However, in the process of transition from a probabilistic to a deterministic finite automaton, the complexity increases.

Rabin[Ra63] showed that for probabilistic automaton with $n$ states, $r$ accepting states, and isolation radius $\delta$ there exists an equivalent deterministic finite automaton with at most $(1 + \frac{r}{\delta})^n$ states. Later this estimate was slightly improved by Paz[Paz66] and Gabbasov and Murtazina[GM79]. However, even after these improvements the estimate remained exponential, only the base of the exponent became smaller.

Construction of concrete languages for which deterministic automata are more complex than probabilistic ones appeared to be quite a complicated problem. Freivalds[Fr82] constructed probabilistic finite automata with $n$ states for which the smallest equivalent deterministic automaton contains $\Omega(2^{\sqrt{n}})$ states. The problem of constructing probabilistic automata with $n$ states such that any equivalent deterministic automaton contains $a^n$ states still remains open.

In this paper we present a language such that

- it is accepted by a probabilistic automaton with $n$ states, and
- any deterministic automaton accepting this language has $\Omega(2^{n/\log n})$ states.

## 2 Results

We shall use the standard definitions of probabilistic and deterministic finite automata[TB73].

**Theorem 1** *There exists a probabilistic finite automaton with n states such that the smallest equivalent deterministic finite automaton contains $\Omega(2^{\frac{n \log \log n}{\log n}})$ states.*

**Proof.** Consider the language $L_m$ having the alphabet $\{a_1, a_2, \ldots, a_m\}$ and consisting of all words that contain each of the letters $a_1, \ldots, a_m$ exactly $m$ times.

**Lemma 1** *If a deterministic finite automaton accepts $L_m$, it has at least $(m + 1)^m$ states.*

**Proof.** The automaton needs to remember the number of $a_1$'s in the part of the input word read, the number of $a_2$'s, $\ldots$, the number of $a_m$'s.

The number of $a_1$'s can assume $(m + 1)$ possible values: $0, 1, \ldots, m$. So, there are $(m + 1)^m$ possible values for the combination of the number of $a_1$'s, the number of $a_2$'s and so on. $\square$

**Lemma 2** *There exists a probabilistic finite automaton with isolated cutpoint which accepts $L_m$ and has $O(m \frac{\log^2 m}{\log \log m})$ states.*

**Proof.** We shall construct an automaton accepting $L_m$ with the probability $3/4$. (For other accepting probabilities the construction is similar.)

In our construction we shall use automata constructed by Freivalds[Fr82] as subroutines. Freivalds[Fr82] considered automata in a one-letter alphabet and proved the following result:

**Lemma 3** *[Fr82] There exists a probabilistic finite automaton with an isolated cutpoint that recognizes whether the word consists of exactly n words and has $O(\frac{\log^2 n}{\log \log n})$ states.*

Let $p$ be the smallest prime larger than $6m$. From prime number distribution theorems we have that $p = 6m + o(m)$.

Let $k = pm^2$. $U_i$ denotes the probabilistic automaton that recognizes (with accepting probability at least $\frac{9}{10}$) whether the word consists of exactly $i$ letters. $s_i$ denotes the number of states in $U_i$ and $s$ denotes $\max(s_1, \ldots, s_k)$.

The states of $U_i$ are denoted by $A_{i,1}, A_{i,2}, \ldots, A_{i,s_i}$. Without the loss of generality, we assume that the starting state of $U_i$ is $A_{i,1}$.

Informally, we shall consider an automaton which, at the beginning, chooses $i \in \{1, \ldots, 6m\}$ equiprobably, and then counts the sum

$$\sum_{j=1}^{m} (i^j \bmod p) n_j$$

where $n_1$ is the number of letters $a_1$ in the word, $n_2$ is the number of $a_2$'s and so on. For the counting of this sum probabilistic automata by Freivalds[Fr82] are used. A more formal desription follows:

Automaton $U$ has $1 + 6ms$ states:

1. Starting state $S$;
2. States $Q_{i,j}$ for $i \in \{1, \ldots, 6m\}$ and $j \in \{1, \ldots, s\}$.

From the starting state $S$, the automaton without reading any input passes to one of the states $Q_{i,1}$ equiprobably (with probability $1/6m$).

If the automaton is in one of states $Q_{i,j}$, it reads the symbol and passes to one of states $Q_{i,1}, \ldots, Q_{i,s}$. The probability of moving from state $Q_{i,j}$ to state $Q_{i,j'}$ after reading $a_t$ is equal to the probability of moving from $A_{i',j}$ to $A_{i',j'}$ after reading $(i^{t-1} \bmod p)$ symbols in automaton $U_{i'}$ where

$$i' = \sum_{j=1}^{m} m(i^{j-1} \bmod p).$$

The state $Q_{i,j}$ is defined to be an accepting state iff the corresponding state $A_{i',j}$ in automaton $U_{i'}$ is accepting.

*Proof of correctness.* We use

**Lemma 4** *If $i_1, \ldots, i_m \in \{1, 2, \ldots, 6m\}$ are pairwise different, then the vectors $v_1, \ldots, v_m$ where*

$$v_j = (i_j^0 \bmod p, i_j^1 \bmod p, \ldots i_j^{m-1} \bmod p)$$

*are linearly independent.*

**Proof.** The proof is similar to the proof of linear independence for Vandermonde's determinant.

Consider the matrix

$$\begin{pmatrix} i_1^0 \bmod p & i_1^1 \bmod p & \ldots i_1^{m-1} \bmod p \\ i_2^0 \bmod p & i_2^1 \bmod p & \ldots i_2^{m-1} \bmod p \\ \ldots & \ldots & \ldots\ldots \\ i_m^0 \bmod p & i_m^1 \bmod p & \ldots i_m^{m-1} \bmod p \end{pmatrix}$$

Rows of this matrix (i.e. $v_j$'s) are linearly independent iff the columns of it are independent.

By the way of contradiction assume that some of columns are dependent. Then $c_1$ multiplied by the first column plus $c_2$ multiplied by the second column and so on, plus $c_m$ multiplied by the last column is equal to $\overrightarrow{0}$ where $c_1, c_2, \ldots, c_m$ are some constants which are not all equal to 0. It follows that, for all $j \in \{1, \ldots, m\}$

$$c_1 + c_2(i_j \bmod p) + c_3(i_j^2 \bmod p) + \ldots + c_m(i_j^{m-1} \bmod p) = 0$$

$$c_1 + c_2 i_j + c_3 i_j^2 + \ldots + c_m i_j^{m-1} = 0 \quad (\bmod p)$$

So, an equation

$$c_1 + c_2 x + c_3 x^2 + \ldots + c_m x^{m-1} = 0 \quad (\bmod p)$$

has $m$ solutions: $i_1, i_2, \ldots, i_m$. By definition, $p > 6m$. Hence all $m$ solutions $i_1, \ldots, i_m$ are different modulo $p$. However, an equation of degree $m-1$ can have only $m-1$ different solutions. Contradiction. $\square$

**Lemma 5** *If the described automaton starts from $Q_{i,1}$, the probability of accepting a word $u$ is*

1. *greater or equal than 9/10 if*

$$\sum_{j=1}^{n} n_j(i^{j-1} \bmod p) = \sum_{j=1}^{n} m(i^{j-1} \bmod p)$$

2. *less or equal than 9/10, otherwise.*

**Proof.** By definition of our automaton, the probabilities of passing from one state to another after reading the symbol $a_j$ are equal to similar probabilities in automaton $U_{i'}$ after reading $(i^{j-1} \bmod p)$ symbols. Hence, the probability of passing from $Q_{i,1}$ to an accepting state after reading the word consisting of $n_1$ symbols $a_1$, $n_2$ symbols $a_2$, ..., $n_m$ symbols $a_m$ is equal to the probability of passing from $A_{i',1}$ to the accepting state in automaton $U_{i'}$ after reading

$$\sum_{j=1}^{m} n_j(i^{j-1} \bmod p)$$

symbols. By definition, $U_{i'}$ accepts the word consisting of

$$i' = \sum_{j=1}^{m} m(i^{j-1} \bmod p)$$

symbols with a probability greater or equal to 9/10 and rejects other words with a probability greater or equal to 9/10. $\square$

Consider two cases:

1. The automaton receives a word $u \in L_m$. Then

$$n_1 = n_2 = \ldots = n_m = m.$$

From Lemma 5 we have that the automaton starting from any of states $Q_{1,1}, \ldots, Q_{6m,1}$ accepts $u$ with a probability 9/10.
Starting from its normal starting state $S$, the automaton equiprobably chooses one of states $Q_{1,1}, \ldots, Q_{6m,1}$ and moves to it. So, it accepts the word $u \in L_m$ with a probability at least 9/10.

2. Automaton receives a word $u \notin L_m$. Then equality

$$\sum_{j=1}^{m} n_j(i^{j-1} \bmod p) = \sum_{j=1}^{m} m(i^{j-1} \bmod p)$$

holds for at most $(m-1)$ different $i \in \{1, \ldots, 6m\}$.
(If this equality holds for $m$ different $i \in \{1, \ldots, 6m\}$ then, from

$$\sum_{j=1}^{m}(n_j - m)(i^{j-1} \bmod p) = 0$$

and from the linear independence of vectors $v_i$ (Lemma 4) it follows that

$$n_j - m = 0$$

for all $j$. This means that $u \in L_m$.)
So, with a probability of at least $1 - \frac{m-1}{6m} > 5/6$ the automaton moves from state $S$ to state $Q_{i,1}$ such that

$$\sum_{j=1}^{m} n_j(i^{j-1} \bmod p) \neq \sum_{j=1}^{m} m(i^{j-1} \bmod p).$$

After starting from such state $Q_{i,1}$ the word is rejected with a probability of at least $9/10$ (Lemma 5). Hence, the probability of rejecting the word $u \notin L_m$ is at least $5/6 * 9/10 = 3/4$.

We have shown that words belonging to $L_m$ are accepted by an automaton with a probability of at least $9/10$ and the words which do not belong to $L_m$ are rejected with a probability of at least $3/4$.

*The number of states in automaton $U$.* The automaton $U$ consists of $1 + 6ms$ states. We have

$$s = O\left(\frac{\log^2 k}{\log\log k}\right), \text{ and } k = pm^2 = (6m + o(m))m^2 = O(m^3).$$

Hence

$$s = O\left(\frac{\log^2 m}{\log\log m}\right).$$

The number of states is

$$1 + 6ms = O\left(m\frac{\log^2 m}{\log\log m}\right).$$

$\square$

Denote

$$n = O\left(m\frac{\log^2 m}{\log\log m}\right).$$

Then, the language $L_m$ can be accepted by a probabilistic finite automaton whith $n$ states (Lemma 2). Any deterministic finite automaton accepting $L_m$ has at least $(m + 1)^m$ states (Lemma 1). We have

$$(m + 1)^m \approx 2^{m\log m} = \Omega\left(2^{\frac{n\log\log n}{\log n}}\right)$$

This proves the theorem. $\square$

**Acknowledgements.** I would like to thank Rūsiņš Freivalds for suggesting the problem and help during this research. Research was supported by Latvia's Science Council Grant No. 93.599.

# References

[Fr82]   R. Freivalds, *On the growth of the number of states in result of determinization of probabilistic finite automata*, Avtomatika i Vicislitelnaja Tehnika, 1982, N.3, 39-42 (in Russian)

[GM79]  N. Z. Gabbasov, T. A. Murtazina, *Improving the estimate of Rabin's reduction theorem*, Algorithms and Automata, Kazan University, 1979, 7-10 (in Russian)

[Paz66]  A. Paz, *Some aspects of probabilistic automata*, Information and Control, 9(1966)

[Ra63]   M. O. Rabin, *Probabilistic automata*, Information and Control, 6(1963), 230-245

[TB73]   B. A. Tracktenbrot, Ya. M. Barzdin', *Finite Automata: Behaviour and Synthesis*. North-Holland, 1973

# Bounded Length UCFG Equivalence

B. Litow

Dept. of Computer Science, James Cook University

**Abstract.** A randomised polylog time algorithm is given for deciding whether or not the sets of words of a given length generated by two unambiguous context-free grammars coincide. The algorithm is in randomised NC4 in terms of the product of the grammar size and the length.

## 1 Introduction

### 1.1 Statement of results

Deciding whether or not two unambiguous context-free grammars (UCFG) generate the same language is not known to be decidable. In this paper we investigate the complexity of the bounded length version of this problem UCFGBEQ. Let $G_1, G_2$ be UCFG with the same terminal alphabet $A$ and let $n$ be a positive integer. $L(G_1), L(G_2)$ are the languages generated by $G_1, G_2$, respectively. UCFGBEQ is the problem of deciding whether or not $L(G_1) \cap A^n = L(G_2) \cap A^n$.

Note that an hypothesis like unambiguousness appears to be essential because the complementary problem, bounded length inequivalence is NP-complete for right linear grammars [3]. An easy way to see this is by reduction from Hamiltonian circuit. Sketching the reduction, the set of length $n$ cycles in a given graph can be presented by a right linear grammar computable in PTIME in $n$ and the graph size. The same can be done for the length $n$ cycles containing repeated nodes. This grammar is probably going to be highly ambiguous, since guessing is used to see whether a node symbol occurs twice. The graph has a Hamiltonian circuit iff the two grammars generate distinct languages.

There has been some effort to investigate the complexity of a variety of language problems. Results are reported in [10], and an extensive study centering on monoid problems and ranking is carried out by Huynh in [6, 5, 7]. More recent work on ranking for UCFG has been carried out in [2] and using a different method in [11]. Both approaches show that UCFG ranking is in DIV, the class of problems NC1 reducible to integer division.

Although several language problems admit NC type algorithms it appears that randomised techniques have not hitherto been used to obtain polylog parallel time solutions to language problems. In this paper, we present a randomised NC (RNC) algorithm for UCFGBEQ.

**Theorem 1.** *UCFGBEQ is in RNC4 in terms of the product of $n$, given in unary notation and the grammar size.*

In general randomised NC algorithms are relatively uncommon. However, the maximum matching algorithm presented in [12] is an example of a randomised NC (RNC2 in fact) algorithm.

## 1.2 Outline of the paper

Section 2 contains preliminaries from context-free language theory and a few facts about related formal series. Section 3 presents the randomisation result, which can be viewed as a kind of sieve. The randomisation is based on an elementary fact from Galois theory, and may have other algorithmic applications. Section 4 gives the proof of Theorem 1.

## 2 Context-free grammar preliminaries

A context-free grammar (CFG) $G$ is a 4-tuple, $G = (A, X, x_0, P)$, where $A$ is the terminal set, $X$ is the set of nonterminals, $x_0 \in X$ is the start symbol, and $P$ is the set of productions. The size of a grammar will be the sum of the lengths of the right hand sides of all its productions. The reader should refer to a standard text, e.g., [4] for other common CFG related terms and definitions.

**Lemma 2.** *Every UCFG can be converted in NC2 (in terms of the grammar size) into a UCFG in Chomsky Normal Form.*

*Proof Sketch.* It is a straightforward that the standard conversion into CNF (e.g. [4]) does not introduce new ambiguity. This conversion can easily be carried out in DLOG, hence certainly in NC2.

Henceforth we will assume that $G_1, G_2$ are UCFG in CNF.

For our purpose a *formal series* $f$ is a mapping from $A^*$ into into the nonnegative integers. $f$ can be viewed as the formal sum $\sum_{w \in A^*} f(w) \cdot w$. A detailed account of the formal series viewpoint in language theory is presented in [8]. A *solution* of a CFG is a list of formal series, $f_0, \ldots, f_r$ corresponding to the nonterminals, $x_0, \ldots, x_r$, such that for each $x_i$ the equation $x_i = \beta_1 + \cdots + \beta_s$ becomes a formal series identity when each nonterminal is replaced by its corresponding series throughout all the equations. $\beta_1, \ldots, \beta_s$ are the right hand sides of the $x_i$ productions.

**Lemma 3.** *A CNF grammar $G$ has a unique solution $(f_0, \ldots, f_r)$ and $f_i(w)$ is the number of leftmost derivations of $w$ from $x_i$.*

*Proof.* This follows from Theorems 14.9 and 14.11 of [8].

**Remark** Note that if $G$ is a UCFG, then $f_i(w) \leq 1$.

Given any CFG $G$ in CNF we define a CFG $G^n$ in CNF such that $L(G^n) = L(G) \cap A^n$. $G^n = (A, X^n, (x_0, 1, n), P^n)$, where

- $X^n = \{(x, i, j) \mid x \in X, 1 \leq i \leq j \leq n\}$.
- - $(x, i, i) \to a \in P^n$ iff $x \to a \in P$.
  - If $i < j$, then $(x, i, j) \to (y, i, k)(z, k+1, j) \in P^n$ iff $x \to yz \in P$.

It is possible that the above construction could introduce useless symbols. The standard algorithm for useless symbol removal, e.g., [4] can be carried out in NC2. We will assume this for $G^n$ in what follows.

**Lemma 4.** $G^n$ is in CNF and has a unique solution $(f_0^n, \ldots)$ such that if $|w| \neq n$, then $f_0^n(w) = 0$, and if $|w| = n$, then $f_0^n(w) = f_0(w)$, where $f_0$ is the unique solution element corresponding to $x_0$.

*Proof Sketch.* It is obvious from its definition that $G^n$ is in CNF. By Lemma 3 both $G$ and $G^n$ have unique solutions. An induction on the difference $j - i$ entirely similar to the proof of correctness of the Cocke-Younger-Kasami algorithm can be used to establish the claims about $f_0^n$.

## 3  A randomised sieve

Let $e(z) = \exp(2\pi\sqrt{-1} \cdot z)$. If $w = w_1 \cdots w_n \in \{0,1\}^n$, then for any odd integer $g$ define $\langle w \rangle_g = \prod_{j=1}^{n} e(g \cdot w_j / 2^{j+1})$.

**Lemma 5.** If $c_w$ is rational for each $w \in \{0,1\}^n$, and $c_w \neq 0$ for at least one $w$, then

$$\sum_{w \in \{0,1\}^n} c_w \cdot \langle w \rangle_g \neq 0$$

*Proof.* It is classical [9] that if $\alpha$ is a primitive complex $m$-th root of unity, then $\alpha$ cannot be a zero of any nonzero rational polynomial of degree below $\phi(m)$. It is clear that $\langle w \rangle_g$ can be written as

$$\langle w \rangle_g = e(g \cdot D / 2^{n+1})$$

such that $0 \leq D \leq 2^n - 1 < \phi(2^{n+1}) = 2^n$. Since $g$ is odd, $e(g/2^{n+1})$ is a primitive $n+1$-st root of unity. By uniqueness of binary notation, and the odd parity of $g$, if $w \neq v$, then $\langle w \rangle_g \neq \langle v \rangle_g$. This means that the above sum can be regarded as the evaluation of a nonzero rational polynomial at $e(g/2^{n+1})$ of degree below $\phi(2^{n+1})$, and the lemma follows.

We need a fact from Galois theory. Let $P(x) \in \mathbb{Z}[x]$, of degree, $d$, and let $\alpha$ be a primitive $m$-th root of unity where $d < \phi(m)$. The norm, $N$, of $P(x)$ is the sum of the absolute values of its coefficients.

**Lemma 6.** *If $S = \{1 \leq j < m \mid GCD(j, m) = 1\}$, then for every $\ell > 0$, for at least $\frac{1}{\ell+1}$ of the $j \in S$, $|P(\alpha^j)| \geq 1/N^{1/\ell}$.*

**Proof** Define

$$\prod = \prod_{j \in S} P(\alpha^j)$$

Recall from elementary Galois theory that

$$\prod \in \mathbb{Z} - \{0\}$$

Indeed this follows from three facts;

- $P(\alpha^j) \neq 0$
- $\prod \in \mathbb{Q}$ since it is fixed under all elements of the Galois group of the field generated by the $\alpha^j$ over $\mathbb{Q}$.
- The $\alpha^j$ are algebraic integers so, by the second item, $\prod$ is in the intersection of the algebraic integers and $\mathbb{Q}$, i.e., $\mathbb{Z}$.

Now if $|P(\alpha^j)| < 1/N^{1/\ell}$ for more than $\frac{1}{\ell+1}$ of the $j$, then since $|P(\alpha^j)| \leq N$ for any $j$, we would have $|\prod| < 1$, which is impossible. $\square$

If $L \subseteq \{0, 1\}^n$, then define $\langle L \rangle_g$ to be

$$\langle L \rangle = \sum_{w \in L} \langle w \rangle_g$$

The next lemma is the key technical result.

**Lemma 7.** *If $L_1, L_2 \subseteq \{0, 1\}^n$, and $L_1 \neq L_2$, then for at least half of the odd $g$, $1 \leq g < 2^{n+1}$,*

$$|\langle L_1 \rangle_g - \langle L_2 \rangle_g| \geq 1/2^n$$

*Proof.* If $L_1 \neq L_2$, then $\langle L_1 \rangle_g - \langle L_2 \rangle_g$ can be regarded as some $P(e(g/2^{n+1}))$, where $P(x)$ is a nonzero polynomial, and with nonzero coefficients in $\{1, -1\}$. In particular, positive terms will correspond to the words in $L_1 - L_2$, and negative terms to those in $L_2 - L_1$. The norm $N$ of $P(x)$ is bounded above by $2^n$. This follows from the fact that the number of terms in $\langle L_1 \rangle_g - \langle L_2 \rangle_g$ is the size of the symmetric difference of $L_1$ and $L_2$, which cannot exceed $2^n$. The result now follows from Lemma 5 and Lemma 6.

# 4  Proof of Theorem 1

## 4.1  Basic error bounds

It will be convenient to work over the alphabet $\{0, 1\}$. If UCFG $G_1, G_2$ have terminal alphabet $A$ such that $2^{k-1} < \text{card}(A) \leq 2^k$, then code the symbols of $A$ by using length $k$ binary strings. We will assume for the rest of the paper that the grammars $G_1, G_2$ to be tested for equivalence have been modified to reflect this coding, which will not materially affect the grammar sizes so far as the complexity bounds are concerned.

Let $G$ be a UCFG in CNF with terminal alphabet $\{0,1\}$. Fix an odd integer $g$ with $1 \leq g < 2^{n+1}$. Define $\beta_i$ to be the sum of the first $3n$ terms of the Taylor series of $e((g \bmod 2^{i+1})/2^{i+1})$. Note that

$$\mathrm{e}((g \bmod 2^{i+1})/2^{i+1}) = \mathrm{e}((g/2^{i+1})$$

However, $0 \leq \frac{g \bmod 2^{i+1}}{2^{i+1}} < 1$, so that we get rapid convergence for the Taylor series.

We are going to associate a system of equations over the complex numbers $[G^n]_g$ with the CFG $G^n$. The set of variables is $\{[(x,i,j)]_g \mid (x,i,j) \in P^n\}$. Next we specify the equations of the system.

- $[(x,i,i)]_g = b_0 + b_1 \cdot \beta_i$ such that $b_0, b_1 \in \{0,1\}$ and $b_0 = 1$ iff $x \to 0 \in P$, and $b_1 = 1$ iff $x \to 1 \in P$.
- If $i < j$, then $[(x,i,j)]_g = \sum [(y,i,k)]_g \cdot [(z,k+1,j)]_g$ where the sum is over all pairs $y, z$ such that $x \to yz \in P$, and over all $i \leq k < j$.

The idea behind this system of equations is to generate for each $w = w_1 \cdots w_n \in L(G^n)$ the number $[w]_g = \prod_{i=1}^{n}[w_i]_g$ where $[w_i]_g = 1$ if $w_i = 0$, and $[w_i]_g = \beta_i$ if $w_i = 1$. Notice the similarity to the bracket notation of Section 3, with the only difference being the appearance of the approximation $\beta_i$. We make this idea precise with the next lemma.

**Lemma 8.** $[G^n]_g$ *has a unique solution namely* $\{[(x,i,j)]_g \mid (x,i,j) \in P^n\}$ *and*

$$|[(x_0,1,n)]_g - \langle L(G^n) \rangle_g| < 1/2^{n+3}$$

*Proof.* Existence and uniqueness of the solution follow directly from Lemma 4. In particular, $[(x_0,1,n)]_g$ is obtained from the formal series $f_0^n$ for $(x_0,1,n)$ by regarding the terminal alphabet $A$ as $\{c_1, \ldots, c_n, d\}$, instantiating these symbols as $c_i = \beta_i$, and $d = 1$ ($d$ stands in for occurrences of the symbol 0 at any position), then evaluating $f_0^n$.

The error of approximation $\delta = |\mathrm{e}(g/2^{i+1}) - \beta_i|$ is certainly bounded above by $1/2^{3n}$. If we let $[w]_g$ designate the value of any $w \in \{0,1\}^n$ after instantiating all occurrences of 0 by 1 and the occurrence of 1 in the $i$-th position by $\beta_i$, and letting $\delta_w = |[w]_g - \langle w \rangle_g|$, we get $\delta_w < 2n\delta$. This is obtained by noting since $|\mathrm{e}(z)| = 1$ for real $z$, that

$$|(\mathrm{e}(z) + \delta)(\mathrm{e}(z') + \delta) - \mathrm{e}(z + z')| < 2\delta + \delta^2 < 3\delta$$

and then iterating this fact $\log n$ times. We conclude from this and the triangle inequality applied at most $2^n$ times (upper bound on $\mathrm{card}(L(G^n))$) that

$$|\langle L(G^n) \rangle_g - [(x_0,1,n)]_g| \leq 2^n \cdot \delta_w < n \cdot 2^{n+1} \cdot \delta < 1/2^{n+3}$$

**Remark** The righmost bound in the last inequality of the proof of Lemma 8 is not intended to be sharp, but it is enough to prove the main result of this subsection.

**Lemma 9.** *If $G_1$ and $G_2$ are UCFG in CNF with start symbols $x_{1,0}$ and $x_{2,0}$, respectively, then*

- *If $L(G_1) \cap \{0,1\}^n = L(G_2) \cap \{0,1\}^n$, then for any $g$,*

$$|[(x_{1,0}, 1, n)]_g - [(x_{2,0}, 1, n)]_g| < 1/2^{n+2}$$

- *If $L(G_1) \cap \{0,1,\}^n \neq L(G_2) \cap \{0,1\}^n$, then for at least half the odd $g$, $1 \leq g < 2^{n+1}$,*

$$|[(x_{1,0}, 1, n)]_g - [(x_{2,0}, 1, n)]_g| > 1/2^{n+1}$$

*Proof.* The first claim follows from Lemma 8 and the triangle inequality. The second claim follows from Lemma 8, the triangle inequality and Lemma 7.

## 4.2 Complexity

We now describe an algorithm which produces the solution to the equation system $[G^n]_g$.

**Lemma 10.** *The solution to $[G^n]_g$ can be computed in NC4 in terms of the problem size which is the product of $n$ and the grammar size.*

*Proof.* It is straightforward that the system $[G^n]_g$ can be constructed in NC4 (in fact certainly NC2 suffices) from $G$ and the binary notation for $n$. This includes computing the complex rational approximations $\beta_i$. If $s$ is the size of $G$, then $[G^n]_g$ has at most $s \cdot n^2$ equations. The solution will be obtained in $\log n$ stages, with each stage involving the solution of a system of linear equations. Each succeeding system is smaller than its predecessor.

The *rank* of the variable $[(x, i, j)]_g$ is defined to be $j - i$. Let $\ell$ be the least integer such that $2^\ell \leq n < 2^{\ell+1}$. For $1 \leq h \leq \ell + 1$ define $R_h$ to be the set of all variables of rank at least $\lfloor n/2^h \rfloor$, but less than $\lfloor n/2^{h-1} \rfloor$. Note that $R_{\ell+1}$ contains the variables $[(x, i, i)]_g$ of rank 0. Next, define $S_h$ to be the set of all equations whose left hand side variable is in $R_h$. Observe that if $h \leq \ell$, then each equation in $S_h$ has at least one variable as a right hand side factor whose rank is less than $n/2^h$.

The solution procedure starts with $H_{\ell+1}$. Each left side variable of an equation in $H_{\ell+1}$ is replaced by the complex rational value on the right side. Next, $H_\ell$ is inspected and for each equation, the right side variables of rank 0 are replaced by the values just determined. By the observation above, at least one of the two right side variables must have rank 0. This modified $H_\ell$ can now be solved as an ordinary linear system.

The situation after $H_{\ell+1}, \ldots, H_m$ have been solved is that all variables of rank less than $\lfloor n/2^{m-1} \rfloor$ have received complex rational values. This collection of variables is precisely $R_{\ell+1} \cup \cdots \cup R_m$. By the observation each right side of an equation in $H_{m-1}$ involves a variable in $R_{\ell+1} \cup \cdots \cup R_m$, and so $H_{m-1}$ can be solved as a linear system. The process terminates on solving $H_1$, where in particular $[(x_0, 1, n)]_g$ will receive a value.

The main cost of this procedure is in solving $\ell + 1$ linear systems, each of size bounded above by $s \cdot n^2$. Since $\ell = O(\log n)$, we get an NC3 bound [1]. We have ignored the numerical computations which involve $\log n$ iterated matrix multiplications. Since the matrices are at most order $s \cdot n^2 \times s \cdot n^2$, and the initial numerical data requires $n^{O(1)}$ bits ( the truncated Taylor series), it is straightforward that the intermediate numbers are also $n^{O(1)}$ bits size. The overall NC4 bound then follows from this and the well known fact e.g. [13] that addition and multiplication are in NC1.

## 4.3 Completion of the proof

The correctness of the algorithm follows from Lemma 9 and its complexity follows from Lemma 10.

# References

1. S. Berkowitz. On computing the determinant in small parallel time using a small number of processors. *Inf. Proc. Lett.*, 18:147–150, 1984.
2. A. Bertoni, M. Goldwurm, and P. Massazza. Counting problems and algebraic formal power series in noncommuting variables. *Inf. Proc. Lett.*, 34:117–121, 1990.
3. Michael R. Garey and David S. Johnson. *Computers and Intractability*. W.H.Freeman and Company, New York, 1979.
4. J. Hopcroft and J. Ullman. *Introduction to Automata Theory, Languages and Computation*. Addison-Wesley, 1979.
5. D. Huynh. The complexity of ranking. In *3rd Structure in Complexity Theory Conf.*, pages 204–212, 1988.
6. D. Huynh. The complexity of deciding code and monoid properties for regular sets. Technical report, Computer science, U. Texas-Dallas, 1990.
7. D.T. Huynh. The complexity of ranking simple languages. *Mathematical Systems Theory*, 23:1–19, 1990.
8. W. Kuich and A. Salomaa. *Semirings, Automata, Languages*. Springer-Verlag, 1986.
9. S. Lang. *Algebra*. Addison-Wesley, 1965.
10. B. Litow. Parallel complexity of the regular code problem. *Inf. and Comp.*, 86,1:107–114, 1990.

11. B. Litow. Numbering unambiguous context-free languages. In *17th Australian Computer Society Conf.*, pages 373–378. University of Canterbury, New Zealand, 1994.

12. K. Mulmuley, U. Vazirani, and V. Vazirani. Matching is as easy as matrix inversion. In *19th Symp. on Theory of Computing (STOC)*, pages 345–354. ACM, 1987.

13. I. Wegener. *The Complexity of Boolean Functions*. Wiley-Teubner, 1987.

# The Steiner Minimal Tree Problem in the
# λ-geometry Plane*

### Abstract

A Steiner Minimal Tree (SMT) for a given set $P$ of points is a shortest network interconnecting the points of $P$ whose vertex set may include some additional points in order to get the minimum possible total length in a metric space. When no additional points are allowed the minimum interconnection network is the well-known minimum spanning tree (MST) of $P$. The Steiner ratio is the greatest lower bound of the ratio of the length of an SMT over that of an MST of $P$. In this paper we study the Steiner minimal tree problem in which all the edges of SMT have fixed orientations. We call it the SMT problem in the λ-*geometry* plane, where λ is the number of possible orientations.

Here is the summary of our results.

1. We show that the Steiner ratio for $|P| \geq 3$ is $\frac{\sqrt{3}}{2}\cos(\pi/2\lambda)$, for $\lambda = 6m + 3$ and integer $m \geq 0$, and is $\sqrt{3}/2$, for $\lambda = 6k$ and integer $k \geq 1$, disproving a a conjecture of Du *et al.*[3] that the ratio is $\sqrt{3}/2$ iff the unit disk in normed planes is an ellipse.

2. We derive the Steiner ratios for $|P| \leq 4$ for all possible λ's and show that for $|P| \geq 3$ there exists an SMT whose Steiner points lie in a *multi-level* Hanan-*grid*, generalizing a result that holds for rectilinear case, i.e., $\lambda = 2$.

These results show that the Steiner ratio is not a monotonically increasing function of λ, as believed by many researchers. We conjecture that the Steiner ratios obtained above ($|P| \leq 4$) are actually true for *all* $|P| \geq 3$.

## 1   Introduction

Let $P$ be a set of $n$ points in a metric space. A Steiner minimal tree (SMT) on $P$ is a shortest network interconnecting $P$ while allowing additional points, called *Steiner points*. Since the problem of finding such a tree has been shown to be NP-hard in Euclidean and rectilinear spaces[6, 7], polynomial time approximation algorithms were proposed. The *performance ratio* of any approximation $\mathcal{A}$ in metric space $\mathcal{M}$ is defined as

$$\rho_\mathcal{M}(\mathcal{A}) = \inf_{P \in \mathcal{M}} \frac{L_s(P)}{L_\mathcal{A}(P)}$$

where $L_s(P)$ and $L_\mathcal{A}(P)$ denote the lengths of a Steiner minimal tree and of the approximation $\mathcal{A}$ on $P$ in space $\mathcal{M}$. One well-studied approximation interconnection network is the minimum spanning tree (MST) in which *no* Steiner points are allowed. When the approximation is the MST, the performance ration is known as the *Steiner ratio*, denoted simply as $\rho$. The following Steiner ratio results are known. In the $L_2$ norm $\rho = \sqrt{3}/2$[1], and in the $L_1$ norm, it is $2/3$[9]. In the $L_2$ norm, the edges connecting $P$ are straight lines of artibrary orientation, while in the $L_1$ norm, they are restricted to be either horizontal or vertical. Let λ denote the number of possible orientations, and the unit disk $D$ in this metric space, called the λ-geometry plane[12] is a centrally symmetric 2λ-gon. Fig. 1 shows the unit disk in various λ. It was shown in [12] that the Steiner ratio, denoted $\rho^\lambda(D)$, satisfies the inequality $\rho^\lambda(D) \geq \frac{\sqrt{3}}{2}\cos(\pi/2\lambda)$. Du *et al.* in [3] showed that in any normed plane where the unit disk is an arbitrary compact convex centrally symmetric domain $\rho(D)$ satisfies the inequality $0.623 < \rho(D) < 0.8686$. They conjectured[3] that $2/3 \leq \rho(D) \leq \sqrt{3}/2$ for any norm and that $\rho(D) = \sqrt{3}/2$ if and only if $D$ is an ellipse and $\rho(D) = 2/3$ if and only if $D$ is a parallelogram. It was shown in [3] that in any normed plane $2/3 \leq \rho(D)$ for $|P| \leq 6$, and in [11] that $\rho(D) \leq \sqrt{3}/2$ for any $L_p$ norms.

We show in this paper that $\rho^\lambda(D) = \frac{\sqrt{3}}{2}\cos(\pi/2\lambda)$, for $\lambda = 6m + 3$ and integers $m \geq 0$. That is, the lower bound derived in [12] is tight for those λ's. When $\lambda = 6k$, $\rho^\lambda(D) = \sqrt{3}/2$, disproving the conjecture that $D$ must be an ellipse. We also derive the ratios $\rho^\lambda(D)$ for $|P| \leq 4$. These results show

* Supported in part by the National Science Foundation under the Grant CCR-9309743, and by the Office of Naval Research under the Grant No. N00014-93-1-0272.

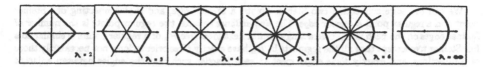

Fig. 1. Examples of unit disks in the λ-*geometry* plane.

that the Steiner ratio is not a monotonically increasing function of $\lambda$, as believed by many researchers.

In the $L_1$ norm ($\lambda = 2$), it is well-known that the Steiner points of any Steiner minimal tree must lie on a grid, known as the *Hanan*-grid, defined by drawing horizontal and vertical lines through the given points. Sarrafzadeh and Wong[12] showed that in the λ-geometry plane, the *Hanan*-grid does not necessarily contain all possible Steiner points in any Steiner minimal tree, but did not offer any characterization of where the Steiner points should lie. We show that their exists a Steiner minimal tree in which the Steiner points can be chosen from a *multi-level* Hanan-grid defined by $P$, generalizing the Hanan-grid result (which holds for $\lambda = 2$) to all $\lambda$'s $< \infty$. Thus this result implies that there is a finite solution for the Steiner minimal tree problem in the λ-geometry plane. In [10] we have reported a similar result for $\lambda = 4$.

The paper is organized as follows. In Section 2 we give the basic definition and terminology along with some preliminary results. Section 3 gives derivations of ratios for smaller point sets. In Section 4 we prove our main result on Steiner ratios and in Section 5 we describe our multi-level grid theorem. Finally we make a conclusion.

## 2 Preliminaries

The given points in $P$ are called *regular points*. The additional points which may exist in an SMT are called *Steiner points*. In the λ-*geometry* plane, each edge in an SMT connecting two points is allowed to make an angle $(i*\pi/\lambda)$ with the positive $x$-axis for integers $0 \leq i < \lambda$. The orientation of angle $(i*\pi/\lambda)$ with the positive $x$-axis is referred to as the $i^{th}$ λ-direction, and the $i^{th}$ and $(i+1st)$ λ-directions are said to be *adjacent*.

The Steiner ratio for a set $P$ of $n$ points in the λ-*geometry* plane $\mathcal{E}^\lambda$ is defined as

$$\rho_n^\lambda = \inf_{P \in \mathcal{E}} \frac{L_{smt}(P)}{L_{mst}(P)}.$$

where $L_{smt}(P)$ and $L_{mst}(P)$ are the lengths of an SMT and MST on $P$ respectively.

From here on it is understood that the point sets are in $\mathcal{E}^\lambda$ and the specification will be omitted. For any two points $p, q$, if the straight line segment $\overline{p, q}$ connecting two points $p$ and $q$ has the $i^{th}$ λ-direction for some $i$, then these two points or the edge connecting them are said to be *straight*. Otherwise they are said to be *nonstraight*. If $p$ and $q$ are nonstraight, then the edge connecting them can be realized by a *canonical path* consisting of exactly *two* segments with orientations equal to two adjacent λ-directions. There are two ways to connect two nonstraight points, and these two nonstraight edges form a parallelogram. One nonstraight edge can be obtained from the other by a *flipping* operation. We use $e(p,q)$ to denote an edge, straight or nonstraight. The distance between $p$ and $q$ is measured by the length of $e(p,q)$.

Let $G_n$ denote a Steiner minimal tree of a set of $n$ regular points. The following are fundamental properties of $G_n$ which are generalizations of that of rectilinear SMT[5, 8, 12].

**Property 1** *The number of Steiner points in $G_n$ is at most $n-2$.*

**Property 2** *The degree of any Steiner point $q$ satisfies $3 \leq deg(q) \leq 4$. When $deg(q) = 4$, these four adjacent points form a cross.*

**Property 3** *There is no acute angle nor straight angle formed by any two edges in $G_n$. Moreover there exists a $G_n$ such that the three angles, $\alpha_1, \alpha_2, \alpha_3$, around a degree-3 Steiner point are:*

$\alpha_1 = \alpha_2 = \alpha_3 = 2\pi/3$, for $2\lambda$ *MOD* $3 = 0$;
$\alpha_1 = \alpha_2 = \lfloor 2\lambda/3 \rfloor (\pi/\lambda)$ *and* $\alpha_3 = \lceil 2\lambda/3 \rceil (\pi/\lambda)$, *for* $2\lambda$ *MOD* $3 = 1$; *or*
$\alpha_1 = \lfloor 2\lambda/3 \rfloor (\pi/\lambda)$ *and* $\alpha_2 = \alpha_3 = \lceil 2\lambda/3 \rceil (\pi/\lambda)$, *for* $2\lambda$ *MOD* $3 = 2$.
*This is referred to as the* **even angle property.**

We call the edge connecting two Steiner points, a *Steiner edge*, and that connecting one Steiner point and a regular point a *regular edge*. The *link distance* between two regular points is the number of Steiner edges on the path in $G_n$ connecting them. A hexagonal tree $\mathcal{H}_n$ of $n$ regular points is similar to a full Steiner tree except that all the segments in $\mathcal{H}_n$ have exactly *three* orientations, not necessarily those of $\lambda$-directions, and exactly *three* edges meet at each *junction* such that the angles between adjacent segments incident to a junction have at most one acute angle. Consider three nonparallel lines such that they intersect pairwise at vertices $A, B$ and $C$. Through vertices $A, B$ and $C$ we draw lines parallel to segments $\overline{B,C}, \overline{C,A}$, and $\overline{A,B}$ respectively. The structure formed by repeating this operation is called a $\Delta$-*lattice* with a base triangle $\Delta(A, B, C)$, and denoted $\mathcal{L}_\Delta$. The vertices of $\mathcal{L}_\Delta$ are called *lattice points*. The lattice whose base triangle is an equilateral triangle of unit length is called a *unit lattice*. The following results were proved by Du *et al.*[3].

**Theorem 1.** *Let $\rho_0$ be a positive number. Then for any norm with unit disk $D$, $\rho(D) \geq \rho_0$ if and only if for all hexagonal norm with unit disk $\mathcal{H}$, $\rho(\mathcal{H}) \geq \rho_0$.*

**Lemma 2.** *Let $C$ be a critical set[a] of $n$ lattice points of a unit lattice such that its MST is of length $n-1$ and let $\mathcal{H}(C)$ be a minimum hexagonal tree on $C$. $\mathcal{H}(C)$ can be transformed into an MST of $C$ by a sequence of translating and flipping operations such that the length of $\mathcal{H}(C)$ is no less than the length of MST of $C$.*

When $G_n$ has exactly $n - 2$ Steiner points, it is a *full* Steiner tree, denoted $G_n^*$. $G_n^*$ must have one of the local structures shown in Fig. 2(a), when $n > 4$. The two regular points that are connected to a common Steiner point are said to form a *regular pair*. It is obvious that $G_n^*, n \geq 4$, must have at least two regular pairs. The upper left diagram in Fig. 2(a) shows two regular pairs with a link distance exactly 2; and the other shows the two regular pairs with a link distance at least 2.

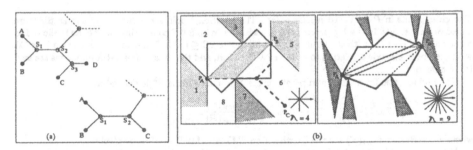

**Fig. 2.** (a) Two local structures in full SMTs. (b) Steiner loci of two regular points.

## 3 Steiner Ratios for Sets of Small Size

### 3.1 Steiner Ratios for $\lambda$-Equilateral Triangles

We show the Steiner ratio for three points that form a $\lambda$-equilateral triangle. Note that $G_3$ may *not* be unique. According to Property 3, a $G_n$ that satisfies the even angle property can always be found and thus is used to prove the Steiner ratio.

Given a $G_3$ for points $A, B, C$, angle $\angle AqB$ is said to be *opposite* to edge $e(q, C)$. We shall use $\lambda = 4$, the 45°-SMT as an example for illustrations only. Similar results can be obtained for other $\lambda > 2$ and the proofs can be carried over with some modifications.

**Lemma 3.** *In $G_3$ at most one nonstraight edge, say edge $e(B, q)$, is incident on $q$. Furthermore the angle opposite to the nonstraight edge equals $135°$ ($\lceil 2\lambda/3 \rceil (\pi/\lambda)$) (Fig. 2(b)). In addition, the segment of the nonstraight edge not incident to $q$ (called the second segment) is not parallel to $e(q, A)$ or $e(q, C)$.*

Lemma 3 is proved by a constructive method. For points $A, B$, we construct two rectangles $R_0(A, B)$ and $R_{+1}(A, B)$ as shown in Fig. 2(b). The boundary of the union of the two rectangles $\mathcal{L}_{A,B} = \partial(R_0(A, B) \bigcup R_{+1}(A, B))$ is the *Steiner locus* of $A, B$.

---

[a] C must satisfy the condition that there exists an MST that divides the interior of the $n$-gon formed by connecting adjacent lattice points, into $n - 2$ equilateral triangles.

When point $C$ lies outside of the region defined by $\mathcal{L}_{A,B}$, the Steiner point must lie on $\mathcal{L}_{A,B}$. If $C$ lies inside, $C$ itself is the Steiner point. From $\mathcal{L}_{A,B}$ we partition the plane into 8 regions as shown in Fig. 2(b), each of which is associated with a boundary segment of $\mathcal{L}_{A,B}$ that contains $q$. Note that for other values of $\lambda$ the Steiner locus is defined by $(\lambda - \lfloor 2\lambda/3 \rfloor)$ parallelograms whose intersection is the parallelogram defined by the two canonical paths connecting $p_A$ and $p_B$.

When $\Delta(A, B, C)$ is an equilateral triangle, $\rho_3^\lambda = \frac{1}{4-2\sqrt{2}}$ for $\lambda = 4$. To see this we assume the Steiner point coincides with the origin $O$ and $e(O, A)$ coincides with the $x$-axis, and edges $e(O, B), e(O, C)$, are symmetric to the $x$-axis, such that they form angles satisfying the angle property. For an arbitrary 3-point set it can be shown that the ratio of its $L_{smt}$ and $L_{mst}$ is always greater than that of a $\lambda$-equilateral triangle.

**Lemma 4.** *If $\lambda$ is even, edge $e(B, C)$ is straight. Otherwise, it is nonstraight and the two segments $e(B, k)$ and $e(C, k)$ are of equal length, and form an angle $\pi - \frac{\pi}{\lambda}$ (Fig. 4).*

We give below the formulae for the ratios.

[1] When $\lambda = \{6m | m \geq 1, m \in I\}$:
$$\rho_3^\lambda = \frac{\sqrt{3}}{2} \qquad \text{(same as the Euclidean Steiner ratio)}$$

[2] When $\lambda = \{6m + 1 | m \geq 0, m \in I\}$:
$$\rho_3^\lambda = \frac{3\cos(\pi/2\lambda) + \frac{\sin(\sigma/\lambda)[\cos(\sigma/2\lambda)\sin(\gamma) - 3\sin(\beta)\sin(\alpha)]}{\sin(\beta)[\sin(\beta) - \sin(5\beta)]}}{4\sin(\alpha)}, \text{ where } \alpha = 2m\pi/\lambda, \ \beta = m\pi/\lambda, \ \gamma = (4m+1)\frac{\pi}{\lambda}$$

In particular, $\rho_3^7 \simeq 0.85824$.

[3] When $\lambda = \{6m + 2 | m \geq 0, m \in I\}$:
$$\rho_3^\lambda = \frac{\frac{\sin(\sigma/\lambda)\cos(\beta) + \sin(\beta) - \sin(\gamma)]\sin(\gamma)}{\sin(\sigma/\lambda)\cos(\beta) + \sin(\beta) - \sin(\gamma)]} + 2}{4\sin(\alpha)}, \text{ where } \alpha = (2m+1)\frac{\pi}{\lambda}, \ \beta = m\pi/\lambda, \ \gamma = (m+1)\frac{\pi}{\lambda}$$

In particular, $\rho_3^2 = 3/4$ and $\rho_3^8 \simeq 0.8603882638$.

[4] When $\lambda = \{6m + 3 | m \geq 0, m \in I\}$:
$$\rho_3^\lambda = \frac{\sqrt{3}}{2}\cos\frac{\pi}{2\lambda}$$

[5] When $\lambda = \{6m + 4 | m \geq 0, m \in I\}$:
$$\rho_3^\lambda = \frac{\frac{\sin(\beta)}{\sin(\gamma)} + \frac{[2\sin(\alpha)\sin(\gamma) - \cos(\gamma)]\sin(\pi/\lambda)}{\sin(\gamma) - \sin(\beta)]\sin(\gamma)} + 2}{4\sin(\alpha)}, \text{ where } \alpha = (2m+1)\frac{\pi}{\lambda}, \ \beta = m\pi/\lambda, \ \gamma = (m+1)\frac{\pi}{\lambda}$$

In particular, $\rho_3^4 = \frac{2+\sqrt{2}}{4} \simeq 0.85355339$ and $\rho_3^{10} \simeq 0.86038826$.

[6] When $\lambda = \{6m + 5 | m \geq 0, m \in I\}$:
$$\rho_3^\lambda = \frac{3\cos(\pi/2\lambda) + \frac{2\sin(\sigma/\lambda)[\cos(\sigma/2\lambda)\cos(\gamma) - \sin(\alpha)]}{\sin(\beta) - \sin(\gamma)}}{4\sin(\alpha)}, \text{ where } \alpha = (2m+2)\frac{\pi}{\lambda}, \ \beta = m\frac{\pi}{\lambda}, \ \gamma = (m+1)\frac{\pi}{\lambda}$$

In particular, $\rho_3^5 = \frac{5 - 2\cos(\pi/5)}{4} \simeq 0.8454$ and $\rho_3^{11} \simeq 0.8621377$.

## 3.2 Steiner Ratios for $n = 4$

It suffices to consider the set $P$ of lattice points in $\mathcal{L}_\Delta$, where $\Delta$ is a $\lambda$-equilateral triangle (cf. Theorem 1). See Fig. 3 for an illustration for $\lambda = 4$. Basically there are two possible critical sets, both defining a parallelogram. When $\lambda = 2m$, there are two topologically different solutions. One has *one* Steiner point

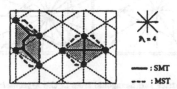

Fig. 3. An example of a unit-lattice for $\lambda = 4$, and two SMT's

of degree 4 (forming a *cross*) and the other has *two* Steiner points, denoted $G_4^{\lambda+}$ and $G_4^{\lambda*}$ respectively. The computation of the SMT's is based on the fact that the Steiner points must be determined by the Steiner loci of two regular pairs and its derivation is purely mechanical. These values all confirm the fact that their ratios are all greater than those for $\lambda$-equilateral triangles. Let $\rho_4^{\lambda+}$ and $\rho_4^{\lambda*}$ denote the Steiner ratios of $G_4^{\lambda+}$ and $G_4^{\lambda*}$ respectively. Below is a summary of the relationship among $\rho_3^{\lambda}$, $\rho_4^{\lambda+}$ and $\rho_4^{\lambda*}$. Except for $\lambda = 2$ (whose ratio is known to be 2/3, and the SMT is a cross), $\rho_3^{\lambda}$ is always a lower bound. In particular, $\rho_4^{\lambda+} = \rho_4^{\lambda*} = \frac{4-\sqrt{2}}{3} > \frac{1}{4-2\sqrt{2}} = \rho_3^{\lambda}$.

$$
\begin{aligned}
&\lambda = 2, && \rho_4^{\lambda+} < \rho_3^{\lambda} < \rho_4^{\lambda*} \\
&\lambda = (6m+2) && \rho_3^{\lambda} < \rho_4^{\lambda*} < \rho_4^{\lambda+} \\
&\lambda = 3, && \rho_3^{\lambda} < \rho_4^{\lambda} \\
&\lambda = (6m+3) && \rho_3^{\lambda} < \rho_4^{\lambda*} \\
&\lambda = 4, && \rho_3^{\lambda} < \rho_4^{\lambda*} = \rho_4^{\lambda+} \\
&\lambda = (6m+4) && \rho_3^{\lambda} < \rho_4^{\lambda*} < \rho_4^{\lambda+} \\
&\lambda = 5, (6m+5) && \rho_3^{\lambda} < \rho_4^{\lambda*} \\
&\lambda = 6, (6m) && \rho_3^{\lambda} < \rho_4^{\lambda*} \\
&\lambda = 7, (6m+1) && \rho_3^{\lambda} < \rho_4^{\lambda*}
\end{aligned}
$$

## 4  Proof of Steiner Ratios when $\lambda$ is Multiple of 3

We briefly review the proof due to Du and Hwang[1, 4] of the Gilbert-Pollak conjecture that the Euclidean Steiner ratio $\rho^\infty = \frac{\sqrt{3}}{2}$.

They first transform the problem into a minimax problem so that one needs only to verify the ratio for a critical set $C$ (Lemma 2) of points drawn from a unit $\Delta$-lattice $\mathcal{L}_\Delta$. They then convert an SMT into a hexagonal tree $\mathcal{H}(C)$. The three orientations of $\mathcal{H}$ are parallel to those of $\Delta$. In doing so, they showed that the length of SMT satisfies:

$$
L_{smt}(C) \geq \frac{\sqrt{3}}{2} * L_{\mathcal{H}}(C),
$$

where $L_{\mathcal{H}}(C)$ denotes the length of $\mathcal{H}(C)$. $\mathcal{H}(C)$ is then transformed so that the junction points of $\mathcal{H}(C)$ coincide with the lattice points, thus producing an MST. In this process they showed that $L_{\mathcal{H}}(C)$ is no less than $L_{mst}(C)$. Therefore

$$
L_{smt}(C) \geq \frac{\sqrt{3}}{2} * L_{\mathcal{H}}(C) \geq \frac{\sqrt{3}}{2} * L_{mst}(C).
$$

$$
\rho = \frac{L_{smt}}{L_{mst}} \geq \frac{\sqrt{3}}{2}.
$$

We follow the same strategy in deriving the Steiner ratio $\rho^\lambda$ for $\lambda = 3m$, and integer $m \geq 1$. To show a lower bound on the Steiner ratio it suffices from Theorem 1 to show that the lower bound holds for a critical set $C$ of $n$ points drawn from a unit lattice. Let $L_\Delta$ denote the unit lattice in the Euclidean sense. $\Delta$, shown in dotted lines, is the same as the actual $\lambda$-equilateral triangle, except for cases when $\lambda = 6m+3$ for some integer $m \geq 0$ (Fig. 4(a)). Since for these $\lambda$'s the Steiner angles in $G_n^*$ are all 120°, it is itself a hexagonal tree. The three orientations of this tree may not be the same as those of $\Delta$, we convert each Steiner edge $e(p, q)$ by at most two segments, $e(p, x), e(x, q)$, parallel to two sides of $\Delta$ (see Fig. 4(b)).

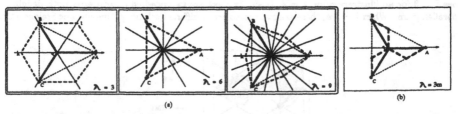

Fig. 4. An example of lattice and hexagonal trees in the $\lambda$-*geometry* plane, where $\lambda = 3m$.

Since $\angle pxq = 120°$, we have

$$L_{e(p,q)} \geq \frac{\sqrt{3}}{2} * (L_{e(p,z)} + L_{e(q,z)})$$

$$L_{smt}(C) = \Sigma\{L_{e(p,q)}|e(p,q) \in G_n^*\} \geq \frac{\sqrt{3}}{2} * \Sigma\{(L_{e(p,z)} + L_{e(q,z)})\}$$

After that we obtain a hexagonal tree which can be transformed into another hexagonal tree $\mathcal{H}'(C)$, whose junctions coincide with the lattice points, i.e., an MST for $C$, of length no greater than that of the hexagonal tree.

$$L_{smt}(C) \geq \frac{\sqrt{3}}{2} * L_{\mathcal{H}'}(C)$$

For $\lambda = 6m$ the length of $\mathcal{H}'(C)$ is the same as $L_{mst}(C)$, we have $\rho^\lambda \geq \frac{\sqrt{3}}{2}$. For $\lambda = 6m + 3$ for some integer $m \geq 0$, $L_{\mathcal{H}'}(C) = L_{mst}(C) * \cos\frac{\pi}{2\lambda}$. Thus we have $\rho^\lambda \geq \frac{\sqrt{3}}{2} * \cos\frac{\pi}{2\lambda}$, and for $\lambda = 3$ we have $\rho^\lambda \geq \frac{3}{4}$. Since the bound is realized by three points, we conclude

**Theorem 5.** *In the $\lambda$-geometry plane, the Steiner ratio is $\rho^\lambda = \frac{\sqrt{3}}{2} * \cos\frac{\pi}{2\lambda}$ for $\lambda = 6m + 3$ for some integer $m \geq 0$, and $\rho^\lambda = \frac{\sqrt{3}}{2}$ for $\lambda = 6m$.*

## 5 Multi-Level Grid Theorem

Given a set $P$ of regular points, the Hanan-grid is the structure obtained by drawing horizontal and vertical lines through each regular point and the intersections of these lines are called *grid* points. It is known that the Steiner points for an SMT of $P$, for $\lambda = 2$, must occur at the the grid points. Sarrafzadeh and Wong[12] showed that in the $\lambda$-geometry plane, the Hanan-grid may not contain all possible candidates of Steiner points when $\lambda > 2$. In fact, Du and Hwang[2] conjectured that the Hanan-grid may not contain all possible candidates of Steiner points if the unit-disk in $d$-dimension space has more than $2d$ extreme points. However, as we outline below (Theorem 6) that for $\lambda = 4$ the Steiner points can be located in a "multi-level grid" so that the number of candidate Steiner points for $G_n$ is finite. This is true for any $\lambda \neq \infty$.

For $\lambda = 4$ we can construct a *multi-level grid* as follows. (For other values of $\lambda$ we do it similarly.) For each regular point $p$ draw four straight lines of slopes $0, +1, -1$ and $\infty$. These $4n$ lines form the *1st-level* grid. In general, the next-*level* grid is formed by repeating the same procedure at each of the new grid points generated at the previous *level*. Let $\mathcal{G}_k$ denote the set of grid points at levels up to $k$, where $k$ is an integer.

**Theorem 6.** *There exists a $G_n$ such that the Steiner points are grid points in $\mathcal{G}_{n-2}$, for $n \geq 3$.*

Theorem 6 is referred to as the *multi-level* grid theorem. From now on, we consider only a full Steiner tree because a non-full Steiner tree can be separated into several full Steiner sub-trees.

Recall from Section 3.1 that for each pair of regular points we can construct its Steiner locus. The Steiner point for a set consisting of this pair and the third point must be on the boundary of the locus. From Lemma 3 among the three regular edges incident to the Steiner point at most one is nonstraight. In other words the Steiner point lies on the 1st level grid determined by the two regular points with straight regular edges. We shall now use this result to show that the theorem is true for point sets of size $n = 4$, and then prove by induction (omitted) that it is true for all $n$.

**Lemma 7.** *There exists a $G_4^*$, such that one of the two Steiner points can be located on the 1st-level grid.*

*Proof.* Consider the full topology of a $G_4^*$ that has two Steiner points. Note that the number of possible topologies of $G_4$ is 5. The relationship between each topology and its loci intersection is shown in Figure 5.

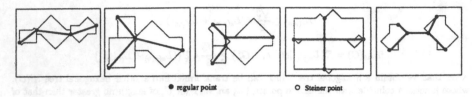

Fig. 5. Five possible topologies (solid lines) of $G_4$ and the relationships of their loci (dotted lines) intersection.

Consider $G_4^a$ defined by two regular pairs $[A, B]$ and $[C, D]$ connected to Steiner points $s_1$ and $s_2$ respectively. (Fig. 6). Suppose that the Lemma is false. That is, each of Steiner points ($s_1$ and $s_2$) has one nonstraight regular edge and the Steiner edge must be straight (cf Lemma 3). It is sufficient to consider the case when the two Steiner loci, $\mathcal{L}_{A,B}$ and $\mathcal{L}_{C,D}$, are disjoint. It is obvious that $s_1$ cannot be inside $\mathcal{L}_{C,D}$ and $s_2$ cannot be inside $\mathcal{L}_{A,B}$. Let us position $e(s_1, s_2)$ on the $z$-axis. There are three possible configurations for the location of $s_1$ and $s_2$ on the Steiner loci $\mathcal{L}_{A,B}$ and $\mathcal{L}_{C,D}$. Assume that $s_1$ lies on the boundary segment of $\mathcal{L}_{A,B}$ of slope +1. Then the possible slopes of the boundary segment of $\mathcal{L}_{C,D}$ on which $s_2$ can lie are +1, −1 and ∞, shown respectively in Fig. 6(a), (b), and (c). Define a *parallel sliding* operation of the Steiner edge $e(s_1, s_2)$ by shifting $e(s_1, s_2)$ parallelly by a distance $\epsilon$ while maintaining the same orientations of the segments incident on the Steiner points and that $s_1$ and $s_2$ lie on $\mathcal{L}_{A,B}$ and $\mathcal{L}_{C,D}$ respectively. We say that these two Steiner trees are *topologically equivalent*. It can be easily verified these two trees in Fig. 6(a) are of the same length. They are said to be *topologically congruent*, while the configurations in Fig. 6(b) and (c) are of different length. Notice that in case (a), if the shifting direction is to reduce the length of the second segment of regular edge $e(A, s_1)$, for instance, by $\epsilon$, the length of the second segment of regular edge on the opposite side (edge $e(C, s_2)$) is increased by the same amount.

Consider the following two cases: (1) The regular points of the two nonstraight edges are either both above the $z$-axis or both below. (2) The regular points of the two nonstraight edges lie on different sides of the $z$-axis.

We can show that in (1) the SMT can always be shortened which is a contradiction (Fig. 6(b)) and that in (2) one of the Steiner points, e.g. $s_1$ (or $s_2$), can be moved to a vertex, a 1st-level grid point, of its locus $\mathcal{L}_{A,B}$ (or $\mathcal{L}_{C,D}$).

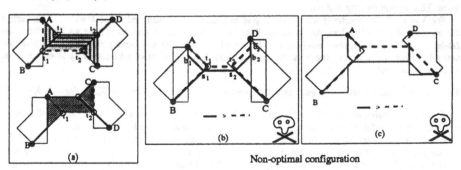

Non-optimal configuration

Fig. 6. Three possible configurations and the parallel sliding operation.

That is, in $G_4^a$, if its solution is unique, both Steiner points are 1st-level grid points. Otherwise, there are infinitely many topologically congruent $G_4^a$'s, and there exist two extreme solutions which contain at least one Steiner point at a 1st-level grid point.

**Lemma 8.** *Suppose the two closest regular pairs in $G_n^a$, $n \geq 4$ are connected by a Steiner path $s_1, ..., s_{k+1}$, $k \geq 1$. Assume that none of the Steiner points on this Steiner path lies on the 1st-level grid. That is, each regular pair has a nonstraight regular edge, and the other regular edges are all straight. We can transform a nonstraight edge by length $\epsilon$ from one end of the Steiner path to the other without changing the total length of SMT, if and only if $k$ is even (resp. odd) and the two nonstraight edges are on the same (resp. different) side of the Steiner path.*

The proof of this lemma can be shown by using parallel sliding operations. If the lemma were false, the sliding operation will either make a regular point coincide with a Steiner point or have the total length reduced. We omit its details.

# 6 Remarks and Conclusion

We have shown that the Steiner ratio for a set of $n \geq 3$ points in the $\lambda$-geometry plane is $\frac{\sqrt{3}}{2}\cos(\pi/2\lambda)$, for $\lambda = 6m + 3$ and integer $m \geq 0$, and is $\sqrt{3}/2$, for $\lambda = 6k$ and integer $k \geq 1$, disproving a a conjecture made by Du et al.[3]. We also have shown that there exists a Steiner minimal tree whose Steiner points must lie on a multi-level Hanan-grid. Our initial study of the Steiner ratios for point sets of size 3 and 4 leads us to conjecture that in the $\lambda$-geometry plane, the Steiner ratio is achieved by 3 points that form a $\lambda$-equilateral triangle. We will focus our future research on settling this conjecture.

Fig. 7 and Table 1 summarize the $\lambda$-geometry Steiner ratio results along with the lower bound result of Sarrafzadeh and Wong[12].

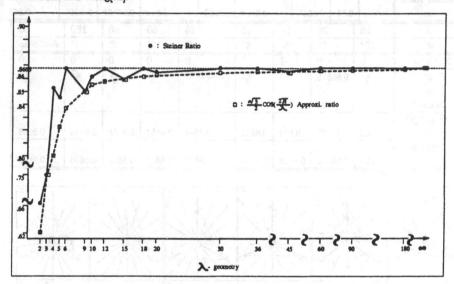

**Fig. 7.** Steiner Ratio and an approximation function in the $\lambda$-geometry plane.

# 7 Acknowledgement

We would like to thank Prof. Ding-Zhu Du for numerous discussions concerning Steiner ratios and providing us references about this problem.

# References

1. D. Z. Du and F. K. Hwang, "A Proof of the Gilbert-Pollak Conjecture on the Steiner Ratio", *Algorithmica* Vol. 7, No. 2/3, 1992, pp. 121-135.
2. D. Z. Du and F. K. Hwang, "Reducing the Steiner Problem in a Normed Space", *SIAM J. Comput.* Vol. 21, No. 6, 1992, pp. 1001-1007.
3. D. Z. Du, Biao Gao, R. L. Graham, Z. C. Liu and Peng-Jun Wan, "Minimum Steiner Trees in Normed Plane," *Discrete Comput. Geom.*, 9, 1993, pp.351-370.
4. D. Z. Du and F. K. Hwang, "State of Art on Steiner Ratio Problems," in Computing in Euclidean Geometry, D. Z. Du and F. K. Hwang, Eds., World Scientific, 1992, 163-191.
5. B. Gao, D. Z. Du and R. L. Graham, "The Tight Lower Bound for the Steiner Ratio in Minkowski Planes", *Proc. tenth Annual Symposium on Computational Geometry*, June, 1994, pp. 183-191.
6. M. R. Garey, R. L. Graham and D.S. Johnson, "The Complexity of Computing Steiner Minimal Trees", *SIAM J. Appl. Math.*, 32, 1977, pp.835-859.
7. M. R. Garey and D. S. Johnson, *Computers and Intractability: a Guide to the theory of NP-completeness,* Freeman, San Francisco, 1979.
8. M. Hanan, "On Steiner's Problem with Rectilinear Distance", *SIAM J. Applied Math.*, Vol 14, No.2, March 1966, pp. 255-265.
9. F. K. Hwang, "One Steiner Minimal Trees with Rectilinear Distance", *SIAM J. AppL. Math.*, Vol 30, No.1, Jan. 1976. pp. 104-114.
10. D. T. Lee, C. F. Shen and C. L. Ding, "On Steiner Tree Problem with 45° Routing", *IEEE international symposium circuit and system*, May, 1995. pp. 1680-1683.
11. Z. C. Liu and D. Z. Du, "On Steiner Minimal Trees with $L_p$ Distance," *Algorithmica*, 7 (1992), 179-191.'
12. M. Sarrafzadeh and C. K. Wong, "Hierarchical Steiner Tree Construction in Uniform Orientations", *IEEE Trans. Computer-Aided Design*, Vol.11, September 1992. pp. 1095-1103.

| λ | 2 | 3 | 4 | 5 | 6 | 9 | 10 | 12 | 15 |
|---|---|---|---|---|---|---|---|---|---|
| θ° | 90° | 60° | 45° | 36° | 30° | 20° | 18° | 15° | 12° |
| 2λ MOD 3 | 1 | 0 | 2 | 1 | 0 | 0 | 2 | 0 | 0 |
| Equil-△ or Equil-◇ | $\frac{3}{4}$ $\frac{2}{3}$ | $\frac{3}{4}$ | $\frac{2+\sqrt{2}}{4}$ | $\frac{5-2\cos(\pi/5)}{4}$ | $\frac{\sqrt{3}}{2}$ | $\frac{\sqrt{3}\cos(\pi/18)}{2}$ | 0.86327 | $\frac{\sqrt{3}}{2}$ | $\frac{\sqrt{3}\cos(\pi/30)}{2}$ |
| $\frac{\sqrt{3}}{2}\cos\left(\frac{\pi}{2\lambda}\right)$ ≃ | 0.623 | 0.75 | 0.80 | 0.823 | 0.836 | 0.852 | 0.855 | 0.858 | 0.861 |
| Steiner Ratio ≃ | 0.666 | 0.75 | 0.8535 | 0.8454 | 0.8666 | 0.852 | 0.86327 | 0.8666 | 0.861 |

| λ | 18 | 20 | 30 | 36 | 45 | 60 | 90 | 180 | ∞ |
|---|---|---|---|---|---|---|---|---|---|
| θ° | 10° | 9° | 6° | 5° | 4° | 3° | 2° | 1° | Euclidean |
| 2λ MOD 3 | 0 | 0 | 1 | 0 | 0 | 0 | 0 | 0 | 0 |
| Equil-△ or Equil-◇ | $\frac{\sqrt{3}}{2}$ | 0.86518 | $\frac{\sqrt{3}}{2}$ | $\frac{\sqrt{3}}{2}$ | $\frac{\sqrt{3}\cos(\pi/90)}{2}$ | $\frac{\sqrt{3}}{2}$ | $\frac{\sqrt{3}}{2}$ | $\frac{\sqrt{3}}{2}$ | $\frac{\sqrt{3}}{2}$ |
| $\frac{\sqrt{3}}{2}\cos\left(\frac{\pi}{2\lambda}\right)$ ≃ | 0.862 | 0.863 | 0.864 | 0.8651 | 0.8654 | 0.8657 | 0.8658 | 0.8659 | 0.8666 |
| Steiner Ratio ≃ | 0.8666 | 0.86518 | 0.8666 | 0.8666 | 0.8657 | 0.8666 | 0.8666 | 0.8666 | 0.8666 |

Table 1: Steiner Ratio and examples of equilateral triangles in the λ-*geometry* plane.

# A Study of the $LMT$-skeleton*

Siu-Wing Cheng[1]    Naoki Katoh[2]    Manabu Sugai[2]

[1] Department of Computer Science, HKUST, Clear Water Bay, Hong Kong
[2] Kobe University of Commerce, Gakuen-Nishimachi, Nishi-ku, Kobe, 651-21 Japan

**Abstract.** We present improvements in finding the $LMT$-skeleton, which is a subgraph of all minimum weight triangulations, independently proposed by Belleville et al, and Dickerson and Montague. Our improvements consist of: (1) A criteria is proposed to identify edges in all minimum weight triangulations, which is a relaxation of the definition of local minimality used in Dickerson and Montague's method to find the $LMT$-skeleton; (2) A worst-case efficient algorithm is presented for performing one pass of Dickerson and Montague's method (with our new criteria); (3) Improvements in the implementation that may lead to substantial space reduction for uniformly distributed point sets.

## 1 Introduction

Given a planar point set $S$, the weight of a triangulation of $S$ is defined to be the sum of Euclidean lengths of the edges. The *minimum weight triangulation problem* is to compute a triangulation of $S$ with minimum weight (denoted by MWT from now on). The complexity of finding a MWT is currently unresolved: it is not known to be in $P$ or $NP$. Nevertheless, many new results are discovered recently concerning its geometric properties [4, 5, 7, 9, 12] and combinatorial properties [2], as well as approximation algorithm [10], fast heuristics [8, 11], and algorithms for finding large subgraphs of a MWT [3, 6].

This paper is a further study on the new subgraph of a MWT independently proposed in [3, 6]. Following [6], we call this subgraph the $LMT$-skeleton. One can repeatedly run the algorithm for finding the $LMT$-skeleton until the subgraph does not grow anymore. The resulting subgraph is called in *extended LMT-skeleton* in [6]. Although it can be proved that the extended $LMT$-skeleton can have linearly many holes in the worst case [4], the experimental results observed so far is encouraging. As indicated by the experimental results in [3, 6], the extended $LMT$-skeleton is often connected and contains a large percentage of the edges in MWT for uniformly distributed point set (up to 250 points are tried by Dickerson and Montague [6] and up to 1000 points are tried by Belleville et al [3]). Unfortunately, finding a extended $LMT$-skeleton currently takes $O(n^4)$ time and $\Theta(n^2)$ space [3] in the worst-case. The quadratic space requirement is a

* Research of the first author is partly supported by the RGC CERG grant HKUST650/95E. Research of the second author is partly supported by the Grant-in-Aid of Ministry of Science, Culture and Education of Japan.

hindrance to experimenting with large point sets. For uniformly distributed point sets, a *diamond test* is proposed in [6] to identify edges which cannot be MWT edges. With this diamond test, it can be proved that the number of possible MWT edges left is *expected* to be $O(n)$ [6]. Then the algorithm for finding the LMT-skeleton can be run on the remaining possible edges and the resulting subgraph is called the *modified LMT-skeleton* in [6]. The total expected running time and expected space usage reduce to $O(n^2)$ and $O(n^{1.5})$, respectively.

In this paper, we pursue improved ways to find a large subgraph of a MWT based on the *LMT*-skeleton. First, we propose a relaxed version of the criteria used in [3, 6] for identifying edges in the *LMT*-skeleton. As in [6], we can apply this relaxed criteria in multiple passes to find a possibly larger subgraph. Our second contribution is a more worst-case efficient algorithm for running one pass, which takes $O(n^3 \log n)$ time and $O(n^2)$ space. Space is one bottleneck for trying large point sets. We suggest improvements in the implementation, which may lead to substantial reduction in space for uniformly distributed point sets. In our experiments, the space usage (in terms of the number of edges need to be recorded) is about five times the point set size. We also run our implementation on some data from natural sources: cross-sections of some human organs. We observe that one pass of our algorithm generates a highly disconnected subgraph for a cross-section of blood vessels. This implies that there exists point sets for which the LMT-skeleton is highly disconnected. This contrasts with the findings for uniformly distributed points observed in this paper and in [6].

In Section 2, we introduce our relaxed criteria and prove its correctness. Section 3 describes our improved algorithm for running one pass. Section 4 describes our implementation, the experiments conducted, and the results obtained.

## 2  A criteria for finding MWT edges

We denote the MWT of a point set $S$ by $MWT(S)$ and the *LMT*-skeleton by $LMT(S)$. The set of all possible edges connecting two points in $S$ is $E(S)$. Let $e$ be an edge in $E(S)$ and $t_1$ and $t_2$ be two empty triangles with vertices in $S$ such that $t_1 \cap t_2 = e$. In [6], $e$ is defined to be *locally minimal* with respect to $t_1$ and $t_2$ if $t_1 \cup t_2$ is not convex, or if $t_1 \cup t_2$ is convex and $|e| \leq |e'|$, where $e'$ is the other diagonal of $t_1 \cup t_2$. A triangulation of $S$ is locally minimal if each edge $e$ in the triangulation is locally minimal with respect to the two triangles containing $e$. In [6], the following method is proposed to identify some edges in the intersection of all locally minimal triangulations, which is $LMT(S)$. (Our description is adapted from [6] and is not identical to that in [6]).

> A list *candEdges* of candidate edges, a list *edgesIn* of edges known to be contained in $MWT(S)$, and a list *deadEdges* of edges known to be not contained in any locally minimal triangulations are maintained. Initially, all edges in $E(S)$ except the convex hull edges are in *candEdges*, *edgesIn* contains the convex hull edges, and *deadEdges* is empty.
> For each unexamined edge $e \in candEdges$,

1. Find all combinations of empty triangles $t_i$ and $t_j$ on the two sides of $e$ such that $t_i$ and $t_j$ are not bordered by any edge in *deadEdge*.
2. Test each combination of $t_i$ and $t_j$ to see if $e$ is locally minimal with respect to $t_i$ and $t_j$. If $e$ is not locally minimal to any such pair $t_i$ and $t_j$, then move $e$ to *deadEdges*. Otherwise, if $e$ intersects no other edge in *candEdges* (examined or unexamined) or *edgesIn*, then move $e$ to *edgesIn*.

Multiple passes of the algorithm can be run until no more edges in *candEdges* can be moved to *edgesIn* or *deadEdges*. The resulting subgraph is the extended *LMT*-skeleton. In the definition of locally minimal edges, if there are neighboring triangles $t_1$ and $t_2$ bordering $e$ such that $t_1 \cup t_2$ is non-convex, then $e$ is locally minimal with respect to $t_1$ and $t_2$. However, the fact that $t_1 \cup t_2$ is non-convex implies that $t_1 \cup t_2$ does not really play any role. This prompts us to define a weaker criteria in the following.

All edges in $E(S)$ and all empty triangles are initially *active*. Edges and empty triangles will become *dead* if they are known to be not contained in any locally minimal triangulation. When each edge is examined, its status will be determined as active, *inactive*, or dead as follows:

Let $\mathcal{T}$ be the set of pairs $\{axb, ayb\}$ of empty active triangles such that $axb \cap ayb = ab$. The edge $ab$ is labelled active if it lies on the boundary of the convex hull, or there exists $\{axb, ayb\} \in \mathcal{T}$ such that $ab \cap xy \neq \emptyset$ and $|ab| \leq |xy|$. Suppose that $ab$ is not labelled active. Then if $\mathcal{T} = \emptyset$ or $ab \cap xy \neq \emptyset$ for all $\{axb, ayb\} \in \mathcal{T}$, then we label $ab$ dead. Otherwise, we label $ab$ inactive.

Whenever an edge $ab$ becomes inactive or dead, we label some of the empty triangles bordering $ab$ dead as follows:

Collect the set $\mathcal{A}$ of all empty active triangles $axb$ satisfying: for all empty active triangles $ayb$ such that $axb \cap ayb = ab$, $ab \cap xy \neq \emptyset$. Label all triangles in $\mathcal{A}$ dead.

After finding the active edges, an active or inactive edge is in $MWT(S)$ if it does not intersect any other active edge. If a MWT edge $e$ is reported using Dickerson and Montague's criteria, then $e$ will also be reported using our criteria. We prove the correctness of our criteria in the following.

**Lemma 1.** *If an empty triangle $t$ is labelled dead, then $t \notin MWT(S)$.*

*Proof.* We prove the lemma by contradiction. Let $axb$ be the first triangle in $MWT(S)$ which is labelled dead and suppose that it is labelled dead because the edge $ab$ becomes inactive or dead. So $ab$ does not lie on the boundary of the convex hull which means that $ab$ is adjacent to another triangle $ayb$ in $MWT(S)$ such that $axb \cap ayb = ab$. Both $axb$ and $ayb$ are active immediately after labeling $ab$ inactive or dead. Since we label $axb$ dead, $ab \cap xy \neq \emptyset$ and $|xy| < |ab|$. Therefore, in the convex quadrilateral $axby$ in $MWT(S)$, we can replace $ab$ by $xy$ to decrease the total length, a contradiction. □

**Lemma 2.** *If $ab \notin MWT(S)$, then $ab$ intersects some active edge.*

*Proof.* Let $ab$ be some edge not in $MWT(S)$. Since $ab \notin MWT(S)$, $ab$ intersects some edge in $MWT(S)$. Let $x_0$ and $y_0$ be the edge in $MWT(S)$ such that $ab \cap x_0 y_0$ is closest to $a$. The triangle $x_0 a y_0 \in MWT(S)$. Since $a$ and $b$ lie on opposite sides of $x_0 y_0$, $x_0 y_0$ is not a convex hull edge. For convenience, we rename $a$ as $v_0$. Let $C_0$ be the cone bounded by the two rays originating from $v_0$ through $x_0$ and $y_0$, respectively. By construction, $b \in C_0$. Since $x_0 y_0$ is not a convex hull edge, there is another triangle $x_0 z y_0$ in $MWT(S)$ adjacent to $x_0 y_0$. See Fig. 1. Assume to the contrary that $ab$ does not intersect any active edge. By Lemma 1, $x_0 v_0 y_0$ and $x_0 z y_0$ were active when $x_0 y_0$ was labelled inactive or dead. So $v_0 z \cap x_0 y_0 = \emptyset$, otherwise, $|v_0 z| < |x_0 y_0|$ and we can flip $x_0 y_0$ to decrease the total length of $MWT(S)$. Therefore, $z \notin C_0$ and $z \neq b$. Without loss of generality, let $z$ lies on the left of $v_0 b$. We rename $x_0$ as $v_1$, $z$ as $x_1$, and $y_0$ as $y_1$. By construction, we discover a new edge $x_1 y_1$ in $MWT(S)$ that intersects $ab$, and $b$ lies inside the cone $C_1$ bounded by the two rays from $v_1$ through $x_1$ and $y_1$. Thus, we can repeat the previous argument again to $x_1 v_1 y_1$ and $x_1 y_1$ to obtain a new edge $x_2 y_2$ that intersects $ab$. In fact, we can repeat this argument indefinitely to obtain an infinite sequence of edges $x_i y_i$, $i > 1$, that intersect $v_0 b$, a contradiction. ◨

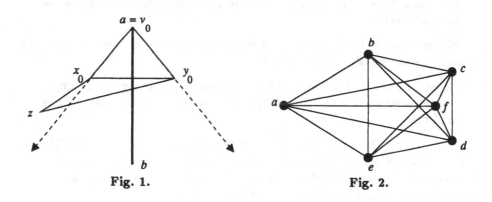

Fig. 1.  Fig. 2.

Lemma 2 implies the correctness of reporting active or inactive edges that do not intersect any other active edges. There is a point set such that in one pass, a larger subgraph of $MWT(S)$ will be reported if our criteria is used instead of the definition of locally minimal edges. Consider the point set $X$ in Fig. 2. Other the convex hull edges, suppose that $E(S)$ is processed in this order $af$, $ac$, $ad$, $ce$, $bd$, $be$, $bf$, $ef$, $cf$, $df$. Using the definition of locally minimal edges, the convex hull edges and the edges $cf$, $df$, $bf$, and $ef$ will be reported. Using our criteria, the edge $be$ will be reported as well.

# 3   A worst-case efficient algorithm

We present an algorithm that applies our criteria to identify MWT edges. One pass of the algorithm runs in $O(n^3 \log n)$ time. The worst-case space usage is $O(n^2)$. To ease the discussion, for the time being, whenever an edge becomes inactive or dead, we will not label any empty active triangle bordering the edge dead. We will discuss how to put this feature back later. Given a direction $\alpha$, we use $S_p(\alpha)$ to denote the sequence of points in $S - \{p\}$ swept by rotating a ray around $p$ in counterclockwise order for an angle $2\pi$ starting from direction $\alpha$. Given any point $x \in S - \{p\}$, we use $S_p(\alpha, x)$ to denote the subset of points that precede $x$ in $S_p(\alpha)$. The following simple lemma enables us to identify the empty triangles adjacent to an edge $ab$.

**Lemma 3.** *Let $\alpha$ be the direction denoted by the ray from point $a$ to point $b$. Given a triangle $axb$ lying on the left of $ab$ with respect to $\alpha$, $axb$ is empty iff $S_a(\alpha, x) \subseteq S_b(\alpha, x)$.*

Let $\alpha$ be the direction from $a$ to $b$. By applying Lemma 3, we can find all the points $x$ such that $axb$ is empty and lies on the left of $ab$ with respect to $\alpha$ in $O(n \log n)$ time by obtaining and scanning $S_a(\alpha)$ and $S_b(\alpha)$. We connect the set $\{x : axb$ is an empty triangle$\}$ and $ab$ to form a simple polygon $L_{ab}$. Similarly, we can construct a simple polygon $R_{ab}$ on the right of $ab$ with respect to $\alpha$. Refer to Fig. 3 for an illustration. Now, it suffices to find the longest diagonal $e$

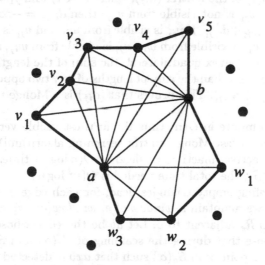

**Fig. 3.** The bold polygonal figure is $L_{ab} \cup R_{ab}$.

of $L_{ab} \cup R_{ab}$ that intersects $ab$. If $e$ exists and $|e| \geq |ab|$, then we label $ab$ active. Suppose that $ab$ is not labelled active. If $L_{ab}$ or $R_{ab}$ is empty, or if the angle at $a$ or $b$ in $L_{ab} \cup R_{ab}$ is not larger than $\pi$, then we label $ab$ dead. Otherwise, we label

$ab$ inactive. The labeling of $ab$ can be done in quadratic time by brute-force. In fact, this is the preferred method whenever the size of $L_{ab} \cup R_{ab}$ is not greater than $\sqrt{n}$ because this incurs less overhead. (This case will apply most of the time if there are not too many empty triangles adjacent to $ab$.) We need a different method if the size of $L_{ab} \cup R_{ab}$ is greater than $\sqrt{n}$.

Name the vertices of $L_{ab}$ on the chain from $a$ to $b$ (excluding $a$ and $b$) as $v_1, v_2, \cdots, v_l$. Name the vertices of $R_{ab}$ on the chain from $b$ to $a$ (excluding $a$ and $b$) as $w_1, w_2, \cdots, w_m$. First, observe that each $w_i$ can see a contiguous subsequence of vertices $\{v_{f(i)}, v_{f(i)+1}, \cdots, v_{g(i)}\}$ on $L_{ab}$. Moreover, the sequences $\{f(i) : 1 \le i \le m\}$ and $\{g(i) : 1 \le i \le m\}$ are non-decreasing. Thus, the "intervals" of visible vertices from all $w_i$'s can be computed by a single traversal of the chain on $L_{ab}$ from $a$ to $b$. To find the longest diagonal of $L_{ab} \cup R_{ab}$ that intersects $ab$, it suffices to find efficiently for each $w_i$, the furthest visible vertex among $\{v_{f(i)}, v_{f(i)+1}, \cdots, v_{g(i)}\}$.

**Lemma 4.** *Given $w_p$ and $w_q$ with $p < q$ and $v_r$ and $v_s$ with $r < s$, if $v_r$ is visible from $w_q$ and $v_s$ is visible from $w_p$, then $v_r$ is visible from $w_p$ and $v_s$ is visible from $w_q$. Moreover, for $1 \le i \le m$ and $1 \le j \le l$, define $d_{ij} = |w_i v_j|$ if $v_j$ is visible from $w_i$ or $-\infty$ otherwise. Then $(d_{ij})$ is a Monge matrix.*

*Proof.* Since $v_r$ is visible from $w_q$ and $v_s$ is visible from $w_p$, $aw_i b \cup av_j b$ is convex for all $w_i$ between $w_p$ and $w_q$ and $v_j$ between $v_r$ and $v_s$. Thus, we conclude that $v_r$ is visible from $w_p$ and $v_s$ is visible from $w_q$. Take any four entries $d_{i_1 j_1}, d_{i_1 j_2}, d_{i_2 j_1}, d_{i_2 j_2}$ of the matrix $(d_{ij})$, where $i_1 < i_2$ and $j_1 < j_2$. If $v_{j_2}$ is not visible from $w_{i_1}$ or $v_{j_1}$ is not visible from $w_{i_2}$, then $d_{i_1 j_2} = -\infty$ or $d_{i_2 j_1} = -\infty$. So $d_{i_1 j_1} + d_{i_2 j_2} \ge d_{i_1 j_2} + d_{i_2 j_1}$. If $v_{j_2}$ is visible from $w_{i_1}$ and $v_{j_1}$ is visible from $w_{i_2}$, then we know that $v_{j_1}$ is visible from $w_{i_1}$, $v_{j_2}$ is visible from $w_{i_2}$, and $w_{i_1} w_{i_2} v_{j_1} v_{j_2}$ is convex. Given any convex quadrilateral, the sum of the lengths of the two diagonals is always no less than the sum of lengths of any two opposite sides. Thus, $d_{i_1 j_1} + d_{i_2 j_2} \ge d_{i_1 j_2} + d_{i_2 j_1}$. This proves that $(d_{ij})$ is a Monge matrix. □

We can then compute in $O(n)$ time the furthest visible vertex on $L_{ab}$ from all $w_i$'s by applying a fast Monge matrix searching algorithm[1]. In all, we can determine if $ab$ is active, inactive, or dead in $O(n \log n)$ time. Since there are $O(n^2)$ edges in $E(S)$, the total time needed is $O(n^3 \log n)$.

To handle labeling empty triangles dead, for each edge $e$ whose status has been determined, we maintain a *window* of $e$, $window(e)$, which consists of the four edges of $L_e \cup R_e$ adjacent to $e$. Let $ab$ be the edge whose status is to be determined. Suppose that during the scanning of $S_a(\alpha)$ and $S_b(\alpha)$ to produce $L_{ab}$, we encounter a point $x$ in $S_a(\alpha)$ such that $axb$ is detected to be empty but $ax$ is inactive or dead. (The symmetric case for $bx$ is handled similarly.) If $ax$ is dead, then $axb$ is dead and we do not include $x$ in forming $L_{ab}$. Suppose that $ax$ is inactive. There are two edges $ac$ and $dx$ in $window(ax)$ for some points $c$ and $d$ that lie on the side of $ax$ opposite to that containing $b$. If $bc \cap ax \ne \emptyset$ and $bd \cap ax \ne \emptyset$, then $axb$ is dead and we do not include $x$ in forming $L_{ab}$. After $L_{ab}$ and $R_{ab}$ are obtained, we process them as described before.

In all, we determine the status of all edges in $E(S)$ in $O(n^3 \log n)$ time if one pass of the algorithm is run. The space needed is $O(n^2)$. The next task is to report all the active and inactive edges that do not intersect any other active edge.

If there are no more than $n^{1.5}\sqrt{\log n}$ active or inactive edges left, then simply do the checking by brute-force. Suppose that there are more than $n^{1.5}\sqrt{\log n}$ alive edges left. We examine each point $p \in S$ and determine for each active or inactive edge $e$ incident to $p$, whether $e$ intersects any active edge not incident to $p$. For each point $p \in S$, we first compute the "envelope" of active edges not incident to $p$ around $p$. Imagine that we emit light rays from $p$ in all directions, we want to find out all the points on active edges (not incident to $p$) hit by these light rays. The situation is basically the same as computing the lower envelope of a set of line segments with at most $n$ distinct endpoints. We describe below an algorithm to compute the lower envelope of $O(n^2)$ line segments with at most $n$ endpoints. This algorithm can be adapted to compute the "envelope" around $p$.

Given $O(n^2)$ line segments with $n$ distinct endpoints, we first sort their endpoints by $x$-coordinates. Then we split the set of endpoints into two equal halves by a vertical splitting line $\ell$. We take out all the line segments that intersect $\ell$. We first recursively find out the lower envelope of line segments on both sides of $\ell$. Then we compute the lower envelope of line segments intersecting $\ell$ and then merge the three envelopes to obtain the final solution. The lower envelope of line segments intersecting $\ell$ can be found as follows. It suffices to focus on the left half of this lower envelope on the left of $\ell$ as the other half can be handled similarly. We sort the line segments intersecting $\ell$ into the order $s_1, s_2, \cdots$ in increasing order of their left endpoints. Suppose that we have already found the left half of the lower envelope for $s_1, s_2, \cdots, s_{i-1}$, which consists of a number disjoint concave chains ordered from left to right. Note that $s_i$ can only intersect the rightmost concave chain $C$ in this envelope. By binary searching, we can determine the the lower envelope of $C$ and $s_i$ in $O(\log n)$ time. (Note the lower envelope of $C$ and $s_i$ may consists of two disjoint concave chains.) This together with the other concave chains form the left half of the lower envelope of $s_1, \cdots, s_i$. Since there are $O(n^2)$ line segments intersecting $\ell$, the total time needed is $O(n^2 \log n)$. Hence, the time to compute the "envelope" around $p$ satisfies the recurrence $T(n) \leq 2T(n/2) + O(n^2 \log n)$ which solves to $O(n^2 \log n)$.

After computing the "envelope" of active edges around $p$ which has size $O(n^2 \alpha(n,n))$, where $\alpha(n,n)$ is the inverse Ackermann function , we connect each vertex of this "envelope" with $p$ to form a set of disjoint wedges around $p$. Now, for each active or inactive edge $e$ incident to $p$, we can perform a binary search to find the wedge intersecting $e$. Then $e$ intersects an active edge not incident to $p$ if and only if $e$ intersects the side of the wedge not incident to $p$. In all, it takes $O(n^2 \log n)$ time to process one point in $S$ and it sums to a total time of $O(n^3 \log n)$.

# 4 Implementation and experiments

We implement our criteria using the programming language C. We only experimented with labeling all empty triangles bordering dead edges dead. We have not experimented with labeling the appropriate empty triangles bordering inactive edges dead. We run only one pass for most of the experiments and the edges in $E(S)$ are examined *in order of decreasing lengths*. We produce the edges in decreasing lengths in $\Theta(n)$ space as follows. First, for each point, we select the longest edge incident to it and output it to a priority queue. Then we extract the longest edge, say $pq$, in the priority queue. Then we select the longest edge incident to $p$ that is shorter than $pq$ and insert it to the priority queue. Perform the same for $q$. The above is repeated until the queue becomes empty. The space needed is $\Theta(n)$ and the running time is $O(n^3)$.

For each edge $ab$ extracted from the priority queue, we first construct $L_{ab} \cup R_{ab}$ as described in Section 3. We compare the length of $ab$ with the lengths of diagonals of $L_{ab} \cup R_{ab}$ crossing $ab$ by brute-force enumeration. In constructing $L_{ab} \cup R_{ab}$, we need to ignore empty triangles that are bordering dead edges as follows. Edges that have not been examined must be active and we remember all the active and inactive edges among those examined so far. Suppose that an empty triangle $t$ is bordering an edge $e$. If $e$ is shorter than the current edge $ab$ being examined, then $e$ is active. Otherwise, if $e$ is not an active or inactive edge remembered, then $e$ must be dead and the triangle $t$ can be ignored. Hence, checking $t$ only takes $O(\log n)$ time. (The "window" idea in Section 3 for checking a triangle bordering an inactive edge has not been experimented.) After we have obtained all active and inactive edges, we compare them pairwise by brute-force to identify those present in $MWT(S)$.

For uniformly distributed point sets, the observed running time of our implementation is roughly $O(n^3)$ and the space usage is proportional to the number of active and inactive edges found, which is observed to be $O(n)$.

We have run *one pass* of our implementation on uniformly point sets. Table 1 shows the average number of connected components in the resulting graph, the average number of active and inactive edges found, and the average percentage of MWT edges found. For point sets of size 2000, it is more often to obtain more than one connected component in one pass (the number of connected component goes up to four in some test case with 2000 points). We would like to emphasize again that this is done *in just one pass*. In all our experiments with uniformly distributed point sets, we observe that the number of active and inactive edges identified is always bounded by five times the size of the point set.

Other than uniformly distributed point sets, we have also run one pass of our implementation on a few two-dimensional cross-sections of some human organs. Fig. 4 shows the MWT edges identified for a cross-section of blood vessels. In Fig. 4, there are 307 points and there are 37 components after one pass. We performed more passes of our implementation until no more MWT edges are found and produce the graph in Fig. 5. Thus, point set of reasonable size exists for which multiple passes are essential. Running our implementation for multiple passes will not increase the space usage because we only need to test the status

**Table 1.** For each size from 100 to 500, 50 random trials were performed. For size 1000, 10 random trials were performed. For size 2000, 5 random trials were performed.

| Size | Aver. # comp. | Aver. # active/inactive edges | Aver. % MWT |
|------|---------------|-------------------------------|-------------|
| 100  | 1.04 | 406.14  | 77.15 |
| 200  | 1.36 | 841.84  | 75.73 |
| 300  | 1.24 | 1298.96 | 74.72 |
| 400  | 1.62 | 1781.46 | 73.62 |
| 500  | 1.34 | 2239.70 | 74.04 |
| 1000 | 1.6  | 4630.9  | 73.38 |
| 2000 | 2.8  | 9422    | 73.88 |

of the edges determined to be active or inactive in the first pass.

# Acknowledgment

We would like to thank Matthew Dickerson, Mordecai Golin, and Jack Snoeyink for helpful discussion. We would also like to thank Kam Hing Lee for passing us the data on two-dimensional cross-sections of human organs. The data was obtained originally from Gill Barequet.

# References

1. A. Aggarwal, M.M. Klawe, S. Moran, P.W. Shor, and R. Wilber: *Geometric applications of a matrix-searching algorithm*, Algorithmica, 2 (1987), pp. 195–208..

2. O. Aichholzer, F. Aurenhammer, S.W. Cheng, N. Katoh, G. Rote, M. Taschwer, and Y.F. Xu, *Triangulations intersect nicely*, manuscript, 1996.

3. P. Belleville, M. Keil, M. McAllister, and J. Snoeyink, *On computing edges that are in all minimum-weight triangulations*, Video Presentation, Symp. Computational Geometry, 1996.

4. P. Bose, L. Devroye, and W. Evans, *Diamonds are not a minimum weight triangulation's best friend*, Proceedings of Canadian Conference on Computational Geometry, 1996. See also technical report 96-01, Dept. of Computer Science, Univ. of British Columbia, January 1996.

5. S.W. Cheng and Y.F. Xu, *Approaching the largest β-skeleton within a minimum weight triangulation*, Proc. Symp. Computational Geometry, 1996, pp. 196–203.

6. M.T. Dickerson and M.H. Montague, *A (usually?) connected subgraph of minimum weight triangulation*, Proc. Symp. Computational Geometry, 1996, pp. 204–213.

7. M. Golin, *Limit theorems for minimum-weight triangulations, other Euclidean functionals, and probabilistic recurrence relations*, Proc. Symp. Discrete Algorithms, 1996, pp. 252–260.

8. L. Heath and S.V Pemmaraju, *New results for the minimum weight triangulation problem*, Algorithmica, 12 (1994), pp. 533–552.

9. J.M. Keil, *Computing a subgraph of the minimum weight triangulation*, Computational Geometry: Theory and Applications, 4 (1994), pp. 13–26.

10. C. Levcopoulos and D. Krznaric, *Quasi-greedy triangulations approximating the minimum weight triangulation*, Proc. Symp. Discrete Algorithms, 1996, pp. 392–401.

11. C. Levcopoulos and D. Krznaric, *A fast heuristic for approximating the minimum weight triangulation*, Proc. Scandinavian Workshop on Algorithmic Theory, 1996.

12. B. Yang, Y. Xu, and Z. You, *A chain decomposition algorithm for the proof of a property on minimum weight triangulation*, in Proc. International Symposium on Algorithms and Computation, 1994, pp. 423–427.

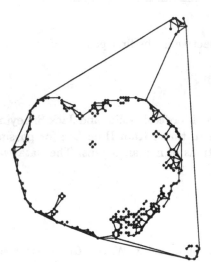

**Fig. 4.** A cross-section of blood vessels.

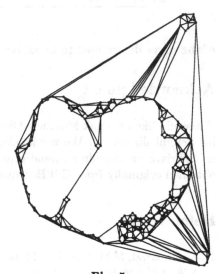

**Fig. 5.**

# A New Subgraph of Minimum Weight Triangulations[1]

Cao An Wang, [2] Francis Chin, [3] and Yin-Feng Xu [4]

## Abstract

In this paper, two sufficient conditions for identifying a subgraph of minimum weight triangulation of a planar point set are presented. These conditions are based on local geometric properties of an identifying edge in the given point set. Unlike the previous known sufficient conditions for identifying subgraphs, such as Keil's $\beta$-skeleton and Yang and Xu's double circles, The local geometric requirement in our conditions is not necessary symmetric with respect to the edge to be identified. The identified subgraph is different from all the known subgraphs including the newly discovered subgraph: so-called the intersection of local-optimal triangulations by Dickerson, Montague, and Keil. An $O(n^3)$ time algorithm for finding this subgraph from a set of $n$ points is presented.

## 1 Introduction

Let $S = \{s_i \mid i = 0, ..., n - 1\}$ be a set of $n$ points in a plane. For simplicity, we assume that $S$ is in general position so that no three points in $S$ are colinear. Let $\overline{s_i s_j}$ for $i \neq j$ denote the line segment with endpoints $s_i$ and $s_j$, and let $\omega(s_i s_j)$ denote the weight of $\overline{s_i s_j}$, that is the Euclidean distance between $s_i$ and $s_j$.

A *triangulation* of $S$, denoted by $T(S)$, is a maximum set of non-crossing line segments with their endpoints in $S$. It follows that the interior of the convex hull of $S$ is partitioned into non-overlapping triangles. The weight of a triangulation T(S) is given by

$$\omega(T(S)) = \sum_{\overline{s_i s_j} \in T(S)} \omega(s_i s_j).$$

A *minimum weight triangulation*, denoted by $MWT$, of $S$ is defined as for all possible $T(S)$, $\omega(MWT(S)) = min\{\omega(T(S))\}$.

$MWT(S)$ is one of the outstanding open problems listed in Garey and Johnson's book [GJ79]. The complexity status of this problem is unknown since 1975 [SH75]. A great deal of works has been done to seek the ultimate solution of the problem. Basically, there are two directions to attack the problem. The first one is to identify edges inclusive or exclusive to $MWT(S)$ [Ke94, YXY94, CX95] and the second one is to construct exact $MWT(S)$ for restricted classes of point set [Gi79, Kl80, AC93, CGT95]. In the first direction, two subdirections have been taken. It is obvious that the intersection of all possible $T(S)$s is a subgraph of $MWT(S)$. Recently, Dickerson and Montague [DM96] have shown that the intersection of all local optimal triangulations of $S$ is a subgraph of $MWT(S)$. (A triangulation $T(S)$ is called $k$-gon local optimal, denoted by $T_k(S)$, if any $k$-sided simple polygon extracted from $T(S)$ is an optimal triangulation for this $k$-gon by those edges of $T(S)$ lying inside this $k$-gon.) Then, if the $MWT(S)$ is unique, then the following inclusion property holds:

$$\bigcap T(S) \subseteq \bigcap T_4(S) \subseteq \cdots \subseteq \bigcap T_{n-1}(S) \subseteq MWT(S)$$

This approach has some flavor of global consideration when $k$ is increased, however, it seems difficult to find the intersections as $k$ is increased.

Gilbert [Gi79] showed that the shortest edge in $S$ is in $MWT(S)$. Yang, Xu and You [YXY94] showed that mutual nearest neighbors are also in $MWT(S)$. Keil [Ke94] presented that the so-called

[1]This work is partially supported by NSERC grant OPG0041629 and HKU funded project HKU 287/95E through RGC firect allocation.

[2]Department of Computer Science, Memorial University of Newfoundland, St.John's, NFLD, Canada A1C 5S7.
[3]Department of Computer Science, University of Hong Kong, Hong Kong.
[4]The school of management, Xi'an Jiaotong University, Xi'an, P.R. China, 710049.

$\beta$-skeleton of $S$ for $\beta = \sqrt{2}$ is a subgraph of $MWT(S)$. Cheng and Xu [CX96] extended Keil's result to $\beta = 1.17682$. The edge identification of $MWT(S)$ seems to be a promising approach and has the following merits. The more edges of $MWT(S)$ being identified, the less disconnected components is $S$. Thus, it is possible that eventually all these identified edges form a connected graph so that an $MWT(S)$ can be constructed by dynamic programming in polynomial time [CGJ95]. Moreover, it has been shown in [XZ96] that the increase of the size of subgraph of $MWT(S)$ could improve the performance of some heuristics.

The second direction is to construct exact $MWT(S)$ for restricted classes of point set $S$. Gilbert and Klinesek [Gi79,Kl80] independently showed an $O(n^3)$ time dynamic programming algorithm to obtain an $MWT(S)$, where $S$ is restricted to a simple $n$-gon. Recently, Anagnostou and Corneil [AC93] gave an $O(n^{3k+1})$ time algorithm to find an $MWT(S)$, where $S$ is restricted on $k$ nested convex polygons. Meijer and Rappaport [MR93] later improved the bound to $O(n^k)$ when $S$ is restricted on $k$ non-intersecting lines. At the same time, it was shown in [CGT95] that if given a subgraph of $MWT(S)$ with $k$ connected components, then the complete $MWT(S)$ can be computed in $O(n^{k+2})$ time.

This paper can be classified as the first direction. The paper is organized as follows. Section 2 surveys the recent results in this direction. Section 3 presents our sufficient conditions. Section 4 proposes an algorithm for finding a subgraph of $MWT(S)$ using the given sufficient conditions. Finally, we conclude our work.

## 2 A review of previous approaches

A trivial subgraph of the $MWT(S)$ is the convex hull of $S$, $CH(S)$, since it exists in any $MWT(S)$. A simple extension of the above idea is the intersection of all possible triangulations of $S$ called *stable line segments* [MWX96], denoted by $SL(S)$, such that

$$SL(S) = \bigcap_{T(S) \in J} T(S),$$

where $J$ denotes the set of all possible triangulations of $S$. The structure properties and the algorithms for finding $SL(S)$ were discussed in [MWX96].

A recent result obtained by [DM96] showed that a subgraph $LOT(S)$ of $\bigcap T_4(S)$ can be found in $O(n^4)$ time and $O(n^3)$ space. The following inclusion relation holds:

$$CH(S) \subseteq SL(S) \subseteq LOT(S) \subseteq \bigcap T_4(S).$$

Another class of subgraphs of $MWT(S)$ was identified using some local geometric properties related to an edge [Ke94, CX96, YXY94]. Keil first pointed out an inclusion condition for an edge in $MWT(S)$, so-called $\beta$-skeleton.

**Fact 1** [Ke94]. *If $x$ and $y$ are the endpoints of an edge in the $\sqrt{2}$-skeleton of $S$, and $p, q, r$, and $s$ are four distinct points in $S$ other than $x$ and $y$ with $p$ and $s$ lying on one side and $q$ and $r$ on the other of the line extending $\overline{xy}$. Assume $\overline{pq}$ and $\overline{rs}$ cross $\overline{xy}$, and $\overline{pq}$ does not intersect $\overline{rs}$. Then, either $| \overline{pq} | > | \overline{qr} |$ or $| \overline{rs} | > | \overline{qr} |$. (Refer to part (a) of Figure 1).*

With the above Fact (called *remote length lemma*), Keil proved that if the shaded disks are empty of points of $S$, $\overline{xy}$ is an edge of any $MWT(S)$. Thus, $\sqrt{2}$-skeleton($S$) is a subgraph of $MWT(S)$ which can be found in $O(n \log n)$ time and $O(n)$ space. The $\beta$-value was strengthened to 1.17682 in [CX96].

Yang, Xu, and You [YXY94] showed that the mutual nearest neighbours in $S$ are subgraph of an $MWT(S)$ and their result can be stated as follows.

**Fact 2** [YXY94]. *If any edge $\overline{pq}$ intersecting $\overline{xy}$ for $p, q, x, y \epsilon S$ satisfies the following inequality:*

$$\omega(xy) \le min\{\omega(px), \omega(py), \omega(qx), \omega(qy)\},$$

*then $\overline{xy}$ is in any $MWT(S)$.*

(Refer to Part (b) of Figure 1 for YXY's condition.)

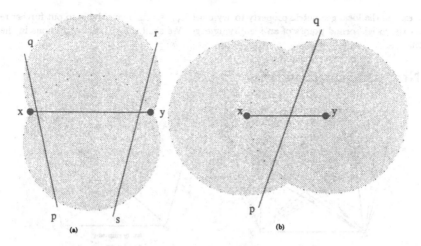

Figure 1: An illustration for the remote lemma of Keil (a) and YXY's double circles (b).

As shown in Figure 1, both Keil's $\sqrt{2}$-skeleton$(S)$ and YXY's double-circle are symmetric with respect to edge $\overline{xy}$. YXY's condition also includes Gilbert's result [Gi79] which stated that the shortest line segment among all the line segments with their endpoints in $S$ belongs to any $MWT(S)$.

In many cases, an edge of $MWT(S)$ may not have symmetric geometric property required by sufficient conditions for identification. However, a simple extending of the known methods to asymmetric can run into difficulty. The difficulty caused by non-symmetric can be easily demonstrated by the following six-point set.

Figure 2 showed an example of six points such that $|\overline{xs}| + |\overline{xb}| \le |\overline{rs}| + |\overline{rb}|$ and $|\overline{xy}| \le |\overline{rb}|$, $|\overline{ra}|$. Edge $\overline{xy}$ satisfies our sufficient conditions, however $\overline{xy}$ cannot be detected by any previous inclusion method. In part (a) of Figure 2, vertex $r$ lies inside the Keil's as well as YXY's empty circles, thus $\overline{xy}$ cannot be identified by these two methods. In part (b) of Figure 2, $\overline{rs}$ is shorter than $\overline{ab}$, thus, they do not swap in quadraliteral ($arbs$). Edge $\overline{ra}$ is shorter than $\overline{xs}$, thus, they do not swap

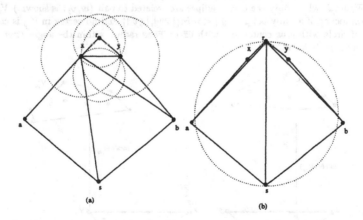

Figure 2: An example of a non-detectable edge for the known sufficient conditions

in quadraliteral ($axrs$). Edge $\overline{rb}$ is shorter than $\overline{ys}$, thus, they do not swap in quadraliteral ($sryb$). Then, $\overline{rs}$ is an edge of a $4-optimal$ triangulation. Hence, $\overline{xy}$ does not belong to the intersection of all $4-optimal$ triangulations and cannot be identified by Dickerson and Montague's method.

To extend the local geometric property to asymmetric, it seems unavoidable to put further restriction on the 'neighboring' points of an identifying edge. We shall show these restrictions in the next section.

## 3 New sufficient conditions

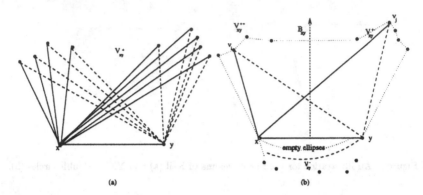

(a)                                        (b)

Figure 3: An illustration for the definitions

**Definition:** Let $E(S)$ denote the set of all line segments with their endpoints in $S$. Let $E_{xy}$ be the subset of $E(S)$, each of which crosses edge $\overline{xy}$ for $x, y \epsilon S$. Let $V_{xy}$ denote the endpoint set of $E_{xy}$, $V_{xy}^+$ denote the subset of $V_{xy}$ on the upper open halfplane bounded by the line extending $\overline{xy}$, and $V_{xy}^-$ denote that on the lower closed halfplane, i.e., $V_{xy}^-$ includes $\{x, y\}$. Then, $V_{xy}^+$ is called **2star-shaped** if the interior of any triangle $\triangle xvy$ for every $v \epsilon V_{xy}^+$ does not contain a vertex in $V_{xy}^+$. Similarly we can define 2*star-shaped* for $V_{xy}^-$. (Refer to part (a) of Figure 3.) Let $V_{xy}^{++}$ denote the subset of $V_{xy}^+$ on the same side of $x$ along the perpendicular bisector $B_{xy}$ of $\overline{xy}$, and let $V_{xy}^{+-}$ denote the subset on the same side as $y$. $V_{x,y}^+$ is called **ellipse-disconnected** if for any $v_i \epsilon V_{xy}^{++}$ and any $v_j \epsilon V_{xy}^{+-}$, $| \overline{xy} | < min\{| \overline{xv_i} |, | \overline{yv_j} |\}$ and no vertex in $V_{xy}^-$ is contained by the portion of ellipses $EL_{v_i, v_j, y} \cup EL_{v_i, v_j, x}$ within the interval bounded by $v_i \dot{\ } y$ and $v_j \dot{\ } x$, where $EL_{v_i, v_j, y}$ denotes the ellipstic area specified by foci $v_i$ and $v_j$ with boundary point $y$. That is, vertices $v_i \epsilon V_{xy}^{++} \cup \{x\}$ and $v_j \epsilon V_{xy}^{+-} \cup \{y\}$ are the foci (except simultaneously $v_i = x$ and $v_j = y$) and $| \overline{yv_i} | + | \overline{yv_j} |$ as the fixed sum of the lengths. (Refer to part (b) of Figure 3, where only the empty ellipse area related to pair $(v_i, v_j)$ is shown.) $V_{xy}^+$ is called **circle-disconnected** if for any $v \epsilon V_{xy}^+$ $| \overline{xy} | < min\{| \overline{xv} |, | \overline{yv} |\}$ and no vertex in $V_{xy}^-$ is contained by the portion of circle with $v$ as center and with $\overline{vx}$ or $\overline{vy}$ as radius within the angle $\angle xvy$. A similar definition can be made for $V_{xy}^-$.

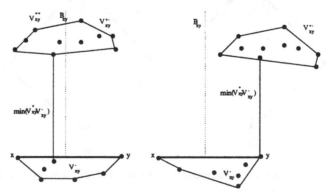

Figure 4: An illustration of a thin set.

**Definition:** The *diameter* of $V_{xy}^-$ ($V_{xy}^+$) is denoted by $d(V_{xy}^-)$ ($d(V_{xy}^+)$). Set $V_{xy}^+$ is called **thin** if $d(V_{xy}^+) < min\{\overline{v_p v_q} \mid v_p \in V_{xy}^+, v_q \in V_{xy}^-\}$. Similarly. define thin for $V_{xy}^-$.
(Refer to Figure 4 for the definition.)

**Theorem 1** *Edge $\overline{xy}$ is in every $MWT(S)$ if (1) $V_{xy}^+$ is 2star-shaped-ellipse-disconnected and (2) $\omega(\overline{xy}) \geq d(V_{xy}^-)$.*

**Proof** By contradiction. Suppose that $\overline{xy}$ does not belong to any $MWT(S)$. Then, there exists an $MWT(S)$ such that some of its edges, denoted by $E_M(\epsilon E_{xy})$, cross $\overline{xy}$. Let $V_{xy}^{+*}$ denote the endpoint set of $E_M$ belonging to $V_{xy}^+$ and $V_{xy}^{-*}$, belonging to $V_{xy}^-$. If we remove $E_M$ from this $MWT(S)$, the resulting non-triangulated area, denoted $R$, is a connected region with $V_{xy}^{+*} \cup V_{xy}^{-*} \cup \{x, y\}$ as vertices. Since $V_{xy}^{+*}$ is a subset of $V_{xy}^+$, $V_{xy}^{+*}$ is also 2star-shaped-ellipse-disconnected, and since $V_{xy}^{-*}$ is a subset of $V_{xy}^-$, $\omega(\overline{xy}) > d(V_{xy}^{-*})$ is hold. Note that the number of vertices contained in $R$ is $\mid V_{xy}^{+*} \mid + \mid V_{xy}^{-*} \mid +2$, which is $\mid E_M \mid +3$. Then, the number of internal edges of $R$ for any triangulation of $R$ is also $\mid E_M \mid$. Now, we shall build a triangulation $T(R)$ such that $T(R)$ contains $\overline{xy}$ as an edge and $\omega(int(T(R)))$ is less than $\omega(E_M)$, where $int(T(R))$ is the internal edges of $T(R)$.

To do so, we add edges $\overline{v_i x}$ clockwisely at $x$ and add edges $\overline{v_j y}$ anti-clockwisely at $y$, where $v_i \epsilon V_{xy}^{+*L}$ and $v_j \epsilon V_{xy}^{+*R}$, and $V_{xy}^{+*L}$ and $V_{xy}^{+*R}$ are the left subset and the right subset of $V_{xy}^{+*}$ along $B_{xy}$, respectively; Add $\overline{xv_j'}$, where $v_j'$ is the last vertex of $V_{xy}^{+*R}$ in anti-clockwise order; Then, add $\overline{xy}$ so that the portion of $R$ above the line extending $\overline{xy}$ is triangulated. This can be realized because the

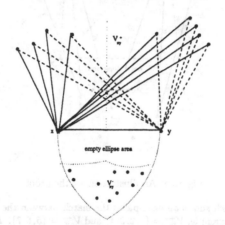

Figure 5: An illustration for sufficient conditions

2star-shaped property of $V_{xy}^{+*}$. We further add edges to triangulate the portion of $R$ below $\overline{xy}$ by any method. Thus, $T(R)$ is completed.

We only need compare the weight of the internal edges of $T(R)$ with the weight of $E_M$ since the two triangulations of $R$ shared the same boundary edges. The number of internal edges of $T(R)$ is equal to that of $E_M$, which allows us to build a match between them. If $\mid V_{xy}^{+*} \mid= 0$, we have done. If $\mid V_{xy}^{+*} \mid= 1$, then by condition (1), the edges of $E_M$ ending at $v_i$ (resp. $v_j$) is longer than $\overline{xv_i}$ (resp. $\overline{yv_j}$), and $\overline{xv_i}$ (resp. $\overline{yv_j}$) is longer than $\overline{xy}$, thus any edge of $E_M$ is longer than $\overline{xy}$. Furthermore, by condition (2) any edge whose both endpoints are in $V_{xy}^-$ is shorter than any edge in $E_M$. Thus, any edge in $int(T(R))$ is shorter than any edge in $E_M$, we have done. If $\mid V_{xy}^{+*} \mid\geq 2$, we shall consider two subcases: (a) one of $V_{xy}^{+*L}$ and $V_{xy}^{+*R}$, say $V_{xy}^{+*L}$, is empty and (b) none of them is empty. In subcase (a), all these edges of $int(T(R))$ above $\overline{xy}$ are $\overline{v_j y}$ for $v_j \in V_{xy}^{+*R}$. By condition (1), every $\overline{v_j y}$ is shorter than these of $E_M$ ending at $v_j$. Since there are $\mid V_{xy}^{+*R} \mid -1$ such edges, the remainder of unmatched $E_M$ is $\mid V_{xy}^{-*} \mid$. There must have $\mid V_{xy}^{-*} \mid$ edges including $\overline{xy}$ to completely triangulate the remaining portion of $R$. Note by condition (2) that each of these added edges is shorter than or equal to any of $E_M$. Thus, $\omega(E_M) > \omega(int(T(R)))$. In subcase (b), there exists an edge of $int(T(R))$ crossing $B_{xy}$. Let it be $\overline{xv_j'}$. Let $v_i'$ be the vertex closest to $B_{xy}$ in the sequence of $V_{xy}^{+*L}$ by angular at $x$. We first match $\overline{xv_i'}$ and $\overline{xv_j'}$ with two edges of $E_M$ ending at $v_i'$ and $v_j'$ and sharing the other endpoint in $V_{xy}^{-*}$. This match is always possible because edge $\overline{v_i'v_j'}$ must be on the boundary of $R$ hence belongs to any $MWT(S)$, and then $\overline{v_i'v_j'}$ must belong to a triangle of $T(R)$ and a triangle of $E_M$ of $MWT(S)$. By the empty ellipse property of $Ed$ in condition (1), $\omega(|\overline{xv_i'}| + |\overline{xv_j'}|)$ is less than that of these two matched edges in $E_M$. For the remaining edges of $int(T(R))$, we match $\overline{v_i x}$ with an edge of $E_M$ ending at $v_i$ and $\overline{v_j y}$ with an edge of $E_M$ ending at $v_j$ (if it possible). It is implied by the $Ed$ property of condition (1) that an edge of $E_M$ ending at $v_i$ (resp. $v_j$) is longer than $\overline{xv_i}$ (resp. $\overline{yv_j}$) because the empty circle area, determined by the circle with $v_i$ ($v_j$) as center and with $\overline{xv_i}$ ($\overline{yv_j}$) as radius, is contained by the empty ellipse area. The remaining edges of $int(T(R))$ are matched with $\overline{xy}$ and those edges whose both endpoints in $V_{xy}^{-*}$. This match is always possible as the same description in subcase (a). By condition (2) and note any edge of $E_M$ is longer than $\overline{xy}$, the weight of these remaining edges of $int(T(R))$ is less than or equal to that of the matched edges in $E_M$. Thus, $\omega(int(T(R)))$ is less than $\omega(E_M)$, a contradiction. □

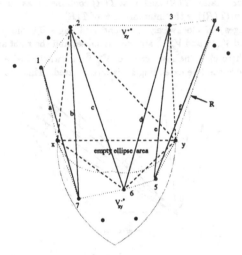

Figure 6: An illustration for the proof

(Refer to Figure 6, which shows an example of the match between the two sets of internal edges of $R$ in the proof. In this example, $V_{xy}^{+*} = \{1, 2, 3, 4\}$ and $V_{xy}^{-*} = \{5, 6, 7\}$. $E_M = \{a, b, c, d, e, f\}$. The dashed line segments are in the new triangulation $T(R)$, they are $\{\overline{x2}, \overline{x6}, \overline{y3}, \overline{y2}, \overline{y6}, \overline{xy}\}$. Note that $|c| + |d| > |\overline{y2}| + |\overline{y3}|$ due to the ellipse empty area property. One possible matching for the rest edges can be: $(\overline{x6}, a)$, $(\overline{y6}, f)$, $(\overline{x2}, b)$, and $(\overline{xy}, e)$.)

**Theorem 2** *Edge $\overline{xy}$ is in every $MWT(S)$ if (1) $V_{xy}^+$ is a Cd-thin set and (2) $\omega(\overline{xy}) \geq d(V_{xy}^-)$.*

**Proof** Let the notations used in the following analysis be the same as those in the proof of Theorem 1. We shall build a new $T(S)$ with $\overline{xy}$ as an edge such that its weight is less than $MWT(S)$. Let $R'$ be the portion of $R$ above $\overline{xy}$. If $|V_{xy}^{+*}| leq 1$, then $\overline{xy}$ is in any $MWT(S)$ obviously. If $|V_{xy}^{+*}| geq 2$, then we consider two subcases: (a) none of $V_{xy}^{+*L}$ and $V_{xy}^{+*R}$ is empty and (b) one of them, say $V_{xy}^{+*R}$, is empty. In subcase (a), we shall traverse the boundary of $R'$, clockwisely starting at $x$, to triangulate the area between the boundary of $R'$ and the convex hull of $V_{xy}^{+*L} \cup \{x\}$. Similarly, for $y$ and $V_{xy}^{+*R}$. (Refer to Part (a) of Figure 7 for the algorithm.). Add edges between the two convex chains to completely triangulate $R'$. In subcase (b), we traverse the boundary of $R'$, clockwisely starting at $y$, to triangulate the area between the boundary of $R'$ and the convex hull of $V_{xy}^{+*R} \cup \{y\}$, then add edges from $x$ to the convex hull of $V_{xy}^{+*R} \cup \{y\}$ to completely triangulate $R'$. (Refer to Part (b) of Figure 7 for the algorithm.). Finally, we use any method to triangulate $V_{xy}^{-*}$. Thus, the area determined by $V_{xy}^{+*} \cup V_{xy}^{-*} \cup \{x, y\}$ is completely triangulated. The above triangulation can always be done by our algorithm since $R'$ is a connected polygonal region that is weakly visible from $\overline{xy}$. Note that the number of edges in $int(T(R))$ is equal to $E_M$. Let us consider the weight of $int(T(R))$ and $E_M$. There are three types edges of $T(R)$ added to $R'$. An edge of the first type has its both endpoints in $V_{xy}^{+*}$, an edge of the second type is either $\overline{xv_i}$ for $v_i \in V_{xy}^{+*L}$ or $\overline{yv_j}$ for $v_j \in V_{xy}^{+*R}$, and an edge of the third type is either $\overline{yv_i}$ for $v_i \in V_{xy}^{+*L}$ or $\overline{xv_j}$ for $v_j \in V_{xy}^{+*R}$. By the thin property of condition (1), an edge of the first type is shorter than $d_{min}(V_{xy}^+, V_{xy}^-)$, which in turn is shorter than $\overline{xy}$, and by the thin

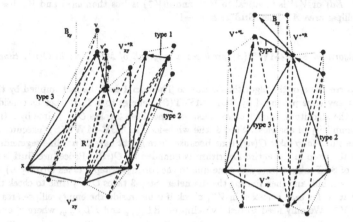

Figure 7: An illustration for the proof, where $(v, v', v'', y)$ is the convex hull of $V_{xy}^{+*L} \cup \{y\}$.

and $Cd$ properties, an edge of the second type or the third type is also shorter than the edge of $E_M$ ending at the same vertex of $V_{xy}^+$. Thus, the weight of $int(T(R))$ above $\overline{xy}$ is less than the weight of these matched edges of $E_M$. By condition (2), the weight of remaining edges of $E_M$ is less than or equal to that of those edges of $int(T(R))$ below $\overline{xy}$ and including $\overline{xy}$. Thus, $\omega(int(T(R)))$ is less than $\omega(E_M)$, and hence the new triangulation $T(S)$ has a weight less than that of the original $MWT(S)$, a contradiction. Thus, $\overline{xy}$ must belong to any $MWT(S)$. $\square$

# 4 The algorithm

**SUB-MWT(S)**

  Input: S (a set of points in general position), $|S| = $ n) and $E(S)$ (the set of all edges in $S$).
  Output: **SUB-MWT(S)** (a subgraph of $MWT(S)$).

  1. **SUB-MWT(S)** $\leftarrow \emptyset$.

  2. Find $GT(S)$ and store the edges in $E_{GT}$ by ascending order.

  3. While $E_{GT} \neq \emptyset$ Do

     (a) $e \leftarrow attract(E_{GT})$;
     (b) Find the subset $E_e$ of $E(S)$ crossed by $e$;
     (c) Call **Suff**$(e, E_e)$;
     (d) If **Suff**$(e, E_e)$=1 Then **SUB-MWT(S)** $\leftarrow e$; Delete $E_e$.

  4. EndDo.

**Procedure Suff$(e, E_e)$**

  1. Let $V^+$ be these vertices of $E_e$ lie on one halfplane bounded by the line extending $e$, and $V^-$ be those vertices of $E_e$ on the other halfplane. Sort $V^+$ by angular clockwisely at the left endpoint

of $e$, and let $V^{+l}$ denote this sequence; Sort $V^+$ by angular anti-clockwisely at the right endpoint of $e$, and let $V^{+r}$ denote this sequence. Sort $V^-$ similarly and let the resulting sequences be $V^{-l}$ and $V^{-r}$. Reverse $V^{+r}$, denote by $\bar{V}^{+r}$, and reverse $V^{-r}$, obtain $\bar{V}^{-r}$.

2. Find the diameters $d(V^+)$ and $d(V^-)$.

3. If $V^{+l}$ is identical to $\bar{V}^{+r}$ and $d(V^-)$ is less than $\omega(e)$ and $V^-$ lies outside the empty ellipse area $A^-$ (i.e., $Ed$) or $V^{-l}$ is identical to $\bar{V}^{-r}$ and $d(V^+)$ is less than $\omega(e)$ and $V^+$ lies outside the empty ellipse area $A^+$ Then **Suff**$(e, E_e) \leftarrow 1$;

**End-Suff.**

**Lemma 1** *Algorithm* **SUB-MWT(S)** *produces a subgraph of* $MWT(S)$ *in* $O(n^3)$ *time and* $O(n^2)$ *space.*

**Proof** The correctness of the algorithm is due to Theorem 1 and the fact implied by the sufficient condition that any edge produced by **SUB-MWT(S)** belongs to $GT(S)$. Let us consider the time complexity of the algorithm. Step 1 takes constant time. Step 2 takes $O(n^3)$ time by a trivial greedy method (for simplicity of analysis). Step 3, the while-loop in **SUB-MWT(S)**, executes $O(n)$ times. Step (b) of the while-loop takes $O(n^2)$ time because there might have $O(n^2)$ line segments in $E_e$. The total time for this step in the entire algorithm is bounded by $O(n^3)$. Procedure **Suff** is called $O(n)$ times. Step 1 of **Suff** takes $O(n \log n)$ time due to the sortings. Step 2 takes $O(n \log n)$ time by first finding the convex hull and then finding the diameter. Step 3 takes $O(n^2)$ time to check the sufficient condition. That is, for each vertex $v$ in $V^-$, check if $v$ lies inside the empty ellipse area determined by vertices in $V^+$. We only need to test two ellipses: $EL_{v_i', y, x}$ and $EL_{v_j', x, y}$, where $v_i'$ and $v_j'$ are the two vertices closest to $B_{xy}$ in the sequence of $V^+_{xy}$. This is because all other empty ellipse areas are contained by these two. It takes $O(n)$ time. Thus, the total time for procedure **Suff** in the entire algorithm is bounded by $O(n^3)$. Step (a) and Step (b) do not exceed $O(n^2)$. The time complexity of the entire algorithm then follows. The Step (b) of Step 3 may yield $O(n^2)$ edges in $E_e$. The space complexity follows. □

Now, let us consider an algorithm for the second sufficient condition. Let the algorithm, denoted by **SUB-1-MWT(S)**, be the same as **SUB-MWT(S)** except the procedure **Suff** $-1(e, E_e)$.

**Procedure Suff-1$(e, E_e)$**

1. Let $V^+$ be these vertices of $E_e$ lie on one open halfplane bounded by the line extending $e$, and $V^-$ be those vertices of $E_e$ on the other closed halfplane. Find $d(V^+)$ and $d(V^-)$.

2. Find $d_{min}(V^+, V^-)$.

3. Test if $V^+$ is $Cd$ with respect to $V^-$ and vice versa.

4. If $d(V^+) < d_{min}(V^+, V^-)$ and $V^+$ is $Cd$ or $d(V^-) < d_{min}(V^+, V^-)$ and $V^-$ is $Cd$, then **Suff-1** $\leftarrow 1$; Else **Suff-1** $\leftarrow 0$

   **End-Suff-1.**

**Lemma 2** *Algorithm* **SUB-1-MWT(S)** *produces a subgraph of* $MWT(S)$ *in* $O(n^3)$ *time and* $O(n^2)$ *space.*

**Proof** The correctness of the algorithm is due to Lemma 2. We only need consider the complexity of **Suff-1**. Let $V = V^+ \cup V^-$. It takes $O(|V|^2)$ to identify $V$. It takes $O(|V| \log |V|)$ time to find the diameters of $V^+$ and $V^-$ and the minimum distance between the two sets. Note that testing the $Cd$ property of $V^+$ and $V^-$ takes at most $O(|V^+| * |V^-|)$. Thus, the entire **Suff-1** takes $O(|V|^2)$ time and space. The Lemma follows. □

## 5  Concluding Remarks

The new subgraph of $MWT$ found in this paper is totally different from the known ones. Since all known subgraphs of $MWT$ can be divided into two classes. (1) edges in all local optimal triangulation. (2) $\beta$-skeleton and mutual nearest neighbors. The first one has some global optimal considerations, hence this method is more closely linked with the whole point set. However, the algorithms for identifying the subgraph is more tricky. The two cases in the second class have *symmetric* inclusion regions. Our conditions given in Section 3, are local and non-symmetric.

# References

[AART95] O. Aichholzer, F. Aurenhammer, G. Rote, and M. Tachwer, Triangulations intersect nicely, Proc. 11th Ann. Symp. on Computational Geometry, Vancouver, B.C., Association for Computing Machinery, pp. 220-229, 1995.

[AC93] E. Anagnostou and D. Corneil, Polynomial time instances of the minimum weight triangulation problem, Computational Geometry: Theory and applications, vol. 3, pp. 247-259, 1993.

[CGJ95] S.-W. Cheng, M. Golin and J. Tsang, Expected case analysis of b-skeletons with applications to the construction of minimum weight triangulations, CCCG Conference Proceedings, P.Q., Canada, pp. 279-284, 1995.

[CX95] S.-W. Cheng and Y.-F. Xu, Constrained independence system and triangulations of planar point sets, in: D.-Z. Du, Ming Li, (eds.), Computing and Combinatorics, Proc. First Ann. Int. Conf., COCOON'95, LNCS 959, Springer-Verlag, pp. 41-50, 1995.

[CX96] S.-W. Cheng and Y.-F. Xu, Approaching the largest $\beta$-skeleton within the minimum weight triangulation, Proc. 12th Ann. Symp. Computational Geometry, Philadelphia, Association for Computing Machinery, 1996.

[DM96] M.T. Dickerson, M.H. Montague, The exact minimum weight triangulation, Proc. 12th Ann. Symp. Computational Geometry, Philadelphia, Association for Computing Machinery, 1996.

[Gi79] P.D. Gilbert, New results in planar triangulations, TR-850, University of Illinois Coordinated science Lab, 1979.

[GJ79] M. Garey and D. Johnson, Computer and Intractability. A guide to the theory of NP-completeness, Freeman, 1979.

[HD90] L. Heath and S. Pemmarajiu, New results for the minimum weight triangulation problem, Virginia Polytechnic Institute and State University, Dept. of Computer Science, TR 92-30, 1992.

[Ke94] J.M. Keil, Computing a subgraph of the minimum weight triangulation, Computational Geometry: Theory and Applications pp. 13-26, 4 (1994).

[Kl80] G. Klinesek, Minimal triangulations of polygonal domains, Ann. Discrete Math., pp. 121-123, 9 (1980).

[Li87] A. Lingas, A new heuristic for the minimum weight triangulation, SIAM Journal of Algebraic and Discrete Methods, pp. 4-658, 8(1987).

[MR92] H. Meijer and D. Rappaport, Computing the minimum weight triangulation of a set of linearly ordered points, Information Processing Letters, vol. 42, pp. 35-38, 1992.

[MWX96] A. Mirzain, C. Wang and Y. Xu, On stable line segments in triangulations, Proceedings of 8th CCCG, Ottawa, 1996, pp.68-73.

[XZ96] Y. Xu, D. Zhou, Improved heuristics for the minimum weight triangulation, Acta Mathematics Applicatae Sinica, vol. 11, no. 4, pp. 359-368, 1995.

[YXY94] B. Yang, Y. Xu and Z. You, A chain decomposition algorithm for the proof of a property on minimum weight triangulations, Proc. 5th International Symposium on Algorithms and Computation (ISAAC'94), LNCS 834, Springer-Verlag, pp. 423-427, 1994.

# Dynamic Tree Routing under the "Matching with Consumption" Model*

GRAMMATI E. PANTZIOU[1], ALAN ROBERTS[2] and ANTONIS SYMVONIS[2]

[1] Computer Technology Institute, P.O. Box 1122, 26110 Patras, Greece
[2] Department of Computer Science, University of Sydney, N.S.W. 2006, Australia

**Abstract.** In this paper we present an extensive study of dynamic routing on trees under the "matching with consumption" routing model. We present an asymptotically optimal on-line algorithm which routes $k$ packets to their destination within $d(k-1) + d \cdot dist$ routing steps, where $d$ is the degree of tree $T$ on which the routing takes place and $dist$ is the maximum distance any packet has to travel. We also present an off-line algorithm that solves the same problem within $2(k-1) + dist$ steps. The analysis of our algorithms is based on the establishment of a close relationship between the matching and the hot-potato routing models.

## 1 Introduction

In a *packet routing problem* on a connected undirected graph $G$ we are given a collection of packets, each packet having an origin and a destination node, and we are asked to route them to their destinations as fast as possible. During the routing, the movement of the packets follows a set of rules. These rules specify the *routing model*. Routing models might differ on the way edges are treated, the number of packets each node can receive/transmit/hold in a single step, the number of packets that are allowed to queue in a node (queue-size), etc.

When all packets are available at the beginning of the routing, we have a *static* routing problem, while, when it is possible to generate packets during the course of the routing we have a *dynamic* routing problem. When each node is the origin of at most $h_1$ packets and the destination of at most $h_2$ packets, we have an $(h_1, h_2)$-*routing* (or *many-to-many* ) problem. In the case where $h_1 = 1$ and $h_2 > 1$ we have a *many-to-one* routing problem (*many* nodes send packets to *one* node); when $h_1 = h_2 = 1$ and the number of packets is (less than or) equal to the number of nodes of the graph we have a *(partial) permutation*.

The *matching model* was defined by Alon, Chung and Graham when they studied the routing of permutations [1]. In the matching model, each node initially holds exactly one packet and the only operation allowed during the routing is the exchange of the packets at the endpoints of an edge. The exchange of the packets at the endpoints of a set of disjoint edges can occur in a single routing step. These edges are said to be *active* during the routing step. When a packet reaches its destination node it is not consumed. Instead, it continues to participate in the routing until the time all the packets in the graph simultaneously reach their destination nodes.

* The work of Dr Pantziou was partly supported by the EEC ESPRIT Projects GEPPCOM (contract No. 9072) and ALCOM IT. The work of Dr Symvonis is supported by an ARC Institutional Grant. Email: pantziou@cti.gr, {alanr,symvonis}@cs.su.oz.au.

Alon, Chung and Graham [1] showed that any permutation on a tree of $n$ nodes can be routed in at most $3n$ steps. Roberts, Symvonis and Zhang [13] reduced the number of steps to at most $2.3n$. Furthermore, for the special cases of bounded degree trees and complete $d$-ary trees of $n$ nodes, they showed that routing terminates after $2n + o(n)$ and $n + o(n)$ steps, respectively. Zhang [14] and Høyer and Larsen [8] subsequently reduced the number of steps required to route a permutation on an arbitrary tree to $2n$. The only work related to on-line routing on trees consists of the study of sorting on linear arrays based on the odd-even transposition method [6].

In this paper, we consider the natural extension of the original model which allows for the consumption of packets. We refer to this routing model as the *matching with consumption model*. Krizanc and Zhang [10] independently considered many-to-one routing under the same model. For $n$-node trees, they showed that any many-to-one routing pattern can be routed in at most $9n$ steps and posed the question whether it is possible to complete the routing for that type of pattern in less than $4n$ steps. In this paper we answer their question to the affirmative.

Consider any $(h_1 - h_2)$-routing problem which has to be routed under the matching model. Even though at most $h_1$ packets originate from any given node $v$, initially at most one of them participates in the routing. The remaining packets which originate at node $v$ are *injected* into the routing at times where $v$ holds no other packet, i.e., at times when either no packet entered $v$ or the packet which did so was consumed at $v$.

Another commonly used routing model is the *hot-potato* (or *deflection*) routing model in which packets continuously move between nodes from the time they are injected into the graph until they are consumed at their destination. This implies that i) at any time instance the number of packets present at any node is bounded by the out-degree of the node, and ii) at any routing step each node must transmit the packets it received during the previous step (unless they were destined for it). Because packets always move, it is not possible to always route all packets to nodes closer to their destination. At any given routing step several packets might be derouted away from their destination. This makes the analysis extremely difficult. Consequently, even though hot-potato routing algorithms have been around for several years [2], no detailed and non-trivial analysis of their routing time was available until recently.

The work of Feige and Raghavan[5] which provided analysis for hot-potato routing algorithms for the torus and the hypercube renewed the interest in hot-potato routing. As a result, several papers appeared with hot-potato routing as their main theme (see [9, 11] and the references therein). Borodin et al [3] formalised the notion of the *deflection sequence*, a nice way to charge each deflection of an individual packet to distinct packets participating in the routing. Among other results, they show that routing $k$ packets in a hot-potato manner can be completed within $2(k-1) + dist$ steps for trees where $dist$ is the initial maximum distance a packet has to travel. A similar result was proven earlier by Hajek [7] and Brassil and Cruz [4] for hypercubes.

Due to space limitations, it is not possible to provide complete proofs for most of our results. Details can be found in [12].

## 2 Preliminaries

A *tree* $T = (V, E)$ is an undirected acyclic graph with node set $V$ and edge set $E$. The nodes of $V$ are supposed to be ordered according to some ordering criteria. Throughout the paper we assume $n$-node trees, i.e., $|V| = n$. An undirected edge connecting nodes

$u$ and $v$ is denoted by $\{u, v\}$, while a directed edge from node $u$ to node $v$ is denoted $(u, v)$. The set of neighbours of node $u$ is defined as $Neighbours(u) = \{v \mid \{u, v\} \in E\}$. The degree of node $u$ is defined as $degree(u) = |Neighbours(u)|$. In a similar way we define the *in-degree* and the *out-degree* of a directed graph. For a graph $G = (V, E)$ and two nodes $u$, $v \in V$, we denote by $dist_T(u, v)$ the distance (i.e., the length of the shortest path) from $u$ to $v$ on $G$.

A static routing problem $\mathcal{R}$ can be defined to be a tuple $\mathcal{R} = (G, S)$ where $G$ is the graph on which the routing takes place and $S$ is the set of packets to be routed. Each packet $p \in S$ can be described by the tuple $p = (orig, dest)$ where $orig$ and $dest$ denote the origin and the destination of packet $p$, respectively. The notation $orig(p)$ and $dest(p)$ is also used to denote the origin and the destination of packet $p$. For simplicity, we assume that for every packet $p \in S$ it holds that $orig(p) \neq dest(p)$.

In the analysis of our algorithms for the matching model we are going to use the *"charging argument"* formulated by Borodin, Rabani, and Schieber [3] for the hot-potato routing model. Consider an arbitrary packet $p$ which, at time $t$, is located at node $v$ and, during the next routing step, moves away from its destination because all edges incident to node $v$ which lead to nodes closer to the destination of $p$ are used for the routing of other packets. In this case, we say that packet $p$ suffers a *deflection* at time $t$ and that any of the packets which move closer to the destination of $p$ is responsible for (or *caused*) that deflection.

Borodin et al [3] defined the notions of the *deflection sequence* and the *deflection path* for a particular deflection as follows: Consider a deflection of a packet $p$ at time $t_1$ and let $p_1$ be the packet which caused the deflection. Follow packet $p_1$ until time $t_2$ where it reaches its destination or it is deflected by packet $p_2$, whichever happens first. In the latter case, follow packet $p_2$ until time $t_3$ where it reaches its destination or it is deflected by packet $p_3$, and so on. We continue in this manner until we follow a packet $p_l$ which reaches its destination at time $t_{l+1}$. The sequence of packets $p_1, p_2, \ldots, p_l$ is defined to be the *deflection sequence* corresponding to the deflection of packet $p$ at time $t_1$. The path (starting from the deflection node and ending at the destination of $p_l$) which is defined by the deflection sequence is said to be the *deflection path* corresponding to the deflection of packet $p$ at time $t_1$.

**Lemma 1. ( Borodin, Rabani, Schieber [3])** *Suppose that for any deflection of packet $p$ from node $v$ to node $u$ the shortest path from node $u$ to the destination of $p_l$ (the last packet in the deflection sequence) is at least as long as the deflection path. Then, $p_l$ cannot be the last packet in any other deflection sequence of packet $p$. Consequently we can associate (or "charge") the deflection to packet $p_l$.*

Lemma 1 is quite useful in the analysis of hot-potato algorithms. Consider for example the case where the routing takes place on an undirected graph and the hot-potato algorithm sends a packet away from its destination only if all edges which lead closer to its destination are used by other packets which advance closer to their destinations. Let $p$ be an arbitrary packet which initially is $dist$ steps away from its destination and assume that $k$ packets participate in the routing (including $p$). According to Lemma 1, each deflection of $p$ can be associated (or charged to) with a distinct packet which also participates in the routing. Therefore, given that the total number of packets is $k$, packet $p$ can be deflected at most $k - 1$ times. So, in the worst case, packet $p$ spends $k-1$ steps moving away from its destination, $k-1$ steps negating the result of the deflections (recall that the graph in this example is undirected), and *dist* steps moving towards its destination. Thus, packet *p* reaches its destination within at most $2(k-1) + dist$ routing steps.

# 3   On-line Routing

In this section we consider on-line routing on *n*-node trees of maximum degree *d*. We prove a lower bound which applies to a natural class of algorithms and we provide an algorithm which matches it (asymptotically).

## 3.1   A Lower Bound

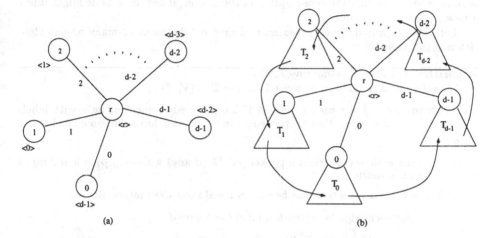

(a)                                                                                              (b)

**Fig. 1.** Worst case permutations for (a) an *n*-node star of degree $d = n - 1$, and (b) a tree of maximum degree *d*, for some constant $d \geq 2$. The numbers in the nodes are node labels, the numbers attached to edges denote the order in which edges are activated and the numbers between angle brackets denote packets with a given destination.

In order to route a pattern under the matching model an on-line algorithm must on each step choose a matching. Once this matching has been chosen for a given step, the packets at the endpoints of each edge of the matching are compared and the decision to swap them is made depending on some rule. The on-line algorithms to which our bound applies are the ones in which the edges of each node are considered in a fixed order throughout the course of the routing. These algorithms repeatedly cycle through a fixed sequence of matchings making swapping decisions based on a deterministic rule.

Consider the permutation shown in Figure 1(a) for a *star* of degree *d*. We assume that the edges become active in increasing order of the labels attached to the edges of the star. Consider an arbitrary packet which originates at a node other than the centre of the star. Observe that any such packet has to spend at least $d-1$ steps at the centre of the star waiting for the edge that leads to its destination to become active. This is because the edge which leads to its destination is activated $d-1$ steps after the time the edge through which the packet arrived at the centre of the star was active. So, each of the $d = n - 1$ packets occupies the centre of the star for at least $d-1$ steps and thus, $\Omega(dn)$ steps are required for the routing of this permutation on the star of degree $d$.

In the above routing problem the maximum degree of the tree is a function of the number of nodes in the tree. It is not difficult to construct a tree of constant degree $d$ and a permutation for which the same bound applies. This is shown in Figure 1(b). Each subtree $T_i$, $0 \le i \le d-1$, has $(n-1)/d$ nodes and the packets in subtree $T_i$ have destinations in subtree $T_{(i-1)\bmod d}$, $0 \le i \le d-1$.

## 3.2 The On-line Algorithm

In the description of the algorithm we assume that, at the end of each routing step, each node examines the packet it holds and if the packet was destined for that node it is consumed. Following, the consumption of the packet, if any, each node might inject a new packet into the routing.

Let $T$ be an $n$-node tree of maximum degree $d$. The many-to-many on-line algorithm is as follows:

---

**Algorithm** *On-Line-Tree-Routing$(T, M)$*
/* $M$ is the set of packets to be routed on tree $T = (V, E)$ */

1. [Preprocessing] For each node $v \in V$ label the edges incident on $v$ with labels in $\{0, \cdots, d-1\}$, so that no two edges incident on $v$ have the same label.

2. $t = 0$

3. For each node $v \in V$ select a packet $p \in M$ (if any) with $orig(p) = v$ and inject it into the routing.

4. While *there are packets that haven't reached their destination* do

   (a) For each edge $\{u, v\}$ with a label $l = t \bmod d$
       do *Update$(u, v)$*.

   (b) Consume packets that reached their destination.

   (c) Inject new packets (if there are any to be injected).

   (d) $t = t + 1$

---

Procedure *Update$(u, v)$* performs a swap of the packets at the endpoints of edge $\{u, v\}$ if and only if both packets will move closer to their destinations. In the description of the procedure, we assume that one packet is present at each endpoint. The procedure can be trivially extended to cover the case where none or only one packet is present at the endpoints of edge $\{u, v\}$. Consider any node $v \in V$ at time $t$. Then, by *packet$(v)$* we denote the packet $p \in M$ (if any) which resides in node $v$ at time $t$.

---

**Procedure** *Update$(u, v)$*

1. $u' = dest_T(packet(u))$

2. $v' = dest_T(packet(v))$

3. if $dist_T(u, v') + dist_T(v, u') < dist_T(u, u') + dist_T(v, v')$ then
       swap the packets at the endpoints of $\{u, v\}$.

---

## 3.3 Analysis of Algorithm *On-Line-Tree-Routing*

The analysis of our on-line algorithm is based on reducing matching routing to hot-potato routing and then applying the general charging scheme that is used for the analysis of hot-potato routing algorithms. Consider the routing problem $\mathcal{R} = (T, M)$ which is routed by algorithm *On-Line-Tree-Routing*. Based on $\mathcal{R} = (T, M)$ and algorithm *On-Line-Tree-Routing*, we define a routing problem $\mathcal{R}' = (G_T, H)$ and the hot-potato Algorithm *On-Line-Simulation* such that, the number of steps required for the routing of problem $\mathcal{R} = (T, M)$ by algorithm *On-Line-Tree-Routing* is a function of the number of steps required for the routing of problem $\mathcal{R}' = (G_T, H)$ by Algorithm *On-Line-Simulation*.

Consider a tree $T$ of maximum degree $d$ and let each edge in $T$ be labelled with an integer $i \in \{0, \cdots, d-1\}$, so that no two edges incident to the same node have the same label. We use $T$ and the labels of its edges to construct a directed graph $G_T$ as follows: For each node $v$ of $T$, we create $d$ nodes $v_j$, $j \in \{0, \cdots, d-1\}$, in $G_T$, and we say that these nodes of $G_T$ *correspond* to node $v$ of $T$. For each edge $\{u, v\}$ of $T$ we create a node $\{u, v\}^i$ in $G_T$, where $i$ is the label of $\{u, v\}$ in $T$. We say that this node of $G_T$ corresponds to edge $\{u, v\}$ of $T$. For each edge $\{u, v\}$ in $T$ with label $i$, we add the following four directed edges in $G_T$: $(u_i, \{u, v\}^i)$, $(\{u, v\}^i, u_{(i+1) \bmod d})$, $(v_i, \{u, v\}^i)$, $(\{u, v\}^i, v_{(i+1) \bmod d})$. Note that, if a node $v$ in $T$ has degree $d' < d$, not all labels in $\{0, \cdots, d-1\}$ appear at the edges incident to it. Consider such a node $v$ and let $l$ be a label that does not appear in an edge incident to $v$. Then we create a node $\{v\}^l$ in in $G_T$ and we add the directed edges $(v_l, \{v\}^l)$, $(\{v\}^l, v_{(l+1) \bmod d})$. For an example of the construction of a graph $G_T$ corresponding to a labelled tree $T$, see Figure 2.

### 3.3.1 Many-to-One Routing

For simplicity, we first analyse Algorithm *On-Line-Tree-Routing* for many-to-one routing problems. In the next section, we extend the analysis to many-to-many routing. So, assume that problem $\mathcal{R} = (T, M)$ is a many-to-one routing problem, that is, $|M| \leq n$ and for every pair of distinct packets $p$ and $q \in M$ it holds that $orig(p) \neq orig(q)$.

We complete the construction of routing problem $\mathcal{R}' = (G_T, H)$ by describing how to construct the set of packets $H$ based on the packets of set $M$. For each packet $p_m \in M$, we create a packet $p_h$ in $H$ and we set its origin and destination nodes as follows. Let $u - origin(p_m)$, $v - dest(p_m)$ and $l$ be the label of the edge that is last in the shortest path from $u$ to $v$ in $T$ (assume that $orig(p_m) \neq dest(p_m)$). Then, for packet $p_h$ we set $origin(p_h) = u_0$ and $dest(p_h) = v_{(l+1) \bmod d}$.

Algorithm *On-Line-Simulation* is the hot-potato algorithm which we use for the routing of problem $\mathcal{R}' = (G_T, H)$. It specifies the rules that each of the nodes of graph $G_T$ uses when it decides which packet to forward (if any) to each of its outgoing edges.

---

**Algorithm** *On-Line-Simulation*

*Rules for nodes of $G_T$ that correspond to nodes of $T$*

[On-line-node-1] If the packet received in the previous step reached its destination it is consumed; otherwise, it is forwarded through the only out-going edge.

*Rules for nodes of $G_T$ that correspond to unused labels around nodes of $T$*

[On-line-label-1] The packet received in the previous step is forwarded through the only out-going edge.

*Rules for nodes of $G_T$ that correspond to edges of $T$*

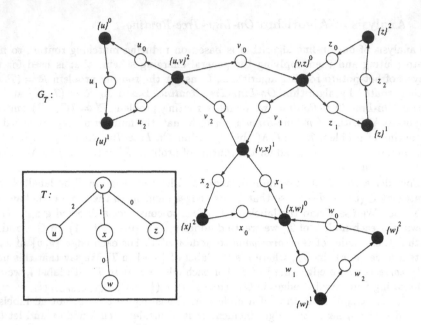

**Fig. 2.** Tree $T$ and the corresponding graph $G_T$ used in the analysis of Algorithm *On-Line-Tree-Routing.*

---

[*On-line-edge-1*] If there is only one packet at the node, the packet is forwarded to the edge that brings it closer to its destination.

[*On-line-edge-2*] In the case that there are two packets in the node, the decision is made as follows: Let $\{u,v\}^i$, $i \in \{0 \cdots d-1\}$, be the node under consideration. Let $p_h$ be the packet that arrived from $u_i$ and $q_h$ be the packet that arrived from $v_i$. Moreover, let $u'$ and $v'$ be the nodes of $T$ which correspond to $dest(p_h)$ and $dest(q_h)$, respectively. If $dist_T(u,v') + dist_T(v,u') < dist_T(u,u') + dist_T(v,v')$ then we forward $p_h$ to $v_{(i+1)\bmod d}$ and $q_h$ to $u_{(i+1)\bmod d}$; Otherwise, $p_h$ is forwarded to $u_{(i+1)\bmod d}$ and $q_h$ is forwarded to $v_{(i+1)\bmod d}$.

---

**Lemma 2.** *Let the many-to-one routing problem $\mathcal{R} = (T, M)$ be routed by Algorithm On-Line-Tree-Routing and the many-to-one routing problem $\mathcal{R}' = (G_T, H)$ by Algorithm On-Line-Simulation. Consider an arbitrary packet $p_m \in M$ and let packet $p_h \in H$ be the packet which corresponds to it. Then,*
*(i) packet $p_m$ is consumed at time $c$ iff packet $p_h$ is consumed at time $2c$, and*
*(ii) at time $t$ packet $p_m$ is at node $u$ iff at time $2t$ packet $p_h$ is at node $u_{t\bmod d}$, $t \le c$.*

**Theorem 3.** *Algorithm On-Line-Tree-Routing routes any many-to-one routing problem $\mathcal{R} = (T, M)$ in at most $d(k-1) + d \cdot dist$ routing steps, where $d$ is the maximum degree of tree $T$, $k = |M|$ is the number of packets to be routed, and dist is the maximum distance that any packet in $M$ has to travel in order to reach its destination.*

*Proof.* Based on Lemmata 1 and 2. ☐

282

## 3.3.2 Many-to-Many Routing

For the purposes of the analysis, we first route problem $\mathcal{R} = (T, M)$ by Algorithm *On-Line-Tree-Routing* and we observe for each individual packet the time at which it is injected into the routing. When the routing of $\mathcal{R} = (T, M)$ terminates, we are ready to fully specify problem $\mathcal{R}' = (G_T, H)$. For each packet $p_m \in M$ which was injected into the matching routing at time $t$, we create a packet $p_h$ in $H$ with $birth(p_h) = 2t$. The origin and the destination nodes of $p_h$ are set as in the analysis of the many-to-one routing.

**Lemma 4.** *Consider the many-to-many routing problem $\mathcal{R} = (T, M)$ which is routed by Algorithm* On-Line-Tree-Routing *and the constructed dynamic routing problem $\mathcal{R}' = (G_T, H)$ which is routed by Algorithm* On-Line-Simulation. *Let $p_m$ be an arbitrary packet in $M$ and let $p_h$ be its corresponding packet in $H$. If Algorithm* On-Line-Tree-Routing *injects packet $p_m$ at time $t$ then Algorithm* On-Line-Simulation *can inject packet $p_h$ at time $2t$.*

**Theorem 5.** *Algorithm* On-Line-Tree-Routing *routes any many-to-many routing problem $\mathcal{R} = (T, M)$ in at most $d(k-1) + d \cdot dist$ routing steps, where $d$ is the maximum degree of tree $T$, $k = |M|$ is the number of packets to be routed, and dist is the maximum distance that any packet in $M$ has to travel in order to reach its destination.*

# 4 Off-line Routing

For our off-line routing algorithms we use some special forms of directed graphs whose underlying undirected structure is that of a tree. More specifically, by *in-tree* we refer to the directed graph that satisfies the following properties: i) its undirected version is a tree, ii) there is a single node of out-degree 0 that is designated as the *root* of the in-tree, iii) all other nodes have out-degree 1. By *1-loop in-tree* we refer to the directed graph that satisfies the following properties: i) its undirected version is a tree, ii) all nodes have out-degree 1, iii) there is a pair of adjacent nodes the outgoing edges of which form a loop, referred as the *1-loop* of the tree. Finally, a node with no incoming and no outgoing edges is referred to as an *isolated* node. Graph $G(T, t)$ in Figure 3 consists of two in-trees rooted at nodes $e$ and $f$, respectively, one 1-loop in-tree with nodes $a$ and $b$ forming the 1-loop, and one isolated node i.e., node $g$.

Consider tree $T$ at time $t$ of the matching routing. Each node of the tree contains at most 1 packet which currently participates in the routing. We construct an auxiliary directed graph $G(T, t) = (V, E^t)$ which is used by our off-line algorithm to determine the set of edges that swap the packets at their endpoints during the next routing step. The directed edge $(u, v)$ is in $E^t$ if and only if at time $t$ there is a packet $p$ at node $u$ and $v$ is the first node in the shortest path from $u$ to $dest(p)$ (of course, $u$ and $v$ are neighbours in $T$). Figure 3 shows the auxiliary graph obtained from tree $T$ at time $t$, assuming that the location of each packet is as described in the figure. The out-degree of each node in graph $G(T, t)$ is at most 1 and thus $G(T, t)$ is a collection of isolated nodes, in-trees, and 1-loop in-trees.

---

**Algorithm** *Off-Line-Tree-Routing(T, M)*
/* $M$ is the set of packets to be routed on tree $T = (V, E)$ */

    1. $t = 0$

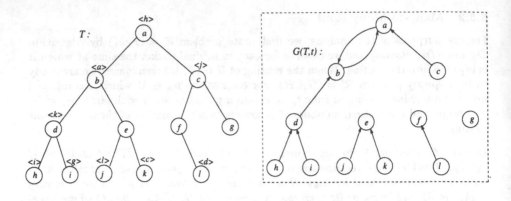

**Fig. 3.** Tree $T$ at step $t$ and the corresponding auxiliary graph $G(T, t)$.

2. For each node $v \in V$ select a packet $p \in M$ (if any) with $orig(p) = v$ and inject it into the routing.

3. While *there are packets that haven't reached their destination* do

   (a) Construct the auxiliary graph $G(T, t)$.

   (b) Denote by $S$ be the set of tree edges which swap the packets at their endpoints during the next routing step. Insert into set $S$:
   –One edge for each 1-loop in-tree. The edge is the one that corresponds to the 1-loop.
   –One edge for each in-tree. Out of the edges which enter the root of the in-tree, select the one which is emanating from the node of lowest order. The tree edge that is inserted in $S$ is the one which corresponds to the selected edge of the in-tree.

   (c) Swap the packets at the endpoints of edges in $S$.

   (d) Consume packets that have reached their destination.

   (e) Inject new packets whenever possible (if any are still to be injected into the routing).

   (f) $t = t + 1$

---

For example, based on the tree $G(T, t)$ of Figure 3 and assuming that the nodes of $T$ are ordered lexicographically, the active edges which swap the packets at their endpoints are $\{a, b\}$, $\{e, j\}$, and $\{f, l\}$.

For the analysis of Algorithm *Off-Line-Tree-Routing* we again employ elements of hot-potato routing. For details see [12].

**Theorem 6.** *Algorithm* Off-Line-Tree-Routing *routes any many-to-one routing problem* $\mathcal{R} = (T, M)$ *in at most* $2(k-1) + dist$ *routing steps, where* $k = |M|$ *is the number of packets to be routed, and dist is the maximum distance that any packet in $M$ has to travel in order to reach its destination.*

Krizanc and Zhang [10] independently showed that any many-to-one problem on an $n$-node tree can be solved under the matching routing model in at most $9n$ steps and they posed the question whether it is possible to complete the routing of any many-to-one pattern in less than $4n$ steps. Algorithm *Off-Line-Tree-Routing* dramatically improves upon the result of Krizanc and Zhang and answers their question to the affirmative.

**Theorem 7.** *Algorithm* Off-Line-Tree-Routing *routes any many-to-many routing problem* $\mathcal{R} = (T, M)$ *in at most* $2(k-1) + dist$ *routing steps, where* $k = |M|$ *is the number of packets to be routed, and* $dist$ *is the maximum distance that any packet in* $M$ *has to travel in order to reach its destination.*

# References

1. Alon, Chung, and Graham. Routing permutations on graphs via matchings. *SIAM Journal on Discrete Mathematics*, 7:513–530, 1994.

2. P. Baran. On distributed communication networks. *IEEE Trans. on Commun. Systems*, CS-12:1–9, 1964.

3. A. Borodin, Y. Rabani, and B. Schieber. Deterministic many-to-many hot potato routing. Technical Report RC 20107 (6/19/95), IBM Research Division, T.J. Watson Research Center, Yorktown Heights, NY 10598, June 1995.

4. J.T. Brassil and R.L. Cruz. Bounds on maximum delay in networks with deflection routing. In *Proceedings of the 29th Allerton Conference on Communication, Control and Computing*, pages 571–580, 1991.

5. U. Feige and P. Raghavan. Exact analysis of hot-potato routing. In *Proceedings of the 33rd Annual Symposium on Foundations of Computer Science (Pittsburgh, Pennsylvania, October 24–27, 1992)*, pages 553–562, Los Alamitos-Washington-Brussels-Tokyo, 1992. IEEE Computer Society Press.

6. N. Haberman. Parallel neighbor-sort (or the glory of the induction principle). Technical Report AD-759 248, National Technical Information Service, US Department of Commerce, 5285 Port Royal Road, Springfieldn VA 22151, 1972.

7. B. Hajek. Bounds on evacuation time for deflection routing. *Distributed Computing*, 5(1):1–6, 1991.

8. P. Høyer and K.S. Larsen. Permutation routing via matchings. Technical Report 16, Dept of Mathematics and Computer Science, Odense University, June 1996.

9. C. Kaklamanis, D. Krizanc, and S. Rao. Hot-potato routing on processor arrays. In *Proceedings of the 5th Annual ACM Symposium on Parallel Algorithms and Architectures, SPAA'93 (Velen, Germany, June 30 – July 2, 1993)*, pages 273–282, New York, 1993. ACM SIGACT, ACM SIGARCH, ACM Press.

10. D. Krizanc and L. Zhang. Packet routing via matchings. *Unpublished manuscript*, 1996.

11. Newman and Schuster. Hot-potato algorithms for permutation routing. *IEEE Transactions on Parallel and Distributed Systems*, 6(11):1168–1176, November 1995.

12. G. Pantziou, A. Roberts, and A. Symvonis. Many-to-many routing on trees via matchings. Technical Report TR-507, Basser Dept of Computer Science, University of Sydney, July 1996. Available from ftp://ftp.cs.su.oz.au/pub/tr/TR96_507.ps.Z.

13. A. Roberts, A. Symvonis, and L. Zhang. Routing on trees via matchings. In *Proceedings of the Fourth Workshop on Algorithms and Data Structures (WADS'95), Kingston, Ontario, Canada*, pages 251–262. Springer-Verlag, LNCS 955, aug 1995. Also TR 494, January 1995, Basser Dept of Computer Science, University of Sydney. Available from ftp://ftp.cs.su.oz.au/pub/tr/TR95_494.ps.Z.

14. L. Zhang. Personal communication, 1996.

# Dimension-Exchange Token Distribution on the Mesh and the Torus

Michael E. Houle[1] and Gavin Turner[2]

[1]Dept. of Computer Science, University of Newcastle
[2]Dept. of Computer Science, Victoria University of Wellington

**Abstract.** *A solution to the token distribution problem is presented for the 2-dimensional mesh and torus, based on the dimension-exchange strategy. The approach is shown to reduce the discrepancy $\Delta$ between maximum and minimum processor loads to $\delta$ in optimal $\Theta((\Delta-\delta) \cdot n)$ time steps, where $2 \leq \delta < \Delta$ for the mesh, and $4 \leq \delta < \Delta$ for the torus.*

## 1  Introduction

One of the fundamental data distribution problems on parallel architectures is that of *token distribution*, a static variant of the well-studied load balancing problem. Each processing element (PE) of the parallel architecture possesses an initial set of *tokens*, each of which represents a task to be performed; the number of tokens stored at a particular PE is called the *load* of that PE. Ideally, one would prefer that the distribution of the tokens over the set of PEs be as even as possible, as imbalances would result in a delay in the time needed to perform all the tasks. The goal of a token distribution algorithm is to redistribute the tokens in such a way that the final loads of the PEs differ as little as possible. Here it is assumed that each token requires only a constant amount of time to send from one PE to an adjacent PE, and that no tokens are created or destroyed before the redistribution is complete.

There are many data distribution methods which achieve a balanced token distribution by gathering and making use of a certain amount of global information. One method that requires no such global information is the so-called *dimension-exchange* method, which is based on the repetitive application of an extremely simple and scalable local exchange protocol. Dimension-exchange algorithms use the edge-colouring of a network to pair processors for data exchange, and have used successfully for solutions to the problems of sorting (for example the well known algorithm of Batcher [1]), and form an integral part of many of the so-called 'hot-potato' routing algorithms [6, 7].

To date, a large body of results exist anlysing the dimension-exchange approach over infinitely-divisible loads [2, 3, 9]; on the other hand, little has been known concerning dimension-exchange for finitely-divisible loads (tokens) on meshes and tori of constant degree. In this paper, we present asymptotically worst-case optimal dimension-exchange algorithms for token distribution on the 2-dimensional mesh and torus. For the $n$-by-$n$ mesh, we prove that if the discrepancy is greater than 3, $16n$ steps of the algorithm suffice to reduce the discrepancy by 1, and if the discrepancy is equal to 3, $22n$ steps suffice. For the $n$-by-$n$ torus ($n$ even), we prove that if the discrepancy is greater than 4,

$14n$ steps of the algorithm suffice to reduce the discrepancy by 1. These results are the first to establish that dimension-exchange techniques lead to optimal solutions for finitely-divisible load balancing on a mesh-connected network of constant degree.

The organization of the paper is as follows: in the next section, we describe the model of computation. In Section 3, we prove a lower bound on the complexity of the token-distribution problem, and propose dimension-exchange algorithms for the mesh and the torus. The notation and preliminary concepts that we use in the analysis of the algorithms is introduced in Section 4. The analysis of the algorithm on the torus appears in Section 5, and in Section 6, the result for the torus is extended to the mesh. Concluding remarks are made in Section 7.

## 2  Model of Computation

One of the simplest and most practical fixed connection networks is the one-port mesh-connected array. In this model, the processing elements (PEs) are arranged in a square grid, and are connected to their neighbours by uni-directional communication links. The PEs of the mesh may send or receive at most one message at any one time. This model is considerably weaker than the MIMD-model, where bi-directional links are assumed and concurrent communication to all the neighbours is allowed. The interconnection network of the 2-dimensional, one-port mesh-connected array can be formally defined as follows.

**Definition 1** *A 2-dimensional mesh-connected array consists of $n^2$ nodes arranged in an n-by-n grid. Each node represents a processor and each edge a communication link. In a toroidal configuration (**torus**) a node at location $[i, j]$, $0 \leq i, j \leq n-1$, is connected through direct communication links to 4 neighbours : $[(i\pm1) \bmod n, j]$ and $[i, (j\pm1) \bmod n]$. A **mesh** configuration is obtained from the torus by removing the links between $[0, j]$ and $[n-1, j]$, and $[i, 0]$ and $[i, n-1]$.*

A torus with side-length four is shown in Figure 1. In the analysis to follow, we assume that $n$ is even.

## 3  Token Distribution Problem and Algorithm

The token distribution problem $TD(A; \Delta, M, \delta)$ was first posed by Peleg and Upfal in [5] and can be stated as follows: Given parallel architecture $A$ containing $n^2$ processors $P_1, ..., P_{n^2}$, with each processor $P_i$ containing a stack of $\mu \leq l(P_i) \leq M$ tokens (for all $1 \leq i \leq n^2$) and for a global discrepancy between loads equal to $\Delta = M - \mu$, distribute the tokens such that at the end the global discrepancy has been reduced to at most $\delta$.

### 3.1  A Lower Bound for the Mesh and the Torus

For a 2-dimensional mesh-connected array $A$ with or without toroidal connections, a lower bound on the time required to solve token distribution problem $TD(A; \Delta, M, \delta)$ is easily derived. The result is stated in Lemma 2.

**Lemma 2** *Let $\mathcal{A}$ be an $n-by-n$ mesh-connected array. Any solution to problem* $TD(\mathcal{A}; \Delta, M, \delta)$ *requires time in* $\Omega((\Delta - \delta) \cdot n)$.

*Proof.* Is a simple bisction width argument and has been omitted in this version due to space limitations.

## 3.2 Dimension-Exchange Token Distribution Algorithm

To implement any dimension-exchange algorithm two choices must be made: the network colouring over which the algorithm will be applied, and a suitable comparison-exchange step. The algorithm presented in this paper, 2DEB (2-d Dimension-Exchange Balancing), is based on the so-called *odd-even* colouring illustrated in Figure 1, in which the *odd* connections $[i, j-1] \leftrightarrow [i, j]$ ($j$ odd) and $[i-1, j] \leftrightarrow [i, j]$ ($i$ odd) are given colours 1 and 2, respectively, and the *even* connections $[i, j-1] \leftrightarrow [i, j]$ ($j$ even) and $[i-1, j] \leftrightarrow [i, j]$ ($i$ even) are given colours 3 and 4, respectively.

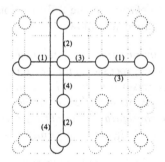

Algorithm 2DEB
FOR $i = 1$ to $c_0 \cdot (\Delta - \delta) \cdot n + c_1 n$
    Apply the elementary step [Plus-Minus1] over all
    network connections having colour ($i$ mod 4)+1;
END

**Fig. 1.** Odd-even colouring of the 2-dimensional torus, and the formal statement of Algorithm 2DEB.

The local comparison-exchange step used by the algorithm can be informally stated as follows:

**[Plus-Minus1]:** Neighbouring PEs compare their loads and, if the loads differ, a *single token* is sent from the heavily-loaded PE to the lightly-loaded PE.

Let $\Delta = M - \mu$ be the initial discrepancy of the network. Algorithm 2DEB, formally stated in Figure ??, is almost exactly the same for both the mesh and the torus, the exception being that the constants appearing below are $c_0 = 16$ and $c_1 = 22$ on the mesh, and $c_0 = 14$ and $c_1 = 0$ on the torus.

In the remainder of the paper, we shall prove two theorems that imply that 2DEB optimally decreases the discrepancy to a final value of no more than $\delta$, where $2 \leq \delta < \Delta$ for the mesh, and $4 \leq \delta < \Delta$ for the torus.

# 4 Notation and Preliminaries

In this section, we present notation and observations needed for the analysis of 2DEB over the torus $\mathcal{T}$. Instead of viewing the positions of tokens as being tied to *absolute* (fixed) locations of $\mathcal{T}$, it will often be convenient to view their positions *relative* to a migrating row and a migrating column. The next subsection establishes the relationship between the absolute and the relative interpretations of Algorithm 2DEB on the torus, and the second subsection examines the behaviour of the tokens under the relative interpretation.

## 4.1 Row and Column Migration

Under the relative interpretation of the algorithm, the torus $\mathcal{T}$ is represented as a collection $(N, S, E, W)$ of four sets of rows and columns: the set of north-migrating rows $N = \{N_0, N_1, \ldots, N_{\frac{n}{2}-1}\}$, the set of south-migrating rows $S = \{S_0, S_1, \ldots, S_{\frac{n}{2}-1}\}$, and the set of east-migrating and west-migrating columns $E = \{E_0, E_1, \ldots, E_{\frac{n}{2}-1}\}$, and $W = \{W_0, W_1, \ldots, W_{\frac{n}{2}-1}\}$ respectively. If $R$ is a row of $N \cup S$, and $C$ is a column of $E \cup W$, the intersection $(R, C)$ is associated with a *pile* of tokens, the number of which may change as each step of the algorithm is executed. The number of tokens of this pile after $t$ steps of 2DEB will be called its *size*, and will be denoted by $\mathcal{T}(R, C)_t$. Letting $\mathcal{T}[i, j]_t$ be the number of tokens at $[i, j]$ at time $t$ under the absolute interpretation, then initially we have $\mathcal{T}(N_i, E_j)_0 = \mathcal{T}[2i, 2j]_0$, $\mathcal{T}(S_i, E_j)_0 = \mathcal{T}[2i+1, 2j]_0$, $\mathcal{T}(N_i, W_j)_0 = \mathcal{T}[2i, 2j+1]_0$ and $\mathcal{T}(S_i, W_j)_0 = \mathcal{T}[2i+1, 2j+1]_0$, for all $0 \le i, j < \frac{n}{2}$.

The analysis to come relies heavily upon a classification of all piles into four *directional sets* — $\mathcal{NE}$, $\mathcal{NW}$, $\mathcal{SE}$ and $\mathcal{SW}$ — that depend on the types of row and column at the intersection of which the pile is stored. $\mathcal{NE}$ is the set of all piles located at the intersection of rows of $N$ and columns of $E$, and $\mathcal{NW}$, $\mathcal{SE}$ and $\mathcal{SW}$ are defined in a similar fashion. The pair of *opposite* sets $\mathcal{NE}$ and $\mathcal{SW}$ are each *orthogonal* to the opposite sets $\mathcal{NW}$ and $\mathcal{SE}$.

If a pile has size that is minimum or maximum over all positions of $\mathcal{T}$, we simply refer to it as a *minimum* or *maximum*. Let $\underline{\mathcal{NE}}_t \subset \mathcal{NE}$ and $\overline{\mathcal{NE}}_t \subset \mathcal{NE}$ be the set of minima and maxima that are members of $\mathcal{NE}$ after step $t$, respectively. Similarly, we define $\underline{\mathcal{NW}}_t$, $\overline{\mathcal{NW}}_t$, $\underline{\mathcal{SW}}_t$, $\overline{\mathcal{SW}}_t$, $\underline{\mathcal{SE}}_t$ and $\overline{\mathcal{SE}}_t$.

With respect to relative locations, the comparison-exchange rule of 2DEB can be reinterpreted. Consider the rows $N_i$ of $N$ and columns $E_j$ of $E$ at step $t > 0$ of 2DEB, for $0 \le i, j < \frac{n}{2}$. If $t$ is odd, for every row $R$ of $N \cup S$, the piles at $(R, E_j)$ and $(R, W_k)$ are compared and tokens are possibly exchanged, where $k = (j + \frac{(t-1)}{2}) \bmod \frac{n}{2}$. If $t$ is even, for every column $C$ of $E \cup W$, the piles at $(N_i, C)$ and $(S_{k'}, C)$ are compared and tokens are possibly exchanged, where $k' = (i + \frac{(t-2)}{2}) \bmod \frac{n}{2}$. The comparison of corresponding elements of columns $E_j$ and $W_k$ will be called a *swap* of $E_j$ and $W_k$; similarly, we shall refer to swaps of rows. Note that in the course of $n$ steps of 2DEB, each column $E_j$ of $E$ swaps with each column $W_k$ of $W$ exactly once, and $k$ increases by 1 with each swap (modulo $\frac{n}{2}$); the same holds true for rows. Accordingly, we shall refer to a sequence of $n$ consecutive steps of 2DEB as an *n-cycle*.

Whenever the piles at $(R, C)$ and $(R', C)$ are compared at time $t$, the new sizes $\mathcal{T}(R, C)_t$ and $\mathcal{T}(R', C)_t$ are recalculated as follows:

if $\mathcal{T}(R,C)_{t-1} > \mathcal{T}(R',C)_{t-1} + 1$ then
$\qquad \mathcal{T}(R,C)_t = \mathcal{T}(R',C)_{t-1} + 1; \ \mathcal{T}(R',C)_t = \mathcal{T}(R,C)_{t-1} - 1;$
else if $\mathcal{T}(R,C)_{t-1} + 1 < \mathcal{T}(R',C)_{t-1}$ then
$\qquad \mathcal{T}(R,C)_t = \mathcal{T}(R',C)_{t-1} - 1; \ \mathcal{T}(R',C)_t = \mathcal{T}(R,C)_{t-1} + 1;$
else
$\qquad \mathcal{T}(R,C)_t = \mathcal{T}(R,C)_{t-1}; \ \mathcal{T}(R',C)_t = \mathcal{T}(R',C)_{t-1};$
endif

A similar set of rules applies when the piles $(R,C)$ and $(R,C')$ are compared during a column swap.

The following lemma establishes the bijective relationship between absolute locations of tokens of $\mathcal{T}$ and the positions of piles relative to migrating rows and columns. The proof is straightforward, and has been omitted.

**Lemma 3** *For all* $0 \le i, j < \frac{n}{2}$,

$$\mathcal{T}(N_i, E_j)_t = \mathcal{T}[2i+t', 2j+t'']_t, \qquad \mathcal{T}(S_i, E_j)_t = \mathcal{T}[2i+1-t', 2j+t'']_t,$$
$$\mathcal{T}(N_i, W_j)_t = \mathcal{T}[2i+t', 2j+1-t'']_t \ \text{and} \ \mathcal{T}(S_i, W_j)_t = \mathcal{T}[2i+1-t', 2j+1-t'']_t$$

*(indices taken modulo $n$), where* $t' = t'' = \frac{t}{2}$ *if $t$ is even, and* $t' = \frac{t-1}{2}$ *and* $t'' = \frac{t+1}{2}$ *if $t$ is odd.*

## 4.2 Maxima and Minima

Consider now the situation in which $\mathcal{T}(R,C)_{t-1} = M$, the maximum possible value, for some row $R$ and column $C$. If at step $t$ $R$ swaps with row $R'$, we distinguish two outcomes. If $\mathcal{T}(R',C)_{t-1} \ge M-1$, then the rules outlined above indicate that after the swap, $\mathcal{T}(R,C)_t = \mathcal{T}(R,C)_{t-1} = M$ and $\mathcal{T}(R',C)_t = \mathcal{T}(R',C)_{t-1}$. Otherwise, if $\mathcal{T}(R',C)_{t-1} < M-1$, the greatest value possible for $\mathcal{T}(R,C)_t$ and $\mathcal{T}(R',C)_t$ is $M-1$. Similarly, if $\mathcal{T}(R,C)_{t-1} = \mu$, we have two cases: if $\mathcal{T}(R',C)_{t-1} \le \mu+1$ then $\mathcal{T}(R,C)_t = \mathcal{T}(R,C)_{t-1} = \mu$ and $\mathcal{T}(R',C)_t = \mathcal{T}(R',C)_{t-1}$; if $\mathcal{T}(R',C)_{t-1} > \mu+1$ then $\mathcal{T}(R,C)_t$ and $\mathcal{T}(R',C)_t$ must be at least $\mu+1$. Noting that the same holds true for column swaps as well as row swaps, we are led to the following observation:

**Observation 4** *Let $M$ and $\mu$ be the maximum size of the piles of $\mathcal{T}$ after step $t_0$. For any row $R$ and column $C$, if $\mathcal{T}(R,C)_{t_0} < M$, then $\mathcal{T}(R,C)_t < M$ for all $t > t_0$. Also, if $\mathcal{T}(R,C)_{t_0} > \mu$, then $\mathcal{T}(R,C)_t > \mu$ for all $t > t_0$.*

In other words, at any given position, maxima and minima can be destroyed but never created.

Consider the pile at $(N_*, E_*)$ of $\mathcal{NE}$. In the course of an $n$-cycle starting after step $t_0$, the pile at $(N_*, E_*)$ is compared exactly once with each of the piles at positions $(S_i, E_*)$ and $(N_*, W_j)$, for $0 \le i, j < \frac{n}{2}$. If $\mathcal{T}(N_*, E_*)_{t_0} < M-1$, then the first maximum of $\overline{\mathcal{SE}}_{t_0} \cap E_*$ or $\overline{\mathcal{NW}}_{t_0} \cap N_*$ to swap with $(N_*, E_*)$ after step $t_0$ will be destroyed: if such a swap does not occur by step $t_0+n$, the implication is that no maxima remain in $\overline{\mathcal{SE}}_{t_0+n} \cap E_*$ or $\overline{\mathcal{NW}}_{t_0+n} \cap N_*$. This follows partly from the observation that for $t > t_0$, $\mathcal{T}(N_*, E_*)_t$ cannot increase to $M-1$ without $(N_*, E_*)$ being swapped with a maximum after step $t_0$. For this reason, we shall say that $(N_*, E_*)$ *threatens* to destroy a maximum of $\overline{\mathcal{SE}}_{t_0} \cap E_*$ or $\overline{\mathcal{NW}}_{t_0} \cap N_*$. Also, if $\mathcal{T}(N_*, E_*)_{t_0} > \mu - 1$, we shall say that $(N_*, E_*)$ threatens a minimum of $\underline{\mathcal{SE}}_{t_0} \cap E_*$ or $\underline{\mathcal{NW}}_{t_0} \cap N_*$. In an analogous manner, we extend the notion of *threats* to maxima and minima to the other three directional sets.

To summarize, every pile of a directional set threatens either a minimum or a maximum (or both) of an orthogonal directional set in the same row or column. This is illustrated in Figure 2.

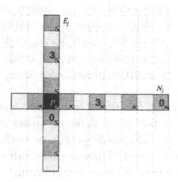

**Fig. 2.** *Pile $p = (N_i, E_j)$ threatens the maxima of $(\overline{\mathcal{NW}}_t \cap N_i) \cup (\overline{\mathcal{SE}}_t \cap E_i)$, the minima of $(\underline{\mathcal{NW}}_t \cap N_i) \cup (\underline{\mathcal{SE}}_t \cap E_i)$, or both. These locations are darkly shaded.*

## 5  Analysis of 2DEB on the Torus

In this section, Algorithm 2DEB is proven to optimally solve token distribution problems $TD(\mathcal{T}; \Delta, M, \delta)$ for tori $\mathcal{T}$, and $\delta \geq 4$. Although it is easily shown that Algorithm 2DEB is unable to solve token distribution problems $TD(\mathcal{T}; \Delta, M, \delta)$ for $\delta < 4$, the proof of this claim has been omitted for the sake of brevity. The optimality of Algorithm 2DEB is stated and proven, in Theorem 10, at the end of this section. We now present several lemmas that are required to complete the proof.

**Lemma 5** *Let $\mathcal{T}$ be an n-by-n torus (n even) whose elements are non-negative integers, and let $M$ and $\mu$ be the maximum and minimum values of these elements, respectively. If $M - \mu > 2$, then after $n$ steps of Algorithm 2DEB on $\mathcal{T}$, no row or column of $\mathcal{T}$ can contain piles $\alpha$ and $\beta$ from different directional sets such that the size of $\alpha$ is $\mu$ and the size of $\beta$ is $M$.*

*Proof.* Without loss of generality, let $\alpha = (N_*, E_*)$ and $\mathcal{T}(N_*, E_*)_0 = \mu$, and let $\beta = (N_*, W_*)$ and $\mathcal{T}(N_*, W_*)_0 = M$. At some step $0 < t \leq n$ within the course of an $n$-cycle, $\mathcal{T}(N_*, E_*)_t$ is compared with $\mathcal{T}(N_*, W_*)_t$. If $\mathcal{T}(N_*, E_*)_{t-1} = \mu$ and $\mathcal{T}(N_*, W_*)_{t-1} = M$, then after the swap, $\mathcal{T}(N_*, E_*)_{t-1} = M - 1$ and $\mathcal{T}(N_*, W_*)_t = \mu + 1$: both the maximum and the minimum are destroyed. Therefore either $\alpha$ can no longer be a minimum after step $n$, or $\beta$ can no longer be a maximum. Since the argument holds true for any choice of $\alpha$ and $\beta$ from different directional sets of the same row or column, the result follows. $\square$

**Lemma 6** *Let $\mathcal{T}$ be an n-by-n torus (n even) whose elements are non-negative integers, and let $M$ and $\mu$ be the maximum and minimum values of these elements, respectively. If $M - \mu > 2$, then after 9n steps of Algorithm 2DEB on $\mathcal{T}$, for each row and column there can be no piles $\alpha$ and $\beta$ from the same directional set such that the size of $\alpha$ is $\mu$ and the size of $\beta$ is $M$.*

*Proof.* Let us assume that the proposition is false; that is, there exists a minimum and a maximum belonging to the same row and directional sets after 9n steps of the algorithm. Lemma 5 implies that this maximum and minimum must belong to different directional sets. Without loss of generality, let us assume that the maximum and minimum belong to $N_* \cap (\overline{\mathcal{NW}}_{9n} \cup \underline{\mathcal{NW}}_{9n})$, for some row $N_*$ of $N$. For the purposes of the analysis, we will consider the behaviour of the algorithm through nine consecutive phases consisting of one full n-cycle each, starting at step 0.

Consider the pile $(N_i, E_j)$, for any row $N_i$ of $N$ and column $E_j$ of $E$, before the first step of 2DEB. The pile must threaten either maxima of $(\overline{\mathcal{NW}}_0 \cap N_i) \cup (\overline{\mathcal{SE}}_0 \cap E_i)$, minima of $(\underline{\mathcal{NW}}_0 \cap N_i) \cup (\underline{\mathcal{SE}}_0 \cap E_i)$, or both. Such a pile is illustrated in Figure 2 for $\mu = 0$ and $M = 3$. Thus in every full n-cycle, every pile of $\mathcal{NE}$ threatens maxima or minima in its row and column.

The total number $\tau$ of minima of $\underline{\mathcal{NW}}_0$ and $\underline{\mathcal{SE}}_0$, and maxima of $\overline{\mathcal{NW}}_0$ and $\overline{\mathcal{SE}}_0$ that can be destroyed by threats is at most

$$\tau < |\mathcal{NW}| + |\mathcal{SE}| < \frac{n^2}{2}.$$

Consider now the set of threats produced at positions in $\mathcal{NE}$ during the first four phases of the algorithm. A subset of these threats (call it $D$) will destroy maxima or minima by the end of the fifth phase. Once all of the piles of $D$ have destroyed their maxima and minima, some of the columns of $E$ may no longer contain maxima of $\overline{\mathcal{SE}}$ and some may no longer contain minima of $\underline{\mathcal{SE}}$. Let $\varphi$ be the set of columns of $E$ containing at least one maximum of $\overline{\mathcal{SE}}_{5n}$ and at least one minimum of $\underline{\mathcal{SE}}_{5n}$ at the end of the fifth phase. Let $\varepsilon$ be the set of columns of $E$ that do not contain both such maxima and minima.

For every column $C$ of $\varphi$, all threats which exist in $C$ during each of the four n-cycles must destroy at least one maximum or a minimum per n-cycle: there are at least $2n$ such threats over the four n-cycles. Therefore

$$2n \cdot |\varphi| \leq |D| \leq \tau < \frac{n^2}{2}, \text{ and } |\varphi| < \frac{n}{4} < \frac{n}{2} - |\varphi| = |\varepsilon|.$$

At the end of the fifth phase, there must be at least $\frac{n}{4}$ columns of $E$ that contain either no minima of $\underline{\mathcal{SE}}$ or no maxima of $\overline{\mathcal{SE}}$, or both.

For phase six of the algorithm, consider any row $N_*$ of $N$ containing a minimum $\alpha$ of $\underline{\mathcal{NW}}_{9n}$, and containing a maximum $\beta$ of $\overline{\mathcal{NW}}_{9n}$. In phase six, by the comparison-exchange rules outlined in the previous section, the comparison of minimum $\alpha$ with each $(N_*, C_i)$, $C_i \in \varepsilon$, generates a pile of size at most $\mu + 1$ at $(N_*, C_i)$. Since $M - \mu > 2$, this pile is a threat to the maxima of $\overline{\mathcal{NW}}_{9n}$ in $N_*$.

Similarly, the comparison of maximum $\beta$ with each $(N_*, C_i)$, $C_i \in \varepsilon$, generates a threat at $(N_*, C_i)$ for the minima of $\overline{\mathcal{NW}}_{9n}$ in $N_*$. Since $C_i$ cannot contain both maxima and minima which can be threatened by $(N_*, C_i)$, one of the two situations will produce a threat that can only destroy a maximum or minimum of $N_*$ — the other can be ignored.

A total of at least $|\varepsilon|$ threats to maxima and minima of $N_* \cap (\overline{\mathcal{NW}}_{9n} \cup \underline{\mathcal{NW}}_{9n})$ are thus produced which cannot destroy maxima or minima outside $N_*$: one per column of $C_i \in \varepsilon$. These threats either each destroy a maximum or minimum by the end of phase seven, or exhaust either the minima or the maxima of $N_* \cap (\overline{\mathcal{NW}}_{9n} \cup \underline{\mathcal{NW}}_{9n})$. Repeating the process of phases six and seven in phases eight and nine, if there still remain both maxima and minima of $N_* \cap (\overline{\mathcal{NW}}_{9n} \cup \underline{\mathcal{NW}}_{9n})$, then at least $2 \cdot |\varepsilon| \geq \frac{n}{2}$ maxima and minima of $N_* \cap \mathcal{NW}$ must have been destroyed — a contradiction. $\square$

**Lemma 7** *Let $\mathcal{T}$ be an $n$-by-$n$ torus ($n$ even) whose elements are non-negative integers, and let $M$ and $\mu$ be the maximum and minimum values of these elements, respectively. If $M - \mu > 2$, then after $11n$ steps of Algorithm 2DEB on $\mathcal{T}$, there can be no elements $\alpha$ and $\beta$ from opposite directional sets such that the value of $\alpha$ is $\mu$ and the value of $\beta$ is $M$.*

*Proof.* Let us assume that the proposition is false; that is, there exists a minimum and a maximum belonging to opposite directional sets after $11n$ steps of the algorithm. Without loss of generality, let us assume that the minimum belongs to $N_* \cap \underline{\mathcal{NE}}_{11n}$, and the maximum belongs to $S_* \cap \overline{\mathcal{SW}}_{11n}$ for some rows $N_*$ of $N$ and $S_*$ of $S$. Row $N_*$ will swap with $S_*$ at some step $t$, $9n < t \leq 10n$. Let $I$ and $J$ be index sets such that for all $i \in I$ and $j \in J$, $N_*$ contains an element of $\underline{\mathcal{NE}}_t$ at $(N_*, E_i)$, and $S_*$ contains an element of $\overline{\mathcal{SW}}_t$ at $(S_*, W_j)$.

Any threat to minima in row $N_*$ must belong to $\mathcal{NW}$ and have size at least $\mu + 2$, and any threat to maxima in row $S_*$ must belong to $\mathcal{SE}$ and have size at most $M - 2$. After the row swap at step $t$, the pile at $(S_*, E_i)$ has size at most $\mu + 1$, and since $M - \mu > 2$, it must threaten maxima in $S_* \cap \overline{\mathcal{SW}}_{9n}$ and $E_j \cap \overline{\mathcal{NE}}_{9n}$; however, since Lemmas 5 and 6 imply that there exists no maximum in $E_i$ after step $9n$, the threat can only destroy a maximum of $S_* \cap \overline{\mathcal{SW}}_{9n}$. Similarly, a threat is created at $(N_*, W_j)$ that can destroy only minima of $N_* \cap \underline{\mathcal{NE}}_{9n}$. The number of such threats created for minima in $N_* \cap \underline{\mathcal{NE}}_{9n}$ is $|S_* \cap \overline{\mathcal{SW}}_{9n}|$, and the number created for maxima in $S_* \cap \overline{\mathcal{SW}}_{9n}$ is $|N_* \cap \underline{\mathcal{NE}}_{9n}|$. As a result, either all minima from row $N_*$ or all maxima from row $S_*$ (whichever is the lesser) must be removed from the distribution by step $t + n$. $\square$

**Lemma 8** *Let $\mathcal{T}$ be an $n$-by-$n$ torus ($n$ even) whose elements are non-negative integers, and let $M$ and $\mu$ be the maximum and minimum values of these elements, respectively. If $M - \mu > 4$, then after $14n$ steps of Algorithm 2DEB on $\mathcal{T}$, there can be no elements $\alpha$ and $\beta$ from the same directional set such that the value of $\alpha$ is $\mu$ and the value of $\beta$ is $M$.*

*Proof.* Omitted in this version due to space limitations.

**Lemma 9** *Let $\mathcal{T}$ be an $n$-by-$n$ torus ($n$ even) whose elements are non-negative integers, and let $M$ and $\mu$ be the maximum and minimum values of these elements, respectively. If $M - \mu > 3$, then after $8n$ steps of Algorithm 2DEB on $\mathcal{T}$, there can be no elements $\alpha$ and $\beta$ from orthogonal directional sets such that the value of $\alpha$ is $\mu$ and the value of $\beta$ is $M$.*

*Proof.* Omitted in this version due to space limitations.

**Theorem 10** *Let $\mathcal{T}$ be an n-by-n torus whose elements are non-negative integers, and let $M$ and $\mu$ be the maximum and minimum values of these elements, respectively. If $M-\mu > 4$, then after 14n steps of Algorithm 2DEB on $\mathcal{T}$, there can be no elements $\alpha$ and $\beta$ such that the value of $\alpha$ is $\mu$ and the value of $\beta$ is $M$.*

*Proof.* The proof follows immediately from Lemmas 3, 7, 8 and 9 which together guarantee that after step 14n there cannot exist such an $\alpha$ or a $\beta$ in identical, in opposite, or in orthogonal directional sets. □

**Corollary 11** *Algorithm 2DEB reduces the discrepancy of the n-by-n torus $\mathcal{T}$ from $\Delta = M-\mu$ to $\delta \geq 4$ in $\mathcal{O}((\Delta-\delta) \cdot n)$ time.*

# 6  Extension of Analysis to the Mesh

In this section, we show how the results of Section 5 for the torus lead directly to prove that Algorithm 2DEB optimally solves token distribution problems $TD(\mathcal{M}; \Delta, M, \delta)$ for mesh $\mathcal{M}$, and $\Delta \geq \delta \geq 2$. The result is obtained via a simulation of the mesh by a torus of twice the sidelength, upon which the results of Section 5 are applied. It can easily be shown that Algorithm 2DEB is unable to solve token distribution problems $TD(\mathcal{T}; \Delta, M, \delta)$ for $\delta < 2$; however, the proof of this claim has been omitted for the sake of brevity.

Unlike the analysis of the previous section for the torus, the analysis for the mesh uses the original, absolute interpretation of the algorithm. Let $\mathcal{M}[a, b]_t$ be the value stored at $[a, b]$ of $\mathcal{M}$ after $t$ steps of Algorithm 2DEB on $\mathcal{M}$, and let $\mathcal{T}[a, b]_t$ be the value stored at $[a, b]$ of $\mathcal{T}$ after $t$ steps of Algorithm 2DEB on $\mathcal{T}$.

Given an $n$-by-$n$ mesh $\mathcal{M}$, we construct a $2n$-by-$2n$ torus $\mathcal{T}$ and initialize it as follows: for all $0 \leq i, j \leq n-1$, we set

$$\mathcal{T}[i, j]_0 = \mathcal{T}[2n-1-i, j]_0 = \mathcal{T}[i, 2n-1-j]_0$$
$$= \mathcal{T}[2n-1-i, 2n-1-j]_0 = \mathcal{M}[i, j]_0.$$

The southeast quadrant of $\mathcal{T}$ contains a copy of $\mathcal{M}$, and the remaining quadrants contain mirror images of $\mathcal{M}$ reflected about imaginary lines between the $(n-1)$st and $n$th rows, and the $(n-1)$st and $n$th columns.

**Lemma 12** *For all $0 \leq i, j \leq n-1$, $\mathcal{T}[i, j]_t = \mathcal{T}[2n-1-i, j]_t = \mathcal{T}[i, 2n-1-j]_t = \mathcal{T}[2n-1-i, 2n-1-j]_t = \mathcal{M}[i, j]_t$.*

*Proof.* Omitted due to space restrictions

**Theorem 13** *Let $\mathcal{M}$ be an n-by-n mesh whose elements are non-negative integers, and let $M$ and $\mu$ be the maximum and minimum values of these elements, respectively. If $M-\mu > 3$, then after 16n steps of Algorithm 2DEB on $\mathcal{M}$, there can be no elements $\alpha$ and $\beta$ such that the value of $\alpha$ is $\mu$ and the value of $\beta$ is $M$. If $M-\mu = 3$, then 22n steps suffice.*

*Proof.* Assume that such elements $\alpha$ and $\beta$ do exist, and that $\alpha = \mathcal{M}[i,j]$ for some $0 \le i,j \le n-1$. Let $\mathcal{T}$ be the $2n$-by-$2n$ torus constructed from $\mathcal{M}$ as described above. The maximum and minimum values of elements of $\mathcal{T}$ are also $M$ and $\mu$. For any integers $n$ and $a$, the difference between $a$ and $2n-1-a$ is an odd number, $2n-1-2a$. This implies that the values stored at $\mathcal{T}[i,j]$, $\mathcal{T}[2n-1-i,j]$, $\mathcal{T}[i,2n-1-j]$ and $\mathcal{T}[2n-1-i,2n-1-j]$ all belong to different directional sets, and the previous lemma implies that the values they store at any time are all equal to $\mu$. One of these four directional sets must be orthogonal to that of an element storing $\beta$, and another two must be opposite. Applying Lemma 9 when $M-\mu > 3$, and Lemma 7 when $M-\mu = 3$, we arrive at a contradiction. $\square$

**Corollary 14** *Algorithm 2DEB reduces the discrepancy of the $n$-by-$n$ mesh $\mathcal{M}$ from $\Delta = M-\mu$ to $\delta \ge 2$ in $\mathcal{O}((\Delta-\delta) \cdot n)$ time.*

# 7 Conclusion

In this paper, we presented a dimension-exchange data distribution algorithm and proved that it is asymptotically-optimal for token distribution on the two-dimensional mesh and torus. The benefits of the dimension-exchange approach, in that it is extremely simple, uses only locally-available information and is completely scalable, cannot be overstated. The analysis shows for the first time that dimension-exchange techniques can lead to optimal solutions for token distribution on mesh-connected networks of constant degree.

# References

1. K. Batcher. Sorting networks and their applications. In *Proceedings AFIPS Spring Joint Conference*, volume 32, pages 307–314, 1968.
2. G. Cybenko. Dynamic load balancing for distributed memory multiprocessors. *Journal of Parallel and Distributed Computing*, 7:279–301, 1989.
3. S.H. Hosseini, B. Litow, M. Malkawi, J. McPherson, and K. Vairavan. Analysis of graph coloring based distributed load balancing algorithm. *Journal of Parallel and Distributed Computing*, 10:160–166, 1990.
4. F. Meyer auf der Heide, B. Oesterdiekhoff, and R. Wanka. Strongly adaptive token distribution. In *Proceedings of the 20th ICALP*, pages 398–409, 1993.
5. D. Peleg and E. Upfal. The generalised packet routing problem. *Theoretical Computer Science*, 53:218–293, 1987.
6. R. T. Plunkett and A. Symvonis. On the hot-potato permutation routing algorithm of borodin, rabani and schieber. In *Proceedings of Computing: The Australasian Theory Symposium (CATS '96)*, Jan 1996.
7. A. Roberts and A. Symvonis. On deflection worm routing on meshes. In *Proceedings of the First IEEE International Conference on Algorithms and Architecture for Parallel Processing (ICA3PP)*, pages 375–378, Apr 1995.
8. G. Turner and H. Schröder. Token distribution on reconfigurable $d$-dimensional meshes. In *Proceedings of the 1st IEEE International Conference on Algorithms and Architectures for Parallel Processing*, volume 1, pages 335–344, 1995.
9. C.Z. Xu and F.C.M. Lau. Analysis of the generalized dimension exchange method for dynamic load balancing. *Journal of Parallel and Distributed Computing*, 16:385–393, 1992.

# Directed Hamiltonian Packing in $d$-dimensional Meshes and Its Application

## (Extended Abstract)

Jae-Ha Lee    Chan-Su Shin    Kyung-Yong Chwa

Dept. of Computer Science, KAIST, Korea,
{jhlee,cssin,kychwa}@jupiter.kaist.ac.kr.

**Abstract.** A digraph $G$ with minimum in-degree $d$ and minimum out-degree $d$ is said to have a directed hamiltonian packing if $G$ has $d$ link-disjoint directed hamiltonian cycles. We show that a $d$-dimensional $N_1 \times \cdots \times N_d$ mesh, when $N_i \geq 2d$ is even, has a directed hamiltonian packing, where an edge $(u, v)$ in $G$ is regarded as two directed links $\langle u, v \rangle$ and $\langle v, u \rangle$. As its application, we design a time-efficient all-to-all broadcasting algorithm in 3-dimensional meshes under the wormhole routing model.

## 1 Introduction

A digraph $G$ with minimum in-degree $d$ and minimum out-degree $d$ is said to have a *directed hamiltonian packing* if $G$ has $d$ link-disjoint directed hamiltonian cycles. Graph theoretical results on this problem are summarized as a famous conjecture [1] that every $k$-diregular tournament has a directed hamiltonian packing. Actually, it has been known to be true for $k \leq 4$ in [1].

In general, the directed hamiltonian packing of popular interconnection networks such as hypercubes, meshes and tori(meshes with wraparound edges) is obtained by finding edge-disjoint undirected hamiltonian cycles and then regarding one hamiltonian cycle as two directed hamiltonian cycles, i.e., clockwise and counter-clockwise cycle. This is based on an assumption that an undirected edge $(u, v)$ consists of two directed links $\langle u, v \rangle$ and $\langle v, u \rangle$ which join $u$ to $v$ and $v$ to $u$, respectively. It has been known that some product graphs including $2k$-dimensional hypercubes and $k$-dimensional tori can be decomposed into $k$ edge-disjoint undirected hamiltonian cycles[2], and so, we can say that they have a directed hamiltonian packing. However, when the degree of hypercubes is odd, it is clear that the directed hamiltonian packing cannot be obtained any more from edge-disjoint hamiltonian cycles.

For $d$-dimensional meshes, no results about the directed hamiltonian packing has been known. In this paper, we present a method to construct a directed hamiltonian packing of $d$-dimensional meshes even if the minimum degree $d$ is odd. Precisely, we show that $d\,(\geq 2)$-dimensional $N_1 \times \cdots \times N_d$ mesh has a directed hamiltonian packing, where $N_i (\geq 2d)$ is even for all $i$.

In practice, directed hamiltonian cycles are useful structures for many communication algorithm in multicomputers under the wormhole routing model [3, 4].

In hypercubes and 2-dimensional meshes, it was shown that multicasting algorithms using hamiltonian paths give a better performance than tree-based and multipath-based ones [5]. If an interconnection network $G$ with degree $k$ has a directed hamiltonian packing, we can easily design reliable algorithms for some important communication problems such as broadcasting, multi-casting, and all-to-all broadcasting ones. Reliability of those algorithms is achieved by executing the same algorithm for each of $k$ directed hamiltonian cycles, independently. This implies that every node in $G$ receives $k$ copies of each message after the algorithm ends. Using such a framework, Lee and Shin [4] presented reliable all-to-all broadcasting algorithms on hypercubes, tori, and hexagonal meshes.

In addition, the directed hamiltonian packing can be used to reduce the execution time of the communication algorithm. This is obtained by partitioning an entire communication problem into $k$ sub-problems, assigning sub-problems to $k$ directed hamiltonian cycles one-to-one, and then solving the sub-problem on each of $k$ cycles, independently. To demonstrate this immediate application of the directed hamiltonian packing in meshes, we, in this paper, design and present a time-efficient all-to-all broadcasting algorithm on wormhole-routed 3-dimensional meshes.

## 2 Preliminaries

A $N_1 \times \cdots \times N_d$ $d$-dimensional mesh $M_d$ is a graph whose node set is $[N_1] \times \cdots \times [N_d]$ and whose edge connects two nodes $(a_1, \cdots, a_d)$ and $(b_1, \cdots, b_d)$ only if $\sum_i |a_i - b_i| = 1$. Here $[N]$ denotes the set of $\{0, 1, \cdots, N-1\}$. Throughout the paper, we assume that an edge $(u, v)$ consists of two directed links $\langle u, v \rangle$ and $\langle v, u \rangle$. If two end-nodes of a link differ in the $k$-th coordinate, the link is said to be $x_k$-*dimensional link*. The *parity* of a node $(a_1, \cdots, a_d)$ is defined to be even if $\sum_i a_i$ is even, otherwise odd.

Let $G$ be a directed graph of $N$ nodes. A *path* of length $k$ in $G$ is a sequence of $k+1$ nodes $\langle v_1, v_2, \cdots, v_{k+1} \rangle$ such that $\langle v_i, v_{i+1} \rangle$ is an edge in $G$ and that $v_i \neq v_j$ if $i \neq j$. A *cycle* of length $k$ in $G$ is a closed path such that $v_1 = v_{k+1}$. A *directed hamiltonian cycle* of $G$ is a cycle of length $N$ that is passing through all nodes of $G$.

**Definition 1** *Let $e$ be a $x_k$-dimensional link in $M_d$. The link $e = \langle (a_1, \cdots, a_k, \cdots, a_d), (a_1, \cdots, a_k+1, \cdots, a_d) \rangle$ is called $U_k$-link if the parity of $(a_1, \cdots, 0, \cdots, a_d)$ is even, otherwise $D_k$-link. The link $e = \langle (a_1, \cdots, a_k, \cdots, a_d), (a_1, \cdots, a_k-1, \cdots, a_d) \rangle$ is called $D_k$-link if the parity of $(a_1, \cdots, 0, \cdots, a_d)$ is even, otherwise $U_k$-link.*

**Definition 2** *Let $v = (a_1, \cdots, a_d)$ be a node in $M_d$. A rotation of $v$ is defined to be the right rotation of its coordinates, i.e., $(a_d, a_1, \cdots, a_{d-1})$. The rotation of a link $e = \langle u, v \rangle$ is $e' = \langle u', v' \rangle$, where $u'$ and $v'$ are the rotated nodes of $u$ and $v$, respectively.*

Suppose that $M_d$ is a $N \times \cdots \times N$ mesh. Then we can observe that the rotation of $U_k(D_k)$-link $e$ in $M_d$ defines $U_{k+1}(D_{k+1})$-link in $M_d$. For example, a rotation of $U_1$-link $e = \langle(0,0),(1,0)\rangle$ is $\langle(0,0),(0,1)\rangle$, that is $U_2$-link. Similarly, we can also define a rotation of a directed hamiltonian cycle in $M_d$.

Let $M_d[S_1,\cdots,S_d]$ denote the induced subgraph of $M_d$ on a node set $V' = \{(x_1,\cdots,x_d) \mid x_i \in S_i, 1 \le i \le d\}$, where $S_i \subseteq [N_i]$. For simplicity of the notation, we omit $S_i$'s in the bracket of $M_d[\cdot]$ if $S_i = [N_i]$. For example, $M_d[S_i = \{0\}]$ represents a $(d-1)$-dimensional submesh in $M_d$ with fixing $i$-th dimension as 0.

## 3 Directed hamiltonian packing of $d$-dimensional meshes

In this section, we construct $d$ link-disjoint directed hamiltonian cycles in $N \times \cdots \times N$ $d$-dimensional mesh $M_d$, where $N \ge 2d$ and is even. Directed hamiltonian packing for meshes with arbitrary sides can be easily obtained from that for $N \times \cdots \times N$ meshes.

Our construction proceeds by induction on $d$. For each $d$, we construct a directed hamiltonian cycle, denoted by $DHC_1^d$, that satisfies some invariants and then making $DHC_i^d$'s for $2 \le i \le d$ by rotating $DHC_1^d$ $(i-1)$ times. Recursively, $DHC_1^d$ in $M_d$ is made from $DHC_1^{d-1}$ in $M_{d-1}$.

### 3.1 Construction in $M_2$

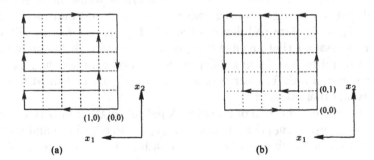

**Fig. 1.** A construction of $DHC_1^2$ and $DHC_2^2$ in $M_2$ (a) $DHC_1^2$ (b) $DHC_2^2$

$DHC_1^2$ of $M_2$ is constructed as shown in Figure 1 (a) so that satisfying the following properties: (1) it uses only links in the subset of the following links: $U_1$-links in $M_2$ and $D_2$-links in $M_2[S_1 = \{0, 1, N-1\}]$. (2) if $N \ge 6 (= 2d+2)$, $DHC_1^2$ contains a path $\langle(2,3),(1,3),(1,4),(2,4)\rangle$. Next, the other directed hamiltonian cycle $DHC_2^2$ is obtained by rotating $DHC_1^2$ once (see Figure 1 (b)).

**Lemma 1** $DHC_1^2$ and $DHC_2^2$ use no common links in $M_2$.

**Proof:** $DHC_1^2$ uses $U_1$-links and $D_2$-links, and $DHC_2^2$ uses $U_2$-links and $D_1$-links by definition of the rotation. Since $U_1$-links and $D_1$-links are disjoint and $U_2$-links and $D_2$-links are disjoint, $DHC_1^2$ and $DHC_2^2$ are also disjoint in $M_2$. □

## 3.2 Construction in $M_d$

Roughly we will explain how to construct $DHC_1^d$ in $M_d$. A submesh $M_d[S_d = \{k\}]$ for fixed constant $1 \le k \le d$ is isomorphic to $(d-1)$-dimensional $N \times \cdots \times N$ mesh $M_{d-1}$. So we can place $DHC_1^{d-1}$ of $M_{d-1}$ into a submesh $M_d[S_d = \{0\}]$ of $M_d$, which is denoted by $DHC_{1,0}^{d-1}$. Then $DHC_{1,j}^{d-1}$ $(j = 0, \cdots, N-1)$ is a translation of $DHC_{1,0}^{d-1}$ from $M_d[S_d = \{0\}]$ to $M_d[S_d = \{j\}]$, in which the direction of all links is reversed if $j$ is odd (See Figure 2 (b)). We call all these $DHC_{1,j}^{d-1}$'s as a *shelf* of $DHC_1^{d-1}$. Next we connect the shelf of $DHC_1^{d-1}$ into a directed hamiltonian cycle in $M_d$, which is $DHC_1^d$. Finally, we obtain $DHC_i^d$ for $2 \le i \le d$ by rotating $DHC_1^d$ $(i-1)$ times. An example of such a construction for $M_3$ is shown in Figure 2.

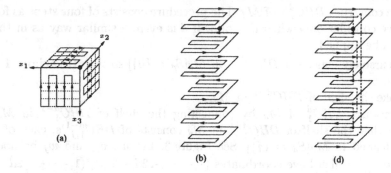

**Fig. 2.** A construction of $DHC_i^3$ for $1 \le i \le 3$ in $M_3$. (a) $DHC_1^2, DHC_2^2, DHC_3^2$. (b) The shelf of $DHC_1^2$. (c) $DHC_1^3$.

Formally, we describe a construction method of $DHC_1^d$ in $M_d$ from $DHC_1^{d-1}$ of $M_{d-1}$. As an induction hypothesis, we suppose that $DHC_1^{d-1}$ in $M_{d-1}$ satisfies the following two invariants:

**Invariant 1:** $DHC_1^{d-1}$ uses the subset of the following links:

$U_1$-links in $M_{d-1}$
$D_2$-links in $M_{d-1}[S_1 = \{0, 1, N-1\}]$
$D_3$-links in $M_{d-1}[S_2 = \{3, 4\}]$
$\cdots$

$D_k$-links in $M_{d-1}[S_{k-1} = \{2k-3, 2k-2\}]$
$\cdots$

$D_{d-1}$-links in $M_{d-1}[S_{d-2} = \{2d-5, 2d-4\}]$

**Invariant 2:** if $(d-1)$ is even and $N \ge 2d$ $(= 2(d-1)+2)$, $DHC_1^{d-1}$ contains the following path in 2-dimensional submesh $M_{d-1}[S_i = \{2i-1\}$ $(i = 2, \cdots, d-2)]$:

$\langle (2, -, \cdots, -, 2d-3), (1, -, \cdots, -, 2d-3), (1, -, \cdots, -, 2d-2), (2, -, \cdots, -, 2d-2) \rangle$.

If $(d-1)$ is odd and $N \ge 2d$, $DHC_1^{d-1}$ contains the path with the reverse direction.

From **Invariant 1**, we can prove that $M_{d-1}$ has a directed hamiltonian packing.

**Lemma 2** $DHC_i^{d-1}$'s of $M_{d-1}$, for $1 \le i \le d-1$, have no common links.

**Proof:** In this proof, we assume that for two integers $a$ and $b$, $a + b$ operation means "modulo+1" operation with respect to $(d-1)$, i.e., $(a+b) \bmod (d-1)+1$. Also, $a - b$ means $(a-b) \bmod (d-1) + 1$. Since $DHC_i^{d-1}$ is obtained by rotating $DHC_1^{d-1}$ $(i-1)$ times, $DHC_i^{d-1}$ uses the subset of the following links:

$$U_i\text{-links in } M_{d-1}$$
$$D_{i+1}\text{-links in } M_{d-1}[S_i = \{0, 1, N-1\}]$$
$$D_{i+k}\text{-links in } M_{d-1}[S_{i+k-1} = \{2k-1, 2k\}], \quad k = 2, \cdots, d-2$$

So, we can easily know that each $U_i$-link is used at most once. It remains to show that each $D_i$-link is used at most once for all $i$. In $DHC_{i-1}^{d-1}$, $D_i$-links are used in $M_{d-1}[S_{i-1} = \{0, 1, N-1\}]$. In $DHC_i^{d-1}$, $D_i$-links are not used, but $U_i$-links are used only. In $DHC_{i+1}^{d-1}$, $D_i$-links are used in $M_{d-1}[S_{i-1} = \{2d-5, 2d-4\}]$. In $DHC_{i+j}^{d-1}$ $(j = 2, \cdots, d-3)$, $D_i$-links are used in $M_{d-1}[S_{i-1} = \{2d-2j-3, 2d-2j-2\}]$. Since $N - 1 \geq 2d - 1$, $D_i$-links used are disjoint. $\square$

Now we construct $DHC_i^d$'s of $M_d$. The procedure consists of four steps as follows. Here, we consider only when $d$ is odd. If $d$ is even, a similar way as in the odd case can be applied.

1. Recursively construct $DHC_1^{d-1}$ in $M_d[S_d = \{0\}]$ so that **Invariant 1** and **2** hold.

2. Make the shelf of $DHC_1^{d-1}$ in $M_d$.

3. Construct $DHC_1^d$ of $M_d$ by connecting the shelf of $DHC_1^{d-1}$ in $M_d$. Recall that the shelf of $DHC_1^{d-1}$ in $M_d$ consists of $DHC_{1,j}^{d-1}$'s, each of which is defined in $M_d[S_d = \{j\}]$. See Figure 3. Let $u_j$, $v_j$, and $w_j$ be nodes on $DHC_{1,j}^{d-1}$ which have coordinates $(1, -, \cdots, -, 2d-3, j)$, $(1, -, \cdots, -, 2d-2, j)$, and $(2, -, \cdots, -, 2d-2, j)$ in $M_d[S_i = \{2i-1\}(i = 2, \cdots, d-2), S_d = \{j\}]$, respectively. By **Invariant 2** in induction hypothesis, $DHC_{1,j}^{d-1}$ must contain a path $\langle u_j, v_j, w_j \rangle$ if $j$ is even and a path $\langle w_j, v_j, u_j \rangle$ if $j$ is odd. Now we construct $DHC_1^d$ as follows: starting at $v_0$ on $DHC_{1,0}^{d-1}$, moving down along $\langle v_j, v_{j+1} \rangle$ until $j = N - 2$ and then entering $v_{N-1}$ (see Figure 3 (a)), moving around $DHC_{1,N-1}^{d-1}$ and then entering $w_{N-1}$, moving up to $w_{N-2}$, moving around $DHC_{1,N-2}^{d-1}$ and then entering $u_{N-2}$, moving up to $u_{N-3}$ (see Figure 3 (b)), moving around $DHC_{1,N-3}^{d-1}$ and then entering $w_{N-3}$, and so on. As a consequence, the path reaches $u_0$ on $DHC_{1,0}^{d-1}$. By including $\langle u_0, v_0 \rangle$, $DHC_1^d$ is completely constructed (See Figure 3 (c)). Note that all links used to connect the shelf of $DHC_1^{d-1}$ are $D_d$-links.

4. Obtain $DHC_i^d$ for $i \geq 2$ by rotating $DHC_1^d$ $(i-1)$ times.

It remains to prove that two invariants for $DHC_i^d$'s also hold.

**Lemma 3** *After Step 3, Invariant 1 and 2 for $DHC_1^d$ are satisfied.*

**Proof: Invariant 1:** all links used to connect $DHC_{1,i}^{d-1}$'s are $D_d$-links in $M_d[S_{d-1} = \{2d-3, 2d-2\}]$. **Invariant 2:** if $N \geq 2d + 2$, the path $\langle u'_{2d}, u_{2d}, u_{2d-1}, u'_{2d-1} \rangle$ is used, where $u'_j$ is $(2, -, \cdots, -, 2d-3, j)$ in $M_d[S_i = \{2i-1\}(i = 2, \cdots, d-2), S_d = \{j\}]$. That path is $\langle (2, -, \cdots, -, 2d), (1, -, \cdots, -, 2d), (1, -, \cdots, -, 2d-1), (2, -, \cdots, -, 2d-1) \rangle$ in $M_d[S_i = \{2i-1\}(2 \leq i \leq d-1)]$. Notice that $d$ is odd. $\square$

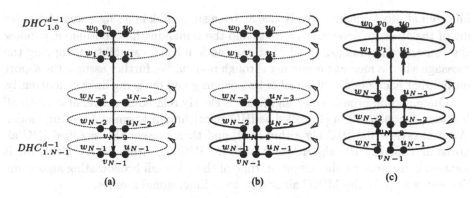

**Fig. 3.** A construction of $DHC_1^d$ by connecting the shelf of $DHC_1^{d-1}$.

**Theorem 1** *For even $N_i \geq 2d$, any $N_1 \times \cdots \times N_d$ mesh has a directed hamiltonian packing.*

For $N_1 \times \cdots \times N_d$ meshes, the only difference from the case of $N \times \cdots \times N$ meshes is that $DHC_i^d$ is not obtained by rotating $DHC_1^d$, but is isomorphic to $DHC_1^d$ of $N_i \times \cdots \times N_d \times N_1 \times \cdots \times N_{i-1}$ mesh.

## 4 All-to-all broadcasting as an application

All-to-all broadcasting addresses a fundamental communication problem in multicomputer systems, in which every node must broadcast its message to all other nodes. This is essential for many important applications such as finite element method[8], $N$-body analysis, clock synchronization[4], and database primitives including parallel joins.

Using the directed hamiltonian packing in 3-dimensional $N_1 \times N_2 \times N_3$ ($N_i \geq 6$, even) mesh $M_3$, we will present an efficient all-to-all broadcasting algorithm of $M_3$ under the wormhole routing model. Three-dimensional meshes have been received considerable attentions in recent years, because of simplicity of its structure, easy embeddability, high scalability and relatively smaller diameter than that of 2-dimensional meshes. There are experimental products such as MIT J-machine and CalTech Mosaic C.

Most of the previous works on all-to-all broadcasting has been devoted to the store-and-forward model. Optimal algorithms whose execution time exactly matches the lower bound are known in hypercubes[8], 2-dimensional meshes[9], and tori. In 2-dimensional meshes, since optimal all-to-all broadcasting algorithm with store-and-forward routing uses two directed hamiltonian cycles, this algorithm can be adapted to wormhole routing. However, no efficient all-to-all broadcasting algorithms are known in 3-dimensional meshes even with store-and-forward model.

In the wormhole routing, a message consists of a sequence of transmission units, called *flits*. During each step, each flit of a message can advance across a link, but at most one flit can advance across a single link in a single step. A header flit of the message determines a route and the remaining flits continuously follow the route. We assume, as in [4], that each node $v$ has a capability of copying the message when a message is passing through node $v$. We further assume the $d$-port model that each node can use all of its incoming and outgoing links concurrently.

Our all-to-all broadcasting algorithm is closely related to the reliable all-to-all broadcasting algorithm [4], called as IHC algorithm, in wormhole routing model. We first review the IHC algorithm in [4] and then present the modified IHC algorithm (called MIHC algorithm). Unlike the IHC algorithm, MIHC algorithm is focused on improving the execution time of the all-to-all broadcasting algorithm. Second, we apply the MIHC algorithm to 3-dimensional meshes.

## 4.1 Reliable all-to-all broadcasting algorithm (IHC algorithm)

IHC algorithm can be implemented in any network $G$ with multiple link-disjoint directed hamiltonian cycles. Let $\gamma$ be the number of link-disjoint directed hamiltonian cycles in $G$, denoted by $DHC_1, \cdots, DHC_\gamma$, and let $N$ be the number of nodes of $G$. In IHC algorithm, same procedure is executed on each $DHC_i$ ($i = 1, \cdots, \gamma$), so we consider only one directed hamiltonian cycle, say $DHC_i$. Let $S_i$ be an arbitrary node on $DHC_i$. We assume that nodes are numbered along $DHC_i$ from 0 to $N - 1$ starting from $S_i$.

The IHC algorithm is performed in $\eta$ phases. The interleaving distance $\eta$ is the spacing between nodes that are initiating messages in one phase. In phase $j$ ($0 \leq j \leq \eta - 1$), every $k$-th node such that $(k \bmod \eta) = j$ is permitted to initiate a message along $DHC_i$. Once messages have been started along $DHC_i$, they keep flowing for $N - 1$ nodes along the cycle. The interleaving distance $\eta$ have to be determined so that no messages are blocked. Consider any two nodes $u$ and $v$ that initiate their messages in the same phase. To guarantee the non-blocking during a phase, a header of $u$'s message should not arrive at $v$ until a tail of $v$'s message leaves $v$. In other words, the distance $\eta$ between $u$ and $v$ must be at least the number of flits $L$ of the message. In fact, $\eta$ have to be determined by several system parameters as well as $L$. For convenience, we assume that $\eta = L$.

> **IHC** algorithm for the $k$-th node along $DHC_i$
> **for** $j = 0$ **to** $\eta - 1$
>      **for** $t = 1$ **to** $N + L - 1$
>          **if** ($1 \leq t \leq L$ **and** $(k \bmod \eta) = j$)
>              send $t$-th flit of a message along $DHC_i$;
>          receive a flit of message from $(k - 1)$-th node;
>          send the flit received from $(k - 1)$-th node to $(k + 1)$-th node;
> **endIHC**

On completion of IHC algorithm, since $\gamma$ copies of every message are delivered to all other nodes through different directed hamiltonian cycles, the reliability of the algorithm increases as $\gamma$ increases.

## 4.2 Time-efficient all-to-all broadcasting algorithm (MIHC algorithm)

When $P = (V_1, \cdots, V_\gamma)$ is defined as a partition of a node set $V$ of $G$, MIHC algorithm broadcasts the message of the nodes in $V_i$ only along $DHC_i$ to all nodes of $G$. So, MIHC algorithm is not reliable but roughly $\gamma$-times faster than IHC algorithm. In order to achieve this speed-up, $G$ must have a "good" partition, which is defined later.

Ideally, if the nodes in $V_i$ are spaced on $DHC_i$ with equi-distance and $|V_i| = \frac{N}{\gamma}$, MIHC algorithm can obtain almost $\gamma$ speed-up of IHC algorithm. However, since $N$ is not exactly divided by $\gamma \ (= 3)$ in 3-dimensional meshes, we should consider approximately equal partitions. For any node $u, v \in V_i$, $dist_i(u, v)$ is the distance between $u$ and $v$ when traversing $DHC_i$. For any node $u, v \in V_i$, $ID_i(u, v)$ is the number of $v$ by renumbering nodes in $V_i$ along $DHC_i$ from $u$ with $0, \cdots, |V_i| - 1$. Note that $ID_i(\cdot, \cdot)$ gives a relative numbering of the nodes in $V_i$ only. For any node $u, v \in V_i$, $offset_i(u, v) = dist_i(u, v) - \gamma \cdot ID_i(u, v)$. The $offset_i^*$ is defined to be $\min_{u,v \in V_i} offset_i(u, v)$.

**Definition 3** *A partition of a node set $V$ into $V_1 \cdots V_\gamma$ is said to be a good partition within $c$ if there exist constant $c \geq 0$ such that $offset_i^* \geq -c$ for all $i = 1, \cdots, \gamma$.*

When $G$ has a good partition within some constant $c$, MIHC algorithm is performed in $\eta'$-phases. Let $S_i$ be an arbitrary node in $V_i$. In each pahse $j$ $(0 \leq j \leq \eta' - 1)$, every node $v \in V_i$ such that $(ID_i(S_i, v) \bmod \eta') = j$ is permitted to initiate a message along $DHC_i$.

**MIHC algorithm for the $k$-th node $v \in V_i$ along $DHC_i$.**
1. Assume that $G$ has a good partition $P = (V_1, \cdots, V_\gamma)$ within a constant $c$.
2. Let $\eta' = \lceil \frac{\eta+c}{\gamma} \rceil$ and let $S_i$ be an arbitrary node in $V_i$.
3. IHC$(\eta')$ with respect to $DHC_i$
    for $j = 1$ to $\eta' - 1$
        for $t = 1$ to $N + L - 1$
            if $(1 \leq t \leq L$ and $v \in V_i$ and $(ID_i(S_i, v) \bmod \eta') = j)$
                send $t$-th flit of a message along $DHC_i$;
            receive a flit of message from $(k - 1)$-th node;
            send the flit received from $(k - 1)$-th node to $(k + 1)$-th node;
    **endMIHC**

**Theorem 2** *During the execution of MIHC algorithm, no messages are blocked.*

**Proof:** Consider the two nodes $u$ and $v$ in $V_i$ that initiate messages in a phase. Then, we can easily know that $ID_i(u, v) \geq \eta'$. As mentioned in the previous section, it suffices to show that the distance between $u$ and $v$, $dist_i(u, v)$, is at least $\eta$. Since $offset_i(u, v) \geq -c$, it is true that $dist_i(u, v) \geq \gamma \cdot ID_i(u, v) - c \geq \gamma \cdot \eta' - c \geq \gamma \cdot \lceil \frac{\eta+c}{\gamma} \rceil - c \geq \eta$. $\square$

In the model of [6], if $c$ is a small constant, MIHC algorithm is about $\gamma$-times faster than IHC algorithm. In the theoretical model of [7], if $c$ is a small constant, we can show that MIHC algorithm in 3-dimensional meshes is asymptotically optimal (with small constant). In the next subsection, we will show that 3-dimensional meshes have a good partition within small $c$.

## 4.3 MIHC algorithm in 3-dimensional meshes

Let $M_3$ be a $N_1 \times N_2 \times N_3$ 3-dimensional meshes, where $N_i (\geq 6)$ is even for $i = 1, 2, 3$. In this subsection, we show that $M_3$ has a good partition within small constant $c$. Our partition method is to divide a node set $V$ of $M_3$ into 3 sets $V_1, V_2, V_3$ such that $V_i = \{(a_1, a_2, a_3) \in V \mid (\sum a_i) \bmod 3 = (i - 1)\}$, and will be shown to be a good partition within $c \leq 18$. As stated in the previous subsection, nodes in $V_i$ is associated with $DHC_i^3$. Since $DHC_i^3$'s are symmetric, we consider only $offset_{i1}^*$ with $V_1$ and $DHC_1^3$. Now we hope to prove the following theorem.

**Theorem 3** *The above partition for $M_3$ is a good partition within $c = 18$.*

By definition of the offset, we can easily know the two properties: (1) if $z$ is on the path from $x$ to $y$ for $x \neq y$ and $x, y, z \in V_1$, $offset_i(x, y) = offset_i(x, z) + offset_i(z, y)$. (2) Since $|V_1| \leq \lceil \frac{N}{3} \rceil$, $offset_i(x, y) + offset_i(y, x) \geq N - 3 \cdot \lceil \frac{N}{3} \rceil \geq -2$ for $x, y$ in $V_1$. Therefore, $offset_i(x, y) \geq -(offset_i(y, x) + 2)$.

Let $s$ be a node in $V_1$ and let $u, v$ be nodes in $V_1$ such that $offset_{i1}^* = offset_i(u, v)$. First we consider a case when $s$ is not on the path from $u$ to $v$. Then, $offset_i(u, v) = offset_i(s, v) - offset_i(s, u) \geq \min_{w \in V_1} offset_i(s, w) - \max_{w \in V_1} offset_i(s, w)$. Second we consider a case when $s$ is on the path from $u$ to $v$. Then, $offset_i(u, v) = offset_i(u, s) + offset_i(s, v) \geq -(offset_i(s, u) + 2) + offset_i(s, v) \geq -\max_{w \in V_1} offset_i(s, w) + \min_{w \in V_1} offset_i(s, w) - 2$. The first inequality is due to property 2. From both cases, we conclude that $offset_{i1}^* \geq \min_{w \in V_1} offset_i(s, w) - \max_{w \in V_1} offset_i(s, w) - 2$. In Lemma 8, we prove that $\min_{v \in V_1} offset_i(s, v) \geq -8$ and that $\max_{v \in V_1} offset_i(s, v) \leq 8$ for some node $s \in V_1$. This directly implies the theorem.

We call the submesh $M_3[S_3 = \{j\}]$ of $M_3$ as $j$-th *layer*. Recall that $DHC_1^3$ is constructed (see Figure 3) by moving down from layer 0 to the last layer $N - 1$, and then repeating the procedure from layer $N - 1$ that is moving around $DHC_{1,j}^2$ in the layer $j$ and moving up to the next layer $j - 1$. Let $a_j(b_j)$ be the firstly(lastly) visited node of $V_1$ in layer $j$ when moving around the layer $j$. The following three basic lemmas are needed for proving Lemma 7, but their proofs are ommited due to the limited space.

**Lemma 4** *For any node $v \in V_i$ in layer $j$, $-3 \leq offset_i(a_j, v) \leq 3$.*

**Lemma 5** *For any $j$, $\sum_{k=0}^{2} offset_i(a_{j+k}, b_{j+k}) = 0$.*

**Lemma 6** *For any $j$ and $l$,*
$\sum_{k=0}^{5} offset_i(b_{j+k}, a_{j+k-1}) = 0$ *and* $-2 \leq \sum_{k=0}^{l} offset_i(b_{j+k}, a_{j+k-1}) \leq 2$.

**Lemma 7** *Let* $s = a_{N_3-1}$. *Then* $\min\limits_{v \in V_1} \text{offset}_i(s,v) \geq -8$ *and* $\max\limits_{v \in V_1} \text{offset}_i(s,v) \leq 8$.

**Proof:** Let $v$ be a node in layer $k$. $\text{offset}_i(s,v) = \sum_{j=N_3-1}^{k+1} \text{offset}_i(a_j, b_j) + \sum_{j=N_3-1}^{k+1} \text{offset}_i(b_j, a_{j-1}) + \text{offset}_i(a_k, v)$. By Lemma 5, $\text{offset}_i(a_k, v) \geq -3$. For consecutive $l \geq 0$ layers from layer $j$, $\sum_{k=0}^{l} \text{offset}_i(a_{j+k}, b_{j+k}) \geq -3$ by Lemma 5 and 6. So, $\sum_{j=N_3-1}^{k+1} \text{offset}_i(a_j, b_j) \geq -3$. By Lemma 7, $\sum_{j=N_3-1}^{k+1} \text{offset}_i(b_j, a_{j-1}) \geq -2$. Therefore, $\min_{v \in V_1} \text{offset}_i(s,v) \geq -8$. With similar arguments, we can also prove that $\max_{v \in V_1} \text{offset}_i(s,v) \leq 8$. $\square$

# 5 Concluding Remarks

In this paper, we construct a directed hamiltonian packing of $d$-dimensional $N_1 \times \cdots \times N_d$ meshes, where $N_i (\geq 2d)$ is even. As its application, we also present a time-efficient all-to-all broadcasting algorithm in 3-dimensional $N_1 \times N_2 \times N_3$ mesh, where $N_i (\geq 6)$ is even. In our current state, we do not know whether $N_1 \times N_2 \times N_3$ mesh has a directed hamiltonian packing when at least one of $N_i$'s is odd. However, we think that the case is practically meaningless because almost multicomputers adopting the mesh topology have sides of length $2^k$.

# References

1. Jujaj Bosak, *Decompositions of graphs*, Kluwer Academic Publishers, 1990.
2. Richard Stong, "Hamilton decompositions of cartesian products of graphs", *Discrete Mathematics 90*, pp.169-190, 1991.
3. Xiaola Lin and Lionel M. Ni "Deadlock-free multicast wormhole routing in multi-computer networks", *Symposium on Computer Architecture*, pp.116-125, 1991.
4. Sunggu Lee and Kang G. Shin, "Interleaved all-to-all reliable broadcast on meshes and hypercubes", *IEEE Trans. Parallel and Dist. Comp.*, vol.5 no.5, pp.449–458, May 1994.
5. Xiaola Lin, Philip K. McKinley and Lionel M. Ni "Performance evaluation of multicast wormhole routing in 2D-mesh multicomputers", *Proc. International Conference on Parallel Processing*, vol.1, pp.435-442, 1991.
6. Lionel M. Ni and Philip K. McKinley, "A survey of wormhole routing technique in direct networks", *IEEE Computer*, pp.62–76, Feb., 1993.
7. S. Felperin, P. Raghavan, and E. Upfal, "A Theory of wormhole routing in parallel computers", *Proc. 33rd IEEE Foundations of Computer Science*, pp.563–572, 1992.
8. S. T. Johnsson and C.-T. Ho, "Optimum broadcasting and personalized communications in hypercubes", *IEEE Trans. on Computers*, 38(9) pp.1249-1268, 1989.
9. D. S. Scott, "Efficient all-to-all communication patterns in hypercube and mesh topologies", Proc. of 6th Distributed Memory Computing Conference, pp398–403, 1991.

# $k$-Pairs Non-Crossing Shortest Paths in a Simple Polygon

Evanthia Papadopoulou

Northwestern University, Evanston, Illinois 60208, USA

**Abstract.** This paper presents an $O(n + k)$ time algorithm to compute the set of $k$ *non-crossing* shortest paths between $k$ source-destination pairs of points on the boundary of a simple polygon of $n$ vertices. Paths are allowed to overlap but are not allowed to cross in the plane. A byproduct of this result is an $O(n)$ time algorithm to compute a balanced geodesic triangulation which is easy to implement. The algorithm extends to a simple polygon with one hole where source destination pairs may appear on both the inner and outer boundary of the polygon. In the latter case, the goal is to compute a collection of non-crossing paths of minimum total cost.

## 1 Introduction

The *k-pairs non-crossing shortest path problem* in a simple polygon is defined as follows. Given a simple polygon $P$ of $n$ vertices and $k$ source-destination pairs of points $(s_i, t_i), i = 1, \ldots, k$, along the boundary of $P$, find the collection of $k$ *non-crossing* shortest paths connecting every pair $(s_i, t_i)$ that lie entirely in $P$. *Non-crossing* paths are allowed to overlap i.e., share vertices or edges, but they are not allowed to cross each other. Note that if the source-destination pairs are arbitrarily located on the polygon boundary there need not be any set of non-crossing paths between them. In the case where $P$ contains a hole and the $k$ source-destination pairs of points may appear in both the outer and the inner boundary of $P$ the problem is to compute the collection of $k$ *non-crossing* paths whose total length is minimum.

The *non-crossing* shortest path problem was introduced by D.T. Lee in [7]. In [12] Takahashi *et al* considered the following problem: Given an undirected plane graph with nonnegative edge lengths, and $k$ terminal pairs on two specified face boundaries, find $k$ *non-crossing* paths in $G$, each connecting a terminal pair, whose total length is minimum. They presented an $O(n \log n)$ time algorithm, where $n$ is the total number of vertices in the graph (including terminal pairs), using divide and conquer. In a subsequent paper [11, 10], Takahashi *et al* consider the non crossing shortest path problem in a polygonal domain of $n$ rectangular obstacles and the $L_1$ metric and provide an $O((n + k) \log(n + k))$ time algorithm. Motivation for non-crossing shortest path problems comes from VLSI layout design (see [7, 10, 12, 11]).

In the case of a simple polygon or a simple polygon with one hole, we can obtain an $O(k + n \log k)$ time algorithm using a divide and conquer approach similar to [12, 11] and the $O(n)$ time algorithm of Lee and Preparata[8] for the single pair shortest path problem in a simple polygon. Alternatively, for a simple polygon without holes, the problem can be solved by answering $k$ shortest path queries using the data structure of Guibas and Hershberger [4]. The data structure of [4] can be constructed in $O(n)$ time and space and can answer shortest path queries between pairs of points in a simple polygon in $O(\log n + l)$ time where $l$ is the number of links of the shortest path.

In this paper we present an $O(n + k)$ time algorithm for the $k$ pairs non-crossing shortest path problem in a simple polygon and a polygon with one hole. The collection of non-crossing shortest paths is organized in a forest that allows shortest path queries between any given pair to be answered in time proportional to the size of the path. We also make an observation that can reduce the time complexity of the algorithm of Takahashi *et al* [11, 10] from $O((n + k)\log(n + k))$ to $O(k + n\log n)$. (Note that $k$ can be large compared to $n$.) A byproduct of our algorithm is a simple $O(n)$ method for computing a *balanced geodesic triangulation* of a simple polygon. A *balanced geodesic triangulation* is a decomposition of a simple polygon into triangle-like regions, called *geodesic triangles*, with the property that any line segment interior to $P$ crosses only $O(\log n)$ geodesic triangles [2, 3]. Such a decomposition finds often application in problems involving simple polygons e.g., ray shooting or shortest path queries [2, 3] and can be computed in $O(n)$ time [2]. The idea is to compute the collection of shortest paths between the following pairs of vertices $(v_1, v_{n/3}), (v_{n/3}, v_{2n/3}), (v_{2n/3}, v_1), (v_1, v_{n/6}), (v_{n/6}, v_{n/3}), (v_{n/3}, v_{n/2})$, etc. until a pair consists of consecutive vertices. Our $O(n)$ time algorithm for computing the collection of non-crossing shortest paths among the above $O(n)$ pairs of vertices gives an alternative method for computing a balanced geodesic triangulation which is simple to implement. In general, a collection of non-crossing shortest paths between pairs of points on the boundary of simple polygon can be used to obtain useful polygon decompositions by including in the decomposition shortest paths between certain boundary points.

## 2 Preliminaries

Let $P$ be a simple polygon and let $\partial P$ be its boundary. Given two points $x, y \in P$, $\pi(x, y)$ denotes the shortest path from $x$ to $y$ that lies entirely in $P$. $\pi(x, y)$ is assumed to be directed from $x$ to $y$. We are given a set of $k$ source-destination pairs along $\partial P$, $\mathcal{ST} = \{(s_i, t_i), 1 \le i \le k\}$. We shall adopt the convention that sources $s_i, 1 \le i \le k$, are ordered clockwise along $\partial P$ and that for any pair, $s_i$ appears before $t_i$ when we traverse $\partial P$ clockwise starting at $s_1$. Source $s_1$ is regarded as the starting point along $\partial P$. Unless otherwise noted, we assume a clockwise traversal of $\partial P$ starting at $s_1$.

Let $\partial P$ be mapped onto a unit circle. Then the $k$ source-destination pairs of points get mapped to $k$ circle chords, $\overline{s_i t_i}, i = 1, \ldots, k$. Figure 1 shows the mapping. If chord $\overline{s_i t_i}$ intersects $\overline{s_j t_j}$ then clearly $\pi(s_i, t_i)$ and $\pi(s_j, t_j)$ must cross each other. We say that pairs $(s_i, t_i)$ and $(s_j, t_j)$ are *non-crossing*, if and only if the corresponding chords in the unit circle $\overline{s_i t_i}$ and $\overline{s_j t_j}$ do not intersect. For example, in figure 1, pairs $(s_1, t_1), (s_2, t_2), (s_3, t_3)$ and $(s_5, t_5)$ are non-crossing while pairs $(s_4, t_4)$ and $(s_5, t_5)$ are crossing. It is not difficult to see that shortest paths between two pairs of points in $\mathcal{ST}$ are non-crossing if and only if the pairs are themselves non-crossing.

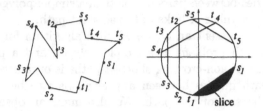

**Fig. 1.** Mapping of $k$ source-destination pairs of vertices into $k$ chords on the unit circle.

**Lemma 1.** *Given a simple polygon $P$ of $n$ vertices and $k$ source-destination pairs along $\partial P$, $(s_i, t_i)$, $1, \ldots, k$, detecting if any two shortest paths connecting $s_i$ to $t_i$ cross each other can be done in $O(n + k)$ time.*

Given the unit circle representing $\partial P$ and a pair $(s_i, t_i)$, the region of the unit circle enclosed by the chord $\overline{s_i t_i}$ and the arc from $s_i$ to $t_i$ is referred to as a *slice* and is denoted as $sl(s_i, t_i)$. (A clockwise traversal is assumed.) For the pair $(s_1, t_1)$ we define two slices, $sl(s_1, t_1)$ and $sl'(s_1, t_1)$, where $sl'(s_1, t_1)$ is the complement of $sl(s_1, t_1)$ i.e., the region of the unit circle enclosed by the chord $\overline{s_i t_i}$ and the arc from $t_1$ to $s_1$. In figure 1, $sl(s_1, t_1)$ is shown shaded. Note that paths $\pi(s_i, t_i)$ and $\pi(s_j, t_j)$ are *non-crossing* if and only if the corresponding slices, $sl(s_i, t_i)$ and $sl(s_j, t_j)$, do not intersect or one of them totally contains the other. In the former case, the two non-crossing pairs are said to be in *series*, and in the latter, they are said to be in *parallel*. The destination points of pairs that are in parallel are in reverse order as the source points. Figure 2 shows an example of a set of non-crossing slices; slices $sl(s_1, t_1), sl(s_3, t_3), sl(s_6, t_6)$ are in series while $sl(s_1, t_1), sl(s_2, t_2)$ are in parallel.

Consider now a tree, $T_{sl}$, representing the set of non-crossing slices where the root corresponds to the unit disk and each node represents a slice. The children of the root are $sl(s_1, t_1)$ and $sl'(s_1, t_1)$. Slices that are totally contained in $sl(s_i, t_i)$ are descendants of $sl(s_i, t_i)$ in the tree. The immediate children of $sl(s_i, t_i)$ (resp. $sl'(s_1, t_1)$) are those slices in series that are totally contained in $sl(s_i, t_i)$ (resp. $sl'(s_1, t_1)$) but are not contained in any other descendent of $sl(s_i, t_i)$. Figure 2 shows the tree representation of the corresponding slices. The nodes are labeled by the slice number; slice 0 refers to the unit disk. Note that $T_{sl}$ may be balanced or completely unbalanced.

Let $R$ be the set of all points in $\mathcal{ST}$. We assume that $R$ is ordered according to a clockwise traversal of $\partial P$ starting at $r_1 = s_1$. Given any two points $x, y$ along $\partial P$, we say that $y$ follows $x$, and denote as $x < y$, if and only if $y$ follows $x$ in a clockwise traversal of the polygon boundary starting at $s_1$. Given $R' \subseteq R$

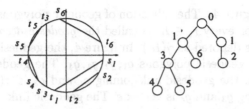

**Fig. 2.** Non-crossing slices and the tree representation.

and $r_j \in R'$, let $n_{R'}(r_j)$ denote the point in $R'$ following $r_j$ and $p_{R'}(r_j)$ denote the point in $R'$ preceding $r_j$. For $R' = R$, $n_R(r_j) = r_{j+1}$ and $p_R(r_j) = r_{j-1}$. Let $R' \subseteq R$ be such that the shortest paths between any two pairs of distinct points are disjoint. The area enclosed by the collection of $\pi(r_j, n_{R'}(r_j))$ for every $r_j \in R'$ is called the *geodesic polygon* induced by $R'$. In figure 3, the shaded area is the geodesic polygon induced by $R' = \{r_1, r_4, r_6, r_7, r_{10}\}$. The last common vertex along $\pi(r_j, n_{R'}(r_j))$ and $\pi(r_j, p_{R'}(r_j))$ is an *apex* of the geodesic polygon, and is denoted as $\alpha(r_j)$. If $r_j$ is the only common point of $\pi(r_j, n_{R'}(r_j))$ and $\pi(r_j, p_{R'}(r_j))$ then $\alpha(r_j) = r_j$. In figure 3, $\alpha(r_4) = a$ and $\alpha(r_1) = r_1$. The common part of $\pi(r_j, n_{R'}(r_j))$ and $\pi(r_j, p_{R'}(r_j))$ (i.e., $\pi(r_j, \alpha(r_j))$) is referred to as the *geodesic link* of $r_j$. Two consecutive apexes of the geodesic polygon, $\alpha(r_j)$ and $\alpha(n_{R'}(r_j))$, are joined by their shortest path, $\pi(\alpha(r_j), \alpha(r_{j+1}))$, which is clearly a convex chain with convexity facing towards the interior of the geodesic polygon.

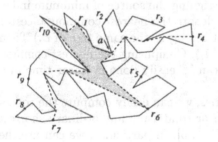

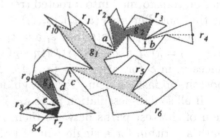

**Fig. 3.** The geodesic polygon induced by $R' = \{r_1, r_4, r_6, r_7, r_{10}\}$.

**Fig. 4.** The geodesic decomposition of $R = \{r_1, \ldots, r_{10}\}$.

If $R'$ contains pairs of distinct points whose shortest paths share some common part, the collection of $\pi(r_j, n_{R'}(r_j))$ induces more than one geodesic polygon in $P$. In figure 4, set $R = \{r_1, \ldots, r_{10}\}$ induces four geodesic polygons $g_1, g_2, g_3, g_4$, where $g_1 = \{r_1, a, r_5, r_6, c, r_{10}\}$, $g_2 = \{a, r_2, r_3, b\}$, $g_3 = \{d, e, r_9\}$, and $g_4 = \{e, r_7, r_8\}$. Two paths $\pi(r_i, n_{R'}(r_i))$ and $\pi(r_j, n_{R'}(r_j))$, $j > i + 1$, that are not disjoint share a subpath of at least one vertex which is referred to as a *geodesic link*. For example in figure 4, segment $\overline{cd}$ is common to $\pi(r_6, r_7)$ and $\pi(r_9, r_{10})$ i.e., $\overline{cd}$ is a geodesic link. Note that a geodesic link may consist of a single vertex in which case the incident geodesic polygons have a common apex (e.g., vertex $a$ in figure 4). The collection of geodesic polygons and links induced by $\pi(r_i, n_{R'}(r_i))$ for every $r_i \in R'$ is called the *geodesic decomposition* of $P$ induced by $R'$ and is denoted as $\gamma(R')$. In figure 4, the geodesic decomposition of $R$ consists of four geodesic polygons $g_1, g_2, g_3, g_4$, The geodesic link joining $g_1$ and $g_2$ is vertex $a$, the geodesic link joining $g_1$ and $g_3$ is $\pi(c, d) = \overline{cd}$ and the geodesic link joining $g_3$ and $g_4$ is vertex $e$. The geodesic link of $r_4$ is $\pi(r_4, b)$. For every $r_j, j \neq 4$, $\alpha(r_j) = r_j$, and $\alpha(r_4) = b$.

Apex $\alpha(r_1)$ is considered to be the root of the geodesic decomposition. This induces a unique directed path from the root to every geodesic polygon which defines a parent-child relation between the geodesic polygons. The apex of a geodesic polygon $g$ incident to its parent is called the *main apex* of $g$ and is denoted as $\alpha(g)$. Given a geodesic polygon $g$ and $r_j \in R'$ let $\alpha(r_j, g)$ be the apex of $g$ that belongs to $\pi(\alpha(g), r_j)$. (If $g$ contains $\alpha(r_j)$ then $\alpha(r_j, g) = \alpha(r_j)$). In figure 4, $\alpha(r_4, g_1) = a$ and $\alpha(r_4, g_3) = d$.

Given a vertex $v$ along a convex chain $C$ and a point $p$, we say that $v$ is *supporting* (with respect to $\overline{pv}$) or that segment $\overline{pv}$ is supporting to $C$ if the line containing $\overline{pv}$ is tangent to $C$.

## 3 The Algorithm

The collection of shortest paths between every pair of points in $\mathcal{ST}$ (given that $\mathcal{ST}$ is non-crossing) forms a forest denoted as $\mathcal{E}$. Our algorithm computes $\mathcal{E}$ in $O(n + k)$ time and processes it to answer shortest path queries between any pair in $\mathcal{ST}$ in time proportional to the length of the path. In particular, any tree of $\mathcal{E}$ can be transformed into a rooted tree by assigning the source of minimum index as the root. Let $nca_{\mathcal{E}}(s_i, t_i), (s_i, t_i) \in \mathcal{ST}$, denote the nearest common ancestor of $s_i$ and $t_i$ in $\mathcal{E}$. (Note that $s_i$ and $t_i$ must belong to the same tree of $\mathcal{E}$.) Then $\pi(s_i, t_i) = \pi(s_i, nca_{\mathcal{E}}(s_i, t_i) \cup \pi(nca_{\mathcal{E}}(s_i, t_i), t_i)$. Computing the nearest common ancestor in $\mathcal{E}$ for every pair $(s_i, t_i) \in \mathcal{ST}$ can be easily done in linear time in a bottom-up fashion using $T_{sl}$ as a guide.

If all source-destination pairs are in series, we can easily compute the collection of shortest paths between them in linear time using any ordinary shortest path algorithm for a single pair of points[8, 1, 5]. In particular, we can use the Lee-Preparata algorithm [8] for each pair independently. The following lemma assures that the time complexity remains linear.

**Lemma 2.** *The collection of non-crossing shortest paths between $k$ source-destination pairs of points on $\partial P$ can be found in linear time if the pairs are in series.*

In general, we compute $\mathcal{E}$ in a bottom-up fashion using the tree of slices, $T_{sl}$, as a guide. In phase 1, we compute the geodesic decomposition of $P$ induced by $R$. This reveals the collection of $\pi(s_i, t_i)$ for every pair $(s_i, t_i)$ at a leaf node of $T_{sl}$. For any pair $(s_i, t_i)$ at the bottom level of $T_{sl}$, we color all edges along $\pi(s_i, t_i)$ red, and add them to $\mathcal{E}$. The remaining edges of the geodesic decomposition are assumed to be white. Let $R^q$, for $1 \leq q \leq h$, where $h$ is the height of $T_{sl}$, denote the set of points appearing at levels 1 to $(h - q + 1)$ in $T_{sl}$. Note that $R^1 = R$ and

$R^h = \{s_1, t_1\}$. In phase $q$, $1 < q \leq h$, we compute the geodesic decomposition of $P$ induced by $R^q$ which reveals the collection of $\pi(s_i, t_i)$ for every pair $(s_i, t_i)$ at level $(h - q + 1)$ of $T_{sl}$. Edges of the decomposition at phase $q$ are either red or white. Red edges are those that have been added to $\mathcal{E}$ in some previous phase. At the end of phase $q$, we add all white edges along $\pi(s_i, t_i)$ to $\mathcal{E}$ and color them red, for every pair $(s_i, t_i)$ at level $(h - q + 1)$ of $T_{sl}$.

At every phase $q$, $1 \leq q \leq h$, the geodesic decomposition of $R^q$ is maintained. Geodesic polygons and links are kept as a doubly linked list of their vertices. In general, an apex is adjacent to two vertices in its geodesic polygon and one vertex in the incident geodesic link. An exception are apexes that are common to two geodesic polygons (e.g. apex $a$ in figure 4), which are incident to two vertices in each of the incident geodesic polygons. Edges of geodesic polygons and geodesic links are marked as red or white, where all red edges have been included in $\mathcal{E}$. In order to have the ability to extract the white edges of $\pi(s_i, t_i)$ for some pair $(s_i, t_i)$ at level $(h - q + 1)$ of $T_{sl}$ without visiting the red edges along $\pi(s_i, t_i)$, we also maintain a doubly linked list of the polygon vertices incident to white edges, referred to as the *white-list*. The white-list includes $\alpha(r_j)$ for every $r_j \in R^q$. Two vertices $v, u$ in the white-list are joined by a link if and only if $v$ and $u$ belong to the shortest path from $\alpha(r_j)$ to $\alpha(n_{R^q}(r_j))$ and either $\overline{vu}$ is a white edge or $\pi(v, u)$ is a maximal red subpath of $\pi(\alpha(r_j), \alpha(n_{R^q}(r_j)))$. Every link of the white-list corresponds either to a white edge of the decomposition, referred to as a *white link*, or to a maximal sequence of red edges, referred to as a *red link*. Note that a white edge along a geodesic link appears twice in the white-list and thus each of the incident vertices also appears twice.

Let's consider the geodesic decomposition of phase 1, i.e., the geodesic decomposition of $R$. Pairs $(r_i, r_{i+1})$, $1 \leq i \leq 2k$, are clearly in series and therefore the collection of $\pi(r_i, r_{i+1})$ can be computed in $O(n + k)$ time (lemma 2). Once $\pi(r_i, r_{i+1})$ for every $i$, $1 \leq i \leq 2k$, have been computed, the geodesic decomposition of $P$ can be easily obtained in linear time. For every pair of points $(s_i, t_i)$ at the bottom level of $T_{sl}$, we color edges along $\pi(s_i, t_i)$ red and add them to $\mathcal{E}$. We also build the white list by a simple scan starting at $\alpha(r_1)$.

To compute the geodesic decomposition of $R^q$, $q > 1$, we use the geodesic decomposition of $R^{q-1}$. To facilitate updating we use the shortest path tree from $s_1$ to all points in $R$. Computing the shortest path tree from $s_1$ can be done in linear time using the algorithm of Guibas et al [5] or the simpler to implement algorithm of Hershberger and Snoeyink [6]. Given two points $r_i, r_j \in R$, the nearest common ancestor of $r_i$ and $r_j$ in the shortest path tree from $s_1$ is denoted as $nca(r_i, r_j)$. Suppose that the geodesic decomposition of $P$ induced by $R^{q-1}$ has been computed, and that we want to compute the geodesic decomposition induced by $R^q$. Let $R'^q = R^{q-1} - R^q$. For brevity, we will use $n(r_j)$ to denote $n_{R^q}(r_j)$, $p(r_j)$ to denote $p_{R^q}(r_j)$, and $\gamma$ to denote the geodesic decomposition during phase $q$. At the end of phase $q$, $\gamma = \gamma(R^q)$.

Consider $r_j \in R^q$ such that $n(r_j) \neq n_{R^{q-1}}(r_j)$. Since $r_j$ and $n(r_j)$ are neighboring points in $R^q$, we need to compute $\pi(\alpha(r_j), \alpha(n(r_j)))$ and update the geodesic decomposition. For this purpose we incrementally compute $\pi(\alpha(r_j), \alpha(r_i))$

for every $r_i \in R'^q \cup \{n(r_j)\}$ with $r_j < r_i \le n(r_j)$, updating at the same time the geodesic decomposition. Suppose that $\pi(\alpha(r_j), \alpha(r_i))$ has been computed for some $r_i \in R'^q, r_j < r_i < n(r_j)$, and that the geodesic decomposition has been updated accordingly. i.e., $\gamma = \gamma(R^q \cup \{r \in R^{q-1}, r \ge r_i\})$. It is enough to update the geodesic polygon $g$ that contains $\alpha(r_i)$ according to $\pi(\alpha(r_j, g), \alpha(r_l, g))$.

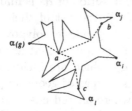
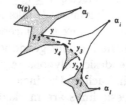

**Fig. 5.** $\gamma(\alpha(g), \alpha_j, \alpha_i, \alpha_l)$        **Fig. 6.** The update of $g$ in case 3.

Let $\alpha_j, \alpha_i$ and $\alpha_l$ denote $\alpha(r_j, g), \alpha(r_i)$ and $\alpha(r_l, g)$ respectively. Note that $\alpha_j$ or $\alpha_l$ may coincide with the main apex of $g$. Consider the geodesic decomposition induced by $\{\alpha(g), \alpha_j, \alpha_i, \alpha_l\}$, denoted as $\gamma(\alpha(g), \alpha_j, \alpha_i, \alpha_l)$. If $\alpha(g) = \alpha_j$ or $\alpha(g) = \alpha_l$, $\gamma(\alpha(g), \alpha_j, \alpha_i, \alpha_l)$ must form a geodesic triangle (i.e., have three apexes). Otherwise, $\gamma(\alpha(g), \alpha_j, \alpha_i, \alpha_l)$ forms either a geodesic quadrilateral (four apexes, figure 5) or two geodesic triangles (figure 7). In any case, let $g(a_i)$ denote the geodesic polygon of $\gamma(\alpha(g), \alpha_j, \alpha_i, \alpha_l)$ containing $\alpha_i$.

Let $a, b$ and $c$ denote the apexes of $\gamma(\alpha(g), \alpha_j, \alpha_i, \alpha_l)$, other than $\alpha_i$, such that $a$ is the last common vertex along $\pi(\alpha(g), \alpha_j)$ and $\pi(\alpha(g), \alpha_l)$, $b$ is the last common vertex along $\pi(\alpha_j, \alpha(g))$ and $\pi(\alpha_j, \alpha_i)$, and $c$ is the last common vertex along $\pi(\alpha_l, \alpha_i)$ and $\pi(\alpha_l, \alpha(g))$. If $\alpha(g) = a_j$ (resp. $\alpha(g) = a_i$) then $a = \alpha(g) = b$ (resp. $a = \alpha(g) = c$). If $g(\alpha_i)$ is a geodesic triangle (see figure 7), let $a'$ denote the main apex of $g(\alpha_i)$. Note that $a' = nca(\alpha_j, \alpha_i)$ or $a' = nca(\alpha_i, \alpha_l)$. Furthermore, if $g(\alpha_i)$ is a geodesic quadrilateral then $a = nca(r_j, r_l)$. To update $g$, we need to compute $\pi(\alpha_j, \alpha_l)$ i.e., it is enough to compute $\pi(b, c)$.

If $g(\alpha_i)$ is a geodesic quadrilateral (see figure 5) we basically need to compute segment $\overline{yz}$ where $y \in \pi(\alpha_i, b) \cup \pi(b, a)$ and $z \in \pi(\alpha_i, c) \cup \pi(c, a)$ such that $y$ and $z$ are supporting and $\overline{yz}$ lies entirely in $P$ (see figure 8). Then $\pi(b, c) = \pi(b, y) \cup \overline{yz} \cup \pi(z, c)$, where $\pi(b, y)$ and $\pi(c, z)$ are both known. There are four possible cases for segment $\overline{yz}$ (see figure 8). *Case 1:* $y \in \pi(\alpha_i, b)$ and $z \in \pi(\alpha_i, c)$ (figure 8(a)). *Case 2:* $y \in \pi(b, a)$ and $z \in \pi(\alpha_i, c)$, $y \ne b$ (figure 8(b)). *Case 3:* $y \in \pi(\alpha_i, b)$ and $z \in \pi(c, a)$, $z \ne c$ (figure 8(c)). *Case 4:* $y \in \pi(b, a)$ and $z \in \pi(c, a)$, $y \ne b$, $z \ne c$ (figure 8(d)).

If $g(\alpha_i)$ is a geodesic triangle then $\pi(\alpha_j, \alpha_l)$ is already known. (Recall that $\pi(\alpha(g), a_j)$ and $\pi(\alpha(g), a_l)$ belong to the shortest path tree from $s_1$. If $a' = nca(\alpha_j, \alpha_i)$, let $z = a'$ and let $y$ be the vertex following $z$ along $\pi(a', b)$. Segment $\overline{yz}$ is supporting to both $\pi(a', b)$ and $\pi(a', \alpha_i)$. Since $y \in \pi(a, b)$ and $z \in \pi(\alpha_i, c)$, this can be regarded as case (2) for $y \ne b$ or case (1) for $y = b$. If $a' = nca(\alpha_l, \alpha_i)$, let $y = a'$ and $z$ be the following vertex along $\pi(a', c)$. Since $y \in \pi(b, \alpha_i)$ and $z \in \pi(c, a)$ this can be regarded as case (3) for $z \ne c$ or case (1) for $z = c$.

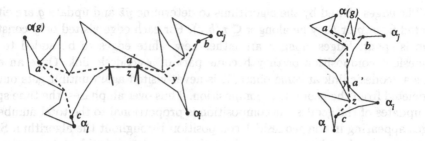

**Fig. 7.** $g(\alpha_i)$ is a geodesic triangle.

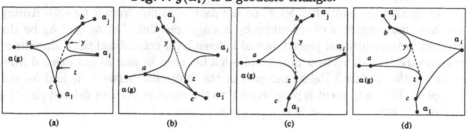

(a)  (b)  (c)  (d)

**Fig. 8.** The four cases for segment $\overline{yz}$.

To determine which of the four cases is the one occurring and determine vertices $y$ and $z$, we advance a variable vertex $v$ along $\pi(\alpha_i, b) \cup \pi(b, a)$ and a variable vertex $u$ along $\pi(\alpha_i, c) \cup \pi(c, a)$ until $v = y$ and $u = z$ ($v$ and $u$ start at $\alpha_i$). The problem is that vertices $a, b$ and $c$ as well as apexes $\alpha_j$ and $\alpha_l$ and $\alpha(g)$ are not known in advance. However, useful information can be obtained while advancing $v$ and $u$ using the shortest path tree from $s_1$. In particular, we can determine that $v = b$ (resp. $u = c$) due to the following property: $v = b$, for $v \in \pi(\alpha_i, \alpha_j)$, (resp. $u = c$, $u \in \pi(\alpha_i, \alpha_l)$) if and only if the predecessor of $v$ (resp. $u$) on the shortest path from $s_1$, denoted as $pred(v)$, is supporting with respect to $\overline{pred(v)v}$ (resp. $\overline{pred(u)u}$) and $pred(v) \notin \pi(\alpha_i, \alpha_j)$ (resp. $pred(u) \notin \pi(\alpha_i, \alpha_l)$). Vertex $a$ needs to be known only in case 4 in which case $a = nca(r_j, r_l)$. Segment $\overline{vu}$ is advanced so that it always lies entirely in the interior of $g(\alpha_i)$. This can be easily enforced due to the following property: $\overline{vu}$ intersects $\pi(a, c)$ (resp. $\pi(a, b)$) if and only if $pred(u)$ (resp. $pred(v)$) lie at the same side of $\overline{vu}$ as $\alpha_i$. If at any point during the advancement it is determined that $\overline{vu}$ intersects $\pi(a, c)$ (resp. $\pi(a, b)$), we have case 3 or case 4 (resp. case 2 or 4).

After segment $\overline{yz}$ has been determined the update of $g$ can be briefly stated as follows: *Case 1*: Remove $\pi(y, \alpha_i) \cup \pi(\alpha_i, z)$ and add $\pi = \overline{yz}$. *Case 2*: Remove $\pi(b, \alpha_i) \cup \pi(\alpha_i, z)$ and add $\pi = \pi(b, y) \cup \overline{yz}$. *Case 3*: Remove $\pi(c, \alpha_i)\pi(\alpha_i, y)$ and add $\pi = \pi(c, z) \cup \overline{yz}$. *Case 4*: Remove $\pi(b, \alpha_i)\pi(\alpha_i, c)$ and add $\pi = \pi(b, y) \cup \overline{yz} \cup \pi(z, c)$. To update $g$ in any of the four cases we walk along the corresponding $\pi \subseteq \pi(b, c)$ starting at vertices $b$ or $c$. (In case 4 we walk along $\pi_1 = \pi(b, y)$ and $\pi_2 = \pi(c, z)$ and then consider segment $\overline{yz}$.) Figure 6 illustrates the update of $g$ for case 3 i.e., $\pi = \pi(c, y)$. Note that edge $\overline{y_2 y_3}$ and vertex $y_4$ become geodesic links.

The edges visited by the algorithms to determine $\overline{yz}$ and update $g$ are either deleted from $g$ or they lie along $\pi \subseteq \pi(b, c)$. For each edge visited only constant time is spent. Edges along $\pi$ are either new white edges to be added to the geodesic decomposition or they become part of a geodesic link. Once an edge joins a geodesic link at some phase, it is never visited again until it gets output or deleted from the geodesic decomposition. Thus over all phases, the time spent for updates of the geodesic decomposition is proportional to the total number of edges appearing in the geodesic decomposition throughout the algorithm. Since no two such edges intersect and since they are always incident to vertices or points in $R$, their number is bounded by $O(n + k)$.

To update the white list, let $\pi'$ be the path obtained from $\pi$ by substituting any maximal sequence of red edges by a single red link. Let $h_1$ and $h_2$ be the vertices in the white-list preceding and following (or coinciding) the endpoints of $\pi$ respectively. (All vertices in the white-list between $h_1$ and $h_2$ got deleted from $\gamma$ during the update.) Delete the part of the white list between $h_1$ and $h_2$ and merge $\pi'$. The time spent is proportional to the number vertices deleted plus the size of $\pi$. Thus,

**Lemma 3.** *The total time spent for updating the geodesic decomposition and the white-list in every phase of the algorithm is $O(n + k)$.*

To complete the update, we need to compute $nca(r_j, r_l)$ since $r_j$ and $r_l$ become neighbors after the deletion of $r_i$. Clearly $nca(r_j, r_l)$ is the one between $nca(r_j, r_i)$ and $nca(r_i, r_l)$ that occurs first along $\pi(s_1, r_i)$ i.e., the one nearest to the root. Assigning a level number to all nodes of the shortest path tree from $s_1$ allows to compute $nca(r_j, r_l)$ from $nca(r_j, r_i)$ and $nca(r_i, r_l)$ in constant time.

After the geodesic decomposition induced by $R^q$ has been computed, we need to extract white edges of $\pi(s_i, t_i)$ for every pair $(s_i, t_i)$ at level $(h - q + 1)$ of $T_{sl}$, add them to $\mathcal{E}$, color them red, and update the white list. Starting at $\alpha(s_i)$ we follow the white list until $\alpha(t_i)$, adding the encountered white edges to $\mathcal{E}$. Note that white edges along $\pi(s_i, t_i)$ that are part of geodesic links appear twice along the white-list and that both occurrences must become red. Finally, substitute $\pi(\alpha(s_i), \alpha(t_i))$ by a single red link between $\alpha(s_i)$ and $\alpha(t_i)$. The time spent is proportional to the number of white edges that get output to $\mathcal{E}$ i.e., become red.

We therefore conclude that $\mathcal{E}$ can be computed in $O(n + k)$ time.

## 4 Concluding Remarks

We have given a simple linear time algorithm for the $k$-pairs non-crossing shortest path problem in a simple polygon $P$. Similarly, to [12, 11, 10], this result can be extended to a simple polygon $P$ with one hole $Q$ where source-destination pairs may appear on both $\partial P$ and $\partial Q$. The main observation is that once a path $\pi_i$ between $s_i \in \partial P$ and $t_i \in \partial Q$ has been routed there is only one way (if any) to route the rest of the pairs. Furthermore, $\pi_i$ can be routed in only two ways: clockwise and counterclockwise around $Q$.

If $P$ is a set of rectangles enclosed by an outer rectangle and points in $\mathcal{ST}$ appear on the outer rectangle and one inner rectangle the shortest non-crossing path problem can be solved in $O((n + k)\log(n + k))$ time as shown in [11, 10] where $n$ is the number or rectangles and $k$ is the number of pairs. Note however that $k$ can be arbitrarily large compared to $n$. Our observation is that $k$ need not play a role in the main part of the algorithm. The idea is as follows: Given $\mathcal{ST}$ and the set of rectangles, derive a set $\mathcal{ST}'$ of $O(\min\{n, k\})$ pairs, solve the problem for $\mathcal{ST}'$ using the algorithm of [11] in $O(n \log n)$ time, and modify the solution in additional $O(k)$ time to derive the set of non-crossing paths for $\mathcal{ST}$. In particular, consider the partition of the outer rectangle into intervals derived by the vertical and horizontal *valid* projections of the inner rectangles. (Valid projection lines do not intersect obstacles.) For $(s_i, t_i) \in \mathcal{ST}$, let $I_{s_i}$ and $I_{t_i}$ denote the intervals on the outer rectangle containing $s_i$ and $t_i$ respectively. Let $a, b$ and $c, d$ be the endpoints of $I_{s_i}$ and $I_{t_i}$. Then $\mathcal{ST}' = \{(I_{s_i}, I_{t_i}), (s_i, t_i) \in \mathcal{ST}\}$, where $(I_{s_i}, I_{t_i}) = \{(a, c), (a, d), (b, c), (b, d)\}$. Computing $\pi(s_i, t_i)$ for $(s_i, t_i) \in \mathcal{ST}$ in [11] can now be substituted by computing $\pi(a, c), \pi(a, d), \pi(b, c)$, and $\pi(b, d)$. The details are left for the full paper.

**Acknowledgment.** I would like to thank professor D.T. Lee for many helpful discussions and comments.

# References

1. B. Chazelle, "A theorem on polygon cutting with applications," *Proc. 23rd Annu. IEEE Sympos. Found. Comput. Sci.*, 1982, pp. 339–349.
2. B. Chazelle, H. Eddelsbrunner, M. Grigni, L. Guibas, J. Hershberger, M. Sharir, and J. Snoeyink, "Ray shooting in polygons using geodesic triangualtions", *Algorithmica*, 12, 1994, 54-68.
3. M. T. Goodrich and R. Tamassia, "Dynamic Ray Shooting and Shortest Paths via Balanced Geodesic Triangualtions", *In Proc. 9th Annu. ACM Sympos. Comput. Geom*, 1993, 318-327.
4. L.J. Guibas and J. Hershberger, "Optimal shortest path queries in a simple polygon", *J. Comput. Syst. Sci.*, 39, 1989, 126-152.
5. L.J. Guibas, J. Hershberger, D. Leven, M. Sharir, and R.E. Tarjan, "Linear-time algorithms for visibility and shortest path problems inside triangulated simple polygons". *Algorithmica*, 2, 209-233, 1987.
6. J. Hershberger and J. Snoeyink, "Computing Minimum Length Paths of a given homotopy class", *Comput. Geometry: Theory and Applications*, 4, 1994, 63-97.
7. D.T. Lee, "Non-crossing paths problems", Manuscript, Dept. of EECS, Northwestern University, 1991.
8. D. T. Lee and F. P. Preparata, "Euclidean Shortest Paths in the Presence of Rectilinear Barriers", *Networks*, 14 1984, 393-410.
9. F.P. Preparata and M. I. Shamos, *Computational Geometry: an Introduction*, Springer-Verlag, New York, NY 1985.
10. J. Takahashi, H. Suzuki, and T. Nishizeki, "Finding shortest non-crossing rectilinear paths in plane regions", *Proc. of ISAAC'93, Lect. Notes in Computer Science, Spinger-Verlag*, 762, 1993, 98-107.
11. J. Takahashi, H. Suzuki, and T. Nishizeki, "Shortest non-crossing rectilinear paths in plane regions", *Algorithmica*, to appear.
12. J. Takahashi, H. Suzuki, and T. Nishizeki, "Shortest non-crossing paths in plane graphs", *Algorithmica*, to appear.

# Minimum Convex Partition of a Polygon with Holes by Cuts in Given Directions

A. Lingas
Dept. Computer Science
Lund University
S-22100 Lund, Sweden
Andrzej.Lingas@dna.lth.se

V. Soltan*
Mathematical Institute
Academy of Sciences
MD-20028 Chişinău, Moldova
17soltan@mathem.moldova.su

### Abstract

Let $\mathcal{F}$ be a given family of directions in the plane. The problem to partition a planar polygon $P$ with holes into a minimum number of convex polygons by cuts in the directions of $\mathcal{F}$ is proved to be NP-hard if $|\mathcal{F}| \geq 3$ and it is shown to admit a polynomial-time algorithm if $|\mathcal{F}| \leq 2$.

## 1 Introduction

The problems of partitioning a planar polygon into convex polygons are well-studied in computational geometry (cf. [4, 11, 13, 20]). They have various applications in computer graphics [19], mesh generation [2], pattern recognition [22, 26], VLSI [1], *etc.*

One of the most studied is the problem of partitioning a polygon into a *minimum number* of convex polygons. If a polygon is simple, i.e., it has no holes, the problem is known to be solvable in time $O(n^3)$, where $n$ is the number of the vertices of the polygon, and it is NP-hard for the case of polygons with holes (see [3] and [15], respectively). As usual, by a *convex partition* of a polygon $P$ (either simple or with holes) we mean a decomposition of $P$ into finitely many closed non-overlapping convex polygons.

There are two variants of this problem, based on certain restrictions on the partition technique. In one of them partitions without introducing additional vertices, called Steiner vertices, are considered (see [7], [10]). This variant of the problem is solvable in time $O(n^3 \log n)$ for the case of simple polygons (see [10]), and it is NP-hard for the case of polygons with holes (see [15]).

The other variant of the problem deals with partitions of polygons by cuts along a given family of directions (e.g., [23, 24]). One of the most known here is the problem of partitioning a rectilinear polygon (with or without holes) into a minimum number of convex pieces by cuts along two orthogonal directions, i.e., partitioning into a minimum number of rectangles. This variant is known to be solvable in time $O(n^{3/2} \log n)$, see [9, 12, 17, 18, 25].

In this connection, one can pose the following questions. Why the minimum convex partition problem by cuts along two directions is polynomially solvable for the case of rectilinear polygons with holes, and the respective problems without restrictions on cutting directions (allowing or disallowing Steiner points) are NP-hard? Does this difference depend on shape of the polygon, or it depends on the family of cutting directions?

To answer these questions, we consider the following problem. Let $\mathcal{F}$ be a given family of non-oriented directions in the plane (in particular, $\mathcal{F}$ can be the family of all directions).

---

*The second author was supported by TFR during his visit to Lund University.

PROBLEM. *Minimum Number Convex Partition in the Directions of* $\mathcal{F}$ (MNCP($\mathcal{F}$) for short). Partition a polygon $P$ with holes into a minimum number of convex polygons by cuts in the directions of $\mathcal{F}$.

Naturally, the decision version of MNCP($\mathcal{F}$) is as follows: given a polygon $P$ with holes and a positive integer $k$, decide whether $P$ can be partitioned into at most $k$ convex polygons by cuts in the directions of $\mathcal{F}$. In what follows, we will also denote the decision version of the above problem by MNCP($\mathcal{F}$). It will be clear from the context which version is considered.

The main result of this paper is described in the following theorem, where the polynomial-time solvability of MNCP($\mathcal{F}$) is treated with respect to the number $n$ of the vertices of a polygon $P$.

THEOREM 1. *MNCP($\mathcal{F}$) is NP-hard if* $|\mathcal{F}| \geq 3$ *and it is solvable in polynomial time if* $|\mathcal{F}| \leq 2$.

A similar problem can be posed for the case of convex partitions disallowing Steiner points, i.e., partitions by diagonal cuts of a polygon parallel to the directions from $\mathcal{F}$. By using the method of the proof of the NP-hardness of MNCP($\mathcal{F}$) in the case $|\mathcal{F}| \geq 3$, we obtain the following corollary.

COROLLARY 1. *Let* $\mathcal{F}$ *be a given family of at least four directions in the plane. The problem of determining whether a polygon with holes can be partitioned into at most* $k$ *convex polygons by diagonal cuts in the directions of* $\mathcal{F}$ *is NP-complete.*

It is well-known (cf. [21]) that the problem of partitioning a three-dimensional polyhedron into a minimum number of convex polyhedra is NP-hard. As it is proved in [5], this problem remains NP-hard even for partitions by cuts parallel to three given mutually orthogonal planes in the case when the faces of the polyhedron are parallel to the cutting planes. From Theorem 1 we can deduce the following corollary.

COROLLARY 2. *For any family* $\mathcal{H}$ *of planes in three-dimensional Euclidean space, which includes three distinct planes having a common parallel line, the problem of determining whether a polyhedron can be partitioned into at most* $k$ *convex polyhedra by cuts parallel to the planes of* $\mathcal{H}$ *is NP-hard.*

The complexity status of MNCP($\mathcal{F}$) in the case $|\mathcal{F}| \leq 2$ is detailed in the following two assertions.

THEOREM 2 ([6]). *MNCP($\mathcal{F}$) has time complexity* $O(n \log n)$ *in the case* $|\mathcal{F}| = 1$.

THEOREM 3. *MNCP($\mathcal{F}$) has time complexity* $O(n^{3/2} \log n)$ *in the case* $|\mathcal{F}| = 2$.

If a polygon $P$ is allowed to be multiple connected and to have degenerate holes, then $P$ will be called a *polygonal domain* (see Section 2 for exact definition).

REMARK 1. As shown in [6] and in Section 8 of this paper, assertions of Theorems 2 and 3 also hold for the case of polygonal domains.

The paper is organized as follows. Section 8 presents the proof of Theorem 3 which implies the second part of Theorem 1. The other sections deal with the proof of Theorem 5, which shows the exact value for the minimum number of pieces in a convex partition of a polygonal domain by cuts in two given directions. Theorem 5 yields the polynomial time solvability of MNCP($\mathcal{F}$) in the case $|\mathcal{F}| = 2$. *Because of space considerations, the part on the NP-hardness of MNCP($\mathcal{F}$) in the case $|\mathcal{F}| \geq 3$ and Corollaries 1, 2, as well as the figures, the majority of lemma proofs and the examples in the remaining parts are omitted. In [16], the reader can find the full version of our paper.*

## 2  Description of Polygonal Domains

This section is a starting point in proving Theorem 3. By applying a suitable affine transformation of the plane, we may consider that the two directions of $\mathcal{F}$ are parallel to the coordinate axes in the plane.

Let $P$ be a polygon, possibly multiply connected, in the plane $E$. The *boundary* bd $P$ of $P$ is the union of finitely many closed non-selfintersecting polygonal contours. We assume that any two of these contours are not interlaced, i.e., they are situated either one inside the other or mutually non-inclusive, that they may have common vertices but no common line segment. In this way, the *interior* int $P$ of $P$ is the set of points each lying inside an odd number of boundary contours. By a *hole* of $P$ we mean any bounded component of the complement $E \setminus P$, and the *exterior* of $P$ is the unbounded component of $E \setminus P$. Clearly, any hole and the exterior of $P$ are open sets.

To extend our considerations for the case of polygonal domains, we need some auxiliary notions. Let us assume that some points $v_1, v_2, \ldots, v_t$ and some closed line segments $s_1, s_2, \ldots, s_u$ are placed inside a polygon $P$ such that the following conditions hold:

1) all $v_1, v_2, \ldots, v_t$ belong to int $P \setminus (s_1 \cup s_2 \cup \cdots \cup s_u)$,

2) the relative interior of every line segment $s_i$, $i = 1, 2, \ldots, u$, lies in int $P$,

3) if some line segments $s_i, s_j$ have a common point, then it is an endpoint for both $s_i, s_j$.

The point-set union of these points and segments is called the *ornament* of $P$ and will be denoted by Or $P$:

$$\text{Or}\, P := \left( \bigcup_{i=1}^{t} v_i \right) \cup \left( \bigcup_{i=1}^{u} s_i \right).$$

In what follows, by a *polygonal domain* we mean a polygon $P$ considered together with a certain ornament Or $P$.

DEFINITION 1. For a given polygonal domain $P$ with the ornament Or $P$, the sets Bd $P :=$ Or $P \cup$ bd $P$  and Int $P := $ int $P \setminus$ Or $P (= P \setminus$ Bd $P)$ are called the *formal boundary* and the *formal interior* of $P$, respectively. A bounded component of the complement $E \setminus$ Int $P$ is called a *formal hole* of $P$, and the unbounded component of $E \setminus$ Int $P$ is called the *formal exterior* of $P$ and is denoted by Ext $P$.

It is easily seen that for any polygonal domain $P$ the sets Bd $P$, Ext $P$ and any formal hole of $P$ are closed, while the formal interior Int $P$ is an open set.

REMARK 2. Observe that the boundary bd $P$ and the interior int $P$ of a polygonal domain $P$ do not depend on the ornament Or $P$. In order to stress this fact, we will call the sets bd $P$ and int $P$ the *topological boundary* and the *topological interior* of $P$, respectively.

A point $x \in P$ is called a *vertex* of $P$ if it is either an endpoint of a line segment forming Bd $P$ or an isolated point in Bd $P$. We denote by Vert $P$ the set of vertices of $P$. A closed line segment $[x, z] \subset$ Bd $P$ is called an *elementary segment* of Bd $P$ if $[x, z] \cap$ Vert $P = \{x, z\}$.

Note that if two elementary segments of Bd $P$ have a common point, then the point is an endpoint for both segments.

DEFINITION 2. A polygonal domain $P$ is said to be *partitioned into convex polygons* $Q_1, Q_2, \ldots, Q_r$ if $\bigcup_{i=1}^{r} \operatorname{int} Q_i \subset \operatorname{Int} P \subset \bigcup_{i=1}^{r} Q_i$, and $\operatorname{int} Q_i \cap \operatorname{int} Q_j = \emptyset$, $i \neq j$.

LEMMA 1. *For a polygonal domain $P$, one has $\alpha_0 - \alpha_1 + c + h' = c' + h$, where*

   $\alpha_0$ *is the number of vertices of $P$, $\alpha_1$ is the number of elementary segments of Bd $P$,*

   $c$ *is the number of components of Int $P$, $c'$ is the number of components of $P$,*

   $h$ *is the number of formal holes of $P$, $h'$ is the number of topological holes of $P$.*

# 3 Measure of Local Nonconvexity

Any vertex $v$ of a polygonal domain $P$, unless it is an isolated point of Bd $P$, is the apex of at least one inner angle of $P$, formed by two elementary segments of Bd $P$. We allow the sides of this angle to coincide (in this case the angle equals $2\pi$). Inner angles of more than $\pi$ will be called *concave* angles.

DEFINITION 3. A point $v \in P$ is called a *point of local nonconvexity* of $P$ provided it is either an isolated point in Bd $P$ or the apex of a concave inner angle of $P$; otherwise $v$ is called a *point of local convexity* of $P$.

LEMMA 2. *Any formal hole of a polygonal domain $P$ contains at least one point of local nonconvexity of $P$.*

LEMMA 3. *A polygonal domain $P$ has no points of local nonconvexity if and only if each component of the formal interior Int $P$ is an open convex polygon.*

LEMMA 4. *Let $Q = \{Q_1, Q_2, \ldots, Q_r\}$ be a partition of a polygonal domain $P$ into simple polygons, and $P_Q$ be the polygonal domain such that $\operatorname{Bd} P_Q := \bigcup_{i=1}^{r} \operatorname{bd} Q_i$ and $\operatorname{Int} P_Q := \operatorname{int} P \setminus \operatorname{Bd} P_Q$. Then all the polygons $Q_1, Q_2, \ldots, Q_r$ are convex if and only if $P_Q$ has no points of local nonconvexity.*

One can associate to any point $x$ in the coordinate plane $E$ four open quadrants $O_1(x)$, $O_2(x)$, $O_3(x)$, $O_4(x)$ with common apex $x = (\xi_0; \eta_0)$ :

$$O_1(x) = \{x\} \cup \{(\xi; \eta) : \xi > \xi_0, \eta > \eta_0\}, \quad O_2(x) = \{x\} \cup \{(\xi; \eta) : \xi > \xi_0, \eta < \eta_0\},$$

$$O_3(x) = \{x\} \cup \{(\xi; \eta) : \xi < \xi_0, \eta < \eta_0\}, \quad O_4(x) = \{x\} \cup \{(\xi; \eta) : \xi < \xi_0, \eta > \eta_0\}.$$

DEFINITION 4. The *measure $m(v)$* (of local nonconvexity) of a polygonal domain $P$ at a point $v \in P$ is defined as follows:

   $m(v) = 0$ if $v$ is a point of local convexity of $P$;

   $m(v) = 1$ if $v$ is a point of local nonconvexity of $P$ and all elementary segments of Bd $P$ with apex $v$ do not lie in the same quadrant $O_i(v)$, $i = 1, 2, 3, 4$;

   $m(v) = 2$ if $v$ is a point of local nonconvexity of $P$ and all elementary segments of Bd $P$ with apex $v$ lie in the same quadrant $O_i(v)$ for some $i = 1, 2, 3, 4$ (in particular, $v$ can be an isolated point in Bd $P$).

REMARK 3. From Definitions 3 and 4 it easily follows that for a point $v \in P$, one has $m(v) > 0$ if and only if $v$ is either an isolated point in Bd $P$ or the apex of an inner concave angle of $P$.

DEFINITION 5. For a polygonal domain $P$, the sum $m(P) := \sum m(v)$ taken over all points of local nonconvexity of $P$, is called the *measure* (of local nonconvexity) of $P$.

# 4 Addition of Segments to the Formal Boundary

Any partition of a polygonal domain $P$ into convex polygons can be considered as a repeated addition of closed line segments to the formal boundary of $P$. Let $[x, z] \subset P$ be a closed line segment, either horizontal or vertical, such that the intersection of $[x, z]$ with $\operatorname{Bd} P$ is a finite, possibly empty, set. We say that the polygonal domain $P'$ is obtained from $P$ by the *addition* of $[x, z]$ to the formal boundary $\operatorname{Bd} P$ provided $\operatorname{Bd} P' := \operatorname{Bd} P \cup [x, z]$, $\operatorname{Int} P' := \operatorname{Int} P \setminus [x, z]$, and $\operatorname{Vert} P' := \operatorname{Vert} P \cup \{x, z\} \cup (\operatorname{Bd} P \cap [x, z])$.

A line segment added to $\operatorname{Bd} P$, does not become, in general, an elementary segment of $\operatorname{Bd} P'$; however, it is the union of elementary segments of $\operatorname{Bd} P'$.

LEMMA 5. *Let $P'$ be the polygonal domain obtained from a polygonal domain $P$ by the addition of a line segment $[x, w] \subset P$, $w \in \operatorname{Vert} P$, to the formal boundary $\operatorname{Bd} P$. Then $m(w) - 1 \leq m'(w) \leq m(w)$, where $m'(w)$ denotes the measure of $P'$ at $w$.*

DEFINITION 6. A line segment $[x, z]$ is called a *simple chord* of a polygonal domain $P$ if the following conditions hold:

1) $[x, z]$ is either horizontal or vertical,

2) $x, z \in \operatorname{Bd} P$,

3) the open interval $]x, z[$ lies in $\operatorname{Int} P$.

From definitions it follows that if a simple chord $[x, z]$ of a polygonal domain $P$ is added to the formal boundary $\operatorname{Bd} P$, then $[x, z]$ becomes an elementary segment of the polygonal domain $P'$ such that $\operatorname{Bd} P' = \operatorname{Bd} P \cup [x, z]$, $\operatorname{Int} P' = \operatorname{Int} P \setminus ]x, z[$, $\operatorname{Vert} P' = \operatorname{Vert} P \cup \{x, z\}$.

LEMMA 6. *For any vertex $v$ of local nonconvexity of a polygonal domain $P$ there is a simple chord $[v, z]$ of $P$ whose addition to $\operatorname{Bd} P$ decreases $m(v)$ by one.*

# 5 Effective Chords

Further we will see that some line segments are of significant importance for a partition of a polygonal domain into a minimum number of convex polygons.

DEFINITION 7. A line segment $[v, w]$ is called an *effective chord* of a polygonal domain $P$ if the following conditions hold:

1) $[v, w]$ is either horizontal or vertical,

2) the open interval $]v, w[$, except a finite (possibly empty) set of points, lies in $\operatorname{Int} P$,

3) $v$ and $w$ are points of local nonconvexity of $P$ and $m'(v) = m(v) - 1$, $m'(w) = m(w) - 1$, where $m'(x)$ is the measure of the polygonal domain $P'$ at a vertex $x \, (\in \{v, w\})$, obtained from $P$ by the addition of $[v, w]$ to $\operatorname{Bd} P$,

4) if $x$ is a point of $\operatorname{Bd} P$ lying in $]v, w[$, then $x$ is a vertex of $P$ with $m(x) = 1$ such that all elementary segments of $\operatorname{Bd} P$ with apex $x$ lie in the same half-plane determined by the line $(v, w)$.

It is easily seen that any two different effective chords of $P$ have at most one common point, and a proper part of an effective chord is not an effective chord.

LEMMA 7. *Let $[v, w]$ be an effective chord of a polygonal domain $P$, and let $x_1, x_2, \ldots, x_r$ be all vertices of $P$ contained in $]v, w[$. The addition of $[v, w]$ to $\operatorname{Bd} P$ decreases the measure of $P$ at each of the vertices $v, w, x_1, x_2, \ldots, x_r$ by one.*

DEFINITION 8. Let $[v, x]$ and $[x, w]$ be two effective chords of a polygonal domain $P$ such that $]v, x[ \cap ]x, w[ = \emptyset$. We will say that $[v, x]$ and $[x, w]$ are in a *special position* if $m(x) = 2$ and either of the conditions holds:

1) $[v, x]$ and $[x, w]$ are collinear,

2) $x$ is not an isolated point in $\operatorname{Bd} P$, the segments $[x, v]$, $[x, w]$ are orthogonal to each other, and all elementary segments of $\operatorname{Bd} P$ with apex $x$ are contained in the same quadrant $O_i(x)$ opposite to the right angle formed by $[x, v]$, $[x, w]$.

DEFINITION 9. A family $\mathcal{G}$ of effective chords of a polygonal domain $P$ is called *admissible* if any two intersecting chords from $\mathcal{G}$ are in a special position. The maximum number of effective chords of $P$ forming an admissible family is named the *effective number* of $P$ and is denoted by $e(P)$.

# 6 Decomposition of the Supplementary Ornament

Assume that a polygonal domain $P$ is partitioned into convex polygons $Q_1, Q_2, \ldots, Q_r$ (see Definition 2). Let $W$ be the union of all vertices of $Q_1, Q_2, \ldots, Q_r$. Denote by $L_1, L_2, \ldots, L_r$ the boundary contours of $Q_1, Q_2, \ldots, Q_r$, respectively, and put $L = L_1 \cup L_2 \cup \cdots \cup L_r$. We may consider $L$ as the formal boundary of the polygonal domain $P_0$ such that

$$\operatorname{Bd} P_0 = L, \quad \operatorname{Int} P_0 = \operatorname{Int} P \setminus L, \quad \operatorname{Vert} P_0 = \operatorname{Vert} P \cup W.$$

Note that $P$ and $P_0$ have the same topological boundary and topological interior, and the components of $\operatorname{Int} P_0$ are the interiors of convex polygons $Q_1, Q_2, \ldots, Q_r$.

As it was mentioned above, the polygonal domain $P_0$ can be obtained from $P$ by the repeated addition of closed line segments to $\operatorname{Bd} P$. In this section we investigate the inverse approach: to obtain $P$ from $P_0$ by the repeated deletion of *open* line intervals from $\operatorname{Bd} P_0$. For this purpose, the "supplementary ornament" $Y := L \setminus \operatorname{Bd} P$ will be studied.

LEMMA 8. *The set* $Y = L \setminus \operatorname{Bd} P$ *is decomposable into finitely many disjoint open line intervals.*

LEMMA 9. *Let* $\alpha_0$ *and* $\alpha_1$ *be the numbers of vertices and elementary segments of* $P$, *respectively, and let* $\beta_0$ *and* $\beta_1$ *be the analogous quantities for the polygonal domain* $P_0$. *If the set* $Y = L \setminus \operatorname{Bd} P$ *is the union of* $k$ *disjoint open intervals* $l_1, l_2, \ldots, l_k$, *then*

$$k = (\beta_1 - \alpha_1) - (\beta_0 - \alpha_0).$$

Let $\mathcal{L} = \{l_1, l_2, \ldots, l_k\}$ be a decomposition of the set $Y = L \setminus \operatorname{Bd} P$ into disjoint open line intervals $l_i = ]x_i, z_i[$, $i = 1, 2, \ldots, k$. We will study below the process of repeated deletion of the intervals $l_1, l_2, \ldots, l_k$ from $\operatorname{Bd} P_0$. Denote by $P_i$, $i = 1, 2, \ldots, k$, the polygonal domain obtained from $P_0$ by deleting $l_1 \cup l_2 \cup \cdots \cup l_i$ from $L$.

REMARK 4. Generally, $P_k$ can be different from $P$, since it may have some new vertices . At the same time $\operatorname{Bd} P_k = \operatorname{Bd} P$, and any vertex $w \in \operatorname{Vert} P_k \setminus \operatorname{Vert} P$ is a point of local convexity of $P$. Hence $m(P_k) = m(P)$.

REMARK 5. Since the set $L_i := L \setminus (l_1 \cup l_2 \cup \cdots \cup l_i)$ can be non-closed for $1 \leq i < k$ (see example below), we cannot assert that $L_i$ is the ornament of $P_i$ (since the ornament of a polygonal domain is a closed set, being the union of isolated points and closed segments).

From Remark 5 it follows that, in general, we cannot apply to $P_i$, $1 \leq i < k$, the notion of measure (of local nonconvexity). Nevertheless, we can avoid this obstacle for vertices of $P$ (but not for vertices of $P_i$ ) in the following way. A set $L_i$ is non-closed if we delete, together with some interval $l_p$, $p \leq i$, an endpoint of an interval $l_q$, $q > i$. Since $l_p \subset Y$ and $Y \cap \operatorname{Bd} P = \emptyset$, this deleted vertex cannot be a vertex of $\operatorname{Bd} P$. Hence for any vertex $w$ in $\operatorname{Vert} P$, we can choose a small square $S$, with vertical and horizontal sides, centered at $w$ such that the intersection of $S$ and $L_i$ is a closed set. Now we define the measure $m_i(w)$ of

$P_i$ at $w$ with respect to that part of Bd $P$ which lies in $S$. Since the notion of measure is local, our approach is correct.

During the process of the repeated deletion of the segments $l_1, l_2, \ldots, l_k$ from Bd $P_0$, we will consider the measures of the polygonal domains $P_0, P_1, \ldots, P_k$. For this purpose, we define the following numbers.

DEFINITION 10. For any decomposition of the set $Y = L \setminus \mathrm{Bd}\, P$ into disjoint open line intervals $l_i = ]x_i, z_i[$, $i = 1, 2, \ldots, k$, define the numbers $\mu(l_i)$, $i = 1, 2, \ldots, k$, as follows:

1) $\mu(l_i) = 0$ if $x_i \notin \mathrm{Vert}\, P$ and $z_i \notin \mathrm{Vert}\, P$,

2) $\mu(l_i) = m_i(x_i) - m_{i-1}(x_i)$ if $x_i \in \mathrm{Vert}\, P$ and $z_i \notin \mathrm{Vert}\, P$,

3) $\mu(l_i) = m_i(z_i) - m_{i-1}(z_i)$ if $x_i \notin \mathrm{Vert}\, P$ and $z_i \in \mathrm{Vert}\, P$,

4) $\mu(l_i) = m_i(x_i) - m_{i-1}(x_i) + m_i(z_i) - m_{i-1}(z_i)$ if $x_i \in \mathrm{Vert}\, P$ and $z_i \in \mathrm{Vert}\, P$,

where $m_j(w)$ denotes the measure of $P_j$ at a vertex $w \in \mathrm{Vert}\, P$.

Due to Lemma 5, we have $0 \leq m_i(x_i) - m_{i-1}(x_i) \leq 1$, $\quad 0 \leq m_i(z_i) - m_{i-1}(z_i) \leq 1$. This gives us the following corollary.

COROLLARY 3. 1) $0 \leq \mu(l_i) \leq 2$ for all $i = 1, 2, \ldots, k$,

2) $\mu(l_i) \leq 1$ if at most one of the vertices $x_i, z_i$ is in $\mathrm{Vert}\, P$,

3) $\mu(l_i) = 0$ if both vertices $x_i, z_i$ are not in $\mathrm{Vert}\, P$. Note that the values $\mu(l_i)$, $i = 1, 2, \ldots, k$, depend on the order in which the intervals $l_1, l_2, \ldots, l_k$ are enumerated.

LEMMA 10. $m(P) = \mu(l_1) + \mu(l_2) + \cdots + \mu(l_k)$.

The remainder of this section is devoted to the proof of the following theorem.

THEOREM 4. Let $\mathcal{L} = \{l_1, l_2, \ldots, l_k\}$ be a decomposition of the set $Y = L \setminus \mathrm{Bd}\, P$ into disjoint open line intervals. The intervals $l_1, l_2, \ldots, l_k$ can be renumbered by the indices $1, 2, \ldots, k$ such that $0 \leq \mu(l_i) \leq 1$ for at least $k - e$ of them, where $e$ denotes the effective number of $P$.

*Proof.* The main idea of the proof is to select in the closure of $Y$ an admissible family of effective chords of maximum cardinality, say $g$, and to show further that under a suitable renumbering of $\mathcal{L}$, each of these chords contains at most one interval $l_i \in \mathcal{L}$ with $\mu(l_i) = 2$. Since $g \leq e$, this proves the theorem.

During the proof we execute some renumberings of the family $\mathcal{L}$.

*First Renumbering.* It is organized for the intervals $l_1, l_2, \ldots, l_k$ according to the following procedure:

1. Put $i = 1$ and $j = 0$.

2. If $\mu(l_i) \leq 1$, put $j = 0$ and go to 3, else go to 4.

3. If $i < k$, put $i = i + 1$ and go to 2, else exit.

4. If $j < k - i - 1$, put $j = j + 1$, renumber the intervals $l_{i+1}, l_{i+2}, \ldots, l_k$, $l_i$ by the indices $i, i+1, \ldots, k$, respectively, and go to 2, else exit.

As a result of First Renumbering we obtain a new sequence $l_1, l_2, \ldots, l_k$ such that:

1) $0 \leq \mu(l_i) \leq 1$ for all $i = 1, 2, \ldots, r - 1$, where $1 \leq r \leq k + 1$,

2) $\mu(l_r) = 2$ for any renumbering of $l_r, \ldots, l_k$ by the indices $r, \ldots, k$ (if $r \leq k$).

An effective chord $[v, w]$ of $P$ is said to be *compatible* with the family $\{l_r, \ldots, l_k\}$ provided $[v, w] \setminus \mathrm{Vert}\, P$ is the union of some intervals from $\{l_r, \ldots, l_k\}$. Denote by $\mathcal{F} = \{t_1, t_2, \ldots, t_g\}$ an admissible family of effective chords of $P$ compatible with $\{l_r, \ldots, l_k\}$, and put $T := t_1 \cup t_2 \cup \cdots \cup t_g$. Obviously, $g \leq e$.

*Second Renumbering.* If $r \leq k$, we renumber the intervals $l_r \ldots, l_k$ such that exactly the intervals $l_s, \ldots, l_k$, $r \leq s \leq k + 1$, form $T$, i.e., $l_s \cup \cdots \cup l_k = T \setminus \mathrm{Vert}\, P$.

LEMMA 11. $r = s$.

*Third Renumbering.* It remains to renumber the intervals $l_s, \ldots, l_k$ in the case $s \le k$. Due to Second Renumbering, these intervals are contained in the set $T = t_1 \cup t_2 \cup \cdots \cup t_g$. Without loss of generality, we may assume that the intervals $l_s, \ldots, l_w$, $w \le k$, form $t_1$, i.e., $]x_s, z_s[\cup \cdots \cup]x_w, z_w[ = t_1 \setminus \text{Vert } P$, such that $z_i = x_{i+1}$, $i = s, \ldots, w-1$ (in the case $s < w$).

From Definition 7 it follows that $m(z_s) = \cdots = m(z_{w-1}) = 1$ and all elementary segments in Bd $P$ incident to $z_i$ lie in the same half-plane determined by the line $(x_s, z_w)$. It is easily seen that $\mu(l_s) = 2$ and $\mu(l_i) = 1$ for all $i = s + 1, \ldots, w$.

Similarly, we renumber repeatedly the intervals $l_{w+1}, \ldots, l_k$ such that for any chord $t_i$, $i = 2, \ldots, g$, there exists exactly one interval $l_{i_r} \subset t_i$, with $\mu(l_{i_r}) = 2$, and $\mu(l_j) = 1$ for all other intervals $l_j \subset t_i$.

Since $g \le e$, the proof of Theorem 4 is complete. $\square$

LEMMA 12. *The number $k$ of disjoint open intervals $l_1, l_2, \ldots, l_k$ decomposing the set $Y = L \setminus \text{Bd}$ is at least $m - e$, where $m$ is the measure of $P$ and $e$ is the effective number of $P$.*

# 7 Minimum Number of Convex Parts

THEOREM 5. *The minimum number of convex polygons obtained by cutting a polygonal domain $P$ in the directions of the coordinate axes in the plane equals $m + c - h - e$, where $m$ is the measure of $P$, $c$ is the number of components of Int $P$, $h$ is the number of formal holes of $P$, $e$ is the effective number of $P$.*

*Proof.* Let $r$ denote the minimum number of convex polygons in the hypothesis of the theorem. First we prove the inequality $r \le m + c - h - e$.

Suppose that $e > 0$, and let $[x_i, z_i]$, $i = 1, 2, \ldots, e$, be effective chords of $P$ forming an admissible family of maximum cardinality. Denote by $P_1$ the polygonal domain obtained from $P$ by the addition of $[x_1, z_1]$ to Bd $P$. Some vertices of $P$ can lie in $]x_1, z_1[$. Denote them by $v_1, v_2, \ldots, v_t$, respectively. Due to Lemma 7, the addition of $[x_1, z_1]$ to Bd $P$ leads to the following changes:

1) the measure of $P$ at each of the points $x_1, v_1, v_2, \ldots, v_t, z_1$ decreases by one,

2) the number $h = h(P)$ of formal holes of $P$ decreases by $p \, (\ge 0)$ and the number $c = c(P)$ of connected components of Int $P$ increases by $q \, (\ge 0)$ such that $p + q = t + 1$,

3) the effective number $e(P)$ decreases by at most one.

In other words, $m(P_1) = m - t - 2$, $c(P_1) = c + q$, $h(P_1) = h - p$, $e(P) \ge e - 1$, and hence $m(P_1) + c(P_1) - h(P_1) - e(P_1) \le m + c - h - e$. Clearly, the chords $[x_i, z_i]$, $i = 2, \ldots, e$, remain to be effective chords for $P_1$.

After a repeated addition of the chords $[x_i, z_i]$, $i = 2, \ldots, e$, to Bd $P$, we obtain the polygonal domain $P_e$ such that Bd $P_e = \text{Bd } P \cup (\bigcup_{i=1}^{e} [x_i, z_i])$, Int $P_e = \text{Int } P \setminus \text{Bd } P_e$, Vert $P_e = \text{Vert } P$. As above, $m(P_e) + c(P_e) - h(P_e) - e(P_e) \le m + c - h - e$ holds.

We claim that $e(P_e) = 0$. Indeed, if there were an effective chord, say $l$, of $P_e$, then, according to Definition 7, the family $\{l, [x_1, z_1], [x_2, z_2], \ldots, [x_e, z_e]\}$ would be an admissible family of cardinality $e + 1$, which is impossible by the assumption. Hence $e(P_e) = 0$ and $m(P_e) + c(P_e) - h(P_e) \le m + c - h - e$. Next assume that $m(P_e) > 0$, and let $u$ be a point of local nonconvexity of $P_e$. By Lemma 6, there is a simple chord $[u, w]$ of $P_e$ whose addition to Bd $P_e$ decreases the measure of $P_e$ at $u$ by one. Denote by $P'$ the polygonal domain obtained from $P_e$ by the addition of $[u, w]$ to Bd $P_e$. This addition either decreases $h(P_e)$ by one or increases $c(P_e)$ by one. So, $m(P') + c(P') - h(P') \le m + c - h - e$ holds. After a repeated addition of simple chords, which delete the local nonconvexity of the polygonal domain $P_e$ at all its vertices of local nonconvexity, we obtain a polygonal domain $\tilde{P}$ without points of local nonconvexity, i.e., with $m(\tilde{P}) = 0$. By Lemma 1, $h(\tilde{P}) = 0$. Similarly to the above, $c(\tilde{P}) = m(\tilde{P}) + c(\tilde{P}) - h(\tilde{P}) \leq m + c - h - e$. Due to Lemma 3, each connected component of $\operatorname{Int}\tilde{P}$ is an open convex polygon. The closures of these polygons partition $\tilde{P}$, and $P$, into $c(\tilde{P})$ convex parts (see Lemma 4). Therefore $r \leq c(\tilde{P}) \leq m + c - h - e$.

Now we are going to prove the opposite inequality. Let $Q = \{Q_1, Q_2, \ldots, Q_r\}$ be a partition of $P$ into convex polygons $Q_1, Q_2, \ldots, Q_r$ by cuts parallel to the directions of the coordinate axes. According to Lemma 4, we may identify this partition with a polygonal domain $R$ obtained from $P$ by the addition of line segments to $\operatorname{Bd} P$ such that $\operatorname{Bd} R$ divides $\operatorname{Int} R$ into $r$ convex parts, i.e., $c(R) = r$. Denote by $\gamma_0$ and $\gamma_1$, respectively, the number of vertices and elementary segments of $R$. By Lemma 1, we have $\alpha_0 - \alpha_1 + c + h' = c' + h$, $\gamma_0 - \gamma_1 + c(R) + h'(R) = c'(R) + h(R)$ for the polygonal domains $P$ and $R$, respectively. It is easily seen that $h(R) = 0$, $c'(R) = c'$, $h'(R) = h'$ holds. Therefore $c(R) = c - h + (\gamma_1 - \alpha_1) - (\gamma_0 - \alpha_0)$ holds. Lemmas 9 and 8 give $(\gamma_1 - \alpha_1) - (\gamma_0 - \alpha_0) = k \geq m - e$. Thus $r = c(R) \geq c - h + m - e$. □

# 8   Proof of Theorem 3

As it was mentioned in Section 2, we assume the two directions of the family $\mathcal{F}$ to be parallel to the coordinate axes in the plane $E$.

Theorem 5 gives not only a formula for a minimum number of convex parts; it also contains an informal description of the respective partition algorithm. Below we describe this algorithm more precisely and show that its time complexity is $O(n^{3/2} \log n)$.

Let $P$ be a polygonal domain in the plane $E$, determined by its vertex-set $V = \operatorname{Vert} P$ (including isolated points of $\operatorname{Bd} P$) and by two linear arrays $S_1$ and $S_2$ of elementary segments of $\operatorname{Bd} P$ forming the topological boundary $\operatorname{bd} P$ and the ornament $\operatorname{Or} P$, respectively. This separate representation of the set $S = S_1 \cup S_2$ of elementary segments of $\operatorname{Bd} P$ enables us to determine, by means of the sweep-line technique, the appertaines of a given horizontal or vertical chord to $\operatorname{int} P$. In what follows, it is assumed that each vertex of $P$ is given by its Cartesian coordinates $(\xi; \eta)$, and each elementary segment of $\operatorname{Bd} P$ is of the form $[x, z]$, $x, z \in V$.

Let $n$ be the number of vertices of $P$. Since every planar graph with $n$ vertices has at most $3n - 6$ edges (cf. [8]), one has $|S| = O(n)$. Similarly, the cardinality of any family of horizontal (or vertical) chords of $P$ of the form $[x, z]$, $x, z \in V$, is at most $O(n)$, whence the family of effective chords of $P$ has at most $O(n)$ elements.

First we organize a preliminary lexicographical sorting of $V$ on coordinates $X, Y$ of vertices (from left-down to right-up corners) and the respective sorting of $S$ by angles of slopes in order to obtain a sorted list $V^*$ of vertices, where each vertex $v_i \in V^*$ is represented in the form

$$v_i = (\xi(i), \eta(i), \alpha_1(i), \alpha_2(i), \alpha_3(i), \alpha_4(i),$$
$$\beta_1(i), \beta_2(i), \beta_3(i), \beta_4(i)), \quad i = 1, 2, \ldots, n,$$

where $\alpha_j(i)$ and $\beta_j(i)$ are defined as follows:

$\alpha_j(i) = 1$ if the open quadrant $O_j(v_i)$ contains at least one elementary segment of $\operatorname{Bd} P$ with apex $v_i$, and $\alpha_j(i) = 0$ otherwise,

$\beta_j(i) = 1$ if there is an elementary segment with apex $v_i$ in the half-line $h_j(v_i)$ and $\beta_j(i) = 0$ otherwise, where the half-lines $h_1(x)$, $h_2(x)$, $h_3(x)$, $h_4(x)$ with common apex $x = (\xi_0; \eta_0)$ are defined as follows:

$$h_1(x) = \{(\xi; \eta_0) : \xi \geq \xi_0\}, \quad h_2(x) = \{(\xi_0; \eta) : \eta \leq \eta_0\},$$

$$h_3(x) = \{(\xi; \eta_0) : \xi \le \xi_0\}, \quad h_4(x) = \{(\xi_0; \eta) : \eta \ge \eta_0\}.$$

Clearly, any isolated point $(\xi; \eta)$ of $\operatorname{Bd} P$ is represented in $V^*$ as $(\xi, \eta, 0, 0, 0, 0, 0, 0, 0, 0)$.

This sorted list $V^*$ gives us a possibility to determine in linear time the vertices of local nonconvexity of $P$, as well as the measure of $P$ at each of them. Moreover, it will help us to construct the effective chords of $P$ and to determine whether two effective chords are in a special position.

MINIMUM PARTITION ALGORITHM:

1. Find the family $\mathcal{N}$ of effective chords of $P$.

2. Select in $\mathcal{N}$ an admissible family $\mathcal{G} = \{s_1, s_2, \ldots, s_e\}$ of maximum cardinality.

3. Form the polygonal domain $P_G$, by adding the set $G := s_1 \cup s_2 \cup \cdots \cup s_e$ to $\operatorname{Bd} P$.

4. Partition $P_G$ into convex polygons by drawing simple chords repeatedly inside $P_G$ deleting the local nonconvexity of $P_G$ at each of its vertices of local nonconvexity.

The description of the implementation of steps 1–4 and their complexity analysis can be found in [16]. In particular to implement step 2, the appropriate $O(n^{3/2} \log n)$-time algorithm from [9] is used. In effect, the whole algorithm can be implemented in time $O(n^{3/2} \log n)$.

# 9 Final Remarks

We have shown that the cardinality of the family $\mathcal{F}$ of cutting directions determines whether $\mathrm{MNCP}(\mathcal{F})$ admits a polynomial-time algorithm or it is NP-hard. We have also proved that the modification of $\mathrm{MNCP}(\mathcal{F})$ disallowing Steiner points is NP-hard if $|\mathcal{F}| > 3$. The problem of whether the aforementioned modification of $\mathrm{MNCP}(\mathcal{F})$ disallowing Steiner points admits a polynomial-time algorithm if $|\mathcal{F}| \le 3$ seems to be open.

# References

[1] T. ASANO, M. SATO, T. OHTSUKI, *Computational geometry algorithms*, in Layout Design and Verification, T. Ohtsuki, ed., North-Holland, Amsterdam, 1986, pp. 295–347.

[2] M. BERN, D. EPSTEIN, *Mesh generation and optimal triangulation*, in Computing in Euclidean Geometry, 2nd ed., D.-Z. Du, F.K. Hwang, eds., World Scientific, Singapore, 1995, pp. 47–123.

[3] B. CHAZELLE, D. P. DOBKIN, *Optimal convex decomposition*, in Computational Geometry, G.T. Toussaint, ed., North-Holland, Amsterdam, 1985, pp. 63–133.

[4] B. Chazelle, L. Palios, *Decomposition algorithms in geometry*, in Algebraic Geometry and Applications, C.L. Bajaj, ed., Springer, New York, 1994, pp. 419–449.

[5] V. J. DIELISSEN V.J., A. KALDEWAIJ, *Rectangular partition is polynomial in two dimensions but NP-complete in three*, Inform. Process. Lett. 38 (1991), pp. 1–6.

[6] A. GORPINEVICH, V. SOLTAN, *Algorithms for the partition of a polygonal region into trapezoids and convex strips in problems of LSI design*, (Russian) Voprosy Kibernet. (Moscow) No. 156 (1991), pp. 45–56.

[7] D. H. GREENE, *The decomposition of polygons into convex parts*, in Advances in Computing Research, F.P. Preparata, ed., JAI Press, London, 1983, pp. 235–259.

[8] F. HARARY, Graph Theory, Addison-Wesley Publ., Massachusetts, 1969.

[9] H. IMAI, T. ASANO, *Efficient algorithm for geometric graph search problems*, SIAM J. Comput. 15 (1986), pp. 478–494.

[10] J. M. KEIL, *Decomposing a polygon into simpler components*, SIAM J. Comput. 14 (1985), pp. 799–817.

[11] J. M. KEIL, J.-R. SACK, *Minimum decomposition of polygonal objects*, in Computational Geometry, G.T. Toussaint, ed., North Holland, Amsterdam, 1985, pp. 197–216.

[12] N. M. KORNEENKO, G. V. MATVEEV, N. N. METEL'SKI, R. I. TYŠKEVIČ, *Partitions of polygons*, (Russian) Vesci Akad. Navuk BSSR. Ser. Fiz.-Mat. Navuk, 1978, no. 2, pp. 25–29.

[13] D. T. LEE, F. P. PREPARATA, *Computational geometry – a survey*, IEEE Trans. Comput. C-33 (1984), pp. 1072–1101.

[14] D. LICHTENSTEIN, *Planar formulas and their uses*, SIAM J. Comput. 11 (1982), pp. 329–343.

[15] A. LINGAS, *The power of non-rectilinear holes*, Lecture Notes Comput. Sci. Vol. 140, 1982, pp. 369–383.

[16] A. LINGAS, V. SOLTAN, *Minimum Convex Partition of a Polygon with Holes by Cuts in Given Directions*, Technical Report , Lund University, 1996 (avaible at http://www.dna.lth.se/Research/Algorithms/Papers.description.html).

[17] W. LIPSKI, *Finding a Manhattan path and related problems*, Networks 13 (1983), pp. 399–409.

[18] W. LIPSKI, *An $O(n \log n)$ Manhattan path algorithms*, Inf. Process. Lett. 19 (1984), pp. 99–102.

[19] W. NEWMAN, R. SPROULL, *Principles of Interactive Computer Graphics*, 2nd ed., McGraw-Hill, New York, 1979.

[20] J. O'ROURKE, *Art Gallery and Algorithms*, Oxford Univ. Press, New York, 1987.

[21] J. O'ROURKE, K. SUPOWIT, *Some NP-hard polygon decomposition problems*, IEEE Trans. Inform. Theory 29 (1983), pp. 181–190.

[22] T. PAVLIDIS, *Structural Pattern Recognition*, Springer, Berlin, 1977.

[23] CH. PRISĂCARU, P. SOLTAN, *On partition of a planar region into d-convex parts and its application*, (Russian) Dokl. Akad Nauk SSSR 262 (1982), pp. 271–273. (English transl. Soviet Math. Dokl. 25 (1982), pp. 53–55.)

[24] V. SOLTAN, *Partition of a planar set into a finite number of d-convex parts*, (Russian) Kibernetika (Kiev), 1984, no. 6, pp. 70–74. (English transl. Cybernetics 20 (1984), pp. 855–860.)

[25] V. SOLTAN, A. GORPINEVICH, *Minimum dissection of rectilinear polygon with arbitrary holes into rectangles*, Discrete Comput. Geom. 9 (1993), pp. 57–79.

[26] G. T. TOUSSAINT, *Computational geometric problems in pattern recognition*, in Pattern Recognition. Theory and Applications, J. Kittler, ed., Reidel Publ. Co., Dordrecht, 1982, pp. 73–91.

# Efficient List Ranking on the Reconfigurable Mesh, with Applications *

Tatsuya Hayashi[1], Koji Nakano[1] and Stephan Olariu[2]

[1] Department of Electrical and Computer Engineering,
Nagoya Institute of Technology, Showa-ku, Nagoya 466, Japan
{hayashi,nakano}@elcom.nitech.ac.jp
[2] Department of Computer Science, Old Dominion University,
Norfolk, Virginia 23529, USA
olariu@cs.odu.edu

**Abstract.** Finding a vast array of applications, the list ranking problem has emerged as one of the fundamental techniques in parallel algorithm design. Surprisingly, the best previously-known algorithm to rank a list of $n$ items on a reconfigurable mesh of size $n \times n$ was running in $O(\log n)$ time. It was open for more than eight years to obtain a faster algorithm for this important problem.

Our main contribution is to provide the first breakthrough: we propose a deterministic list-ranking algorithm that runs in $O(\log^* n)$ time as well as a randomized one running in $O(1)$ expected time, both on a reconfigurable mesh of size $n \times n$. Our results open the door to an entire slew of efficient list-ranking-based algorithms on reconfigurable meshes.

## 1 Introduction

Recent years have seen a flurry of activity in the area of bus-based architectures [7]. Among these, the reconfigurable mesh (RMESH, for short) has emerged as a very attractive and powerful architecture. In essence, a RMESH of size $m \times n$ consists of $mn$ identical SIMD processors positioned on a rectangular array with $m$ rows and $n$ columns. As usual, it is assumed that the processors know their coordinates within the mesh: we let $PE(i, j)$ denote the processor in row $i$ and column $j$, with $PE(1, 1)$ in the north-west corner of the mesh.

Each processor is connected to its four neighbors in a mesh-like fashion and has 4 ports denoted by N, S, E, and W in Figure 1. Local connections between these ports can be established, under program control, creating a powerful *bus system* whose configuration changes dynamically to accommodate various computational needs. In general, the bus system comprises a number of disjoint *subbuses*. The results in this paper assume a model that allows at most two connections to be set in each processor at any one time. Furthermore, these two connections must involve disjoint pairs of ports as illustrated in Figure 1. We

* Work supported in part by NSF grant CCR-9522093, by ONR grant N00014-95-1-0779, and by Grant-in-Aid for Encouragement of Young Scientists (08780265) from Ministry of Education, Science, Sports, and Culture of Japan

note that some models proposed in the literature allow processors to fuse an arbitrary number of ports [4]. In accord with other workers [4, 5] we assume that broadcasts along subbuses take O(1) time. Although inexact, recent experiments with the YUPPIE, the GCN, and the PPA reconfigurable multiprocessor system [4] seem to indicate that this is a reasonable working hypothesis.

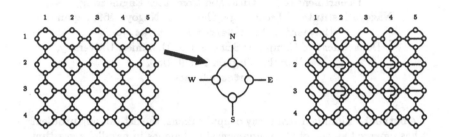

**Fig. 1.** *A reconfigurable mesh of size 4 × 5 and several subbuses*

The problem of list ranking is to determine, in parallel, the *rank* of every element in a given linked list, defined to be one larger than the number of elements following it in the list. The weighted list ranking problem is similar: here, every element of a linked list has a weight. The problem is to compute, for every element of the list the sum of the weights of the elements following it in the list. Since the two problems are solved using the same technique, we shall only discuss the unweighted version. List ranking turned out to be one of the fundamental techniques in parallel processing, playing a crucial role in a vast array of important parallel algorithms [3]. Based on the pointer jumping technique, the list ranking problem can be solved in logarithmic time on the PRAM [2, 3].

The problem of list ranking has been addressed on the RMESH before. An instance of size $n$ of this problem involves a linked list $L$ stored in one row or column of a RMESH of size $n \times n$. An $O(\log n)$ time algorithm for list ranking on a RMESH of size $n \times n$ follows immediately from ad-hoc arguments and is also implicit in [5,9]. Later, using a complicated algorithm, Olariu *et al.* [9] have shown that the problem can be solved in $O\left(\frac{\log n}{\log m}\right)$ time on a RMESH of size $mn \times n$. In particular, the result in [9] implies that for every fixed $\epsilon > 0$, an arbitrary instance of size $n$ of the list ranking problem can be solved in $O(1)$ time on a RMESH of size $n^{1+\epsilon} \times n$. In spite of this result, not much progress was made in the last eight years in solving the list ranking problem in less than $O(\log n)$ time on a RMESH of size $n \times n$.

Our main contribution is to present the first breakthrough: we propose a deterministic list-ranking algorithm running in $O(\log^* n)$ time as well as a randomized one running in $O(1)$ expected time, both on a RMESH of size $n \times n$. Our results open the door to a large number of list-ranking-based efficient algorithm on the RMESH. Due to page limitations, only the Euler-tour will be discussed in this paper. Other applications will be presented in the journal version of the work.

## 2 List ranking – a preview

Our algorithms use the following basic results.

**Lemma 1.** [10] *The task of computing the sum or the prefix-sums of $n$ $O(\log n)$-bit integers stored in one row or column of a RMESH of size $n \times n$ can be performed in $O(1)$ time.* □

The *packing* problem is defined as follows: Let $X$ be an arbitrary $n$-element set and let $P_1, P_2, \ldots, P_m$, $(m \leq n)$, be a collection of predicates such that each element in $X$ satisfies exactly one of the predicates. The packing problem asks to permute the elements of $X$ in such a way that for every $i$, $(1 \leq i \leq m)$, the elements satisfying $P_i$ occur consecutively. It is clear that packing is strictly weaker than sorting. Olariu *et al.* [10] proved the following result.

**Lemma 2.** *The task of solving an instance of size $n$ of the packing problem, stored in one row or column of a RMESH of size $n \times n$ can be performed in $O(1)$ time.* □

For our purposes, a linked list $L$ of nodes $1, 2, \ldots, n$ is specified by a mapping $l$ such that for every $i$, $(1 \leq i \leq n)$, node $l(i)$ follows $i$ in the list. Accordingly, the list $L$ is stored as a collection of $n$ ordered pairs of the form $(i, l(i))$. We assume that $l(i) = i$ only if $i$ is last in the list.

It is natural to map a linked list to a bus that captures the "topology" of the list. More precisely, for a given linked list $L = \{(i, l(i)) | 1 \leq i \leq n\}$, the corresponding bus configuration on a RMESH of size $n \times n$ is obtained by assigning node $i$, $(1 \leq i \leq n)$, to the diagonal processor $PE(i, i)$. The fact that $l(i)$ follows $i$ in the list is captured by a subbus connecting $PE(i, i)$ and $PE(l(i), l(i))$. This subbus proceeds vertically from $PE(i, i)$ to $PE(l(i), i)$ and then horizontally from $PE(l(i), i)$ on to $PE(l(i), l(i))$. Setting up the bus configuration is easy: in the first step, each processor $PE(i, i)$ broadcasts the ordered pair $(i, l(i))$ vertically; all the processors in column $i$ between $PE(i, i)$ and $PE(l(i), i)$ maintain their N–S connection. In the second step, processor $PE(l(i), i)$ broadcasts the ordered pair $(i, l(i))$ horizontally; as a result, all the processors in row $l(i)$ between $PE(l(i), i)$ and $PE(l(i), l(i))$ maintain their E–W connection. Finally, each processor $PE(l(i), i)$ connects ports W–S or N–E, depending on whether or not $l(i) < i$. At the same time, the diagonal processors $PE(i, i)$ connect an appropriate set of ports to join the subbuses corresponding to $(i, l(i))$ and $(l^{-1}(i), i)$. We shall refer to the bus constructed at the end of these three steps as the *bus-embedding* of $L$.

Figure 2 illustrates the bus-embedding of the list $L = \{(1, 4), (2, 5), (3, 8), (4, 2), (5, 6), (6, 6), (7, 1), (8, 7)\}$.

By using the bus-embedding in conjunction with the algorithm of Lemma 1, the rank of a single node $i$ of list $L$ can be computed as follows:

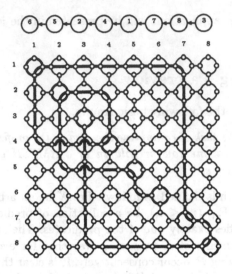

**Fig. 2.** *Embedding a list into a RMESH of size n × n*

---

**Step 1.** Find the bus-embedding of $L$; processor $PE(i, i)$ removes its local connection, disconnecting the subbuses corresponding to $(i, l(i))$ and $(l^{-1}(i), i)$;
**Step 2.** Processor $PE(i, i)$ sends a signal on the subbus corresponding to $(i, l(i))$;
**Step 3.** Each diagonal processor $PE(j, j)$ that receives the signal writes a 1 in
 a local register; all other diagonal processors write a 0;
**Step 4.** Compute the sum of the values written in Step 3 to obtain the rank of
 node $i$ in $L$.

---

The reader should have no difficulty confirming that this simple algorithm correctly computes the rank of node $i$ in O(1) time. On a RMESH of size $n^2 \times n$
this simple algorithm can be run, in parallel, for every node in the list. Thus, we
have the following result.

**Lemma 3.** *An instance of size n of the list ranking problem can be solved in
O(1) time on a RMESH of size $n^2 \times n$.*                                    □

The weighted version of the list ranking problem can also be solved in O(1)
time, provided that the weights are logarithmic; in Step 3, processors receiving
the signal write the weight of the corresponding node instead of 1. Thus we have
the following result.

**Lemma 4.** *An instance of size n of the weighted list ranking problem can be
solved in O(1) time on a RMESH of size $n^2 \times n$, provided that the weights are
logarithmic.*                                                                    □

    The remainder of the paper is devoted to reducing the size of RMESH to
$n \times n$. We begin by showing that the list ranking problem can be solved in O(1)

time on a RMESH of size $n \times n$ if a good ruling set (that we are about to define) in $L$ is available. Consider a subset $S$ of $L$. The *S-shortcut* of $L$ is a linked list involving the nodes in $S$ only. Specifically, the ordered pair $(p, q)$ belongs to $S$-shortcut if both $p$ and $q$ belong to $S$ and the path in $L$ from $p$ to $q$ contains no other node from $S$. If $(p, q)$ belongs to $S$-shortcut, the *q-sublist* is the sublist of $L$ starting at $l(p)$ and ending at $q$. A subset $S$ of nodes in $L$ is termed a *good ruling set* if 1) $S$ includes the last node of $L$, 2) $S$ contains at most $n^{2/3}$ nodes, and 3) each $p$-sublist $(p \in S)$ contains at most $\sqrt{n}$ nodes.

Now suppose that a good ruling set $S$ is available. The following algorithm solves the list ranking problem in $O(1)$ time on a RMESH of size $n \times n$. We refer the reader to Figure 3, where the various steps of the algorithm are illustrated on the list $L = \{(1, 3), (2, 1), (3, 9), (4, 6), (5, 2), (6, 6), (7, 4), (8, 5), (9, 7)\}$ and the ruling set $S = \{2, 6, 9\}$.

**Step 1.** By using the bus-embedding, broadcast $p$ to the processors storing nodes in the $p$-sublist. Using packing move each $p$-sublist to consecutive columns of the RMESH.

**Step 2.** Solve the list ranking problem for each $p$-sublist in the corresponding consecutive columns: since none of these sublists contains more than $\sqrt{n}$ nodes, we can rank all the sublists in parallel using Lemma 3. Note that, the number of nodes in each $p$-sublist is also computed by this procedure.

**Step 3.** Identify the $S$-shortcut as follows: find the bus-embedding of $L$. Each $PE(p, p)$ with $p \in S$ removes its local connection, disconnecting the subbuses corresponding to $(p, l(p))$ and $(l^{-1}(p), p)$. In two broadcasts on the resulting subbuses, each processor $PE(p, p)$ with $p \in S$ finds its predecessor and successor in $S$-shortcut. Use the weighted list ranking algorithm of Lemma 4 to compute the actual rank in $L$ of each node in $S$, where the weight of a node $p$ in $S$ is the number of nodes in $p$-list.

**Step 4.** By broadcasting the actual rank of each node $p$ $(p \in S)$ to the processors in each node in $p$-sublist, every node in $L$ finds its rank in $L$.

Since each step can be performed in $O(1)$ time, we have the following result.

**Theorem 5.** *An instance of size $n$ of the list ranking problem can be solved in $O(1)$ time on an $n \times n$ RMESH, provided that a good ruling set is available.* □

The crux of our approach to list ranking is an algorithm to determine a good ruling set in a given list. In Section 3 we present a deterministic algorithm solving this task in $O(\log^* n)$ time. In Section 4 we present a randomized algorithm that finds a good ruling set in $O(1)$ expected time. In conjunction with Theorem 5, these results yield, respectively, a deterministic list ranking algorithm running in $O(\log^* n)$ time, as well as a randomized one running in $O(1)$ expected time.

## 3 A deterministic ruling set algorithm

The main goal of this section is to show that a good ruling set can be obtained in $O(\log^* n)$ time on a RMESH of size $n \times n$. As a first step in this direction, we

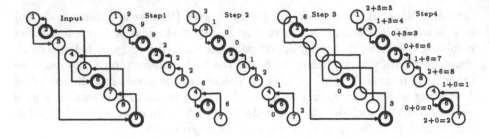

**Fig. 3.** *Illustration of list ranking algorithm of Theorem 5*

exhibit an algorithm for this task running in $O(\log\log n)$ time. We then show that the running time can be reduced to $O(\log^* n)$.

A *k-sample* of a linked list $L$ is a set of nodes obtained by retaining every $k$-th node in $L$, that is, nodes whose ranks in $L$ are $1, k+1, 2k+1, 3k+1, \ldots$. Let $L/k$ denote the ($k$-sample)-shortcut. By definition, the nodes in $L/k$ are a good ruling set whenever $n^{1/3} \le k \le n^{1/2}$. Our goal is to find such a $L/k$.

Figure 4 shows an abstract bus configuration to find a $k$-sample of a list $L$. The $k$ subbuses are laid in $k$ layers, and each column corresponds to a node in $L$. By setting the local connections appropriately, the $k$ subbuses are shifted cyclically in each column, as illustrated in Figure 4. In this bus configuration, a signal is sent through the topmost subbus in the column corresponding to the last node in $L$. By construction, the signal passes through the topmost layer of every $k$-th subsequent column. By selecting the corresponding nodes in $L$, we obtain the desired $k$-sample.

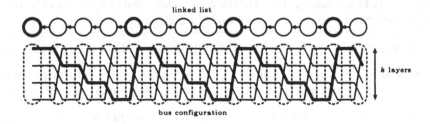

**Fig. 4.** *Illustrating the abstract bus configuration for finding a k-sample*

Next, having obtained the list $L/k$, we can get the $k$-sample list of $L/k$, that is, the ($k^2$-sample)-shortcut, $L/k^2$, of $L$ by using the wide bus-embedding defined next. The *wide bus-embedding* assigns a "bundle" of $k$ subbuses to connect two nodes, while a single subbus corresponds to a link between two nodes in the bus-embedding. Figure 5 shows an example of the wide bus-embedding. With each node $i$ in $L/k$, we assign a submesh of size $k \times k$. Moreover, $k$ rows and $k$

columns are used to connect nodes $i$ and $l(i)$, the successor of $i$ in $L/k$. The wide bus-embedding can be configured in $O(1)$ time as follows: Let $i_1, i_2, \ldots, i_{n/k}$ be nodes in $L/k$ such that $1 \leq i_1 < i_2 < \cdots < i_{n/k}$. We assume that processor $PE(i_j, i_j)$ is in charge of the ordered pair $(i_j, l(i_j))$, $(1 \leq j \leq n/k)$, where $l(i_j)$ denotes the successor of $i_j$ in $L/k$. Note that $PE(i_j, i_j)$ need not know the value of $j$: by using the prefix sums algorithm of Lemma 1, $PE(i_j, i_j)$ can learn the value of $j$. This enables $PE(i_j, i_j)$ to sends $(i_j, l(i_j))$ to $PE(jk, jk)$ and, as a consequence, rows and columns $(j-1)k+1$ through $jk$ are being dedicated to $(l^{-1}(i_j), i_j)$ and $(i_j, l(i_j))$, respectively. By broadcasting $(i_j, l(i_j))$ on these rows and columns, the corresponding wide bus-embedding can be configured in $O(1)$ time, in the same manner as the bus-embedding. By using the resulting wide bus-embedding, the abstract bus configuration for finding the $k$-sample can be embedded in a RMESH of size $n \times n$. Thus, if $L/k$ is available, $L/k^2$ can be computed in constant time.

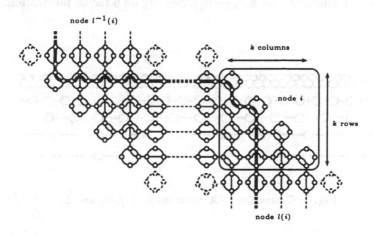

**Fig. 5.** *Illustrating the embedding of the abstract bus configuration for finding a k-sample*

**Lemma 6.** *A good ruling set can be selected in $O(\log \log n)$ time on a RMESH of size $2n \times 2n$.*

*Proof.* Assume that a RMESH of size $2n \times 2n$ is given. As discussed, $L/2$ can be computed in $O(1)$ time. As soon as $L/2$ is available, $L/2^2$ can be computed in $O(1)$ time. In general, if $L/2^{2^i}$ is available, $L/2^{2^{i+1}}$ can be computed in $O(1)$ time. Therefore, $L/2^{2^i}$ can be computed in $O(i)$ time by iterating the procedure above. We stop when $n^{1/4} \leq 2^{2^i} \leq \sqrt{n}$. After that, we compute a $\sqrt{n}/2^{2^i}$-sample of $L/2^{2^i}$, that corresponds to $L/\sqrt{n}$. Clearly, since $\sqrt{n}/2^{2^i} \leq 2^{2^i}$, this takes $O(1)$ time. Since the number of iteration is $O(\log \log n)$, and each iteration takes $O(1)$ time, the good ruling set $L/\sqrt{n}$ can be computed in $O(\log \log n)$ time.

Next, we will show how reduce the size of the RMESH to $n \times n$. By implementing the EREW-PRAM algorithm that computes a 3-ruling set of $L$ in $O(\log^* n)$ time [2] to the $n \times n$ RMESH, $L$ can be reduced such that it has at most $n/2$ nodes. By Lemma 6, the list ranking problem for the reduced linked list can be solved in $O(\log \log n)$ time on a RMESH of size $n \times n$. After that, by obvious broadcasting, the rank of each node in $L$ can be computed. Consequently, the computing time is still $O(\log \log n)$ on the $n \times n$ RMESH.

We now show how to reduce the time to compute a good ruling set to $O(\log^* n)$. The idea is based on the parallel prefix-remainder computation [8]. Let $p_1(= 2), p_2(= 3), p_4(= 5), \ldots, p_q$ be the first $q$ prime numbers. For a list of $L$, $q$ linked lists $L/p_1, L/p_2, \ldots, L/p_q$ can be computed by $p_1 + p_2 + \cdots + p_q$ layers in the abstract bus configuration. Then, we retain in the sample only elements common to all the $q$ linked lists $L/p_1, L/p_2, \ldots, L/p_q$. In other words, a node $p$ such that $r(p) \bmod p_1 = r(p) \bmod p_2 = \cdots = r(p) \bmod p_q = 1$ is selected, where $r(p)$ is the rank of $p$ in $L$. For such $p$, $r(p) \bmod p_1 p_2 \cdots p_q = 1$ holds. Thus, the nodes selected constitute $L/(p_1 p_2 \cdots p_q)$. See Figure 6 for an illustration.

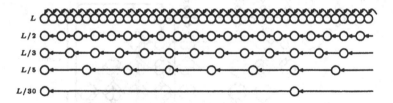

**Fig. 6.** *Computing $L/30$ by using $L/2$, $L/3$, and $L/5$*

To implement this idea, we begin by identifying all the prime numbers not exceeding $n$. For an integer $i$, $(i \leq n^2)$, the task of determining whether or not $i$ is prime can be solved in $O(1)$ time by using $n$ processors: all we need is check divisibility by $2, 3, 4, \ldots, \sqrt{i}$. Therefore, on a RMESH of size $n \times n$ one can find all the primes not exceeding $n$ in $O(1)$ time. In fact, the prime numbers less than $O(\log n)$ are sufficient for the list ranking algorithm. Based on the idea above, we have

**Lemma 7.** *A good ruling set of a list $L$ of size $n$ can be obtained in $O(\log^* n)$ time on a RMESH of size $2n \times 2n$.*

*Proof.* As in Lemma 6, the resulting sample is a good ruling set. By the prime number theorem, $p_q = \Theta(q \log q)$. Therefore, $p_1 + p_2 + \cdots + p_q = \Theta(q^2 \log q)$ and $p_1 p_2 \cdots p_q = (q \log q)^{\Theta(q)}$ hold. Let $k = p_1 + p_2 + \cdots + p_q$. Then, $p_1 p_2 \cdots p_q = 2^{\Theta(k)}$. Thus, if $L/k$ is available, $L/(2^{\Theta(k)})$ can be computed in $O(1)$ time. Assume that $L/K(t)$ is computed in the $t$-th iteration. Then, $K(t+1) = 2^{\Theta(K(t))}$ and

$K(0) = 1$ hold. Since $K(t) \geq \sqrt{n}$ is sufficient for termination, the number of iteration is $O(\log^* n)$.

As before, we can reduce the size of RMESH to $n \times n$ by using a 3-ruling set.

**Lemma 8.** *A good ruling set of a list $L$ can be obtained in $O(\log^* n)$ time on a RMESH of size $n \times n$.* □

Lemma 8 and Theorem 5, combined, imply the main result of this section.

**Theorem 9.** *An instance of size $n$ of the list ranking problem can be solved deterministically in $O(\log^* n)$ time on a RMESH of size $n \times n$.* □

## 4 A randomized ruling set algorithm

The purpose of this section is to exhibit a randomized algorithm for returning a good ruling set in a given linked list in $O(1)$ expected time. The details are spelled out as follows:

**Step 1.** Each $PE(i, i)$ randomly selects node $i$ as a sample in $S$ with probability $n^{-5/12}$.

**Step 2.** Check whether $S$ is a good ruling set. If $S$ is a good ruling set, then the algorithm terminates, otherwise repeat Step 1.

Notice that the list ranking algorithm of Theorem 5 enables us to determine in $O(1)$ time whether $S$ is a good ruling set. To wit, in Step 2 one can check whether $|S| \leq n^{2/3}$; in Step 3 it can be checked whether each $p$-sublist ($p \in S$) has at most $\sqrt{n}$ nodes. Therefore, Steps 1 and 2 of the algorithm above can be completed in $O(1)$ time. Next, we will evaluate the expected number of trials to terminate this algorithm. Step 1 selects $n^{7/12}$ nodes on the average, and the expected size of each $p$-sublist is $n^{5/12}$. Note that to be a good ruling set, $|S| \leq n^{2/3}$ and each $p$-sublist must have at most $\sqrt{n}$ nodes. Therefore, a margin of a factor of $n^{1/12}$ is allowed for the numbers of nodes in $S$ and every $p$-sublist. This margin is sufficient to ensure that a good ruling set is found in a constant expected number of trials. The detailed proof will be presented in the journal version of the work.

Having obtained a good ruling set, the list ranking problem can be solved in $O(1)$ time by using Theorem 5.

**Theorem 10.** *The task of ranking a linked list of size $n$ can be performed in $O(1)$ expected time on a RMESH of size $n \times n$.*

## 5 Applications

List ranking plays a crucial role in a vast array of important parallel algorithms. Due to stringent page limitations, among the many applications of list ranking we present one that is used time and again in devising parallel algorithms. The

*Euler tour* technique developed in [11] allows one to compute a number of tree functions by reducing them to list ranking. We shall now present the details of a variant of this technique. Let $T = (V, E)$ be an arbitrary $n$-node rooted tree. For each node $v \in V$, we fix a certain ordering on the set of vertices adjacent to $v$ – say, $adj(v) = \langle u_0, u_1, \ldots, u_{d-1} \rangle$, where $d$ is the degree of $v$. The next of each arc $\langle u_i, v \rangle$, $l(\langle u_i, v \rangle)$ is defined to be $\langle v, u_{(i+1) \bmod d} \rangle$. As an exception, let $l(\langle u_{d_r}, r \rangle) = \langle u_{d_r}, r \rangle$, where $r$ is the root of $T$ and $d_r$ is its degree. The $l$ thus defined is a linked list ended at $\langle u_{d_r}, r \rangle$ called the *Euler tour* [3]. Since the Euler tour of the $n$-node rooted tree has $2n$ node, the list ranking problem for the Euler tour can be solved on the $2n \times 2n$ RMESH in $O(\log^* n)$ time and in $O(1)$ expected time.

After computing the rank of each node in the Euler tour, a large number of tree-functions can be computed by reducing them to list ranking. A few of these functions are listed next: 1) computing a preorder and/or postorder numbering of the nodes of $T$, 2) computing the number of descendants of every node in $T$, 3) computing the level of every node in $T$, 4) computing the nodes of the (unique) path joining nodes $u$ and $v$ in $T$, 5) computing the lowest common ancestor of nodes $u$ and $v$ in $T$, 6) computing a Breadth-First traversal of $T$, 7) computing the center of $T$, and 8) computing a level-order encoding of $T$.

# References

1. R. J, Anderson and G. L. Miller, A simple randomized algorithm for list-ranking, *Information Processing Letters*, 33, (1990), 269-273.
2. R. Cole and U. Vishkin, Deterministic coin tossing with applications to optimal parallel list ranking, *Information and Control*, 70, (1986), 32–53.
3. J. JáJá, *An introduction to parallel algorithms*, Addison-Wesley, Reading, Massachusetts, 1991.
4. H. Li and M. Maresca, Polymorphic-torus network, *IEEE Transactions on Computers*, 38, (1989), 1345–1351.
5. R. Miller, V. K. P. Kumar, D. Reisis, and Q. F. Stout, Parallel computations on reconfigurable meshes, *IEEE Transactions on Computers*, 42, (1993), 678–692.
6. R. Motwani and P. Raghavan, *Randomized Algorithms*, Cambridge University Press, 1995.
7. K. Nakano, A bibliography of published papers on dynamically reconfigurable architectures, *Parallel Processing Letters*, 5, (1995), 111-124.
8. K. Nakano and K. Wada, Integer summing algorithms on reconfigurable meshes, *Proc. First IEEE International Conference on Algorithms And Architectures for Parallel Processing*, vol. 1, 187–196, 1995.
9. S. Olariu, J. L. Schwing, and J. Zhang, Fundamental algorithms on reconfigurable meshes, *Proc. 29-th Annual Allerton Conf. on Communication, Control, and Computing*, 1991, 811–820.
10. S. Olariu, J. L. Schwing, and J. Zhang, Integer problems on reconfigurable meshes, with applications, *Journal of Computer and Software Engineering*, 1, (1993), 33–46.
11. R. E. Tarjan and U. Vishkin, Finding biconnected components and computing tree functions in logarithmic parallel time, *SIAM Journal of Computing*, 14, (1985) 862–874.

# Periodic Merging Networks*

Mirosław Kutyłowski[1], Krzysztof Loryś[2], Brigitte Oesterdiekhoff[1]

[1] Heinz Nixdorf Institute and Department of Mathematics & Computer Science, University of Paderborn, D-33095 Paderborn, Germany, {mirekk,brigitte}@uni-paderborn.de,
[2] Institute of Computer Science, University of Wrocław, and Dept. of Computer Science, University of Trier, D-54286 Trier, Germany, lorys@TI.Uni-Trier.DE

**Abstract.** We consider the problem of merging two sorted sequences on constant degree networks using comparators only. The classical solution to the problem are the networks based on Batcher's Odd-Even Merge and Bitonic Merge running in $\log(2n)$ time. Due to the obvious $\log n$ lower bound for the runtime, this is time-optimal.

We present new merging networks that have a novel property of being periodic: for some (small) constant $k$, each processing unit of the network performs the same operations at steps $t$ and $t + k$ (as long as $t + k$ does not exceed the runtime.) The only operations executed are compare-exchange operations, just like in the case of the Batcher's networks. The architecture of the networks is very simple, easy to be laid out. The runtimes achieved are $c \cdot \log n$, for a small constant $c$.

## 1 Introduction

*Merging* is the following problem: *given sorted sequences $A$ and $B$ of $n$ keys each, arrange all $2n$ elements of $A$ and $B$ in one sorted sequence.*

Merging is one of the most fundamental problems in computer science and has been intensively studied from theoretical and practical point of view. There have been a lot of efforts to construct optimal merging algorithms in different models. In the parallel setting we have to consider at least two very different situations. For one of them we have a parallel machine performing a non-oblivious computation, for instance a shared memory machine (PRAM). Fascinating sublogarithmic time algorithms have been developed for this model, see for instance [4]. Another possibility is to consider fixed interconnection networks performing oblivious computations. The algorithms developed in this setting might be better suited for implementation, for instance on VLSI circuits. Therefore we are mainly interested in the network model.

### 1.1 The Model

We consider the following architecture: a comparator network algorithm has an *underlying graph* $G = (V, E)$, where $V$ is the set of processing units (serving as registers

---

* Partially supported by KBN grants 8 S503 002 07, 2 P301 034 07, DFG-Sonderforschungsbereich 376 "Massive Parallelität", DFG Leibniz Grant Me872/6-1 and EU ESPRIT Long Term Research Project 20244 (ALCOM-IT); this research was partially done while the second author visited University of Paderborn

as well) and $E$ is the set of the links connecting the processing units. The elements of $E$ will be later called *comparators* due to the function they perform. We assume that $G$ is a graph of a **constant** degree. We assume that during execution of the algorithm every processing unit stores exactly one element at each moment. The set $V$ is ordered, the goal of the merging algorithm is to relocate the input elements, so that they form a nondecreasing sequence according to the ordering of $V$. The convention for initial placing two sorted sequences of input elements might be arbitrary as long it is simple.

During a run of the algorithm the elements may be exchanged by neighboring processing units, that is, those processing units that are connected by an comparator in $E$. A comparator $(P_i, P_j) \in E$, where $P_i < P_j$ in the ordering of $V$, acts as follows: the elements held in $P_i$ and $P_j$ are compared, then the smaller element is placed in $P_i$ and the bigger one is placed in $P_j$. The processing unit $P_i$ ($P_j$) is said to be on the minimum side (on the maximum side) of the comparator $(P_i, P_j)$.

A comparator network algorithm consists of *parallel compare-exchange steps*. A parallel compare-exchange step uses a matching in the graph $G$, that is a set of edges $E' \subseteq E$, such that no element of $V$ belongs to two edges in $E'$. During the parallel step all comparators from $E'$ act in parallel.

We say that a comparator network algorithm is **periodic** with period $k$ if for every moment $t$ of the computation, the parallel compare-exchange steps $t$ and $t + k$ use the same matching, that is, the same groups of comparators. The steps $(i - 1) \cdot k + 1$ through $i \cdot k$ are called the $i$th round of the algorithm.

Comparator network algorithms are often described in a different way. The processing units are replaced by wires. Then step $t$ of a compare-exchange algorithm corresponds to level $t$ of comparators in the model with wires. If the algorithm is not periodic, such an interpretation is necessary in order to implement the algorithm on a network of a fixed degree.

## 1.2 Previous Work

Merging by comparator networks has been studied for a long time. The most famous merging networks are the Odd-Even Merger and the Bitonic Merger of Batcher [1], and the Balanced Merger of Dowd et al. [5]. They run in time $\log(2n)$ and use $n \cdot \log n + O(n)$ comparators. There has been a lot of research on these networks and they have been modified in many ways (see [13], [11], [3], [2], [10]). Milterson et al. [9] prove that the minimal number of comparators needed for merging two $n$-element sequences is at least $n \cdot \log n - O(n)$, improving the previous results (see Knuth [7]). It follows that the networks by Batcher are asymptotically optimal. This result also shows that merging two $n$-element sequences on a comparator network requires $\log n - O(1)$ time.

Recently, periodic comparator network algorithms for sorting have been intensively studied. After a number of algorithms with a non-constant period ([5], [2], [10] and [3]) finally some fast periodic sorting networks of constant period have been found:[12], [6] and [8]. The main result of the last paper mentioned is the so called "periodification scheme". It says that any sorting comparator network algorithm may be rebuilt so that we obtain a constant period algorithm. This modification causes a slowdown of factor $\log n$ in the case of fast algorithms.

In a certain sense we expand the ideas of [8]. However, the periodification scheme cannot yield a time-optimal merging algorithm due to the slowdown inherent in the periodification scheme. Therefore we develop a direct solution.

### 1.3 The New Results

**Theorem 1.** *There is a periodic comparator network of period* 4 *that merges two sorted sequences of $n$ items in time* $10 \cdot \log_3 n \approx 6.3 \cdot \log n$.

It is not difficult to find an example showing that $10 \cdot \log_3 n - O(1)$ is really the worst case runtime of our algorithm. In a forthcoming journal version of this paper we present further variants of Theorem 1, where we reduce the period to 3 or reduce the runtime to about $2.84 \log n$. (However, they require a more tedious analysis.) The last result differs only by a small constant factor from the lower bound for merging in comparator networks, although the restrictions imposed on our network are very strong.

## 2 Construction of the Network

In this section we describe the techniques used by our algorithm. We explain the underlying ideas before we go into (rather technical) details.

It is well known that comparator networks that sort all inputs consisting of 0's and 1's sort correctly arbitrary input sequences (so called 0-1-Principle [7]). A similar phenomenon holds for merging (and can be proved in an analogous way):

**Proposition 2.** *If a comparator network algorithm merges two sorted sequences consisting solely of 0's and 1's, then it correctly merges any two sorted input sequences.*

By Proposition 2, for the rest of the paper we may consider only the input sequences consisting of 0's and 1's.

Our merging algorithm uses a network $M$ the processing units of which are arranged in a $p \times q$-rectangle, where $p = 3^q$, $q$ is even. Let $P_{i,j}$ denote the processing unit of $M$ in the row $i$ and the column $j$, for $i \in \{1, 2, \ldots, p\}$ and $j \in \{1, 2, \ldots, q\}$. Let $C_j = \{P_{i,j} \mid i \leq p\}$ denote the $j$th column of $M$. We order the nodes of $M$ by the snake-like ordering: $P_{i,j} \prec P_{i',j'}$ if and only if

$$(i > i') \quad \vee \quad (i = i' \wedge j < j' \wedge i \text{ odd}) \quad \vee \quad (i = i' \wedge j > j' \wedge i \text{ even}).$$

The algorithm has a strange property that at each step of the computation if an element moves from column $C_i$ to $C_j$, then every element stored in $C_i$ moves to $C_j$. (This property holds only for the input sequences consisting of 0's and 1's, and under the assumption that the comparators may switch equal elements. Thereby our analysis collapses if we do not use the 01-Principle. This is an interesting example how powerful this principle is.) So we may interpret that the algorithm moves the columns around the network. The purpose of these movements is that at different positions different sets of comparators are applied inside a column. The input allocation to this network has the property that the number of 1's at each column is the same up to one. Since the columns move separately, this property will be preserved all the time. After sorting each column, the contents of the network is sorted except of at most one row. Some additional rounds suffice to sort this row.

## 2.1 Input Allocation.

**Definition 3.** We say that a column $C_i$ is of the form $1^d(01)^e0^*$ if $d$ bottom positions in $C_i$ are occupied by 1's, the $p - 2e - d$ top positions in $C_i$ are occupied by 0's, and in the remaining $2e$ positions there are interchangeably 0's and 1's, starting with a 0 at position $d + 1$.

It is crucial that the input allocation fulfills the following property:

**Property 1** *After loading the input sequences*
 - *the number of 1's in the columns of M may differ by at most 1,*
 - *each column has the form $1^d(01)^e0^*$ for some e and d.*

Property 1 is obviously fulfilled for instance for the following easy scheme: The elements of $A$ are loaded into the odd rows, and the elements of $B$ into the even rows; elements of $A$ ($B$) are placed according to the snake-like ordering.

Our algorithm will be designed in such a way that Property 1 will hold during the whole computation.

## 2.2 Horizontal Steps

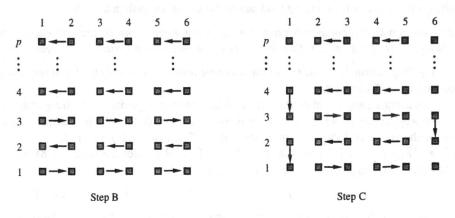

Step B        Step C

**Fig. 1.** The parallel Steps B and C

The algorithm of Theorem 1 executes periodically 4 different steps, called A, B, C, and D. The Steps A and D, called *vertical*, affect each column separately. The Steps B and C, called *horizontal*, exchange elements between consecutive columns in the same row (see Fig. 1). (An exception are the comparators in the leftmost and rightmost columns at Step C.) More precisely, Step B uses comparators $(P_{i,j}, P_{i,j'})$, $i \le p, j \le q$, where

$$j' = j + 1 \quad \text{if } i, j \text{ are odd}, \qquad j' = j - 1 \quad \text{if } i, j \text{ are even}.$$

Step C uses comparators $(P_{i,j}, P_{i,j'})$, where

$$j' = j + 1, \quad \text{if } i \text{ is odd}, j \text{ is even}, j < q \qquad j' = j - 1, \quad \text{if } i \text{ is even}, j \text{ is odd}.$$

Additionally, at $C_1$ there are comparators $(P_{i,1}, P_{i-1,1})$ for each even $i \leq p$, and at $C_q$ there are comparators $(P_{i,q}, P_{i-1,q})$ for each odd $i \leq p$, $i \neq 1$.

**Definition 4.** A column with the contents of the form $1^d(01)^e 0^*$ is classified as an *E-column* if $d$ is even, and $e > 0$, or as an *O-column* if $d$ is odd and $e > 0$. If $e = 0$, then the column is said to be an *S-column*. ("S" stands for "sorted", "O" for "odd", and "E" for "even".)
The lowest $d$ positions in this column form the *foot* of the column, the next $2e$ positions form the *01-region*.
We say that a column is an E/S-column if it is an E- or S-column, and we call it an O/S-column if it is an O- or S-column.

In the following lemma we show that under certain circumstances the columns exchange their contents during horizontal steps. This is a key to our construction.

**Lemma 5.** *Consider the columns $C_i$ and $C_{i+1}$ of $M$ and assume that $C_i$ is an O/S-column, $C_{i+1}$ is an E/S-column and that the number of $1$'s in them differ by at most $1$. Assume further that during Step B or C comparators connect the columns $C_i$ and $C_{i+1}$. Then during this step:*

(a) *$C_i$ and $C_{i+1}$ exchange their items provided that at least one of them is not an S-column.*

(b) *$C_i$ and $C_{i+1}$ sort together according to the snake-like ordering of $M$ provided that they are both S-columns.*

**Proof.** Point (b) is obvious. For (a) assume that $C_i$ is an O-column and $C_{i+1}$ is an E-column. Let $C_i$ be of the form $1^d(01)^e 0^*$ and $C_{i+1}$ be of the form $1^{d'}(01)^{e'} 0^*$. Suppose that $d < d'$. Then the $d$ bottom positions in $C_i$ and $C_{i+1}$ are occupied by $1$'s and this does not change at the horizontal step. So we may say that within the first $d$ rows $C_i$ and $C_{i+1}$ exchange their contents. In the next $2e$ rows the $1$'s of $C_i$ are at the minimum sides of the comparators, whereas the $0$'s are at the maximum sides of the comparators. Thus the comparators move these $0$'s and $1$'s to column $C_{i+1}$ no matter what are the contents of $C_{i+1}$ (equal items are supposed to be switched.) Now let us consider the positions above the row $d + 2e$ in both columns. $C_i$ contains no $1$'s there, so we have to show that all $1$'s from $C_{i+1}$ in this part are moved to $C_i$ at the horizontal step. Column $C_i$ contains $d + e$ ones, hence $C_{i+1}$ contains at most $d + e + 1$ ones. Assume that $C_{i+1}$ contains at least one $1$ above the row $d + 2e$. Since $d$ ones are contained in the first $d$ rows, the next $2e$ rows contain at least $e$ ones, there is exactly one $1$ above the row $d + 2e$ in $C_{i+1}$. Since $C_{i+1}$ is an E-column, its last one is at an even row and therefore must be moved into $C_i$ during this step. So we see that also above the row $d + 2e$ the columns $C_i$ and $C_{i+1}$ exchange their $1$'s.
The other cases can be checked in a similar way. □

## 2.3 Step A and Movements of the Columns

Lemma 5 shows that O- and E-columns exchange their contents at Steps B and C. For our eye observing the computation it looks like switching positions by the columns. So for a computation consisting of Steps B and C, only, we would see only the columns moving around the network. We shall call them *moving columns*. Our algorithm performs also vertical Steps D and A. They influence each of the moving columns separately. How the vertical steps sort the moving columns is an important issue, but we require that immediately before Step B the odd column are O/S-columns, and the even columns are E/S-columns. (Otherwise we could not use Lemma 5.) For this purpose, Step A applies the comparators $(P_{i,j}, P_{i-1,j})$, where $i, j$ are of different parity (see Fig. 2). By the

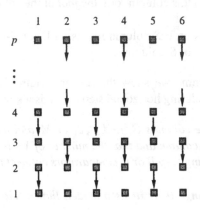

**Fig. 2.** The parallel Step A

definition we get immediately the following property:

**Lemma 6.** *(a)    Suppose that each column is an E-, O- or S-column. Then by performing Step A, each odd column of M becomes an O/S-column, each even column becomes an E/S-column.*
*(b)    At Step C the leftmost column becomes an O/S-column and the rightmost column becomes an E/S-column.*

By Lemma 6 we see that immediately before Step B the O- and E-columns are in the right positions in order to apply Lemma 5. So

   *immediately before Step C each odd column is an E/S-column and each even column is an O/S-column.*

Hence again Lemma 5 can be applied and together with Lemma 6(b) we get:

   *immediately before Step D each odd column is an O/S-column and each even column is an E/S-column.*

By the above observations it is easy to see that the sequence of the steps has been arranged in such a way that:
(i)    During Steps B and C the E-columns move to the left, the O-columns move to the right. The only exception are the leftmost and the rightmost columns at Step C – they cannot move any further. Instead they change their status: the leftmost column becomes an O/S-column and the rightmost column becomes an E/S-column.

(ii)    During Step D some changes within unsorted columns are made. For this reason an E-column (O-column) may loose its status of an E-column (O-column), but at the next Step A it recovers the proper status.

(iii)   Every column that is not sorted *moves* back and forth through $M$, the changes of the direction of the movement occur only at the moment when it reaches the rightmost or the leftmost column. An immediate consequence is that we sort the elements of each column separately.

## 2.4   Construction of Step D

Now we describe Step D responsible for sorting the columns. In a column $C_i$ it uses all comparators $(P_{j,i}, P_{j-3^{w(i)},i})$, for $j > 3^{w(i)}$ having the same parity as $i$. The choice of the parameter $w(i)$ will be discussed in the next section. The comparators $(P_{i,j}, P_{i,j-3^{w(i)}})$ will be called *jump comparators* and $3^{w(i)}$ the *size of the jump*.

Note that all jump comparators within a column have the same length and either originate in odd rows and point to even rows, or originate in even rows and point to odd rows. Thereby, the comparators have no common endpoints and the definition is sound. For the inputs considered, immediately before Step D odd columns are O/S-columns and even columns are E/S-columns. Thus in the odd (even) columns the jump comparators originate at odd (even) rows where we may expect to find 1's of the 01-region. The idea is that with some luck some of these 1's jump into the places occupied previously by 0's of the 01-region and thereby reduce the size of the 01-region. Of course, the effect depends on the relative size of the 01-region and the jump size:

**Lemma 7.** *Suppose that immediately before Step D a column $C_i$ has the form $1^d(01)^e 0^*$, where $d$ has the same parity as $i$. Let $\ell$ be the jump size at column $C_i$. Then after Step D:*

(a) *For $e < \ell/2$, the contents of $C_i$ remains unchanged.*
(b) *For $\ell/2 \le e < \ell$, column $C_i$ gets the form $1^{d+2e-\ell+1}(01)^{\ell-1-e} 0^*$.*
(c) *For $e \ge \ell$, column $C_i$ gets the form $1^{d+\ell}(01)^{e-\ell} 0^*$.*

**Proof.**   Of course, performing Step D leaves the old foot intact, and no 1 can move above the 01-region. So the only part that may change is the 01-region.
*Case (a)*: If $\ell > 2e$, then each jump comparator has either the maximum side in the foot, or the minimum side above the 01-region. Such a comparator does not change the contents of its endpoints, hence performing Step D has no affect.
*Case (b)* (see Fig. 3): As already seen, the jump comparators originating between rows $d+1$ and $d+\ell$ compare two 1's and thereby do not change the contents of the column. The jump comparators with the minimum side at rows $d+\ell+1$ up to $d+2e$ have 1's on the minimum side and 0's on the maximum side. Therefore their 1's jump down. There are $e - (\ell-1)/2$ such comparators, so these ones move into rows $d+1, d+3, \ldots, d+2e-\ell$. Since $2e - \ell < \ell$, all these 1's are moved into the part of the 01-region inside which the ones remain on their places. Thereby, the foot grows up to row $d+2e-\ell+1$, and there is a new 01-region containing the remaining $\ell - 1 - e$ ones that have not been moved. So the column gets the form $1^{d+2e-\ell+1}(01)^{\ell-1-e} 0^*$.
*Case (b)* (see Fig. 3): The difference to the previous case is that not all 1's that jump down fit into the $\ell$ lowest rows of the old 01-region – there are too many jumping 1's.

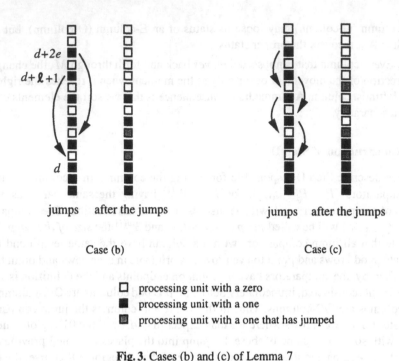

$d+2e$

$d+\ell+1$

$d$

jumps    after the jumps

jumps    after the jumps

Case (b)

Case (c)

☐ processing unit with a zero
■ processing unit with a one
▨ processing unit with a one that has jumped

**Fig. 3.** Cases (b) and (c) of Lemma 7

The 1's from the 01-region that remain on their places and the 1's jumping into the lowest $\ell$ rows of the old 01-region cause the foot to grow by $\ell$. Above the row $d+\ell$ there are only 1's that have been moved, they form a new 01-region with $e-\ell$ ones. So the column gets the form $1^{d+\ell}(01)^{e-\ell}0^*$.    □

### 2.5 Jump Sizes at Step D

We have still to determine the sizes of the jumps so that the columns are sorted fast. In order to sort a column it suffices to reduce gradually the size of the 01-region at Step D. The size of the jumps is essential: if they are too long, then the size is not reduced (see Lemma 7(a)); if they are too short, then the decrease of the size is not substantial (see Lemma 7(c)). A lucky situation is when the size of the 01-region is $2e = 3\ell - 1$, where $\ell$ is size of the jumps at this column. Then by Lemma 7(c), at step D the size of the 01-region is reduced to $2(e-\ell) = \ell - 1$, that is, to about one third of the original size. Thus in at most $\log_3 p$ rounds we could reduce the size of a 01-region to 0.

These ideas can be implemented on a periodic network as follows: For a given column at some initial position we first let it move to $C_1$. Eventually every moving column reaches $C_1$ (see Subsection 2.3). In the meantime, the Steps D and A may reduce the size of the 01-region of this column, but we do not rely on this improvement. Once a moving column arrives at $C_1$ the real game starts.

Between consecutive Steps D two horizontal moves are executed, so a moving column will be at the following positions while executing Step D:
$C_1, C_3, C_5, \ldots, C_{q-1}, C_q, C_{q-2}, \ldots, C_2 \ldots$. Note that immediately before Step D

at $C_1$, $C_3$, $C_5$, ..., $C_{q-1}$ this moving column will be an O/S-column, and at $C_q$, $C_{q-2}$, ..., $C_2$ this column column will be an E/S-column. So we define the jumps of Step D as follows:

- at odd (even) columns the jump comparators of Step D originate at odd (even) rows,
- at columns $C_1$, $C_3$, $C_5$, ..., $C_{q-1}$ the sizes of the jumps are equal, respectively, to $p/3, p/9, \ldots, p/3^{q/2}$,
- at columns $C_q$, $C_{q-2}$, ..., $C_2$ the sizes of the jumps are equal, respectively, to $p/3^{q/2+1}, p/3^{q/2+2}, \ldots, 1$.

The points of origin of the jump comparators (in even or odd rows) are chosen so that Lemma 7 can be applied, and a moving column starting its movement at $C_1$ will be affected by jumps of size $p/3, p/9, \ldots, 1$, as expected.

## 3 Runtime Analysis

Conceptually the computation consists of two phases: during the first phase all columns become sorted, but may be they are still in a wrong order causing one row to be unsorted. During the second phase the order of the columns is corrected.

**Lemma 8.** *Assume that at a given column the comparators of Step D have length $\ell$ and the 01-region has size $2e$ so that $2e \leq 3\ell - 1$. Then Step D reduces the size of the 01-region at this column to at most $\ell - 1$.*

**Proof.** Let $x$ denote the size of the 01-region after Step D. We apply Lemma 7 in three separate cases: If $e < \ell/2$, then $x = 2e < \ell$ by Lemma 7(a). If $\ell/2 \leq e < \ell$, then by Lemma 7(b), $x = 2(\ell - 1 - e) \leq \ell - 2$. For $e \geq \ell$ but satisfying $2e \leq 3\ell - 1$, by Lemma 7(c) we get $x = 2(e - \ell) \leq \ell - 1$. □

**Corollary 9.** *Consider a moving column immediately $i$ rounds after the moment it has reached $C_1$ before Step D for the first time. Then its 01-region has size at most $p/3^i - 1$.*

**Proof.** The proof is by induction on $i$. For $i = 0$, it is obvious, since a 01-region may not be bigger than $p - 1$, because $p$ is odd and the size of the 01-region is even. Assume that the claimed property holds for $i$ and consider the next round. Step A may only decrease the size of the 01-region. Performing Steps B and C moves the column to the next position where Step D performs jumps of length $p/3^{i+1}$. By Lemma 8, Step D reduces the size of the 01-region in this column to at most $p/3^{i+1} - 1$. □

By Corollary 9, a moving column that arrives back at $C_1$ has a 01-region of size at most $p/3^q - 1 = 0$. Thus the column is sorted. Let us count how much time elapses until this moment:

(i) In the worst case, in order to reach $C_1$ immediately before Step D the column needs to move to the right border, change its direction (at Step C), and move back to the left border. Together, it takes at most $2q$ horizontal steps, that is, at most $4q$ steps.

(ii) After arriving at $C_1$ the moving column has to get to the right border and then back to the left border in order to become sorted. It requires $2q$ horizontal steps, that is, $4q$ steps.

We conclude that during the first $8q$ steps every column becomes sorted.

By Lemma 7, once all columns are sorted we only reorder the columns. Steps A and D have no affect anymore, Steps B and C mimic Odd-Even Transposition Sort of $q$ items. Since $q$ steps of Odd-Even Transposition Sort suffice to sort $q$ items [7], relocating the sorted columns requires at most $q$ horizontal steps, that is, $2q$ steps of our algorithm.

Concluding, the algorithm needs together at most $10q$ steps. Since $n = p \cdot q/2 = 3^q \cdot q/2$, we have $q \leq \log_3 n$ and therefore the runtime of the algorithm is bounded by $10 \cdot \log_3 n$ completing the proof of Theorem 1.

### Acknowledgment

Some elements of our construction (jump comparators) have been considered earlier by Grzegorz Stachowiak in the context of sorting networks. The second author would like to thank F. Meyer auf der Heide for inviting him to Paderborn, where this research has started.

## References

1. Kenneth E. Batcher. Sorting networks and their applications. In *AFIPS Conf. Proc. 32*, pages 307–314, 1968.
2. Ronald I. Becker, David Nassimi, and Yehoshua Perl. The new class of g-chain periodic sorters. In *Proc. 5th ACM-SPAA*, pages 356–364, 1993.
3. Gianfranco Bilardi. Merging and sorting networks with the topology of the Omega network. *IEEE Transactions on Computers*, 38:1396–1403, 1989.
4. Allan Borodin and John E. Hopcroft. Routing, merging, and sorting on parallel models of computation. *J. Comput. Syst. Sci.*, 30(1):130–145, 1985.
5. Martin Dowd, Yehoshua Perl, Larry Rudolph, and Michael Saks. The periodic balanced sorting network. *Journal of the ACM*, 36:738–757, 1989.
6. Marcin Kik, Mirosław Kutyłowski, and Grzegorz Stachowiak. Periodic constant depth sorting networks. In *Proc. 11th STACS*, pages 201–212, 1994.
7. Donald E. Knuth. *The Art of Computer Programming, Volume 3: Sorting and Searching*. Addison-Wesley, Reading, MA, 1973.
8. Mirosław Kutyłowski, Krzysztof Loryś, Brigitte Oesterdiekhoff, and Rolf Wanka. Fast and feasible periodic sorting networks of constant depth. In *Proc. 35th IEEE-FOCS*, pages 369–380, 1994.
9. Peter Bro Miltersen, Mike Paterson, and Jun Tarui. The asymptotic complexity of merging networks. In *Proc. 33rd IEEE-FOCS*, pages 236–246, 1992.
10. David Nassimi, Yehoshua Perl, and Ronald I. Becker. The generalized class of g-chain periodic sorting networks. In *Proc. 8th IPPS*, pages 424–432, 1994.
11. Yehoshua Perl. Better understanding of Batcher's merging networks. *Discrete Applied Mathematics*, 25:257–271, 1989.
12. Uwe Schwiegelshohn. A shortperiodic two-dimensional systolic sorting algorithm. In *International Conference on Systolic Arrays*, pages 257–264, 1988.
13. Robert Sedgewick. Data movement in odd-even merging. *SIAM J. Comput.*, 7:239–272, 1978.

# Minimizing Wavelengths in an All-Optical Ring Network

Gordon Wilfong

Bell Laboratories, Murray Hill NJ 07974, USA

**Abstract.** In an optical network, wavelength-division multiplexing permits multiple messages to be sent along the same link in the network by transmitting each message using a distinct wavelength. In such a network, the action of establishing a connection between two nodes entails choosing a route in the network between the nodes and then assigning to the connection, a wavelength for each link in the chosen route. We will consider problems concerning the minimum number of wavelengths sufficient to establish a set of connections. In a wavelength-selective (WS) network, the assigned wavelengths for any given connection must be the same throughout the chosen route. In contrast, for a wavelength-interchanging (WI) network the assigned wavelengths for each link in the route may differ. We show that for an important class of networks, namely ring networks of $n$ nodes with clockwise and counterclockwise links between each neighboring pair of nodes on the ring, the minimum number, $k_n$, of wavelengths that suffice to allow simultaneously all possible connections is the same for either case. In particular, for both WS and WI networks, it is shown that $k_n = (n^2 - 1)/8$ if $n$ is odd, $k_n = n^2/8$ if $n$ is even and divisible by 4 and $k_n = n^2/8 + 1/2$ if $n$ is even but not divisible by 4.

## 1 Introduction

Wavelength-division multiplexing appears to be an important and effective solution for increasing the bandwidth within an optical fiber [6]. However, while multiwavelength transmission allows for increased bandwidth along optical fibers, the processing, switching and buffering delays at the nodes can result in bottlenecks. A solution to this problem was offered by the idea of "wavelength routing" [3, 10]. That is, the wavelength of the transmission, used without further processing, determines the routing of the signal. Such networks are called "all-optical networks" [2] and they require circuit-switching. Thus for a given connection between two nodes to be achieved, a route through the network must be determined and a wavelength allocated to the connection for each link along the route. The resulting route and allocation of wavelengths is referred to as an "optical path" [9].

There are two general types of networks depending on how optical paths are assigned. In a wavelength-selective (WS) network [4], an optical path for a given connection is formed by assigning the same wavelength along each link in the chosen route. In a wavelength-interchanging (WI) network [7, 8], the wavelength of the transmission is chosen independently along each link. The advantage of a WS network over a WI network is that there is no need to perform wavelength conversions at each node along the route. However a key advantage to a WI network over a WS network is that it could potentially allow for the establishment of an optical path that would be impossible with a WS network. The simplicity of a WS network therefore could result in lower bandwidth or equivalently could require more capacity in terms of needing additional wavelengths to accommodate desired connections.

Wavelength assignments in WS networks have been referred to as "light paths" [4] but we will call them *monochromatic assignments* to emphasize the fact that the wavelength ("color" of the transmitting light) along each link is the same. It has been shown [4] that given a set of desired connections along with some fixed routing for each such connection the question as to whether it is possible to achieve monochromatic assignments along each route for a fixed number of available wavelengths is NP-complete. The proof uses a reduction from Graph Coloring [5]. Thus in this general case, even if the number of wavelengths is sufficient to support connections in a WI network it is difficult to determine whether or not a solution exists using monochromatic assignments. However, in the case where the network topology is a tree, the number of wavelengths required for all monochromatic assignments is the same as if general optical paths are permitted [4]. In the Open Problems section of [1] it is stated that "... (an) interesting and practically useful question is getting a good bound on the number of wavelengths required for a given network and a given message pattern". In this paper we show just such a result for an optical network that consists of a ring of $n$ nodes such that between each pair of neighboring nodes on the ring there are two links, one directed clockwise and the other directed counterclockwise and the set of connections to be established consists of a connection between each pair of nodes in the network. For such a network we determine $k_n$, the minimum number of wavelengths that allow for the establishment of general (i.e. not necessarily monochromatic) optical paths for connections between every pair of nodes. This computation is constructive in the sense that a method for routing each connection is given that allows for optical paths for all connections and uses at most $k_n$ wavelengths. Then we show that even if monochromatic assignments are required, $k_n$ wavelengths remain sufficient to establish optical paths for all connections. Again, this is done constructively so that not only is the method for choosing routes for each connection given but also an algorithm for choosing the wavelengths assigned to each route is given and proven correct. These results can be compared to the general results in [1] that imply that for the case under consideration, $O(\sqrt{n}k_n)$ wavelengths are sufficient.

## 2  Definitions and Problem Descriptions

Consider a ring network $\mathcal{R}_n$ consisting of $n$ nodes $v_0, v_1, \ldots, v_{n-1}$ in clockwise order. We assume that $n$, the size of the network, is fixed and so we use the notation $i \oplus j$ to denote $i + j \pmod{n}$ and similarly $i \ominus j$ is used to mean

$i - j \pmod{n}$. For each $i$, $0 \le i \le n-1$, there is a clockwise link, denoted by $\alpha_i$, from node $v_i$ to node $v_{i \oplus 1}$. Similarly, there is a counterclockwise link, denoted by $\beta_i$ from node $v_i$ to node $v_{i \ominus 1}$. Establishing a connection from one node $v_i$ to another $v_j$, that is, determining an optical path from $v_i$ to $v_j$, entails choosing a route from $v_i$ to $v_j$ and assigning a wavelength along each link in the route. Thus in order to be able to establish a given set of such connections, there needs to be an available wavelength along each link for each route traversing the link. This implies there is a lower bound on the number of wavelengths that would allow all such connections to be established. This work is concerned with the problem of determining such lower bounds and then showing how to route connections and assign wavelengths along the routes using the minimum number of wavelengths. These problems are considered for both WI and WS networks.

Each link will be assumed to have the same number of wavelengths and we refer to each wavelength along a link as a *channel* of the link. That is, each connection will require one channel along each link in its route and each link is thought of as having some number of channels one of which is assigned to each connection passing through the link. Then the first problem of interest is determining the minimum number of channels each link needs in order to be capable of establishing connections between each pair of nodes. Each channel represents a wavelength (or "color" of light) and so we think of each channel as being uniquely labeled by one of the $k$ colors $c_1, c_2, \ldots, c_k$. The second problem considered is to determine how many different colors are required so that each connection can be established using channels all of the same color.

More formally, the action of establishing a connection from *starting node* $v_i$ to *destination node* $v_j$ requires choosing a *route* $R_{i,j}$ where a route consists of clockwise links $\alpha_i, \alpha_{i \oplus 1}, \ldots, \alpha_{j \ominus 1}$ or counterclockwise links $\beta_i, \beta_{i \ominus 1}, \ldots, \beta_{j \oplus 1}$. The links in a route are said to be *traversed* by the route. The number of links traversed by a route is called the *length* of the route. Given a route for a connection we sometimes refer to the *length of the connection* and by this we mean the length of the route chosen for the connection. The second step in establishing a connection between nodes $v_i$ and $v_j$ is to assign a *free* channel from each of the links traversed by the route $R_{i,j}$ where a channel is said to be free if it has not been assigned to any other connection. Thus the problem of determining the minimum number of wavelengths sufficient to allow establishing all connections in a WI network can be stated as follows.

**Problem 1** *Suppose we wish to be able to establish connections from each node to every other node at the same time. For a n-node ring $\mathcal{R}_n$, what is the minimum number $k_n$ such that if each link has $k_n$ channels then all required connections can be established. That is, what is the minimum number of channels for each link sufficient to guarantee that for each connection a route can be chosen and free channels can be assigned for each link of the chosen route.*

Now we consider the problem of determining the minimum number of wavelengths sufficient to allow establishing all connections in a WS network. That is, for each route $R_{i,j}$ all channels assigned to $R_{i,j}$ must not only be free but must all have the same color. Such an assignment of channels to a route is said to be *monochromatic*.

**Problem 2** *Suppose we wish to be able to establish connections from each node to each other node at the same time. For a n-node ring $\mathcal{R}_n$, where each link has $k_n$ channels, is it possible to choose routes for each required connection and to make monochromatic assignments of channels for each such route? That is, are $k_n$ channels for each link sufficient to allow all assignments of channels to be chosen to be monochromatic?*

## 3   Minimum Number of Channels

We begin by considering **Problem 1**. That is, we wish to determine $k_n$, the minimum number of channels per link in $\mathcal{R}_n$ that guarantees that all required connections can be established in a WI network. An outline of the argument used to determine $k_n$ is provided.

Suppose all connections are routed using routes of minimum length. Then no route is longer than $n/2$ for even $n$ or $(n-1)/2$ for odd $n$. Notice that $t_n$, the total number of channels used by all the connections for this type of routing, is no more than would be used by any other routing strategy and $t_n = n(n^2-1)/4$ if $n$ is odd and $t_n = n^3/4$ if $n$ is even. Since $t_n$ is the least number of links that can be traversed for any choice of routing, if a routing scheme results in having the routes evenly distributed so that each of the $2n$ links is traversed by exactly $t_n/2n$, the average number of routes per link, then this routing scheme is optimal in the sense that any other routing scheme will have some link with at least $t_n/2n$ routes traversing it. In other words, $k_n \geq t_n/2n$.

Thus in the case of odd $n$ there would be $(n-1)(n+1)/8$ routes per link and for even $n$ there would be $n^2/8$ routes per link. Notice that for odd $n$, since both $(n-1)$ and $(n+1)$ are divisible by 2 and one must be divisible by 4 the value of $(n-1)(n+1)$ is divisible by 8 and hence $(n-1)(n+1)/8$ is an integer. Therefore if we could show a method for routing the connections such that each link was traversed by exactly $(n-1)(n+1)/8$ routes then this would constructively show that $k_n = (n-1)(n+1)/8$. Similarly for even $n$ where $n$ is divisible by 4, the amount $t_n/2n = n^2/8$ is an integer and so our goal will be to describe a routing method that results in each link being traversed by exactly $n^2/8$ routes enabling us to conclude that $k_n = n^2/8$ in this case. For even $n$ where $n$ is not divisible by 4, the value $n^2/8$ is not an integer and so a routing technique to evenly distribute the routes over the links is not possible. However this says that for any routing there is at least one link that is traversed by at least $n^2/8+1/2$ routes. Therefore we wish to show that there is a routing method such that no link is traversed by more than $n^2/8 + 1/2$ and so conclude that in this case $k_n = n^2/8 + 1/2$.

It can be shown that when $n$ is odd, the unambiguous routing strategy of choosing the shortest route for each connection results in each link being traversed by exactly $(n^2-1)/8$ routes. Since $(n^2-1)/8 = t_n/2n$, we conclude that $k_n = (n^2-1)/8$.

We then consider the case where $n$ is even. As before we choose to route connections via a shortest route but in the case of routes of length $n/2$ this decision is ambiguous. This is corrected by choosing the clockwise route for each connection of length $n/2$ starting at an even numbered node (that is, those nodes $v_i$ where $i$ is even) and choosing the counterclockwise route for each connection of length $n/2$ starting at an odd numbered node. For the case where $n$ is not only even but also divisible by 4 it can be shown that the routing strategy given results in each link being traversed by $n^2/8$ routes and hence $k_n = n^2/8$. In the case where $n$ is even but not divisible by 4 it can be shown that the routing scheme results in no link being traversed by more than $n^2/8 + 1/2$ routes and so we conclude that in this case $k_n = n^2/8 + 1/2$.

Therefore collecting all the above results we can write the value of $k_n$, the least number of channels that permit all connections to be established simultaneously, as

$$k_n = \begin{cases} \frac{n^2-1}{8} & \text{if } n \text{ is odd} \\[2mm] \frac{n^2}{8} & \text{if } n \text{ is even and divisible by 4} \\[2mm] \frac{n^2}{8} + \frac{1}{2} & \text{if } n \text{ is even but not divisible by 4} \end{cases} \tag{1}$$

and it is the case that if all connections are routed via a shortest route (where in the case of even $n$ the connections of length $n/2$ are routed clockwise if they start at an even numbered node and counterclockwise otherwise) then this results in each link being traversed by at most $k_n$ routes thus achieving the lower bound.

## 4 Monochromatic Assignment of Channels to Routes

Next we consider **Problem 2** and in fact show that all required connections can be achieved using routes that allow for monochromatic assignments when only $k_n$ colors are available. A proof outline is provided while most of the technical proofs are omitted in this abstract.

It will first be necessary to provide a routing strategy and then show how channels (colors) should be assigned in order to prove that monochromatic assignments are possible using $k_n$ colors. As in the previous problem, we route all connections via shortest paths. We describe later how to resolve the question as to how to route the length $n/2$ connections when $n$ is even. In any case, once the choice of routes has been fixed, the remaining problem is that of finding monochromatic assignments of channels for each route.

### 4.1 Monochromatic Assignments for Odd $n$

Since the shortest routes for connections in the case where $n$ is odd are unambiguous we are left with only the problem of assigning colors to these routes. We consider the problem of assigning colors to the clockwise routes since the counterclockwise problem is independent and can be solved in an analogous fashion.

In this case, the results of the previous section show that the number of routes traversing channels of a given link, namely $k_n = (n^2 - 1)/8$, is the same for each link, and so the problem can be stated as follows. Partition the clockwise routes into $k_n$ collections $S_1, S_2, \ldots, S_{k_n}$ so that for each $i$, $1 \leq i \leq k_n$, all routes in $S_i$ are assigned the same color, say $c_i$ and each $S_i$ must be such that the set of routes in $S_i$ together traverse every clockwise link exactly once. That is, for each $\alpha_i$ and each $S_j$ there is exactly one route in $S_j$ that traverses $\alpha_i$.

Thus the problem of assigning colors to the routes can be summarized as follows for the case of odd $n$. For each color $c_i$, $1 \leq i \leq k_n$, find an ordered list of routes $S_i = (R_i(0), R_i(1), \ldots, R_i(m_i - 1))$ such that the destination vertex of $R_i(j)$ is the starting vertex of $R_i(j + 1 \pmod{m_i})$, $0 \leq j < m_i$, each clockwise link is traversed by exactly one $R_i(j)$ and each route appears in exactly one such list. We call such an ordered list of routes a *tour* since following each one in turn constitutes a non-overlapping traversal of the ring. Thus the $S_i$'s partition the set of routes into $k_n$ tours. Such an assignment of colors to routes (or routes to lists) is said to be a *valid assignment*. Therefore the goal is to find a valid assignment of colors to routes, thus proving that $k_n$ colors is sufficient to achieve monochromatic assignments for all connections.

A convenient way to view this problem will be via a matrix representation as described next. Consider the matrix $M$ with the entry in the $i^{\text{th}}$ row and $j^{\text{th}}$ column denoted by $M(i, j)$ where $0 \leq i < n$ and $1 \leq j \leq (n-1)/2$. Each $M(i, j)$ will be assigned a color $c_t$ meaning that the route beginning at node $v_i$ of length $j$ is assigned color $c_t$ or equivalently is placed in the list $S_t$.

```
(0):     proc color_odd(n) {    /*n is the number of nodes*/
(1):         h := (n - 1)/2;
(2):         r := 1;                    /*c_r is the current color*/
(3):         for(vertex = 0, 1, ..., h - 1) {
(4):             for(column = h - vertex, h - vertex - 1, ..., 1) {
(5):                 M(vertex, column) := c_r;
(6):                 row := vertex + column;
(7):                 jump := h - vertex;
(8):                 while(row + jump < n) {
(9):                     M(row, jump) := c_r;
(10):                    row := row + jump;
(11):                }
(12):                jump := n - row + vertex;
(13):                M(row, jump) := c_r;
(14):                r := r + 1;
(15):            }
(16):        }
(17):    }
```

**Fig. 1.** Procedure *color_odd* for assigning colors to routes

In Fig. 1 we give a procedure, *color_odd*, written in pseudo-code that we claim assigns a color to every route and for any color, the set of resulting routes assigned that color forms a tour. Thus by the argument above, such a procedure is a constructive proof that $k_n$ colors are sufficient for achieving monochromatic assignments.

We now outline the proof that the procedure *color_odd* satisfies

(*i*) the set of clockwise routes assigned a given color forms a tour and
(*ii*) every clockwise route is assigned exactly one color.

In the following discussions we will need to refer to the value of various variables in the procedure *color_odd* at various stages in the execution of *color_odd*. We use the notation $Val(var)$ to denote the value of variable *var* for the indicated stage in the execution. We begin by showing that the routes assigned any particular color form a tour.

**Lemma 4.1** *The collection of routes assigned a given color by the procedure color_odd forms a tour.*

**Proof: (Sketch)** Notice that the color assigned to routes is fixed for fixed values of the variables *vertex* and *column* in the procedure. Consider a color $c_r$ and let $Val(vertex) = i_r$ and $Val(column) = j_r$ be the values of variables *vertex* and *column* respectively during the execution of the procedure when routes are assigned the color $c_r$. Let $jmp = Val(h) - i_r$. Let $R_r(0), R_r(1), \ldots$ be the routes that are assigned color $c_r$ where the routes are listed in the order in which they are assigned colors by the procedure. The starting vertex of $R_r(1)$ and the destination vertex of $R_r(0)$ clearly have index $i_r + j_r$. For $s = 1, 2, \ldots$, let $t_s = i_r + j_r + s * jmp$. A loop invariant argument shows that route $R_r(s)$ has starting node $v_{t_{s-1}}$ and its destination node is $v_{t_s}$ for $s \geq 1$. The last route $R_r(m_r - 1)$ assigned color $c_r$ occurs at line (13) and it can be shown that its starting node is the destination node of $R_r(m_r - 2)$ and has index given by the value of *row*.

Next we need to show that the destination node of $R_r(m_r - 1)$ is the starting node $v_{i_r}$ of $R_r(0)$. Clearly, the index of the destination node of $R_r(m_r - 1)$ is given by the sum of the values of the variables *row* and *jump* which by the definition of *jump* is simply $n \oplus i_r = i_r$. Consider $Val(row)$, the value of the variable *row* at the time that $R_r(m_r - 1)$ is assigned color $c_r$. We must be somewhat careful here to make sure that a route of length $n - Val(row) + i_r$ is feasible. That is, it must be shown that $1 \leq n - Val(row) + i_r \leq (n-1)/2$. But the loop condition being false means exactly that $Val(row) + (n-1)/2 - i_r \geq n$ and so rewriting this gives $n - Val(row) + i_r \leq (n-1)/2$. If $Val(row) \geq n$ then the loop would have terminated before the previous iteration and so it must be that $Val(row) < n$ and hence $n - Val(row) + i_r > i_r \geq 1$. A monotonicity argument shows that no link is traversed by more than one route $R_r(s)$. ∎

The last problem to complete the proof that *color_odd* produces a valid assignment is to show that every route is assigned exactly one color. For each $i_p$, $0 \leq i_p < (n-1)/2$, define the set $A(i_p)$ to be the set of all entries in $M$ that are assigned a color in *color_odd* while the value of variable *vertex* is $i_p$.

We will show that each $A(i_p)$ consists of three regions of the matrix $M$, namely a horizontal strip $H(i_p)$, a vertical strip $V(i_p)$ and a diagonal strip $D(i_p)$ where

$$H(i_p) = \{M(i_p, j) : 1 \le j \le (n-1)/2 - i_p\},$$
$$V(i_p) = \{M(i, (n-1)/2 - i_p) : i_p < i \le (n-1)/2 + i_p\} \text{ and}$$
$$D(i_p) = \{M(i, j) : i + j = n + i_p, (n+1)/2 + i_p \le i \le n - 1\}.$$

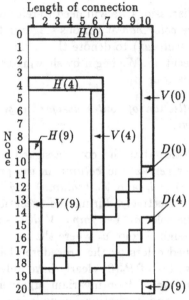

**Fig. 2.** Regions $H(i_p), V(i_p)$ and $D(i_p)$ for $i_p = 0, 4, 9$.

Given this characterization of $A(i_p)$, it can be shown that the $A(i_p)$'s form a partition of the elements of $M$ and so our goal will be to show that for each fixed $p$, each element of $A(i_p)$ is assigned a color exactly once. Figure 2 shows some examples of the partition of the matrix $M$ into regions $H(i_p), V(i_p)$ and $D(i_p)$ for various values of $i_p$ where $n = 21$.

Throughout the remainder of this section, let $i_p$ be some fixed value of the variable *vertex* where $0 \le i_p < (n-1)/2$. It is easy to check that $H(i_p)$, $V(i_p)$ and $D(i_p)$ form a partition of $A(i_p)$. Therefore it is sufficient to show that each element of $H(i_p)$, $V(i_p)$ and $D(i_p)$ is assigned a color exactly once by the procedure. It is straightforward to show that line (5) assigns a color to each element of $H(i_p)$ exactly once. Suppose we could prove the following relationships between the elements assigned colors by execution of line (9) of *color_odd*:

($i_V$)  the elements of $M$ assigned colors in line (9) of the procedure are all elements of $V(i_p)$,

($ii_V$)  two distinct executions of line (9) of the procedure assign colors to distinct elements and

$(iii_V)$ every element of $V(i_p)$ is assigned a color by some execution of line (9) of the procedure

Also, suppose we could prove analogous relationships $(i_D)$, $(ii_D)$ and $(iii_D)$ between the elements assigned colors by execution of line (13) of *color_odd* and the elements of $D(i_p)$. Then we have shown that each element of $H(i_p)$, $V(i_p)$ and $D(i_p)$ is assigned a color exactly once by the procedure.

The following lemma (the proof of which is omitted) establishes property $(i_V)$ and it can be shown that the next two lemmas suffice to establish properties $(ii_V)$ and $(iii_V)$.

**Lemma 4.2** *If element $M(i,j)$ is assigned a color at line (9) of color_odd then $M(i,j) \in V(i_p)$.*

**Lemma 4.3** *For $1 \leq j, j' \leq (n-1)/2 - i_p$, if $i_p + j + k * ((n-1)/2 - i_p) = i_p + j' + k' * ((n-1)/2 - i_p)$ then $j = j'$ and $k = k'$.*

**Lemma 4.4** *For each $\hat{\imath}$, $i_p + 1 \leq \hat{\imath} \leq (n-1)/2 + i_p$, there are values $j$ and $k$ such that $i_p + j + k((n-1)/2 - i_p) = \hat{\imath}$ where $k \geq 0$ and $1 \leq j \leq (n-1)/2 - i_p$.*

**Proof:** Since $\hat{\imath} \geq i_p + 1 > i_p$ we can write $\hat{\imath} - i_p = q + t\left(\frac{n-1}{2} - i_p\right)$ for some integer $t$, $t \geq 0$ and some integer $q$, $0 \leq q < (n-1)/2 - i_p$. Suppose $q > 0$. Then take $j = q$ and $k = t$ and we are done. Suppose $q = 0$. If $t = 0$ then this implies that $\hat{\imath} = i_p$ thus contradicting the fact that $\hat{\imath} \geq i_p + 1 > i_p$. Thus $t \geq 1$ and so we write $\hat{\imath}$ as

$$\hat{\imath} = i_p + \left(\frac{n-1}{2}\right) + (t-1)\left(\frac{n-1}{2} - i_p\right)$$

and this shows that we can take $j = (n-1)/2$ and $k = t - 1 \geq 0$. ∎

The above results allow us to conclude that properties $(i_V)$, $(ii_V)$ and $(iii_V)$ are true. A similar series of lemmas (omitted) establishes properties $(i_D)$, $(ii_D)$ and $(iii_D)$.

Thus each element of $M$ is assigned a color exactly once. These lemmas together with Lemma 4.1 showing that the collection of routes assigned any given color form a tour, allows us to conclude that the procedure *color_odd* produces a valid monochromatic assignment of colors to all routes using $k_n = (n^2 - 1)/8$ colors.

## 4.2 Monochromatic Assignments for Even $n$

For connections of length less than $n/2$ the choice of shortest route is fixed and so we can discuss the problem of assigning colors to each such connection. Consider the procedure *color_even* that is exactly the same as the procedure *color_odd* in Fig. 1 except that the variable $h$ is initially assigned the value $n/2 - 1$ rather than $(n-1)/2$. Then arguments analogous to those in Sect. 4.1 show that *color_even* monochromatically assigns colors to all connections of length less than $n/2$ and uses exactly $n(n-2)/8$ colors to achieve this assignment.

We then show that the connections of length $n/2$ can be routed so that it is possible to monochromatically assign colors to each such route using only $k_n - n(n-2)/8$ additional colors. Let $C_{i,i^*}$ be the length $n/2$ connection from $v_i$ to $v_{i^*}$. That is, $i^* = i \pmod{n/2}$. Then the $n$ length $n/2$ connections can be partitioned into $n/2$ pairs $P_0, P_2, \ldots, P_{n/2-1}$ where $P_i = (C_{i,i^*}, C_{i^*,i})$. If $i$ is even then route the two connections in $P_i$ clockwise and otherwise route them counterclockwise. Then the two routes corresponding to the two connections of $P_i$ form a tour and so the two routes can be assigned one color. It is shown that no more than $k_n - n(n-2)/8$ of the $P_i$'s are routed clockwise and similarly no more than $k_n - n(n-2)/8$ of the $P_i$'s are routed counterclockwise thus we conclude that the routes can be monochromatically assigned channels using no more than $k_n$ colors.

# References

1. A. Aggarwal, A. Bar-Noy, D. Coppersmith, R. Ramaswami, B. Schieber, and M. Sudan. Efficient routing and scheduling algorithms for optical networks. In *Proceedings of SODA*, pages 412–423, 1994.
2. S. B. Alexander, R. S. Banderant, D. Byrne, V. W. S. Chan, S. G. Finn, R. Gallager, B. S. Glance, H. A. Haus, P. Humblet, R. Jain I. P. Kaminow, M. Karol, R. S. Kennedy, A. Kirby, H. Q. Le, A. M. Saleh, B. A. Schofield, J. H. Shapiro, N. K. Shankaranarayanan, R. E. Thomas, R. C. Williamson, and R. W. Wilson. A precompetitive consortium on wide-band all-optical networks. *Journal of Lightwave Technology*, 11(5/6):714–735, May/June 1993.
3. M. C. Brain and P. Cochrane. Wavelength routed optical networks using coherent transmission. In *Proceedings of ICC*, volume 1, pages 26–31, 1988.
4. I. Chlamtac, A. Ganz, and G Karmi. Lightpath communications: An approach to high bandwidth optical wan's. *IEEE Transactions on Communications*, 40(7):1171–1182, July 1992.
5. M. R. Garey and D. S. Johnson. *Computers and Intractability: A Guide to the Theory of NP-Completeness*. W.H. Freeman and Co., San Francisco, CA, 1979.
6. P. E. Green. *Fiber-Optic Networks*. Prentice-Hall, Cambridge, MA, 1992.
7. G. Jeong and E. Ayanoglu. Comparison of wavelength-interchanging and wavelength-selective cross-connects in multiwavelength all-optical networks. Technical Report BLO113430-950927-29TM, AT&T Bell Laboratories, 1995.
8. K-C. Lee and V. O. K. Li. Routing and switching in a wavelength convertible optical network. In *Proceedings of IEEE INFOCOM*, volume 2, pages 578–585, 1993.
9. K. Sato, S. Okamoto, and H. Hadama. Network performance and integrity enhancement with optical path layer technologies. *IEEE Journal on Selected Areas of Communications*, 12(1):159–170, January 1994.
10. D. W. Smith and G. R. Hill. Optical processing in future coherent networks. In *Proceedings of GLOBECOM*, volume 2, pages 678–683, 1987.

# Competitive Analysis of
# On-Line Disk Scheduling*

Tzuoo-Hawn Yeh, Cheng-Ming Kuo, Chin-Laung Lei, and Hsu-Chun Yen

Department of Electrical Engineering
National Taiwan University, Taipei, Taiwan, ROC

**Abstract.** We give a detailed *competitive analysis* of a popular on-line disk scheduling algorithm, namely, *LOOK*. Our analysis yields a tight competitive ratio for the *LOOK* algorithm. By comparing and contrasting the competitive ratio of *LOOK* with the lower bounds of *FCFS* and *SSTF* (which are equally popular in disk scheduling), our results provide strong evidence to suggest that when the workload of disk access is heavy, *LOOK* outperforms *SSTF* and *FCFS*. As a by-product, our analysis also reveals (quantitatively) the role played by the degree of *look-ahead* in disk scheduling.

## 1 Introduction

*Disk scheduling* is a problem that is of practical importance and theoretical interest in the study of computer systems (in particular, in the areas of databases and operating systems) [9]. The goal of disk scheduling is to devise a policy for servicing disk requests in an on-line fashion so as to minimize the total disk access time.

Previous works on the study of the disk scheduling problem resort to *probabilistic analysis* [3,7,11,12], *simulation* [6–8,11,12], or *amortized analysis* [1]. As an alternative to the aforementioned analytical techniques, in this paper we study various disk scheduling algorithms through the use of the so-called *competitive analysis* technique. Competitive analysis has its origin in a paper by Sleator and Tarjan [10]. In this approach, the efficiency of an on-line algorithm is compared with that of an optimal off-line algorithm. An on-line algorithm is *competitive* if its cost is within a constant factor of that of the optimal off-line algorithm for all request sequences. To be more precise, an on-line algorithm $A$ is said to be $\alpha$-*competitive* if there exists a constant $\beta$ such that for any request sequence $\sigma$, $C_A(\sigma) \leq \alpha \cdot C_{OPT}(\sigma) + \beta$, where $OPT$ is an optimal off-line algorithm, and $C_A(\sigma)$ (respectively, $C_{OPT}(\sigma)$) is the cost of algorithm $A$ (respectively, $OPT$) over request sequence $\sigma$. The constant $\alpha$ is called the *competitive ratio* of algorithm $A$. For more about competitive analysis, see, e.g., [2,4,5]

Given a sequence of requests on a disk equipped with a single read-write head and a *waiting queue* of a fixed size, the goal of disk scheduling is to design a strategy for picking requests (in an on-line fashion) from the waiting queue to service so that the total cost of service is minimized. The waiting queue is capable of holding a fixed amount of requests from which the next request to

* This work was supported in part by the National Science Council of the Republic of China under Grant NSC-84-2213-E-002-002.

service is chosen. Once a request enters the waiting queue, it is entitled to be serviced next even though the waiting queue might contain requests that arrived earlier. (That is, requests in the waiting queue are serviced in a random order.) Conversely, a request can be serviced only if it resides in the waiting queue. In our model, disk scheduling algorithms are differentiated mainly by the order in which the next request to service is chosen, as well as by the way future requests enter the waiting queue. (To a certain degree, the waiting queue offers some sort of a look-ahead capability.)

In this paper, we give a detailed competitive analysis for a popular scheduling algorithm, namely, the $LOOK$ algorithm [9]. As it turns out, $LOOK$ is $\frac{WT+D}{WT+D/(2W-1)}$-competitive, where $W$ is the size of the waiting queue, $D$ is the number of cylinders on the disk less one, and $T$ is a constant capturing the seek startup time, the latency time, and the data transfer time needed to complete a disk access. We also show the above bound to be tight. By viewing the size of the waiting queue as the degree of look-ahead, our derived bound precisely reveals how look-ahead affects the competitive ratio of $LOOK$. Interestingly, previous study on the $k - server$ problem suggests that look-ahead plays a negligible role as far as the algorithm's competitive ratio is concerned. Aside from the detailed competitive analysis for $LOOK$, we also compare the lower bound of $LOOK$ with that of other popular scheduling policies such as *first-come-first-served (FCFS)* and *shortest-seek-time-first (SSTF)*. In practice, the size of the waiting queue can also be interpreted as the average workload of the disk. By comparing and contrasting the competitive ratio of $LOOK$ with that (lower bounds) of $FCFS$ and $SSTF$, our analysis provides strong evidence to suggest that when the workload of disk access is heavy, $LOOK$ seems to outperform $SSTF$ and $FCFS$. (The interested reader is referred to [9] for a comparative analysis of the above scheduling policies from a probabilistic viewpoint.)

The remainder of this paper is organized as follows. Section 2 gives the definitions of the disk model, the cost measure, and a number of disk scheduling algorithms found in the literature. Section 3 concerns itself with deriving the lower bounds of the competitive ratios of the $FCFS$, $SSTF$, and $LOOK$ algorithms. In Sections 4 and 5, we derive the upper bound of the competitive ratio of the $LOOK$ algorithm in detail. Discussion and concluding remarks are given in Section 6.

## 2 Preliminaries

The *disk scheduling problem* considered in this paper is that of, given an online sequence of disk requests and a waiting queue of a fixed size, designing a strategy for picking requests from the waiting queue to service so that the total cost (including the seek time, the latency time, and the transfer time) is minimized.

For ease of explanation, a disk with $D + 1$ cylinders can be viewed as a line segment along which cylinder 0 (respectively, $D$) is the leftmost (respectively, rightmost) cylinder. A *request* on cylinder $x$ is denoted as $x$. A *request sequence* is a sequence of requests waiting to be serviced. The waiting queue, which will be referred to as the *window* throughout the rest of the paper, is capable of holding a fixed number of requests from which the next request to be serviced is chosen. Once a request resides in the window, the request is entitled to be serviced next even though the window might contain earlier requests. Conversely, a request must be in the window before it can be serviced. The order in which the next request to service is selected from the window and the way future requests are brought into the window upon the completion of servicing a request depend on the underlying on-line algorithm, and will be elaborated later in conjunction with the definition of the algorithm.

Given a seek distance $y$ (measured in terms of the number of cylinders traveled by the disk head), the seek time is a function of the form $f(y) = s + ky$, where $s$ is the *seek startup time* and $k$ is the seek time per cylinder [6,11,12]. The latency time, as suggested in the literature, is difficult to be modeled precisely; hence, the so-called "average latency time," a constant which is usually given in the disk specification, is employed to approximate the latency time. The transfer time is usually a constant if the size of the data to be transferred is fixed. In this paper, the cost of servicing a request is defined to be a constant (capturing the seek startup time, the average latency time and the transfer time) plus the distance the disk head travels.

The notations used throughout this paper are listed below:

$D$: cylinders are numbered from 0 to $D$.

$W$: a constant representing the size of the window.

$T$: a constant which captures the sum of the seek startup time, the latency time, and the transfer time of a request.

$\sigma = \langle x_1 \cdots x_n \rangle$: a request sequence. $\langle \rangle$ denotes an empty request sequence.

$h_A$: $(0 \le h_A \le D)$ the cylinder at which the disk head of algorithm $A$ locates.

$C_A(\sigma)$: the cost of algorithm $A$ for servicing request sequence $\sigma$.

$l(\sigma)$ (respectively, $r(\sigma)$): the leftmost (respectively, rightmost) request in $\sigma$, where $\sigma$ is either a request sequence or a multiset of requests.

$|\sigma|$: the length of request sequence $\sigma$, i.e., the number of requests in $\sigma$.

Given a request sequence $\sigma$ and an algorithm, the *schedule*, denoted as $\langle\langle x_1, x_2, \ldots, x_n \rangle\rangle$, is the servicing sequence, that is, $x_i \in \sigma$ is the $i$-th serviced request. The *path*, which is a sequence of cylinders, records the cylinders the algorithm travels during its move. The cost of the algorithm is $d + m \cdot T$, where $d$ is the total distance traveled and $m$ is the number of serviced requests.

In this paper, we consider the following disk scheduling algorithms:

*FCFS*: This algorithm always selects the first request in the remaining request sequence to service.

*LOOK*: Using this policy, requests are grouped into blocks, each of which (except perhaps the last one) is of size $W$. All the requests in the window have to be serviced completely before the next $W$ requests are brought into the window. Let $\delta$ be the current block of requests, and $h$ be the current position of the disk head. If $h \le l(\delta) \le r(\delta)$, then the disk head sweeps to cylinder $r(\delta)$, servicing all the requests in the window. The case where $l(\delta) \le r(\delta) \le h$ is symmetric. If $l(\delta) < h < r(\delta)$ and $h$ is closer to $l(\delta)$ than to $r(\delta)$, then the disk head first moves to cylinder $l(\delta)$ and then sweeps (to the right) across the disk until reaching $r(\delta)$, servicing all the requests in between. The case when $r(\delta)$ is closer is symmetric. For simplicity, we assume that the disk head moves to $l(\delta)$ then sweeps to $r(\delta)$ when $r(\delta) - h = h - l(\delta)$.

*SSTF*: This algorithm always chooses the request in the window that is closest to the disk head to service. (For simplicity, break tie by servicing the request to the left first.) We assume that the window for $SSTF$ is full at all times unless no further requests can be brought into the window.

*ADV* can be thought of as a generic scheduling algorithm against which the on-line scheduling algorithms (such as $FCFS$, $LOOK$, and $SSTF$) will be compared. Since the role played by $ADV$ is to maximize the ratio of the costs incurred by the on-line algorithms and $ADV$, it is natural to give as much freedom (or power) to $ADV$ as possible. Having this in mind, in our subsequent discussion we assume that the window for $ADV$ is full at all times unless no further requests can be brought into the window.

## 3 Lower Bounds

In this section, the lower bounds of the competitive ratios for algorithms $FCFS$, $SSTF$, and $LOOK$ are derived. Initially, the disk head is assumed to be at cylinder 0. For ease of expression, we let $\sigma^k$ represent $\sigma$ repeated $k$ times, where $\sigma$ is a request or a sequence of requests.

**Theorem 1.** $\alpha_{FCFS} \geq \frac{T+D}{T+D/(2W-1)}$.

*Proof.* Consider the request sequence $\sigma = \langle (D0)^{(2W-1)n} \rangle$, where $n$ is a non-negative integer. $FCFS$ services $\sigma$ using the schedule $\langle\langle (D0)^{(2W-1)n} \rangle\rangle$, and $C_{FCFS} = 2n(2W-1)(T+D)$. One can service $\sigma$ using the schedule $\langle\langle 0^{W-1} (D^{2W-1}0^{2W-1})^{n-1} D^{2W-1} 0^W \rangle\rangle$, so $C_{ADV} \leq 2n(2W-1)T + 2nD$. Therefore, we have $\alpha_{FCFS} \geq \frac{C_{FCFS}(\sigma)-\beta}{C_{ADV}(\sigma)} \geq \frac{2n(2W-1)(T+D)-\beta}{2n(2W-1)T+2nD}$, which approaches $\frac{T+D}{T+D/(2W-1)}$ when $n$ is large. $\qquad\square$

**Theorem 2.** $\alpha_{LOOK} \geq \frac{WT+D}{WT+D/(2W-1)}$.

*Proof.* Consider the following request sequence $\sigma = \langle ((0^{W-1} D0^W)^{W-1} D^W (D^{W-1}0D^W)^{W-1} 0^W)^n \rangle$, where $n$ is a non-negative integer. $LOOK$ services $\sigma$ using the schedule $\langle\langle \sigma \rangle\rangle$, and $C_{LOOK} = 2nW(2W-1)T + 2n(2W-1)D$. One can service $\sigma$ using the schedule $\langle\langle (0^{(W-1)(2W-1)} D^{2W-1} \cdot D^{(W-1)(2W-1)}0^{2W-1})^n \rangle\rangle$, so $C_{ADV} \leq 2nW(2W-1)T + 2nD$. Therefore, we have $\alpha_{LOOK} \geq \frac{C_{LOOK}(\sigma)-\beta}{C_{ADV}(\sigma)} \geq \frac{2n(2W-1)(WT+D)-\beta}{2n(W(2W-1)T+D)}$, which approaches $\frac{WT+D}{WT+D/(2W-1)}$ when $n$ is large. $\qquad\square$

**Theorem 3.** $\alpha_{SSTF} \geq \frac{T+\lfloor\frac{D}{2}\rfloor}{T+\lfloor\frac{D}{2}\rfloor(2W-1)}$.

*Proof.* Consider the request sequence: $\sigma = \langle D^{W-1}(\lfloor\frac{D}{2}\rfloor 0)^{n(2W-1)+W} \rangle$, where $n$ is a non-negative integer. $SSTF$ services $\sigma$ using the schedule $\langle\langle (\lfloor\frac{D}{2}\rfloor 0)^{n(2W-1)+W} D^{W-1} \rangle\rangle$, and $C_{SSTF} = (2n(2W-1)+3W-1)T + 2(n(2W-1)+W)\lfloor\frac{D}{2}\rfloor + D$. One can service $\sigma$ using the schedule $\langle\langle D^{W-1}\lfloor\frac{D}{2}\rfloor^W (0^{2W-1}\lfloor\frac{D}{2}\rfloor^{2W-1})^n 0^W \rangle\rangle$, so $C_{ADV} \leq (2n(2W-1)+3W-1)T + 2D + 2n\lfloor\frac{D}{2}\rfloor$. When $n$ is large, we have $\alpha_{SSTF} \geq \frac{T+\lfloor\frac{D}{2}\rfloor}{T+\lfloor\frac{D}{2}\rfloor/(2W-1)}$. $\qquad\square$

# 4 The Behavior of *ADV* Against *LOOK*

In this section, we characterize a class of adversaries, called *aggressive* adversaries, that are capable of enforcing the maximum competitive ratios against on-line algorithms. An adversary $ADV$ is aggressive if it always services the request in the window that is *adjacent* to the disk head. (A request $x$ is adjacent to $h_{ADV}$ if $x = h_{ADV}$ or there is no request $x'$ in the window such that $h_{ADV} \leq x' < x$ or $x < x' \leq h_{ADV}$.)

In what follows, we argue that for any adversary $ADV$ and request sequence $\sigma$, there exists an aggressive adversary $ADV'$ such that $C_{ADV'}(\sigma) \leq C_{ADV}(\sigma)$. To prove this, we first construct an $ADV''$ (for $ADV$) which follows exactly the same path as that of $ADV$, and, whenever the traversal of $ADV''$ from one location to another encounters a request already resides in the window, such a request will be serviced immediately, even though the request is serviced later on the original path of $ADV$. It is obvious that $C_{ADV''}(\sigma) = C_{ADV}(\sigma)$. Due to the behavior of $ADV''$ (which follows the same path of $ADV$), it is likely that $ADV''$ might travel to locations at which the associated requests have been serviced already. (Recall that $ADV''$ follows $ADV$ while servicing requests at a possibly faster pace.) $ADV'$ is constructed from $ADV''$ by removing all disk head movements that are unnecessary as far as servicing requests is concerned. Trivially, the cost of $ADV'$ is no more than that of $ADV''$. It is easy to see that $ADV'$ is aggressive.

Based on the above discussion, we have:

**Theorem 4.** *For every adversary ADV and request sequence $\sigma$, there is an aggressive adversary ADV' such that $C_{ADV'}(\sigma) \leq C_{ADV}(\sigma)$.*

The notion of a *phase* is critical in our subsequent discussion about *LOOK*. The original request sequence $\sigma$ is partitioned into blocks of size $W$ each (except perhaps the last block) and each block is called a *phase of requests*. Formally, $\sigma$ is partitioned as $\sigma = \delta_1 \delta_2 \cdots \delta_n$, where $|\delta_k| = W$ ($\forall k, 1 \leq k \leq n-1$) and $|\delta_n| \leq W$. $\delta_k$ is called the $k$-th phase of requests. The period for *LOOK* to service the requests in $\delta_k$ is called the $k$-th phase of *LOOK*.

The definition of a phase for $ADV$ (assuming that $ADV$ is aggressive) is more complicated, for $ADV$ is capable of holding a request in the window while continuously servicing an arbitrary number of future requests. Consider a request sequence $\sigma = \langle x_1, x_2, \ldots, x_n \rangle$. In the first phase, $ADV$ services exactly one requests. From the second phase on, $ADV$ services $W$ requests in each phase, except perhaps the last phase. In the last phase, $ADV$ services the remaining requests. The number of requests serviced in the last phase is less than or equal to $W$.

Consider an $ADV$ and a request sequence $\sigma = \delta_1 \cdots \delta_n$, where $\delta_k, 1 \leq k \leq n$, is the $k$-th phase of requests. Let $h_A^k$ be the disk head position of an algorithm $A$ at the end of phase $k$. $\delta_k$ is called *canonical* with respect to $ADV$ if the following two conditions are satisfied:

**Canonical Condition 1:** For any replacement $\delta'$ of $\delta_k$, if $|\delta'| = |\delta_k|$ and the paths of *LOOK* on $\delta_1 \cdots \delta_k \cdots \delta_n$ and $\delta_1 \cdots \delta' \cdots \delta_n$ are identical, then we have $\sum_{x \in \delta_k} |x - h_{ADV}^{k-1}| \leq \sum_{x \in \delta'} |x - h_{ADV}^{k-1}|$.

| Case | Disk configuration | $\delta'_k$ | Case | Disk configuration | $\delta'_k$ |
|------|--------------------|-------------|------|--------------------|-------------|
| 1 | $L \leq A \leq r, L \leq l$ | $A^{|\delta_k|-1}r$ | 1' | $r \leq L, l \leq A \leq L$ | $A^{|\delta_k|-1}l$ |
| 2 | $A < L \leq l \leq r$ | $L^{|\delta_k|-1}r$ | 2' | $l \leq r \leq L < A$ | $L^{|\delta_k|-1}l$ |
| 3 | $L \leq l \leq r < A$ | $r^{|\delta_k|}$ | 3' | $A < l \leq r \leq L$ | $l^{|\delta_k|}$ |
| 4 | $A \leq l < L < r$ | $l^{|\delta_k|-1}r$ | 4' | $l < L < r \leq A$ | $r^{|\delta_k|-1}l$ |
| 5 | $l < A < r, l < L < r$ | If the first request on $l$ appears before the first request on $r$ in $\delta_k$, then $\delta'_k = A^{|\delta_k|-2}lr$, otherwise, $\delta'_k = A^{|\delta_k|-2}rl$. | | | |

**Table 1.** The construction of $\sigma_k$ from $\sigma_{k-1}$, $ADV_{k-1}$, and $LOOK$. ($A = h^{k-1}_{ADV_{k-1}}$, $L = h^{k-1}_{LOOK}$, $l = l(\delta_k)$, and $r = r(\delta_k)$.)

```
procedure canonical(σ, ADV);
  σ₀ = σ; ADV₀ = ADV;
  for k = 1 to n do
  begin
     Construct δ'_k using the rules provided in Table 1;
     (It is easy to see that the paths of LOOK on δ'_k and δ_k are identical.)
     Let σ_k = δ'₁ ⋯ δ'_{k-1}δ'_kδ_{k+1} ⋯ δ_n;
     Let ADV''_k (on σ_k) follow the same path of ADV_{k-1} (on σ_{k-1});
     Convert ADV''_k into an aggressive ADV_k (on σ_k)—guaranteed by Theorem 4.
  end
  σ' = σ_n; ADV' = ADV_n;
```

**Fig. 1.** The procedure for constructing a canonical request sequence.

**Canonical Condition 2:** For any $x, y \in \delta_k$, if $y < x \leq h^{k-1}_{ADV}$ or $h^{k-1}_{ADV} \leq x < y$, then $x$ proceeds $y$ in $\delta_k$.

A request sequence $\sigma$ is called *canonical* with respect to $ADV$ if all its phases are canonical with respect to $ADV$.

**Theorem 5.** *Given a request sequence $\sigma$ and adversary $ADV$, there exist a request sequence $\sigma'$ and an aggressive adversary $ADV'$ such that $\sigma'$ is canonical with respect to $ADV'$, $C_{ADV'}(\sigma') \leq C_{ADV}(\sigma)$, and $C_{LOOK}(\sigma') = C_{LOOK}(\sigma)$.*

*Proof.* Without loss of generality we assume that $ADV$ is aggressive. Let $\sigma = \delta_1 \cdots \delta_n$, where $\delta_k$, $1 \leq k \leq n$, is the $k$-th phase of requests. $\sigma'$ and $ADV'$ are constructed using the procedure in Figure 1. In what follows, we show the following: (C1) $ADV''_k$ is capable of servicing $\sigma_k$; (C2) $ADV_k$ is aggressive; (C3) $\delta'_k$ is canonical with respect to $ADV_k$; and (C4) $C_{ADV_k}(\sigma_k) \leq C_{ADV_{k-1}}(\sigma_{k-1})$, for all $k$, $1 \leq k \leq n$.

(C2) follows from the construction of $ADV_k$. (See the procedure in Figure 1.) By the procedure in Figure 1, the paths of $ADV_{k-1}$, $ADV''_k$, and $ADV_k$ are identical in the first $k - 1$ phases. Thus, at the beginning of phase $k$, the head positions of them are of the same, i.e., $h^{k-1}_{ADV_{k-1}} = h^{k-1}_{ADV''_k} = h^{k-1}_{ADV_k}$, and their windows contain the same requests at the beginning of phase $k$, except the new request that comes from the first request in the $k$-th phase of requests. In what follows, we first prove (C1) and (C3). (For ease of expression, let $A = h^{k-1}_{ADV_{k-1}}$, $L = h^{k-1}_{LOOK}$, $l = l(\delta_k)$, and $r = r(\delta_k)$ as in Table 1 in the following case analysis.)

**Case 1:** In this case, $L \leq A \leq r$ and $L \leq l$, and we let $\delta'_k = A^{|\delta_k|-1}r$. Suppose $\delta'$ is an arbitrary sequence such that $|\delta'| = |\delta'_k|$ and $LOOK$ follows the same path on $\delta'$ and $\delta'_k$. Then $\delta'$'s requests must lie in the interval $[L, r]$ and request $r$ must be in $\delta'$. Clearly, $\sum_{x \in \delta'_k} |x - A| (= |r - A|) \leq \sum_{x \in \delta'} |x - A|$. Hence, Canonical Condition 1 is met for $\delta'_k$ with respect to $ADV_k$. (Notice that $h^{k-1}_{ADV_k} = h^{k-1}_{ADV_{k-1}}$.) Canonical Condition 2 clearly holds. Thus, (C3) holds. It remains to show that $ADV''_k$ is capable of processing $\delta'_k$ while following the same path of $ADV_{k-1}$ on $\delta_k$. At the beginning of this phase, $ADV''_k$ can service $A^{|\delta_k|-1}$. And $ADV''_k$ can service $r$ when $ADV_{k-1}$ services $r$. The remaining sequence $\langle \delta_{k+1} \cdots \delta_n \rangle$ can be serviced by $ADV''_k$ at the same time when $ADV_{k-1}$ services them. So (C1) holds.

The proof technique of Cases 2—5 is similar to Case 1. The details can be found in [13].

In the following, we prove (C4). Since $ADV''_k$ and $ADV_{k-1}$ follow that same path, we have $C_{ADV''_k}(\sigma_k) = C_{ADV_{k-1}}(\sigma_{k-1})$. By Theorem 4, $C_{ADV_k}(\sigma_k) \leq C_{ADV''_k}(\sigma_k)$. Thus, $C_{ADV_k}(\sigma_k) \leq C_{ADV_{k-1}}(\sigma_{k-1})$, for any $k$, $1 \leq k \leq n$.

By the procedure in Figure 1, we have constructed an aggressive adversary $ADV'$ and a request sequence $\sigma'$ that is canonical with respect to $ADV'$. And the cost of $ADV'$ on $\sigma'$ is no more than the cost of $ADV$ on $\sigma$. (By (C4), $C_{ADV'}(\sigma') = C_{ADV_n}(\sigma_n) \leq C_{ADV_{n-1}}(\sigma_n) \leq \cdots \leq C_{ADV_0}(\sigma_0) = C_{ADV}(\sigma)$.) Since $LOOK$ follows an identical path on $\sigma$ and $\sigma'$ is not changed, the equality $C_{LOOK}(\sigma') = C_{LOOK}(\sigma)$ holds trivially. $\square$

## 5 Competitive Analysis of $LOOK$

In this section we will show that $LOOK$ is competitive and the derived competitive ratio matches its lower bound. Our analysis is based on the potential technique. Let $\sigma$ be an arbitrary request sequence, $n = \lceil |\sigma|/W \rceil$ and $\Phi_k$, $1 \leq k \leq n$, be a potential function that maps the configurations of $LOOK$ and $ADV$ at the end of phase $k$ to a real number. Also, let $\Delta\Phi_k = \Phi_k - \Phi_{k-1}$ be the change of potential during phase $k$, and $C^k_{ADV}$ and $C^k_{LOOK}$ be the costs of phase $k$ for $ADV$ and $LOOK$, respectively. For a constant $\alpha$, if we can show that $C^k_{LOOK} + \Delta\Phi_k \leq \alpha \times C^k_{ADV}$ for any $k$, $1 < k < n$, then,

$$C_{LOOK}(\sigma) = \sum_{k=1}^{n} C^k_{LOOK}$$
$$= \sum_{k=2}^{n-1} \left( C^k_{LOOK} + \Phi_k - \Phi_{k-1} \right) + \left( C^1_{LOOK} + C^n_{LOOK} + \Phi_1 - \Phi_{n-1} \right)$$
$$\leq \alpha \times \sum_{k=2}^{n-1} C^k_{ADV} + \left( C^1_{LOOK} + C^n_{LOOK} + \Phi_1 - \Phi_{n-1} \right)$$
$$\leq \alpha \times C_{ADV}(\sigma) + \underbrace{\left( C^1_{LOOK} + C^n_{LOOK} + \Phi_1 - \Phi_{n-1} \right)}_{\text{additive constant } \beta},$$

That is to say, if (1) we can bound the cost of $LOOK$ plus the change of the potential in every phase (except the first and the last phase) within a factor of $\alpha$ to the cost of $ADV$ in the same phase, and (2) $C^1_{LOOK} + C^n_{LOOK} + \Phi_1 - \Phi_{n-1}$ is bounded by some constant $\beta$, then $LOOK$ is $\alpha$-competitive. The first and last phases are excluded from the discussion because $ADV$ services only one request in the first phase, whereas $LOOK$ may service less than $W$ requests in the last phase. These situations are awkward to analyze in a systematic way and are thus grouped into the additive constant $\beta$.

Let $\sigma = \delta_1\delta_2\cdots\delta_n$ be the request sequence under consideration, where $\delta_k$, $1 \leq k \leq n$, is the $k$-th phase of requests. Define the *delayed set* of $ADV$ at the end of phase $k$, $B_k$, to be the requests in $\delta_1\delta_2\cdots\delta_k$ that have not yet been serviced by $ADV$ at the end of phase $k$. Clearly, $B_k$ is exactly $ADV$'s window content at the end of phase $k$ minus the first request in $\delta_{k+1}$ that is just brought into the window. Let $h^k_{ADV}$ and $h^k_{LOOK}$ be the head positions of $ADV$ and $LOOK$, respectively, at the end of phase $k$. The potential at the end of phase $k$ is defined as $\Phi_k = M \times (\phi_1 + \phi_2 + \phi_3 + \phi_4)$, where

$$M = \frac{WT + D}{(2W - 1)WT + D},$$

$$\phi_1 = \left|h^k_{ADV} - h^k_{LOOK}\right|,$$

$$\phi_2 = -2 \times \sum_{q \in B_k} \left|h^k_{ADV} - q\right|,$$

$$\phi_3 = \begin{cases} 2\left|h^k_{ADV} - h^k_{LOOK}\right|, & \text{if } h^k_{ADV} < h^k_{LOOK} < q \text{ or } q < \\ & h^k_{LOOK} < h^k_{ADV}, \text{ for all } q \in B_k, \\ 0, & \text{otherwise,} \end{cases}$$

$$\phi_4 = K\left(\frac{1}{M} - 1\right), \text{ where } K = \min(h^k_{LOOK}, D - h^k_{LOOK}).$$

Notice that $0 \leq M \leq 1$, and $-2M(W - 1)D \leq \Phi_k \leq 3MD + \lfloor\frac{D}{2}\rfloor(1 - M)$, and $WT \leq C^k_{LOOK} \leq \lceil 3D/2\rceil + WT$, for any $k$, that is, the potential function is bounded from below and above.

Throughout the remainder of this section, we will focus on one phase of operations of $ADV$ and $LOOK$, and we call this phase the "current" phase. Let $\delta$ be the current phase of request, $l = l(\delta)$ and $r = r(\delta)$. Let $C_{ADV}$ and $C_{LOOK}$ be the costs of $ADV$ and $LOOK$, respectively, in the current phase. Let $B$, $A$, $L$, $\Phi$, and $\phi_i$, for $1 \leq i \leq 4$, be the delayed set, head position of $ADV$ and $LOOK$, and the potential and its four components at the beginning of the current phase, respectively; and $B'$, $A'$, $L'$, $\Phi'$, and $\phi'_i$, for $1 \leq i \leq 4$ be the corresponding items at the end of the current phase. Let $\Delta\Phi = \Phi' - \Phi$ and $\Delta\phi_i = \phi'_i - \phi_i$, for $1 \leq i \leq 4$.

By Theorems 4 and 5, we can restrict our analysis to aggressive adversaries and canonical phases of requests. The main result of this section is stated in the following theorem:

**Theorem 6.** *The LOOK algorithm is $M(2W - 1)$-competitive.*

First, we consider the case where $T = 0$. Notice that in this case, $M = 1$, and $\phi_4$ is always 0 and can be omitted in the discussion of the following lemma.

**Lemma 7.** *LOOK is $(2W - 1)$-competitive when $T = 0$. In particular, we have $C_{LOOK} + \Delta\Phi \leq (2W - 1)C_{ADV}$.*

*Proof.* The proof of the lemma is proceeded according to the classification of Table 1. The proof is rather complicated; hence, the details are omitted due to space limitation. The reader is referred to [13] for details. □

Next, we will show how to generalize the result to the case where $T = t > 0$. For ease of expression, let $C_{ADV}^{T=t}$, $C_{LOOK}^{T=t}$, and $\Delta\Phi^{T=t}$ be $C_{ADV}$, $C_{LOOK}$, and $\Delta\Phi$, respectively, when $T = t$.

**Lemma 8.** *If* $C_{LOOK}^{T=0} + \Delta\Phi^{T=0} \leq (2W-1) \times C_{ADV}^{T=0}$ *when* $T = 0$, *then* $C_{LOOK}^{T=t} + \Delta\Phi^{T=t} \leq m(2W-1) \times C_{ADV}^{T=t}$ *when* $T = t > 0$, *where $m$ is the value of $M$ when* $T = t$.

*Proof.* Let $\Delta K$ be the difference of $K$ in $\phi_4$ after and before $LOOK$ services a phase of requests $\delta$. We show that $\Delta K + C_{LOOK}^{T=0} \leq D$ for any phase of requests $\delta$. In the following, we consider cases where $L \leq L'$. Those cases where $L' \leq L$ are omitted. Let $u = \lfloor \frac{D}{2} \rfloor$ denote the middle cylinder of the disk. Consider the case $L \leq L' \leq u$: The disk head of the $LOOK$ algorithm can first move left $x$ cylinders and then it must sweep to cylinder $L'$. Notice that $x + L' - L \leq \lfloor \frac{D}{2} \rfloor$. In this case, we have: $\Delta K = L' - L$ and $C_{LOOK}^{T=0} = 2x + L' - L$. Then, $\Delta K + C_{LOOK}^{T=0} = 2 \times (x + L' - L) \leq D$. The other cases (i.e., $L \leq u < L'$ and $u < L < L'$) are similar. See [13] for details. In all the cases above, we get $\Delta K + C_{LOOK}^{T=0} \leq D$. As a result,

$$C_{LOOK}^{T=t} + \Delta\Phi^{T=t} = C_{LOOK}^{T=0} + Wt + m\left(\Delta\Phi^{T=0} + \Delta K(\frac{1}{m} - 1)\right)$$

$$= m\left(C_{LOOK}^{T=0} + \Delta\Phi^{T=0}\right) + m(2W-1)Wt$$

$$+ (1-m) \times C_{LOOK}^{T=0} + \Delta K(1-m) + Wt(1 - m(2W-1))$$

$$\leq m(2W-1) \times C_{ADV}^{T=0} + m(2W-1)Wt$$

$$+ (1-m) \times C_{LOOK}^{T=0} + \Delta K(1-m) + Wt(1 - m(2W-1))$$

$$= m(2W-1) \times C_{ADV}^{T=t} \frac{2(W-1)Wt}{(2W-1)Wt + D} \times \left(\Delta K + C_{LOOK}^{T=0} - D\right)$$

$$\leq m(2W-1) \times C_{ADV}^{T=t},$$

for any phase and any $t > 0$. □

By Lemmas 7 and 8, we conclude that $LOOK$ is $M(2W-1)$-competitive.

# 6 Discussions and Conclusions

In this section, we discuss the impact of the window size on the competitive ratio of a disk scheduling algorithm with respect to a given disk drive. The results will make it possible for a system programmer to implement a much more efficient disk scheduling algorithm by selecting a suitable window size according to the characteristics of the given disk drive in use. Different disk drives may have different number of cylinders and value of $T$. The $D/T$ ratios of the disk drives discussed in [6,8,11,12] range from 1 to 4. We plot the competitive ratio versus the window size for different $D/T$ ratios in Figure 2. We can see from the figures that different $D/T$ ratios (different disk drives) result in different competitive ratios. But, for $W > 2$, the competitive ratio of $LOOK$ is always lower than the lower bounds of $FCFS$ and $SSTF$ regardless of the $D/T$ ratio.

Summarizing the results in this paper, we have given a detailed competitive analysis of the $LOOK$ algorithm under a cost function that takes the seek

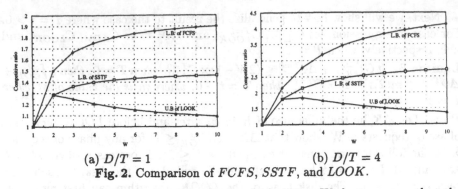

(a) $D/T = 1$          (b) $D/T = 4$

**Fig. 2.** Comparison of $FCFS$, $SSTF$, and $LOOK$.

startup, latency, and transfer times into consideration. We have proven that the competitive ratio of $LOOK$ is $\frac{WT+D}{WT+D/(2W-1)}$, where $D$ is the number of cylinders less one, $W$ is the window size, $T$ is a constant capturing the seek startup time, the latency time and the transfer time. The competitive ratio of $LOOK$ matches its lower bound. We have also shown that $LOOK$ is better than $SSTF$ and $FCFS$ in the competitive sense when the window size $W > 2$. Our results match previous results that $LOOK$ performs better than $SSTF$ and $FCFS$ when the work load of the disk is heavy [9].

## References

1. T. Chen, W. Yang, and R. Lee. Amortized analysis of some disk scheduling algorithms: SSTF, SCAN, and $n$-step SCAN. *BIT*, 32:546–558, 1992.
2. M. Chrobak and L. Larmore. An optimal on-line algorithm for $k$-servers on trees. *SIAM Journal on Computing*, 20:144–148, 1991.
3. E. Coffman, Jr. and M. Hofri. On the expected performance of scanning disks. *SIAM Journal on Computing*, 11(1):60–70, February 1982.
4. A. Fiat, D. Foster, H. Karloff, Y. Rabani, Y. Ravid, and N. E. Young. Competitive paging algorithms. *Journal of Algorithms*, 12:685–699, 1991.
5. A. Fiat, Y. Rabani, and Y. Ravid. Competitive $k$-server algorithms. *J. Computer and System Sciences*, 48:410–428, 1994.
6. R. Geist and S. Daniel. A continuum of disk scheduling algorithms. *ACM Transactions on Computer Systems*, 5(1):77–92, February 1987.
7. M. Hofri. Disk scheduling: FCFS vs. SSTF revisited. *Comm. of the ACM*, 23(11):645–653, November 1980.
8. M. Seltze, P. Chen, and J. Ousterhout. Disk scheduling revisited. In *USENIX Technique Conference*, 1990.
9. A. Silberschatz, J. Peterson, and P. Galvin. *Operating System Concepts*. Addison Wesley, third edition, 1991.
10. D. Sleator and R. Tarjan. Amortized efficiency of list update and paging rules. *Comm. of the ACM*, 28(2):202–208, February 1985.
11. T. Teorey and T. Pinkerton. A comparative analysis of disk scheduling policies. *Comm. of the ACM*, 15(3):177–184, March 1972.
12. N. Wilhelm. An anomaly in disk scheduling: a comparison of FCFS and SSTF seek scheduling using an empirical model for disk accesses. *Comm. of the ACM*, 19(1):13–17, January 1976.
13. T. Yeh, C. Kuo, C. Lei, and H. Yen. Competitive analysis of on-line disk scheduling. Technical report, Dept. of Electircal Engineering, National Taiwan University, 1995.

# Scheduling Interval Ordered Tasks with Non-Uniform Deadlines

Jacques Verriet

Department of Computer Science, Utrecht University,
P.O. Box 80.089, 3508 TB Utrecht, The Netherlands.
E-mail: jacques@cs.ruu.nl

**Abstract.** Garey and Johnson defined an algorithm that finds minimum-lateness schedules for arbitrary graphs with unit-length tasks on two processors. Their algorithm can be easily generalised to an algorithm that constructs minimum-lateness schedules for interval orders on $m$ processors. In this paper, we study the problem of scheduling interval orders with deadlines without neglecting the communication costs. An algorithm is presented that finds minimum-lateness schedules. Like the algorithm by Garey and Johnson, it first computes modified deadlines; these are used to assign a starting time to every task. Unlike the algorithm by Garey and Johnson, calculating a modified deadline for every individual task is not sufficient: in order to fully use the knowledge of the precedence constraints and the communication delays, the algorithm has to compute deadlines for pairs of tasks. The algorithm constructs minimum-lateness schedules in $O(n^2)$ time.

## 1 Introduction

Much of the complexity of parallel computing is due to communication. During the execution of a parallel program on a distributed memory computer, large delays occur between the execution of dependent tasks on different processors. These delays are used to send the result of the computation of a task from one processor to another. Classical scheduling problems do not take these communication delays into account and hence do not capture the complexity of parallel programming.

With the introduction of data transfer costs, scheduling becomes much more complicated. Without communication delays, finding a minimum-length schedule for a tree on an arbitrary number of processors takes polynomial time [4]. The same holds for scheduling arbitrary precedence graphs on infinitely many processors. Both problems become NP-complete, when communication delays are taken into account [5,3]. For interval orders, however, Ali and El-Rewini [1] defined an algorithm that considers communication delays and finds minimum-length schedules in polynomial time.

In this paper, we consider the problem of scheduling unit-length tasks with non-uniform deadlines. The objective is finding schedules in which each task is completed before its deadline. Each task has to be executed on one processor without interruption. A precedence constraint between tasks $u_1$ and $u_2$ represents a data dependency: to execute $u_2$, the result of the computation of $u_1$ must be known. If $u_1$ and $u_2$ are executed on different processors, it is necessary to send data from one processor to another.

This takes unit time during which the sending and the receiving processor can execute another task. If $u_1$ and $u_2$ are executed by the same processor, then no delay occurs.

The problem of finding a schedule in which no task violates its deadline is closely related to the problem of scheduling with the objective of finding a minimum-length schedule. On one hand, the schedule length coincides with a uniform deadline. Hence scheduling with deadlines is a more general problem. On the other, if some tasks have already been assigned a starting time, then the precedence constraints can be used to compute a deadline for the remaining tasks. So scheduling with deadlines is also a sub-problem of the problem of finding a minimum-length schedule.

In earlier work, I presented two algorithms for scheduling unit-length tasks with non-uniform deadlines subject to unit-length communication delays [7]. These algorithms have the same structure as the one by Garey and Johnson [2] for scheduling without communication delays: first the individual deadlines are modified, and second the tasks are scheduled by a list scheduling algorithm applied to the set of tasks ordered by non-decreasing modified deadlines. The precise deadline modification of a task $u$ depends on the subgraph of successors of $u$: if $u$ has sufficiently many successors that have to be completed at time $d$, then the deadline of $u$ can be decreased, since a task has to be scheduled before its successors. The deadline modification does not depend on the predecessors of $u$. For the case of two processors without communication delays [2], this turns out to be sufficient: the algorithm of Garey and Johnson finds minimum-lateness schedules for arbitrary graphs on two processors. Also, a straightforward generalisation of this algorithm finds minimum-lateness schedules for interval ordered tasks on an arbitrary number of processors. In case of two processors with communication delays, I was only able to solve the problem for graphs satisfying the least urgent parent property [7]. This restriction is closely related to the two basic patterns in which communication delays have an impact on scheduling.

**Successor pattern** If $u$ has immediate successors $v_1, \ldots, v_k$, then at most one of these can be executed immediately after $u$.

**Predecessor pattern** If $u$ has immediate predecessors $w_1, \ldots, w_l$, then only one can be executed immediately before $u$.

Restriction to (series-parallel) graphs with the least urgent parent property answers the question associated with the predecessor pattern at the beginning. The question corresponding to the successor pattern is dealt with in the deadline modification stage.

The algorithm that will be presented in this paper takes both patterns into account. This is accomplished by a different approach to computing the modified deadlines: a deadline will be defined to every pair of tasks $(u_1, u_2)$. At least one of the tasks $u_1$ and $u_2$ has to be completed before this deadline. Like the individual deadlines, the deadline of $(u_1, u_2)$ depends on the successors of $u_1$ and $u_2$: if $u_1$ and $u_2$ have sufficiently many common successors that have to be executed before time $d$, then the deadline of $(u_1, u_2)$ can be decreased. The deadline of $(u_1, u_2)$ may be smaller than the individual deadlines of $u_1$ and $u_2$, because the Predecessor pattern is taken into account. If the deadline of $(u_1, u_2)$ is strictly smaller than the deadlines of $u_1$ and $u_2$, then $u_1$ or $u_2$ must be executed earlier than required by its individual deadline.

In Section 4, an algorithm is presented that constructs schedules for interval ordered tasks with non-uniform deadlines. It first computes modified deadlines. These are used to assign a starting time to every task. Both parts can be implemented in $O(n^2)$ time. It will be shown that the algorithm finds minimum-lateness schedules for interval orders.

## 2 Preliminary Definitions

The tasks and the data dependencies between them are represented by a directed acyclic graph or precedence graph. Throughout this paper, $G$ will denote a precedence graph in which every task $u$ has an original deadline $D_0(u)$. These are assumed to be non-negative integers. $G$ contains $n$ nodes and $e$ edges (precedence constraints). Let $u_1, u_2$ be nodes of $G$. If $(u_1, u_2)$ is an edge of $G$, $u_2$ is called a child of $u_1$ and $u_1$ a parent of $u_2$. If there is a directed path from $u_1$ to $u_2$, then $u_1$ is a predecessor of $u_2$ and $u_2$ a successor of $u_1$. This is denoted by $u_1 \prec u_2$. A node $u$ is a successor (child) of a set of nodes $V$ if some task of $V$ is a predecessor (parent) of $u$. The set of successors of $u$ is denoted by $Succ(u)$. A source is a node without predecessors, a task without successors is called a sink.

A schedule for a graph $G$ is a list of subsets of $G$ which are called time slots. A schedule $S = (S_0, \ldots, S_{l-1})$ is valid for $G$ on $m$ processors, if

1. $\bigcup_{t=0}^{l-1} S_t = G$;
2. $S_t \cap S_{t'} = \emptyset$ for all $t \neq t'$;
3. $|S_t| \leq m$ for all $t$;
4. if $u \prec v$, $u \in S_t$ and $v \in S_{t'}$, then $t < t'$;
5. if $u \in S_t$, then $S_{t+1}$ contains at most one child of $u$; and
6. if $u \in S_{t+1}$, then $S_t$ contains at most one parent of $u$.

Note that these properties do not contain information about the assignment of processors. It is easy to assign a processor to every task: if $S$ is a valid schedule for $G$ on $m$ processors, a correct assignment can be found in $O(\min\{mn, n + e\})$ time.

Let $S$ be a schedule on $m$ processors for a graph $G$. If $u \in S_t$, $u$ is said to be scheduled at time $t$. $u$ meets its deadline if its execution is finished at time $D_0(u)$. So $u$ meets its deadline if $t + 1 \leq D_0(u)$. If $D_0(u) \leq t$, $u$ is called late and its lateness is $t + 1 - D_0(u)$. The lateness of a task meeting its deadline is 0. The lateness of $S$ equals the maximum lateness of a task in $G$. $S$ is optimal if no other schedule for $G$ on $m$ processors has lateness smaller than $S$. It is called 0-optimal if no task violates its deadline.

A partial schedule for a graph $G$ on $m$ processors is a valid schedule $S$ for an induced subgraph of $G$. Let $S = (S_0, \ldots, S_{l-1})$ be a partial schedule of a graph $G$ on $m$ processors. Suppose $S$ is a valid schedule for an induced subgraph $G'$ of $G$. A node $u$ of $G \backslash G'$ is called available at time $t$ (with respect to $S$) if $(S_0, \ldots, S_{t-1}, S_t \cup \{u\}, S_{t+1}, \ldots, S_{l-1})$ is a valid schedule for $G' \cup \{u\}$.

## 3  Interval Orders

In this paper, we will consider a special class of graphs: the interval orders. Papadimitriou and Yannakakis [6] defined these in the following way. An interval order is a partial order $(V, \prec)$, for which every element $v$ of $V$ can be assigned a closed interval $I(v)$ in the real line such that for all $v_1, v_2$ in $V$

$$v_1 \prec v_2 \text{ if and only if } x < y \text{ for all } x \in I(v_1), y \in I(v_2).$$

Interval orders are a natural class of graphs for scheduling. A schedule for a graph with arbitrary task lengths defines an interval order: the interval corresponding to a task $u$ with starting time $s_u$ and completion time $t_u$ is $[s_u, t_u - \epsilon]$, where $\epsilon$ is a small positive number. Hence in this interval order, $u \prec v$ if and only if $t_u \leq s_v$. Clearly, every edge in the original graph is an edge in the interval order. Hence the problem of scheduling a graph $G$ can be considered as the problem of finding the best interval order that contains all edges of $G$ and finding an optimal schedule (without communication delays) for it.

Interval orders have a very nice property. Consider an assignment of intervals for interval order $(V, \prec)$. Suppose $I(v_1) = [x_1, y_1]$ and $I(v_2) = [x_2, y_2]$. Assume $y_1 \leq y_2$. From the definition of this assignment, $Succ(v) \subseteq Succ(u)$. Consequently, $Succ(u) \subseteq Succ(v)$ or $Succ(v) \subseteq Succ(u)$ for all $u, v$ in $V$. This result can be generalised.

**Proposition 1.** *Let $(V, \prec)$ be an interval order. Let $U$ be a non-empty subset of $V$. There is a task $u$ in $U$ such that*

$$Succ(u) = \bigcup_{v \in U} Succ(v).$$

## 4  The Algorithm

In this section, I will present an algorithm that finds optimal schedules for interval orders with non-uniform deadlines. It is similar to the one presented by Garey and Johnson [2] for scheduling with deadlines without communication delays. It consists of two steps. First it computes (smaller) deadlines $D(u)$, that are consistent with the precedence constraints. These modified deadlines are used by a list scheduling algorithm that assigns a starting time to every task.

In earlier work, I defined an algorithm for scheduling unit-length tasks with deadlines subject to unit-length communication delays [7]. This algorithm does not use all knowledge of the deadlines and the structure of 0-optimal schedules: its deadline modification part only considers a task and its successors. Consequently, a task with modified deadline $d$ has at most one successor with modified deadline $d + 1$. However, since the predecessors of a task are not considered, a task with modified deadline $d + 1$ can have several parents with modified deadline $d$. Because of the asymmetric deadline modification, it is possible to construct an interval order for which this algorithm does not construct a 0-optimal schedule even if such a schedule exists.

To define an algorithm that finds 0-optimal schedules, extra information is needed. Let $u$ be a task of a graph that has to be scheduled on $m$ processors. Suppose $v_1, \ldots, v_k$ are successors of $u$ with deadlines $D(v_1) \leq \ldots \leq D(v_k)$. Because of communication delays, only one successor of $u$ can be executed immediately after $u$. So, in order to meet every deadline, $u$ has to be completed at time $D(v_k) - 1 - \lceil \frac{k-1}{m} \rceil$.

Suppose $u_1$ and $u_2$ have $l = km + 1$ common successors $v_1, \ldots, v_l$ with deadlines $D(v_1) \leq \ldots \leq D(v_l)$. In order to meet its deadline, $u_1$ must be completed at time $\leq D(v_l) - 1 - \lceil \frac{l-1}{m} \rceil = D(v_l) - 1 - k$. The same holds for $u_2$. If both tasks are scheduled at time $D(v_l) - 2 - k$, one of their common successors violates its deadline, because the first task of $v_1, \ldots, v_l$ cannot be executed until time $D(v_l) - k$. So one of the tasks $u_1$ and $u_2$ has to be scheduled at time $\leq D(v_l) - 3 - k$.

In order to use this knowledge, we will consider pairs of tasks as well as individual tasks. A pair of (not necessarily different) tasks $(u_1, u_2)$ will be assigned a deadline $D(u_1, u_2)$. $(u_1, u_2)$ meets its deadline if $u_1$ or $u_2$ is completed before time $D(u_1, u_2)$.

The following definitions are used to compute the modified deadlines. $N(u_1, u_2, d)$ denotes the number of common successors of $u_1$ and $u_2$ with deadlines at most $d$. $P(u_1, u_2, d) = \max\{0, |V| - 1\}$, where $V$ is a maximum-size subset of $Succ(u_1) \cap Succ(u_2)$, such that $D(v_1) \geq d + 1$ and $D(v_1, v_2) \leq d$ for all $v_1 \neq v_2$ in $V$. Note that in a schedule in which all deadlines are met, at most one task of such a set $V$ is executed at time $\geq d$. Hence at least $N(u_1, u_2, d) + P(u_1, u_2, d)$ common successors of $u_1$ and $u_2$ are completed at time $d$ in every 0-optimal schedule. For individual tasks, we use shorthand notations: $N(u, d) = N(u, u, d)$ and $P(u, d) = P(u, u, d)$.

Let $G$ be a graph with deadlines $D(u_1, u_2)$. $G$ is said to have modified deadlines, if

1. $D(u_1, u_2) \leq \min\{D(u_1), D(u_2)\}$;
2. if $N(u, d) + P(u, d) \geq 1$, then $D(u) \leq d - 1 - \lceil \frac{1}{m}(N(u, d) + P(u, d) - 1) \rceil$;
3. if $N(u_1, u_2, d) + P(u_1, u_2, d) \geq km + 1$, then $D(u_1, u_2) \leq d - 2 - k$; and
4. in any 0-optimal schedule for $G$, no deadline $D(u_1, u_2)$ is violated.

The following property of graphs with modified deadlines is not difficult to prove. It will be used to define the algorithm.

**Lemma 2.** *Let $G$ be a graph with modified deadlines. Let $u_1, u_2, v$ be three tasks of $G$. If $D(u_1) = D(u_2) = D(v) = D(u_1, u_2) + 1$ and $Succ(u_1) \cap Succ(u_2) \subseteq Succ(v)$, then $D(u_1, v) = D(u_2, v) = D(u_1, u_2)$.*

### 4.1 Finding 0-Optimal Schedules

If we are given an interval order $G$ with modified deadlines, we can construct a 0-optimal schedule for $G$ using a straightforward list scheduling algorithm, if such a schedule exists. Algorithm *List scheduling* is shown in Fig. 1.

It is easy to see that Algorithm *List scheduling* finds valid schedules for arbitrary precedence graphs in $O(n^2)$ time.

**Theorem 3.** *Let $G$ a graph. Let $L$ be a list containing all tasks of $G$. Using $L$, Algorithm List scheduling constructs a valid schedule for $G$ in $O(n^2)$ time.*

**Algorithm** *List scheduling*
**Input:** A graph $G$ and a list $L = (u_1, \ldots, u_n)$ containing all tasks of $G$.
**Output:** A valid schedule for $G$ on $m$ processors.
1.    $t := 0$
2.    **while** $L$ contains unscheduled tasks
3.        **do for** $i := 1$ **to** $n$
4.             **do if** $u_i$ is unscheduled and available at time $t$
5.                 **then** $S_t := S_t \cup \{u_i\}$
6.        $t := t + 1$

**Fig. 1.** The list scheduling algorithm

For interval orders, we use the following notion of priority. $u_1$ has a higher priority than $u_2$, if $D(u_1) < D(u_2)$ or $D(u_1) = D(u_2)$ and $Succ(u_1) \supsetneq Succ(u_2)$. A list of tasks ordered by non-increasing priority will be called a priority list. We can prove that, for interval orders, Algorithm *List scheduling* constructs 0-optimal schedules using a priority list $L$, if such a schedule exists.

**Theorem 4.** *Let $G$ be an interval order with modified deadlines. Let $L$ be a priority list of $G$. Let $S$ be the schedule for $G$ on $m$ processors found by Algorithm List scheduling using $L$. If a 0-optimal schedule for $G$ on $m$ processors exists, then in $S$, every pair of tasks meets its deadline.*

*Proof.* Let $G$ be an interval order with modified deadlines. Let $L$ be a priority list for $G$. Suppose there is a 0-optimal schedule for $G$ on $m$ processors. Let $S$ be the schedule for $G$ on $m$ processors constructed by Algorithm *List scheduling* using $L$. Suppose not every pair of tasks meets its deadline. Assume $t$ is the first time such that $S_t$ contains a task in a pair that violates its deadline. Let $u_1$ be this task and let $(u_1, u_2)$ be this pair. $(u_1, u_2)$ violates its deadline. So $D(u_1, u_2) \leq t$. There are three possibilities: $D(u_1) \leq t$, $D(u_2) \leq t$ or $D(u_1, u_2) = t$ and $D(u_1) = D(u_2) = t + 1$. The last is the most complicated. The other two cases can be proved in a similar way. Hence we will assume $D(u_1, u_2) = t$ and $D(u_1) = D(u_2) = t + 1$.

Suppose $Succ(u_1) \subseteq Succ(u_2)$. The priority of $u_2$ is at least as high as that of $u_1$. Let $U$ be the set of tasks with priority as least as high as $u_1$. Let $v_1, v_2$ be two tasks of $U$. Clearly, $D(v_1), D(v_2) \leq D(u_1) = t + 1$. If $D(v_1) \leq t$ or $D(v_2) \leq t$, then $D(v_1, v_2) \leq t$. We will assume $D(v_1) = D(v_2) = t + 1$. Since the priority of $v_1$ and $v_2$ is at least as high as that of $u_1$, $Succ(u_1) \cap Succ(u_2) \subseteq Succ(v_1), Succ(v_2)$. By applying Lemma 2 twice, we obtain $D(v_1, v_2) = t$. So, in order to meet every deadline, at most one task of $U$ can be scheduled at time $\geq t$. Since a 0-optimal schedule for $G$ exists, $U$ contains at most $mt + 1$ tasks. Therefore there is a time before $t$ at which at most $m - 1$ tasks with priority as least as high as $u_1$ are scheduled. Let $t' - 1$ be the last such time before $t$. Define $H = \bigcup_{i=t'}^{t-1} S_i \cup \{u_1, u_2\} \cup \{v \in \bigcup_{i>t} S_i \mid v \prec u_2\}$. $H$ contains at least $m(t - t') + 2$ tasks and every pair of different tasks of $H$ has deadline $\leq t$. $S_{t'-1}$ contains at most $m - 1$ tasks with priority as least as high as $u_1$, so no task of $H$ was available at time $t' - 1$. Let $u$ be a source of $H$. $u$ was not available at time $t' - 1$, because (at least) one of the following three conditions is satisfied.

1. $S_{t'-1}$ contains a parent of $u$.
2. $S_{t'-2}$ contains two parents of $u$.
3. $S_{t'-2}$ contains a parent $u'$ of $u$ and $S_{t'-1}$ contains another child of $u'$.

Thus every task in $H$ has a predecessor in $S_{t'-2}$ or $S_{t'-1}$.

**Case 1.** Every task in $H$ is a successor of $S_{t'-1}$.
Define $Q = S_{t'-1} \cap \{v \in G \mid D(v) \leq t\}$. Since every task in $H$ has deadline at most $t+1$, every task in $H$ is a successor of $Q$. $S$ is a valid schedule, so at most $|Q| \leq m-1$ successors of $Q$ are executed at time $t'$. Therefore $t = t'$. From Proposition 1, $Q$ contains a predecessor $w$ of both $u_1$ and $u_2$. Hence $N(w, t+1) \geq 2$. Consequently, $D(w) \leq (t+1) - 2 = t - 1 = t' - 1$. So $w$ violates its deadline. Contradiction.

**Case 2.** Not every source of $H$ has a parent in $S_{t'-1}$.
Define $V = \{w \in S_{t'-2} \cup S_{t'-1} \mid w \text{ is a parent of } H\}$. With Proposition 1, $V$ contains a task $w_1$, such that every task of $H$ is a successor of $w_1$. Obviously, $w_1$ is scheduled at time $t' - 2$. Let $V' = V \backslash \{w_1\}$.

**Case 2.1.** Every task in $H$ is a successor of $V'$.
From Proposition 1, $V'$ contains a task $w_2$ with $H \subseteq Succ(w_2)$. Hence every task in $H$ is a common successor of $w_1$ and $w_2$. Define $H_1 = \{v \in H \mid D(v) \leq t\}$ and $H_2 = \{v \in H \mid D(v) = t+1\}$. Clearly, $N(w_1, w_2, t) \geq |H_1|$ and $P(w_1, w_2, t) \geq |H_2| - 1$. So $N(w_1, w_2, t) + P(w_1, w_2, t) \geq m(t-t') + 1$. As a result, $D(w_1, w_2) \leq t - 2 - (t - t') = t' - 2$. So $(w_1, w_2)$ violates its deadline. Contradiction.

**Case 2.2.** Not every task in $H$ is a successor of $V'$.
Let $v$ be a task of $H$ that has no predecessor in $V'$. Assume $v$ is a source of $H$. $V'$ does not contain a parent of $v$. So $w_1$ is a parent of $v$. Since $v$ was not available at time $t' - 1$ and $S_{t'-2}$ contains only one parent of $v$, $S_{t'-1}$ contains another child $v'$ of $w_1$. $v'$ is scheduled before $v$, so $v'$ occurs before $v$ in $L$. Therefore $D(v') < D(v)$ or $D(v') = D(v)$ and $Succ(v') \supseteq Succ(v)$. If $D(v') < D(v)$, then $D(v') \leq t$ and every pair consisting of two different tasks of $H \cup \{v'\}$ has deadline $\leq t$. Otherwise, using Lemma 2, every pair of different tasks of $H \cup \{v'\}$ has deadline $\leq t$. Let $H_1 = \{v \in H \cup \{v'\} \mid D(v) \leq t\}$ and $H_2 = \{v \in H \cup \{v'\} \mid D(v) = t+1\}$. Then $N(w_1, t) \geq |H_1|$ and $P(w_1, t) \geq |H_2| - 1$. Therefore $N(w_1, t) + P(w_1, t) \geq m(t - t') + 2$. So $D(w_1) \leq t' - 2$ and $w_1$ is not completed before its deadline. Contradiction.

$\square$

Hence Algorithm *List scheduling* finds 0-optimal schedule for interval orders with modified deadlines, if such schedules exist.

## 4.2 Efficient Deadline Modification

In this section, I will present an algorithm that computes modified deadlines for interval orders. Since Algorithm *List scheduling* does not use the modified deadlines of the pairs of tasks, the deadline modification algorithm does not need to consider all pairs.

The complexity of the deadline modification depends mainly on the computation of $P(u_1, u_2, d)$ and the number of pairs that have to be taken into account. It is not difficult to prove that we only need to consider pairs of tasks $(u_1, u_2)$ with $D(u_1) = D(u_2)$ and $N(u_1, d) + P(u_1, d) = N(u_2, d) + P(u_2, d) = (d - D(u_1) - 1)m + 1$ for some $d$.

**Lemma 5.** *Let $G$ be a graph with modified deadlines. Let $u_1$ and $u_2$ be two tasks of $G$. If $D(u_1, u_2) < \min\{D(u_1), D(u_2)\}$, then $D(u_1) = D(u_2) = D(u_1, u_2) + 1$ and $N(u_1, d) + P(u_1, d) = N(u_2, d) + P(u_2, d) = N(u_1, u_2, d) + P(u_1, u_2, d) = (d - D(u_1) - 1)m + 1$ for some $d$.*

This result allows us to restrict the number of deadlines of pairs of tasks that have to be computed: we only need to compute deadlines $D(u_1, u_2)$ for tasks $u_1$ and $u_2$ with $D(u_1) = D(u_2)$ and $N(u_1, d) + P(u_1, d) = N(u_2, d) + P(u_2, d) = (d - D(u_1) - 1)m + 1$. The other deadlines can be set to the minimum of the deadlines of the individual tasks.

$P(u_1, u_2, d)$ is defined in terms of sets of common successors of $u_1$ and $u_2$. For an arbitrary precedence graph $G$, computing $P(u_1, u_2, d)$ coincides with finding a maximum clique in a subgraph of $G$ containing the common successors of $u_1$ and $u_2$ with deadlines $d + 1$ and edges between tasks $v_1$ and $v_2$ if $D(v_1, v_2) = d$. Consequently, this definition does not allow an efficient method of determining $P(u_1, u_2, d)$. For interval orders, an alternative definition can be derived. This formulation allows us to compute $P(u_1, u_2, d)$ in linear time. Define $D_{\min}(u_1)$ as the minimum deadline $D(u_1, u_2)$ for a task $u_2$ with $D(u_2) = D(u_1)$. Using Lemma 2, we can derive a definition of $P(u_1, u_2, d)$ in terms of $D_{\min}(u)$.

**Lemma 6.** *Let $G$ be an interval order with modified deadlines. Let $u_1, u_2$ be tasks of $G$. $P(u_1, u_2, d) = \max\{0, |V| - 1\}$, where $V = \{v \in Succ(u_1) \cap Succ(u_2) \mid D(v) = d + 1 \wedge D_{\min}(v) = d\}$.*

*Proof.* Let $G$ be an interval order with modified deadlines. Let $u_1, u_2$ be tasks of $G$. If $P(u_1, u_2, d) = 0$, then it is easy to prove that $u_1$ and $u_2$ have at most one common successor $v$ with $D(v) = d + 1$ and $D_{\min}(v) = d$. Hence we may assume $P(u_1, u_2, d) = |V_P| - 1 \geq 1$, where $V_P$ is a maximum-size subset of $Succ(u_1) \cap Succ(u_2)$ with $D(v_1) = d + 1$ and $D(v_1, v_2) = d$ for all $v_1 \neq v_2$ in $V_P$. Suppose $V_P = \{v_1, \ldots, v_k\}$ with $Succ(v_1) \subseteq \ldots \subseteq Succ(v_k)$. Define $V = \{v \in Succ(u_1) \cap Succ(u_2) \mid D(v) = d + 1 \wedge D_{\min}(v) = d\}$. Clearly, $V_P \subseteq V$. Suppose $V_P \neq V$. Assume $v \in V \backslash V_P$.

**Case 1.** $Succ(v_1) \subseteq Succ(v)$.
$Succ(v_1) \cap Succ(v_i) \subseteq Succ(v)$, so, from Lemma 2, $D(v_i, v) = d$ for every $i$. Hence $V_P \cup \{v\}$ is a subset of $Succ(u_1) \cap Succ(u_2)$ with $D(w_1) = d + 1$ and $D(w_1, w_2) = d$ for all $w_1 \neq w_2$ in $V_P \cup \{v\}$. Contradiction.

**Case 2.** $Succ(v) \subseteq Succ(v_1)$.
There is a task $v'$ with $D(v') = d + 1$ and $D(v, v') = d$. Since $Succ(v) \cap Succ(v') \subseteq Succ(v_i)$ and $D(v_i) = d + 1$, $D(v_i, v) = d$ for all $i$. So $V_P$ is not of maximum size. Contradiction.

$\square$

Using the new definition of $P(u_1, u_2, d)$, we can present the deadline modification algorithm. Algorithm *Deadline modification* is shown in Fig. 2. The following notations are used. $L_d$ denotes the set of tasks with modified deadline $d$. $L_{d,d'}$ is a subset of $L_d$ containing tasks $u$ with $N(u, d') + P(u, d') = (d' - d - 1)m + 1$. Algorithm *Deadline modification* starts with computing the modified deadline and the $D_{\min}$-value for the tasks with the largest original deadline, which is denoted by $D$. $d_{\max}$ denotes the maximum $d$, such that there is a task $u$ in $L_d$ with $D_{\min}(u)$ is unknown.

**Algorithm** *Deadline modification*
**Input:** An interval order $G$ with original deadlines $D_0(u)$.
**Output:** Interval order $G$ with modified deadlines $D(u)$.

1.    **for** $u \in G$
2.        **do** $D(u) := D_0(u)$
3.    $d_{\max} := \max_{u \in G} D(u)$
4.    $P := G$
5.    **while** $P \neq \emptyset$
6.        **do let** $u$ be a sink of $P$ with maximum $D(u)$
7.            **while** $D(u) < d_{\max}$
8.                **do for** $v \in L_{d_{\max}}$
9.                    **do** $D_{\min}(v) := d_{\max}$
10.                    **for** $d := d_{\max}$ **to** $D$
11.                        **do if** $N(v, d) + P(v, d) = (d - d_{\max} - 1)m + 1$
12.                            **then** $L_{d_{\max},d} := L_{d_{\max},d} \cup \{v\}$
13.                    **for** $d := d_{\max}$ **to** $D$
14.                        **do if** $|L_{d_{\max},d}| \geq 2$
15.                            **then for** $v \in L_{d_{\max},d}$
16.                                **do** $D_{\min}(v) := d_{\max} - 1$
17.                $d_{\max} := d_{\max} - 1$
18.            **for** $d := d_{\max}$ **to** $D$
19.                **do if** $N(u, d) + P(u, d) \geq 1$
20.                    **then** $D(u) := \min \left\{ D(u), d - 1 - \lceil \frac{1}{m}(N(u, d) + P(u, d) - 1) \rceil \right\}$
21.            $L_{D(u)} := L_{D(u)} \cup \{u\}$
22.            **for** all parents $v$ of $u$
23.                **do** $D(v) := \min\{D(v), D(u) - 1\}$
24.        $P := P \backslash \{u\}$

**Fig. 2.** The deadline modification algorithm

We will assume that, in the beginning, $D(u_1) \leq D(u_2) - 1$ if $u_1 \prec u_2$. This way, Algorithm *Deadline modification* always chooses a task $u$ whose successors have a modified deadline that is larger than $D(u)$, because after each iteration of the outer while loop, the successors of a sink $u$ of $P$ have modified deadlines that are larger than $D(u)$. Suppose sink $u$ of $P$ is chosen. Since the successors of $u$ have a larger deadline than $u$ and $D(u)$ is maximum, the $D_{\min}$-values of the tasks with modified deadline greater than $D(u)$ can be computed. Hereafter, the modified deadline of $u$ is determined and the deadline of each parent $v$ of $u$ is adjusted, such that $D(v) < D(u)$.

Because Algorithm *Deadline modification* starts with $D(u) = D_0(u)$, the resulting interval order has modified deadlines.

$N(u, d)$ can be computed by traversing the successors of $u$ and a prefix sum operation. From Lemma 6, computing $P(u, d)$ can be done by looking at all successors of $u$. This takes $O(n)$ time in a transitive closure for all $d$. The inner while loop is executed for every $d$ in $1, \ldots, D$. Since $L_{d,d'}$ is a subset of $L_d$, such an iteration takes $O(|L_d| n)$ time. Hence $O(\sum_d |L_d| n) = O(n^2)$ time used to execute the inner while loop. The outer while loop is executed $n$ times. Computation of $N(u, d) + P(u, d)$ takes $O(n)$ for each task, so $O(n^2)$ time in total. The adjustment of the deadlines of the parents of the chosen tasks takes $O(e^+)$ time, where $e^+$ is the number of edges in the transitive closure, which can be found in $O(n + e^+)$ time for interval orders. Hence computing modified deadlines takes $O(n^2)$ time.

**Theorem 7.** *Algorithm Deadline modification computes modified deadlines for interval orders in $O(n^2)$ time.*

So Algorithms *Deadline modification* and *List scheduling* find 0-optimal schedules for interval orders in polynomial time. If no such schedules exist, they construct minimum-lateness schedules.

**Theorem 8.** *Let $G$ be an interval order with deadlines. Algorithms Deadline modification and List scheduling construct a minimum-lateness schedule for $G$ in $O(n^2)$ time.*

*Proof.* Let $G$ be an interval order with deadlines $D_0(u)$. Assume the lateness of a minimum-lateness schedule for $G$ equals $l$. Let $G'$ be the interval order $G$ with original deadlines $D'_0(u) = D_0(u) + l$. Algorithms *Deadline modification* and *List scheduling* find a 0-optimal schedule for $G'$. Let $D(u)$ and $D'(u)$ be the modified deadlines for $G$ and $G'$ respectively. It is not difficult to see that $D'(u) = D(u) + l$. Hence any priority list for $G'$ is a priority list for $G$. Consequently, Algorithms *Deadline modification* and *List scheduling* construct a schedule with lateness $l$ for $G$. ∎

# References

1. H.H. Ali and H. El-Rewini. An optimal algorithm for scheduling interval ordered tasks with communication on N processors. *Journal of Computer and System Sciences*, 51(2):301–307, October 1995.
2. M.R. Garey and D.S. Johnson. Scheduling tasks with nonuniform deadlines on two processors. *Journal of the ACM*, 23(6):461–467, July 1976.
3. J.A. Hoogeveen, J.K. Lenstra and B. Veltman. Three, four, five, six, or the complexity of scheduling with communication delays. *Operations Research Letters*, 16:129–137, 1994.
4. T.C. Hu. Parallel sequencing and assembly line problems. *Operations Research*, 9(6):841–848, 1961.
5. J.K. Lenstra, M. Veldhorst and B. Veltman. The complexity of scheduling trees with communication delays. *Journal of Algorithms*, 20(1):157–173, January 1996.
6. C.H. Papadimitriou and M. Yannakakis. Scheduling interval-ordered tasks. *SIAM Journal on Computing*, 8(3):405–409, August 1979.
7. J. Verriet. Scheduling UET, UCT dags with release dates and deadlines. Technical Report UU-CS-1995-31, Department of Computer Science, Utrecht University, September 1995.

# Cryptographic Weaknesses in the Round Transformation Used in a Block Cipher with Provable Immunity Against Linear Cryptanalysis

## (Extended Abstract of ISAAC'96)

Kouichi SAKURAI [‡]          Yuliang ZHENG [†]

[‡] Dept. of Computer Science and Communication Engineering,
Kyushu University, 812-81, JAPAN
e-mail: sakurai@csce.kyushu-u.ac.jp

[†] School of Comp. & Info. Tech., Monash University
McMahons Road, Frankston, Melbourne, VIC 3199,AUSTRALIA
e-mail: yzheng@fcit.monash.edu.au

**Abstract.** MISTY is a data encryption algorithm recently proposed by M. Matsui from Mitsubishi Electric Corporation. This paper focuses on cryptographic roles of the transform used in the MISTY cipher. Our research reveals that when used for constructing pseudorandom permutations, the transform employed by the MISTY cipher is inferior to the transform in DES, though the former is superior to the latter in terms of strength against linear and differential attacks. More specifically, we show that a 3-round (4-round, respectively) concatenation of transforms used in the MISTY cipher is *not* a pseudorandom (super pseudorandom, respectively) permutation. For comparison, we note that with three (four, respectively) rounds, transforms used in DES yield a pseudorandom (super pseudorandom, respectively) permutation.
Another contribution of this paper is to show that a 3-round concatenation of transforms used in (the preliminary version of) the MISTY cipher has an algebraic property, which may open a door for various cryptanalytic attacks.

## 1  Introduction

The Data Encryption Standard (DES) [NBS77] is the most widely used cipher over the world. It has been nearly a quarter of a century since DES was published in the 1970's. Due to rapid advances in cryptanalysis as well as computing technology over the past 20 years, especially the recent discovery of differential cryptanalysis by Biham and Shamir [BS93] and linear cryptanalysis by Matsui [Ma94], the cryptographic strength of DES is being questioned by an increasing number of researchers as well as practitioners. Structurally DES can be viewed as being obtained by the iteration of a basic transform which was first proposed by Feistel [F73, FNS75] and will be called a DES-like transform in this paper.

Not all the design criteria for DES have been made public by its designers. Recent work by some researchers, however, shows that based on the iteration of DES-like transforms, it is possible to construct a block cipher that is provably secure against differential cryptanalysis [NK95, Nyb94].

Based on these observations, Matsui has proposed a new block cipher (encryption algorithm) called MISTY [Ma96]. A preliminary version of the MISTY cipher appears in [Ma96], where it is shown that the MISTY cipher is provably more robust than DES in terms of its resistance against linear or differential cryptanalysis. In studying the security of a cipher, however, we should bear in mind that there is in general no inclusive relationship in the power of cryptanalytic attacks. In particular, a cipher secure against linear and differential cryptanalysis may be insecure against other (seemingly weaker) types of cryptanalysis. One example of such an algorithm can be found in [Nyb93]. In this context the MISTY cipher deserves special attention, as it employs a new transform that is different from a DES-like one. It is quite natural for one to expect that a cryptanalytic attack not applicable to DES may be used for breaking the MISTY cipher, which is precisely the major motivation of this research.

The cryptographic soundness of DES-like transforms has been theoretically studied by Luby and Rackoff [LR86]. In particular they proved that a 3-round concatenation of DES-like transforms yields a pseudorandom permutation. In proving the result they assumed that truly random and independent functions were used in the three round transforms. As the function used in a DES-like transform is far from being random, their result does not form a proof for the security of DES.

It is important to note that Luby and Rackoff [LR86] also proved that a 2-round concatenation of DES-like transforms never gives a pseudorandom permutation, as the resulting permutation is breakable by a chosen plaintext attack when the permutation is regarded as a cipher. From this result one can say that the approach taken by Luby and Rackoff is of fundamental importance to any basic transform used in a cryptographic algorithm. This can be further demonstrated by recent studies on the security of message authentication codes [BKR94, BGR94], and Kerberos-like key distribution [BR95].

When the preliminary version of the MISTY cipher was published in [Ma96], the soundness of a transform used in the MISTY cipher, which will be called a MISTY-like transform hereafter, was not examined in the context of Luby and Rackoff's approach. Hence, the focus of this paper is on the construction of pseudorandom permutations from MISTY-like transforms, with the aim of comparing a MISTY-like transform against a DES-like one. We show that a 3-round concatenation of MISTY-like transforms does not yield a pseudorandom permutation. This should be compared with DES-like transforms: as mentioned earlier, a concatenation of the same number of DES-like transforms does result in a pseudorandom permutation. This contrast also shows that pseudorandomness and resistance against linear or differential cryptanalysis are incomparable. Hence it provides an answer to the second open problem in the last section of [SP92].

More importantly we show that a 3-round concatenation of MISTY-like transforms proposed in [Ma96], has an algebraic invariance property. As 3-round concatenations are recursively used as basic building blocks for each round in a preliminary version of the MISTY cipher, this algebraic property would open a large door for various cryptanalytic attacks, and hence could be a potentially critical weakness of a preliminary version of the MISTY cipher. These facts clearly show that MISTY-like transforms are inferior to DES-like ones.

We have also examined under which conditions MISTY-like transforms would yield a pseudorandom permutation. In particular we have considered cases where similar concatenations of DES-like transforms would result in pseudorandom permutations. Our research in this direction shows that, in every case we considered, MISTY-like transforms fail to produce pseudorandom permutations. To put it in another way, in all these cases DES-like transforms are superior to MISTY-like ones.

## 2 Preliminary

### 2.1 Basic Notation

The set of positive integers is denoted by $\mathcal{N}$. For each $n \in \mathcal{N}$, let $I_n$ be the set of all $2^n$ binary strings of length $n$, i.e., $\{0,1\}^n$. For $s_1, s_2 \in I_n$, $s_1 \oplus s_2$ stands for the bit-wise exclusive-or of $s_1$ and $s_2$, and $s_1 \bullet s_2$ denotes the bit-wise product of $s_1$ and $s_2$.

Denote by $H_n$ the set of all functions from $I_n$ to $I_n$, which consists of $2^{n2^n}$ in total. The composition of two functions $f$ and $g$ in $H_n$, denoted by $f \circ g$, is defined by $f \circ g(x) = f(g(x))$, where $x \in I_n$. And in particular, $f \circ f$ is denoted by $f^2$, $f \circ f \circ f$ by $f^3$, and so on.

By $x \in_R X$ we mean that $x$ is drawn randomly and uniformly from a set $X$.

### 2.2 DES-like Transforms

Associate with each $f \in H_n$ a function

$$\delta_{2n,f}(L, R) = (R \oplus f(L), L)$$

for all $L, R \in I_n$. In cryptography, the function $f$ used in $\delta_{2n,f}$ is commonly referred to as *the F-function* of $\delta_{2n,f}$. Note that $\delta_{2n,f}$ is a permutation in $H_{2n}$, and it is commonly called a DES-like transform associated with $f$ [F73, FNS75, NBS77]. Furthermore, for $f_1, f_2, \ldots, f_s \in H_n$, define $D(f_s, \ldots, f_2, f_1) = \delta_{2n,f_s} \circ \cdots \circ \delta_{2n,f_2} \circ \delta_{2n,f_1}$ as an $s$-round concatenation of DES-like transforms.

Various generalizations of DES-like transforms, together with their cryptographic applications, were studied in [ZMI89, ZMI89, Zhe90].

### 2.3 Notion of Pseudorandomness

Let $n \in \mathcal{N}$. An oracle circuit $T_n$ is an acyclic circuit which contains, in addition to ordinary AND, OR, NOT and constant gates, also a particular kind of gates – oracle gates. Each oracle gate has an $n$-bit output, and it is evaluated using a function from $H_n$. The output of $T_n$, a single bit, is denoted by $T_n[f]$ when a function $f \in H_n$ is used to evaluate all the oracle gates in $T_n$. The size of $T_n$ is the total number of connections in it. Note that we can regard an oracle circuit as a circuit without any input or as a circuit with inputs to which constants are assigned.

A family of oracle circuits $T = \{T_n | n \in \mathcal{N}\}$ is called a statistical test for functions if there is a polynomial $Q(n)$ such that the size of each $T_n$ is not larger than $Q(n)$.

Assume that $S_n$ is a set composed of functions from $H_n$. Let $S = \{S_n | n \in \mathcal{N}\}$ and $H = \{H_n | n \in \mathcal{N}\}$. We say that $T$ is a distinguisher for $S$ if there is a polynomial $P(n)$ such that for infinitely many $n$, we have

$$|Pr[T_n[s] = 1 - Pr[T_n[h] = 1]| \geq 1/P(n),$$

where $s \in_R S_n$ and $h \in_R H_n$. We say that $S$ is pseudorandom if there is no distinguisher for it. (See also [GGM86, LR86]).

## 3  Previous results

This section summarizes some of the currently known results on pseudorandomness of DES-like transforms. We note that only those directly related to this research have been shown below.

**Theorem 1 [LR86].** $\{D(g, f) | g, f \in H_n, n \in \mathcal{N}\}$ *is not a pseudorandom permutation generator.*

**Theorem 2 [LR86].** $\{D(h, g, f) | h, g, f \in H_n, n \in \mathcal{N}\}$ *is a pseudorandom permutation generator.*

**Theorem 3 [Pie90].** $\{D(f^2, f, f, f) | f \in H_n, n \in \mathcal{N}\}$ *is a pseudorandom permutation generator.*

We note that in the above theorems, $f$, $g$ and $h$ are functions drawn from $H_n$ independently.

## 4  Some Facts on MISTY-like Transforms

The MISTY cipher employs a transform different from a DES-like one. This section reviews the definition of the new transform, as well as relevant results on it.

### 4.1  Definition of MISTY-like Transforms

Associate with $f \in H_n$, a function

$$\mu_{2n,f}(L, R) = (R, f(L) \oplus R)$$

for all $L, R \in I_n$. $\mu_{2n,f}$ called a MISTY-like transform associated with $f$ [Ma96]. Similarly to a DES-like transform $\delta_{2n,f}(L, R) = (R \oplus f(L), L)$, we call the function $f$ used in $\mu_{2n,f}$ the F-function of $\mu_{2n,f}$. Comparing $\mu_{2n,f}$ with $\delta_{2n,f}$, two differences between them are apparent: the first is the position where the F-function is placed, and the second is that unlike $\delta_{2n,f}$, $\mu_{2n,f}$ forms a permutation over $I_{2n}$ only when $f$ is also a permutation over $I_n$.

In addition, for $f_1, f_2, \ldots, f_s \in H_n$, we define $M(f_s, \ldots, f_2, f_1) = \mu_{2n,f_s} \circ \cdots \circ \mu_{2n,f_2} \circ \mu_{2n,f_1}$ as an $s$-round concatenation of MISTY-like transforms.

In [Ma96] Matsui observes that unlike DES-like transforms, a 3-round concatenation of MISTY-like transforms allows partial parallel computation. This suggests that a cipher based on MISTY-like transforms would be more suitable for hardware implementation than those based on DES-like transforms. In the next section we turn to a more important issue, that is the resistance of a MISTY-like transform to cryptanalytic attacks, especially linear and differential attacks.

## 4.2 Immunity against differential and linear cryptanalysis

Nyberg and Knudsen [NK95] introduced a measure of security of block ciphers against differential cryptanalysis and showed that DES-like transforms yield block ciphers with provably security against differential attacks. Furthermore, Nyberg [Nyb94] extends the argument into the case of linear cryptanalysis.

The following measures are formulated in [Ma96].

**Definition 4 [Ma96].** For $f \in H_n, \Delta x, \Gamma x \in I_n$ and $\Delta y, \Gamma y \in I_n$, define

$$DP(f) = \underset{\Delta x \neq 0, \Delta y}{MAX} \frac{\#\{x \in X | S(x) \oplus S(x \oplus \Delta x) = \Delta y)\}}{2^n},$$

$$LP(f) = \underset{\Gamma x, \Gamma y \neq 0}{MAX} (\frac{\#\{x \in X | x \bullet \Gamma x = S(x) \bullet \Gamma y)\}}{2^{n-1}} - 1)^2.$$

Using this definition, a result in [NK95, Nyb94] can now be stated as follows:

**Theorem 5 [NK95, Nyb94].** *For an s-round concatenation ($s \geq 3$) of DES-like transforms $D(f_s, \ldots, f_2, f_1)$, assuming that $DP(f_i) \leq p$, we have*

$$DP(D(f_s, \ldots, f_2, f_1)) \leq 2p^2.$$

*Similarly, assuming that $LP(f_i) \leq p$, we have*

$$LP(D(f_s, \ldots, f_2, f_1)) \leq 2p^2.$$

*Remark.* Nyberg [Nyb93] showed that a DES-like transform based on a function $f(x, k) = (x \oplus k)^{-1}$ on $GF(2^n)$ achieves high resistance against differential attacks. Note, however, we can easily crack such a cipher by solving a set of low degree polynomial equations derived from known plaintext/ciphertext pairs. Thus, the measures introduced in Definition 4 are not sufficient for the security of a block cipher. This conclusion is further supported by extended differential attacks proposed in [Lai94, Knu94]. In particular, the higher order differential cryptanalysis discussed in [Knu94] breaks a 6-round version of an example cipher proposed in [NK95], despite of the fact that this example cipher has been proven to be resistant against ordinary differential attacks. These successfully extended attacks could be helpful in refining the security measures in Definition 4.

A key result in [Ma96] is the following which was served as evidence that MISTY-like transforms would have an advantage over DES-like transforms, in terms of resistance against differential and linear cryptanalysis.

**Theorem 6 [Ma96].** *For an s-round concatenation ($s \geq 3$) of MISTY-like transforms*
$M(f_s, \ldots, f_2, f_1)$, *where each $f_i$ is a permutation, assuming that $DP(f_i) \leq p$, we have*

$$DP(M(f_s, \ldots, f_2, f_1)) \leq p^2.$$

*Similarly, assuming that $LP(f_i) \leq p$, we have*

$$LP(M(f_s, \ldots, f_2, f_1)) \leq p^2.$$

*Remark.* Recently, Aoki and Ohta [AO96] reported that the inequalities in Theorem 5 can be improved to the following:

$$DP(D(f_s, \ldots, f_2, f_1)) \le p^2 \quad (LP(D(f_s, \ldots, f_2, f_1)) \le p^2, \text{respectively})$$

under the assumption that each function $f_i$ is a permutation. This result disproves Matsui's conjecture on the advantages of MISTY-like transforms over DES-like transforms with respect to immunity against differential and linear cryptanalysis.

## 5  Our results

We now investigate (non-)randomness of permutations obtained from MISTY-like transforms in order to compare the security of the MISTY cipher with that of DES.

### 5.1  Non-randomness of MISTY-like transforms

The following are results we have obtained so far regarding conditions under which MISTY-like transforms do not generate pseudorandom permutations. For all conditions shown in Theorems 7 - 10, except that in Theorem 8 which is currently being investigated by the authors, it is known that DES-like transforms give pseudorandomness permutations.

**Theorem 7.** $\{M(h, g, f) | h, g, f \in H_n, n \in \mathcal{N}\}$ *is not a pseudorandom permutation generator.*

**Proof:** Let $M_3 = M(h, g, f)$. Then we have $M_3(L, R) = (R \oplus f(L) \oplus g(R), *)$, where $L$ and $R$ are arbitrary vectors from $I_n$ and $*$ denotes a string we do not care. Now we further assume that neither $L$ nor $R$ is 0. Then we have the following:

$$M_3(0, 0) = (f(0) \oplus g(0), *)$$
$$M_3(L, 0) = (f(L) \oplus g(0), *)$$
$$M_3(0, R) = (f(0) \oplus g(R) \oplus R, *)$$
$$M_3(L, R) = (f(L) \oplus g(R) \oplus R, *)$$

Adding together the left halves of the right-hand sides of the above four equations must give us 0. These observations indicate that we can construct an oracle circuit for $\{M(h, g, f) | h, g, f \in H_n\}$. The oracle circuit uses only four (4) oracle gates. When $M_3$ is used in the oracle circuit for function evaluation, the oracle circuit always outputs a bit 1. On the other hand, when a random function from $H_{2n}$ is used in the oracle circuit, the probability for the oracle circuit to produce a bit 1 is $1/2^n$. This completes the proof. $\quad\square$

In our plain language, Theorem 7 states that a 3-round concatenation of MISTY-like transforms is not a pseudorandom permutation, even if the F-function in each round is chosen independently at random from $H_n$.

*Remark.* Theorem 7 is also implied by a more general result by Ohnishi [Ohn88] which states that the family of functions with a depth of at most one (e.g. $f(L) \oplus g(R) \oplus h(L \oplus R)$) is not a pseudorandom function generator. Thus, for pseudorandom function generation, functions must have a depth of at least two (e.g. $f(g(R))$).

For a 4-round concatenation of MISTY-like transforms, we have the following two results.

**Theorem 8.** $\{M(f, f^2, f, f)|h, g, f \in H_n, n \in \mathcal{N}\}$ *is not a pseudorandom permutation generator.*

**Proof:** To prove this theorem, we construct a oracle circuit for $\{M(f, f^2, f, f)|f \in H_n\}$ that uses two oracle gates. The detailed arrangement of the two oracles is obtained through the following two observations:

1. $M(f, f^2, f, f)$ always translates $(0, 0)$ into $(f^3(0), f^3(0) \oplus f(0))$. Adding up the two halves gives us $f(0)$.
2. Now we have $(0, f(0))$, which will be translated by $M(f, f^2, f, f)$ into $(0, *)$.

The oracle circuit based on the above observations outputs a bit 1 with certainty when its two oracles are evaluated using $M(f, f^2, f, f)$, but only with a probability of $1/2^n$ when using a truly random function from $H_{2n}$. $\square$

**Theorem 9.** $\{M(f^{i+j}, f^j, f^i, f^i)|f \in H_n, n \in \mathcal{N}\}$ *is not a pseudorandom permutation generator, where $i$ and $j$ are integers larger than 0.*

A proof for this theorem can be easily obtained by noting the fact that $M(f^{i+j}, f^j, f^i, f^i)$ always translates $(0, 0)$ into $(*, 0)$, where $*$ as before means a string we do not care.

Based on Theorem 9, the following result on a 5-round concatenation of MISTY-like transforms can be obtained:

**Theorem 10.** $\{M(g, f^{i+j}, f^j, f^i, f^i)|g, f \in H_n, n \in \mathcal{N}\}$ *is not a pseudorandom permutation generator, where $i$ and $j$ are integers larger than 0.*

The proof for Theorem 10 is surprisingly simple: $M(g, f^{i+j}, f^j, f^i, f^i)$ always translates $(0, 0)$ into $(0, *)$, even if $f$ and $g$ are chosen independently at random.

It remains an interesting topic to see whether the above techniques can be generalized to other cases, including $M(g, f, f, f)$, $M(f, g, f, f)$, $M(g, f, f, f, f)$, etc.

Finally we study the super pseudorandomness of MISTY-like transforms. Super pseudorandomness is a slightly stronger notion than that of pseudorandomness. It is defined by allowing an oracle circuit to contain also gates that computes the inverse of a permutation. The reader is directed to [LR86] for the precise definition of super pseudorandomness.

In the case of DES-like transforms, Luby and Rackoff showed the following result.

**Theorem 11 [LR86].** $\{D(f_4, f_3, f_2, f_1)|f_4, f_3, f_2, f_1 \in H_n, n \in \mathcal{N}\}$ *is a super pseudorandom permutation generator.*

Our following result shows that, in the case of MISTY-like transforms, a 4-round concatenation is not adequate for achieving super pseudorandomness, although whether it yields a pseudorandom permutation generator still remains open,

**Theorem 12.** $\{M(f_4, f_3, f_2, f_1)|f_4, f_3, f_2, f_1 \in H_n, n \in \mathcal{N}\}$ *is not a super pseudorandom permutation generator.*

**Proof:** Let $M_4 = M(f_4, f_3, f_2, f_1)$ and $M_4^{-1}$ the converse of $M_4$, namely, $M_4^{-1}(M_4(A, B)) = (A, B)$ for all vectors $A, B \in I_n$. Given two vectors $A$ and $B$, first compute $(X, Y) = M_4(A, B)$, then set $(C, D) = M_4^{-1}(X \oplus Z, Y \oplus Z)$ for an arbitrary $Z \in I_n$. Next compute $(U, V) = M_4(A, D)$ and $(S, T) = M_4(C, B)$. It is easy to check that the relation $U \oplus V = S \oplus T$ holds. This proves the theorem. $\square$

## 5.2 Practical consequences of non-randomness

Though the argument of non-randomness in the previous section is theoretical, the way the distinguishers work could suggest potential attacks on MISTY and related block ciphers.

Consider a 3-round concatenation of MISTY-like transforms $M(h, g, f)$. As we have shown in Theorem 7, it is not a pseudorandom permutation. Set $t(L, R)$ be the left half of output of $S_3(L, R)$. Then, the following relation holds:

$$t(L, R) = t(0, 0) \oplus t(L, 0) \oplus t(0, R).$$

More importantly, the following general relation holds:

$$t(L, R) = t(A, B) \oplus t(L, B) \oplus t(A, R)$$

This implies that the left half $t$ has an algebraic structure which allows $t(L, R)$ to be computed from three encrypted data items $t(A, B), t(L, B)$ and $t(A, R)$ for any $A, B, L, R \in I_n$. Based on this property, one may launch a known-plaintext attack against a 3-round concatenation of MISTY-like transforms.

For a cipher to be secure, the above algebraic relation must be avoided. To see this point, we note that by using Luby and Rackoff's argument for Theorem 1, a cipher placed in the public domain which was based on 2-round DES-like transforms, has been shown to be insecure against a known-plaintext attack similar to the one described above. (For details see Pages 351-352 of [Sch95].)

Though such an algebraic structure could disappear in the concatenation of four or more MISTY-like transforms, the F-function of a round (which is a MISTY-like transform) in the preliminary version of the MISTY cipher is defined as the concatenation of three smaller MISTY-like transforms. In other words, an "outer" F-function in the MISTY cipher is recursively constructed from three smaller MISTY-like transforms. Hence, the "outer" F-function has the algebraic structure explained above.

This structure could be used by a cryptanalyst and hence cast very serious doubts on the security of the MISTY cipher. Indeed the MISTY cipher may be immune against the differential or linear attack, however, the fact that the MISTY cipher adopts the concatenation of three smaller MISTY-like transforms in its "outer" F-functions could render the cipher vulnerable to other (chosen plaintext) attacks.

## 6 Future Research

Topics for future research include (1) to identify under what conditions MISTY-like transforms would yield a pseudorandom permutation, and (2) to compare the MISTY cipher with DES from the perspectives of other security criteria. In particular, concerning the first topic the following two concrete questions remain to be tackled:

(Q1) Is a 4-round concatenation of MISTY-like transforms $M(f_4, f_3, f_2, f_1)$ a pseudorandom permutation ?

(Q2) Is a 5-round concatenation of MISTY-like transforms $M(f_5, f_4, f_3, f_2, f_1)$ a super pseudorandom permutation ?

Concerning on the second topic, an interesting question is if there are any security criteria which would indicate the advantage of a MISTY-like transform over a DES-like transform.

A final remark is that a detailed description of the original algorithm in the preliminary version of the MISTY cipher [Ma96] uses a 8-round concatenation of MISTY-like transforms, though no design criteria on this decision have been disclosed.

## Acknowledgment

The authors would like to thank Toshiya Itoh who pointed out an error in an early version of Theorem 8.

## References

[AO96] Aoki, K. and Ohta, K., "Stricter evaluation for the maximum average of differential probability and the maximum average of linear probability," Proc. of the 1996 SCIS'96, Japan (1996).

[BS93] Biham, E. and Shamir, A., "Differential cryptanalysis of the Data Encryption Standard," Springer-Verlag, New York, (1993).

[BGR94] Bellare, M., Guérin, R., and Rogaway, P., "XOR MACs: New methods for message authentication using finite pseudorandom functions," in Advances in Cryptology – Crypto'95, Lecture Notes in Computer Science 963, pp.14-28, *Springer-Verlag*, Berlin (1995).

[BKR94] Bellare, M., Kilian, J., and Rogaway, P., "The security of cipher block chaining," in Advances in Cryptology – Crypto'94, Lecture Notes in Computer Science 839, pp.341-358, *Springer-Verlag*, Berlin (1994).

[BR93] Bellare, M. and Rogaway, P., "Entity authentication and key distribution," in Advances in Cryptology – Crypto'93, Lecture Notes in Computer Science 773, pp.232-249, *Springer-Verlag*, Berlin (1994).

[BR94] Bellare, M. and Rogaway, P., "Optimal Asymmetric Encryption," in Advances in Cryptology – EUROCRYPT'94, Lecture Notes in Computer Science 950, pp.92-111, *Springer-Verlag*, Berlin (1995).

[BR95] Bellare, M. and Rogaway, P., "Provably secure session key distribution — The three party case," Proc. of STOC'95.

[F73] H. Feistel: "Cryptography and computer privacy," in Scientific American, Vol.228, pp.15-23 (1973).

[FNS75] H. Feistel, W.A. Notz and J.L. Smith: "Some cryptographic techniques for machine-to-machine data communications," in Proceedings of IEEE, Vol.63, No. 11, pp.1545-1554 (1975).

[GGM86] Goldreich, O., Goldwasser, S., and Micali, S., "How to construct random functions," in JACM, Vol.33, No.4, pp.792-807 (1986).

[Knu94] Knudsen, L., "Truncated and higher order differentials," Proc. of 2nd Fast Software Encryption, LNCS 1008, pp.197-211, *Springer-Verlag*, Berlin (1995).

[Lai94] Lai, X., "Higher order derivatives and differential cryptanalysis," Proc. of Comm. Coding and Cryptography, (Feb.1994).

[LR86] Luby, M. and C. Rackoff, "How to construct pseudorandom permutations from pseudorandom functions," STOC'86 (also in SIAM-COMP.1988).

[Ma94] Matsui, M., "Linear cryptanalysis method for DES cipher," in Advances in Cryptology – EUROCRYPT'93, LNCS 756, pp.386-397, *Springer-Verlag*, Berlin (1994).

[Ma95a] Matsui, M., "On provably security of block ciphers against differential and linear cryptanalysis," Proc. of SITA'95 (1995).

[Ma96] Matsui, M., "New structure of block cipher with provable security against differential and linear cryptanalysis," in 3rd Fast Software Encryption, Cambridge, U.K., Lecture Notes in Computer Science 1039, pp.205-218, *Springer-Verlag*, Berlin (1996).

[NBS77] National Bureau of Standards, NBS FIPS PUB 46, "Data Encryption Standard," U.S.Department of Commerce (Jan. 1977).

[NK95] Nyberg, K. and Knudsen, L.R., "Provable security against a differential attacks ," J. Cryptology, Vol.8, pp.27-37 (1995).

[Nyb93] Nyberg, K. "Differentially uniform mappings for cryptography," in Advances in Cryptology – EUROCRYPT'93, LNCS 765, pp.55-64, *Springer-Verlag*, Berlin (1994).

[Nyb94] Nyberg, K. "Linear approximation of block ciphers," in Advances in Cryptology – EUROCRYPT'94, Lecture Notes in Computer Science 950, pp.439-444, *Springer-Verlag*, Berlin (1995).

[Ohn88] Ohnishi, Y. "A study on data security," Master Thesis (in Japanese), *Tohoku University*, Japan (March, 1988).

[Pie90] Pieprzyk, J. "How to construct pseudorandom permutations from single pseudorandom functions," in Advances in Cryptology – EUROCRYPT'90, Lecture Notes in Computer Science 473, pp.140-150, *Springer-Verlag*, Berlin (1995).

[Sch95] Schneier, B. "Applied Cryptography (2nd Edition)," *John Wiley & Sons, Inc.*, (1995).

[SP92] Sadeghiyan,B., and Pieprzyk, J. "A construction for pseudorandom permutations from a single pseudorandom function," in Advances in Cryptology – EUROCRYPT'92, Lecture Notes in Computer Science 658, pp.267-284, *Springer-Verlag*, Berlin (1995).

[ZMI89] Zheng, Y., Matsumoto, T. and Imai, H. "Impossibility and optimality results on constructing pseudorandom permutations," in Advances in Cryptology – EUROCRYPT'89, Lecture Notes in Computer Science 434, pp.412-422, *Springer-Verlag*, Berlin (1990).

[ZMI89] Zheng, Y., Matsumoto, T. and Imai, H. "On the construction of block ciphers provably secure and not relying on any unproven hypotheses," in Advances in Cryptology – CRYPTO'89, Lecture Notes in Computer Science 435, pp.461-480, *Springer-Verlag*, Berlin (1990).

[Zhe90] Zheng, Y. "Principles for designing secure block ciphers and one-way hash functions," Ph.D Thesis, *Yokohama National University*, Japan (Dec. 1990).

# The Multi-variable Modular Polynomial and Its Applications to Cryptography

Tsuyoshi Takagi    Shozo Naito

NTT Software Laboratories
3-9-11 Midori-cho Musashino-shi Tokyo, 180
{ttakagi,naito}@slab.ntt.jp

**Abstract.** We prove the extension of the Håstad algorithm to the multi-variable modular polynomial. Although the Håstad attack is one of the strongest known attacks on RSA-type cryptosystems, the original Håstad attack is generally not applicable to multi-variable cryptosystems, where the plain text space (the encryption domain) is multi-variable. As an application of our extension, we attack RSA-type cryptosystems over elliptic curves and show the critical number of encrypted texts.

## 1 Introduction

We consider the integer polynomial computational problem: when we have several modular polynomials having the same single solution, can we find the real-valued polynomial with the solution in polynomial time? In the one-variable case, Håstad showed an algorithm for solving the problem using LLL reduction [4]. [1]
And in the same paper, he indicated a cryptological application of the algorithm: RSA type public-key cryptosystems can be broken without knowing the secret key if one plain text is sent to several people. This Håstad attack is only applicable to one-variable equations, so it cannot be used directly to attack cryptoschemes with more than one variable.
In this paper, we show that Håstad's algorithm can be extended to polynomials with more than one variable, including the original algorithm as a special case. We can apply this new attack to RSA-type cryptoschemes whose plain text space (the domain of encryption) is multi-variable without reducing them to single-variable cryptosystems. One of the most famous two-variable cryptosystems is the cryptoscheme over elliptic curves [5]. Since the elliptic cryptoscheme can be reduced from a

---

[1] Recently, Coppersmith, Franklin, Patarin and Reiter reported a class of low-exponent attacks against RSA cryptoscheme [2] [3]. By the attack, the plaintexs with polynomial relation under a *single* public key can be recovered. But the attack is not applicable to the case where a single message is encrypted under *multiple* different keys discussed in this paper.

two-variable encryption function to a one-variable function using the definition equation, we can apply the original Håstad attack to it [6] [7]. We show that the number of encripted texts necessary for our new attack can be reduced by comparing the number with the original Håstad attack. On the other hand, it is generally not known whether that all multi-variable encryption functions can be converted to one-variable ones.

Section 2 extends the Håstad attack to multi-variable cryptosystems. Section 3 applies the new attack to RSA-type cryptosystems over elliptic curves and compares the results with the original Håstad attack.

## 2 Extension of the Håstad attack

**Theorem 1** *Consider an integer coefficient equivalent k polynomial with l variables not greater than e-th degree, such as*

$$\sum_{j_1,j_2,\ldots,j_l=0}^{j_1+j_2+\cdots+j_l \le e} a_{i,j_1,\ldots,j_l} x_1^{j_1} x_2^{j_2} \ldots x_l^{j_l} \equiv 0 \pmod{n_i}, \quad i = 1, 2, \ldots, k$$

(1)

*where $x_1, x_2, \ldots, x_l < n$, $n = \min n_i$. Then if the following relation holds, we can get the real-valued equation in $x_1, x_2, \ldots, x_l$ in time polynomial in $e, k,$ and $\log n_i$.*

$$N > n^f (k+g)^{\frac{k+g}{2}} 2^{\frac{(k+g)^2}{2}} g^g,$$

(2)

*where $N = \prod_{i=1}^{k} n_i$, $g$ is the number of different terms including the constant term, and $f$ is the sum of the degrees in $x_1, x_2, \ldots, x_l$ through all of the different terms. The maximum numbers of $f$ and $g$ are*

$$f = \sum_{m=1}^{e} m \, _{l+m-1}C_m, \quad g = \sum_{m=0}^{e} {}_{l+m-1}C_m$$

(3)

**Remark 1** *For Theorem 1, in the case of one variable, it is possible to solve the real-valued equation within polynomial time [4]. In the case of more than one variable, it is an open problem to solve the multi-variable algebraic equation. Some methods for solving the generic multi-variable algebraic equation are known. One method is to calculate the Gröbner bases and another is to eliminate those equations using the resultant [1]. And we note that we can construct a real-valued multi-variable polynomial that can be solved in polynomial time.*

**Remark 2** *Theorem 1 includes the original Håstad attack as a special case by restricting the number of variables to one.*

**Proof:** Let $u_j \equiv \delta_{ij} \pmod{n_i}$, where $\delta_{ij}$ is Kronecker's delta. And $s_1, s_2, \ldots, s_k$ are particular integers determined later. Using the Chinese remainder theorem, we obtain the following equation equivalent to equations (1).

$$0 \equiv \sum_{\substack{j_1, j_2, \ldots, j_l = 0}}^{j_1 + j_2 + \ldots + j_l \leq e} x_1^{j_1} x_2^{j_2} \ldots x_l^{j_l} \sum_{i=1}^{k} a_{i, j_1, \ldots, j_l} s_i u_i \qquad (4)$$

$$\equiv \sum_{\substack{j_1, j_2, \ldots, j_l = 0}}^{j_1 + j_2 + \ldots + j_l \leq e} x_1^{j_1} x_2^{j_2} \ldots x_l^{j_l} c_{j_1, j_2, \ldots, j_l} \pmod{N}. \qquad (5)$$

In $n$ dimensional real space $\mathbf{R}^n$, we define lattice $L$ spanned by linearly independent vectors $b_1, b_2, \ldots, b_n$ such that

$$L = \{y \mid y = \sum_{i=1}^{n} a_i b_i, \ a_i \in \mathbf{Z}\}.$$

From Minkowski's theorem, let the shortest norm vector including lattice $L$ be $\lambda_1$, thus

$$\lambda_1 \leq \gamma_n^{\frac{1}{2}} (\det(L))^{\frac{1}{n}},$$

where $\gamma_n$ is called Hermite's constant, which is always less than $n$. And from LLL algorithm, within polynomial time we can find the vector satisfying

$$\|b\| / \lambda_1 \leq 2^{\frac{n}{2}}.$$

Therefore, for $n$−dimensional lattice $L$, we can find vector $b$ satisfying

$$\|b\| \leq n^{\frac{1}{2}} 2^{\frac{n}{2}} \det(L)^{\frac{1}{n}} \qquad (6)$$

within polynomial time [4].

Next, for equation (5), the $_{l+j-1}C_j$ terms, where the sum of the degrees in $x_1, x_2, \ldots, x_l$ is $j$, are as follows:

$$x_1^0 x_2^0 x_3^0 \ldots x_{l-2}^0 x_{l-1}^0 x_l^j$$
$$x_1^0 x_2^0 x_3^0 \ldots x_{l-2}^0 x_{l-1}^1 x_l^{j-1}$$
$$x_1^0 x_2^0 x_3^0 \ldots x_{l-2}^0 x_{l-1}^2 x_l^{j-2}$$
$$x_1^0 x_2^0 x_3^0 \ldots x_{l-2}^1 x_{l-1}^1 x_l^{j-2}$$
$$\ldots$$
$$x_1^{j-1} x_2^1 x_3^0 \ldots x_{l-2}^0 x_{l-1}^0 x_l^0$$
$$x_1^j x_2^0 x_3^0 \ldots x_{l-2}^0 x_{l-1}^0 x_l^0$$

Then we can get $g(= \sum_{j=0}^{e} {}_{l+j-1}C_j)$ terms at most in equation (5). So we associate the following vector with each term, where the number of elements is ${}_{l+j-1}C_j$ at most and the degree in $x_1, x_2, \ldots, x_l$ is $j$:

$$\mathbf{b}_{i,j} = (a_{i,0,0,\ldots,0,j} u_i, a_{i,0,0,\ldots,1,j-1} u_i, \ldots, a_{i,j,0,\ldots,0} u_i). \tag{7}$$

Here, we generate the following $k + g$ vectors at most, where each vector includes the above vector among its elements.

$$\mathbf{b}_1 = (\mathbf{b}_{1,0}, n\mathbf{b}_{1,1}, n^2 \mathbf{b}_{1,2}, \ldots, n^e \mathbf{b}_{1,e}, \frac{N}{n_1 g}, 0, \ldots, 0)$$

$$\mathbf{b}_2 = (\mathbf{b}_{2,0}, n\mathbf{b}_{2,1}, n^2 \mathbf{b}_{2,2}, \ldots, n^e \mathbf{b}_{2,e}, 0, \frac{N}{n_2 g}, \ldots, 0)$$

$$\ldots$$

$$\mathbf{b}_k = (\mathbf{b}_{k,0}, n\mathbf{b}_{k,1}, n^2 \mathbf{b}_{k,2}, \ldots, n^e \mathbf{b}_{k,e}, 0, 0, \ldots, \frac{N}{n_k g})$$

$$\mathbf{b}_{k+1} = (N, 0, 0, \ldots, 0, 0, 0, \ldots, 0)$$

$$\mathbf{b}_{k+2} = (0, nN, 0, \ldots, 0, 0, 0, \ldots, 0)$$

$$\ldots$$

$$\mathbf{b}_{k+g} = (0, 0, 0, \ldots, n^e N, 0, 0, \ldots, 0)$$

Next, we consider the $(k + g)$-dimensional lattice $L$ that is generated by these vectors. Any point in $L$ can be written as follows:

$$\sum_{i=1}^{k+g} s_i \mathbf{b}_i$$

$$= (\sum_{i=1}^{k} s_i \mathbf{b}_{i,0} + s_{k+1} N, \sum_{i=1}^{k} s_i \mathbf{b}_{i,1} + s_{k+2} nN, \ldots, \frac{N s_1}{n_1 g}, \frac{N s_2}{n_2 g}, \ldots, \frac{N s_k}{n_k g})$$

$$\equiv (c_{0,0,\ldots,0,0}, n c_{0,0,\ldots,0,1}, n c_{0,0,\ldots,1,0}, \ldots, \frac{N s_1}{n_1 g}, \frac{N s_2}{n_2 g}, \ldots, \frac{N s_k}{n_k g}) \pmod{N},$$

where $s_1, s_2, \ldots, s_k$ are arbitrary integers. Therefore, from equation (6), we can find vector **b** satisfying the following condition in polynomial time.

$$\|\mathbf{b}\| \leq (k+g)^{\frac{1}{2}} 2^{\frac{k+g}{2}} \operatorname{Det}(L)^{\frac{1}{k+g}}$$

The following equation holds:

$$\operatorname{Det}(L) = n^f N^{k+g-1} g^{-k},$$

where $f = \sum_{j=1}^{e} j \, {}_{l+j-1}C_j$. Finally we can find vector **b** satisfying the following condition within polynomial time.

$$\|\mathbf{b}\| \leq (k+g)^{\frac{1}{2}} 2^{\frac{k+g}{2}} (n^f N^{k+g-1} g^{-k})^{\frac{1}{k+g}} \tag{8}$$

In equation (5), if the relation

$$|c_{j_1,j_2,\ldots,j_l}| \leq \frac{N}{g n^{j_1+j_2+\cdots+j_l}} \tag{9}$$

holds, then noticing $x_1, x_2, \ldots, x_l < n$, we obtain

$$\left| \sum_{\substack{j_1,j_2,\ldots,j_l=0}}^{j_1+j_2+\cdots+j_l \leq e} x_1^{j_1} x_2^{j_2} \cdots x_l^{j_l} c_{j_1,j_2,\ldots,j_l} \right|$$

$$\leq \sum_{\substack{j_1,j_2,\ldots,j_l=0}}^{j_1+j_2+\cdots+j_l \leq e} n^{j_1+j_2+\cdots+j_l} |c_{j_1,j_2,\ldots,j_l}| < N.$$

Therefore we can obtain equation (5) as the real-valued equation. Now consider how we should get the vector **b** satisfying equation (9). If the condition

$$\|\mathbf{b}\| \leq \frac{N}{g} \tag{10}$$

is satisfied, then equation (9) is also satisfied, while if the norm of the vector **b** is greater than $N/g$, then equation (9) is never satisfied.

Therefore, the vector satisfying both the equations (10) and (8) is the vector we are seeking, this means that

$$N > n^f (k+g)^{\frac{k+g}{2}} 2^{\frac{(k+g)^2}{2}} g^g \tag{11}$$

holds. Here, $f$ and $g$ are numbers associated with terms in the equation as described in the theorem.

After determining the vector **b**, we can find the coefficients of the vector **b** spanned by the basis $b_1, b_2, \ldots, b_{K+g}$. Then the integers $s_1, s_2, \ldots, s_k$ are also determined. And it is clear that the vector we choose here is not zero. $\qquad\qquad\square$

# 3   Attack on RSA-type cryptoschemes over elliptic curves

## 3.1   RSA-type cryptoschemes

We can construct RSA-type cryptoschemes over ellilptic curves [5]. We state only the information needed for the attack. The elliptic curve over real field **R** is

$$E(a,b) := \{(x,y) \mid y^2 = x^3 + ax + b\} \cup P_\infty, \tag{12}$$

where $P_\infty$ is a point at infinity; $a, b \in \mathbf{R}$; and $4a^3 + 27b^2 \neq 0$. Then, $E(a,b)$ has the structure of an abelian group. Actually, for any pair of points $P = (x_1, y_1)$ and $Q = (x_2, y_2)$, we define $P + Q = (x_3, y_3)$ as follows:

$$\lambda = \begin{cases} \dfrac{y_2 - y_1}{x_2 - x_1} & if \quad x_1 \neq x_2 \\[3mm] \dfrac{3x_1^2 + a}{2y_1} & if \quad x_1 = x_2 \end{cases}$$

$$x_3 = \lambda^2 - x_1 - x_2$$
$$y_3 = \lambda(x_1 - x_3) - y_1.$$

We can also define the same operation over the ring $\mathbf{Z}_n$.

Now, we calculate $e$ times the value of any point $P = (x, y)$ in $E(a, b)$ i.e., $e \star P = (e \star x, e \star y)$.

From the group law, $e \star P$ is generally a rational function of $x$ and $y$. So, letting $f_x(x, y), f_y(x, y), g_x(x, y), g_y(x, y)$ be polynomials in $x$ and $y$, then we have

$$e \star x = \frac{g_x(x, y)}{f_x(x, y)}, \quad e \star y = \frac{g_y(x, y)}{f_y(x, y)}.$$

From these rational functions, we obtain the following two-variable polynomial equations in $x$ and $y$, which are the ones theorem 1 is applicable to.

$$e \star x f_x(x, y) - g_x(x, y) = 0$$
$$e \star y f_y(x, y) - g_y(x, y) = 0$$

Furthermore, in the elliptic curve cryptosystem, by investigating the division polynomials and using the definition equation $y^2 = x^3 + ax + b$, we can reduce the above two-variable polynomials into one-variable polynomials with $x$ as the only variable. Therefore we can also apply the original Håstad attack to the reduced one-variable polynomials. This method was first pointed out by Kurosawa-Tujii [6], and the number of polynomials necessary for the attack was accurately calculated by Kuwakado-Koyama [7]. This prominent property of the elliptic functions that $e \star x$ can be described with only $x$ with the definition equation comes from the fact that the division polynomials of $e \star x$ are polynomials in terms of $y^2$.

Next, we compare the number of $e \star x$ polynomials necessary for applying the two types of attacks, that is, attacks for one- or two-variable polynomials. To apply theorem 1, we have to calculate the numbers $f$ and $g$, where $g$ is the number of different terms in the $e \star x$ polynomial including the constant term, and $f$ is the sum of the degrees of all the different terms. In the next section, by investigating division polynomials of $e\star$ operations, we get general formulas describing $f$ and $g$ for one- and two-variable $e \star x$ polynomials. On the other hand, because the Håstad attack cannot work for polynomials with large $e$ [7], we investigate $f$ and $g$ for ones with small $e$ in detail.

## 3.2 Division polynomials

Division polynomials $\psi_e \in \mathbf{Z}[a, b, x, y]$ are inductively defined as follows:

$$\psi_1 = 1, \quad \psi_2 = 2y,$$
$$\psi_3 = 3x^4 + 6ax^2 + 12bx - a^2,$$
$$\psi_4 = 4y(x^6 + 5ax^4 + 20bx^3 - 5a^2x^2 - 4abx - 8b^2 - a^3),$$
$$\psi_{2e+1} = \psi_{e+2}\psi_e^3 - \psi_{e-1}\psi_{e+1}^3 \ (e \geq 2),$$
$$2y\psi_{2e} = \psi_e(\psi_{e+2}\psi_{e-1}^2 - \psi_{e-2}\psi_{e+1}^2) \ (e \geq 3).$$

It is easy to check that the $\psi_{2e}$'s are polynomials. We further define polynomials $\phi_e$ and $\omega_e$ by

$$\phi_e = x\psi_e^2 - \psi_{e+1}\psi_{e-1}$$
$$4y\omega_e = \psi_{e+2}\psi_{e-1}^2 - \psi_{e-2}\psi_{e+1}^2.$$

$\psi_e, \phi_e, y^{-1}\omega_e$ (for $e$ odd) and $(2y)^{-1}\psi_e, \phi_e, \omega_e$ (for $e$ even) are polynomials in $\mathbf{Z}[a, b, x, y^2]$. Hence replacing $y^2$ by $x^3 + ax + b$, we can treat them as polynomials in $\mathbf{Z}[a, b, x]$.

As polynomials in $x$, we can show that

$$\phi_e(x) = x^{e^2} + lower\ order\ terms$$
$$\psi_e(x)^2 = e^2 x^{e^2-1} + lower\ order\ terms.$$

Let $P = (x, y) \in E(a, b)$. Then using these polynomials, we can describe $e \star P$ as follows:

$$e \star P = \left( \frac{\phi_e(P)}{\psi_e(P)^2}, \frac{\omega_e(P)}{\psi_e(P)^3} \right).$$

Therefore we obtain the following theorem describing $f$ and $g$ in one-variable $e \star x$ polynomials. (See, for example, [8])
Here, we investigate the number and degree of the different terms corresponding to the changes of $e$. Then we have the next lemma.

**Lemma 1** *Let $f_1(e)$ be the sum of the degrees of all different terms and $g_1(e)$ be the number of different terms including a constant term in $e \star x$ one-variable polynomials. Then the following holds:*

$$f_1 = \frac{e^2(e^2 + 1)}{2} \tag{13}$$

$$g_1 = e^2 + 1 \tag{14}$$

On the other hand, replacing $x^3$ by $y^2 - ax - b$, we can treat division polynomials with not more than two degrees for variable $x$. In this treatment, division polynomials $\psi_k$ are inductively defined as follows:

$$\psi_1 = 1, \quad \psi_2 = 2y,$$
$$\psi_3 = 3xy^2 + 3ax^2 + 9bx - a^2,$$
$$\psi_4 = 4y(y^4 + 3axy^2 + 18by^2 - 9a^2x^2 - 27abx - 27b^2 - a^3),$$
$$\psi_{2e+1} = \psi_{e+2}\psi_e^3 - \psi_{e-1}\psi_{e+1}^3 (e \geq 2),$$
$$2y\psi_{2e} = \psi_e(\psi_{e+2}\psi_{e-1}^2 - \psi_{e-2}\psi_{e+1}^2)(e \geq 3).$$

From these polynomials, it is easy to obtain the following theorem.

**Theorem 2** *Let $f_2(e)$ be the sum of the degree of all different terms and $g_2(e)$ be the number of different terms including a constant term in $e \star x$ two-variable polynomials. Then the following holds:*

$$f_2 = \begin{cases} 3l^2 + 2l & \text{if } e = 0 \bmod 3 \\ 3l^2 + 4l + 1 & \text{otherwise} \end{cases} \tag{15}$$

$$g_2 = e^2 + 1, \tag{16}$$

*where $l = \lfloor \frac{e^2}{3} \rfloor$.*

We note that $g_1$ in lemma 1 and $g_2$ in theorem 2 have the same value. For $f_1$ and $f_2$, the difference is calculated as follows:

$$f_1 - f_2 = \begin{cases} \dfrac{3l^2 - l}{2} & \text{if } e = 0 \bmod 3 \\ \dfrac{3l^2 + l}{2} & \text{otherwise} \end{cases} \tag{17}$$

From this equation, we find that $f_2$ is always smaller than $f_1$, and the difference increases with the square of $l$.

Next, let $e$ be the exponent of the encryption. And when lemma 1 and theorem 2 are applied to encryption function, we obtain the following $f_1, g_1, f_2$, and $g_2$ table for small exponent $e$.

| exponent $e$ | 2 | 3 | 4 | 5 | 6 | 7 |
|---|---|---|---|---|---|---|
| one variable: $g_1, f_1$ | 5, 10 | 10, 45 | 17, 136 | 26, 338 | 37, 703 | 50, 1408 |
| two variable: $g_2, f_2$ | 5, 8 | 10, 32 | 17, 96 | 26, 225 | 37, 456 | 50, 833 |

Table 1. $f$ and $g$ values of one- and two-variable polynomials

## 3.3  Attacks to the cryptosystem

Using the $f$ and $g$ values obtained in the above section, we can calculate the number of $e \star x$ polynomials necessary to obtain a real-valued equation. The numbers are listed in Tables 2 and 3 for 512-bit and 1024-bit encryption keys respectively. From these tables, we can observe that the number of two variables is always smaller than the number of one variables, and the difference increases as the exponent of $e$ increases. Here,

| exponent $e$ | 2 | 3 | 4 | 5 | 6 | 7 $\geq$ |
|---|---|---|---|---|---|---|
| one-variable | 11 | 49 | 173 | * | * | * |
| two-variable | 9 | 36 | 114 | 464 | * | * |

Table 2. Number of polynomials (512-bit means)

we have a question: Is the real-valued two-variable polynomial solvable? The answer is affirmative, because $e \star x$ polynomials of two variables are polynomials in $Z[a, b, x, y^2]$, so if we replace $y^2$ by $x^3 + ax + b$, we can

| the exponent $e$ | 2 | 3 | 4 | 5 | 6 | 7 $\geq$ |
|---|---|---|---|---|---|---|
| one-variable | 11 | 47 | 151 | 428 | * | * |
| two-variable | 9 | 35 | 104 | 269 | 814 | * |

Table 3. Number of polynomials (1024-bit means)

treat them as polynomials in $Z[a, b, x]$. After this treatment, we can apply the LLL algorithm to the real-valued polynomial in one variable of $x$. Therefore, we can recover the plain text. From this discussion, theorem 1 means that the two-variable attack is stronger than the original one-variable Håstad attack.

**Theorem 3** *The RSA-type cryptoscheme with low exponent e over the elliptic curve is not secure when sending a number of messages encrypting the same plaintext equal to or greater than the number listed in tables 3.*

## 4  Conclusion

We extended the Håstad attack to multi-variable polynomial1 and applied this new attack to RSA-type cryptosystems over elliptic curves. This new attack is stronger than the original attack, which is applied to the one-variable polynomials obtained by using the definition polynomial of the elliptic curve.

## References

1. B. Buchberger, "Application of gröbner bases in non-linear computer science," Lecture Notes in Computer Science, Vol.296, (1987), pp.52–80.
2. D. Coppersmith, M. Franklin, J. Patarin, M. Reiter, "Low-exponet RSA with related messages," Advances in Cryptology – EUROCRYPT '96, LNCS 1070, (1996), pp.1–9.
3. D. Coppersmith, "Finding a small root of a univariate modular equation," Advances in Cryptology – EUROCRYPT '96, LNCS 1070, (1996), pp.155-165.
4. J. Håstad, "Solving simultaneous modular equations of low degree," SIAM J. Computing, Vol.17, No.2, (1988), pp.336-341.
5. K. Koyama, U. M. Maurer, T. Okamoto and S. A. Vanstone, "New public-key schemes based on elliptic curves over the ring $Z_n$," Advances in Cryptology – CRYPTO '91, LNCS 576, (1992), pp.252-266.
6. K. Kurosawa, K. Okada, S. Tsujii, "Low exponent attack against elliptic curve RSA," Information Processing Letters, 53, (1995), pp.77-83.
7. H. Kuwakado, K. Koyama; "Security of RSA-type cryptosystems over elliptic curves against the Håstad attack," Electronics Letters, 30, No.22, (1994), pp.1843-1844.
8. J. H. Silverman, "The arithmetic of elliptic curves," GTM106, Springer-Verlag, Berlin, 1986.

# Bounds and Algorithms for a Practical Task Allocation Model (Extended Abstract)

Tsan-sheng Hsu[1] and Dian Rae Lopez[2]

[1] Institute of Information Science, Academia Sinica, Nankang
Taipei 11529, Taiwan, ROC ***
[2] Division of Science and Mathematics, University of Minnesota, Morris
Morris, Minnesota 56267, USA †

**Abstract.** A parallel computational model is defined which addresses I/O contention, latency, and pipe-lined message passing between tasks allocated to different processors and can be used for modeling parallel task-allocation on a network of workstations or on a multi-stage inter-connected parallel machine. To study the performance bounds more closely, basic properties are developed for when the precedence constraints form a directed tree. Approximation algorithms are presented for scheduling directed one-level task trees on an unlimited number of processors and also on a fixed number of $k$, $k > 1$, processors, where the approximation ratios are 3 and $3 + \frac{k-2}{k}$, respectively. The case of equal task execution times is also considered.

## 1 Introduction

Models for scheduling tasks on a parallel MIMD architecture have usually included a communication cost associated with the sending of data between tasks which are located on different processors. Early work on this problem used graph theoretic techniques such as network flow and/or enumeration techniques [9, 16, 19]. Later work concentrated on approximation algorithms [1, 12, 14, 18]. Research then evolved to more restricted models which allowed an infinite number of processors in the system. Polynomial algorithms were found for the cases where the precedence constraints form a tree under certain constraints [2, 3, 4, 7, 15]. A good review of models and algorithms developed for this problem can be found in [2, 6, 13, 17]. Most of this work was very theoretical in nature, i.e., the models were too simplistic for practical application to real machines. More recently, Valiant's BSP Model [20, 21] provided a general framework with which to study more practical algorithms in an asynchronous distributed memory parallel architecture. The LogP model [8] and the QRQW model [11] attempted to further bridge the theoretical and practical models.

This paper uses a practical and realistic model based on Valiant's asynchronous distributed memory architecture while taking into consideration the

*** Research supported in part by NSC Grant 85-2213-E-001-003.
† Research supported in part by Academia Sinica, Taipei, the University of Minnesota, Morris and the University of Minnesota China Center, Minneapolis.

read/write contention of the QRQW model, the latency/overhead time of the LogP model, and the pipe-lined message sending cost which is proportional to the message size. It can be used for a loosely-coupled parallel architecture where communication times are small but still significant. Our model is also general enough to represent a communication network of computers or workstations each with its own memory and microprocessor. The growth of such networks mandate more study into the efficient use of their parallel computing power. More importantly, our model is general enough to be used for any algorithm which can be represented as a set of tasks which communicate with each other and whose execution and communication costs are known or can be estimated. An example where such an algorithm would be helpful is a network of computers using PVM parallel software [10]. In today's environment, PVM program tasks are either scheduled by the programmer or, more often, they are arbitrarily allocated to processors. The work of this paper is designed to allow the compiler and/or operating system to perform such tasks.

Our studies show that when the precedence constraints of a set of tasks form a one-level directed tree, the scheduling of these tasks is an NP-hard problem. Previous work [5] had shown that scheduling a two-level directed precedence tree with communication costs on a simpler and more theoretical model where message sending time is the only cost for communication is NP-hard. These results show that we must either put constraints on the task set or develop approximation algorithms with good performance bounds. We show a linear-time algorithm which restrains the executions costs of all tasks to be equal and give approximation algorithms for intractable cases. Due to space limitations, proofs of lemmas and theorems whose statements are terminated with □ are omitted. Our results are summarized in Table 1.

## 2 The Communication Model

Let $J = \{t_0, t_1, \ldots, t_n\}$ be a set of tasks whose precedence constraints form a directed graph PC($J$). In a precedence graph for a set of tasks, the weight on a directed edge $(u \rightarrow v)$ which points from $u$ to $v$ represents the communication time needed for $u$ to send data to $v$ if $u$ and $v$ are allocated to different processors. The weight on each node represents its execution time. In this model, we consider

Table 1. Summary of algorithms presented in this paper.

| | | arbitrary task execution times | | equal task execution times | |
|---|---|---|---|---|---|
| | | $n$ PE's | $k$ PE's | $n$ PE's | $k$ PE's |
| directed one-level task trees w/ $n + 1$ tasks | approx. ratio | 3 (NP-hard) | $3 + (k - 2)/k$ (NP-hard) | optimal | 3 (NP-hard) |
| | running time | $O(n)$ | $O(n \cdot \log n)$ | $O(n)$ + sorting | $O(n)$ |

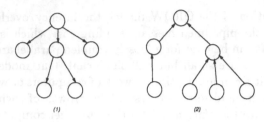

**Fig. 1.** In (1), a directed out-tree is illustrated. In (2), a directed in-tree is illustrated.

the case when all processors in the system are identical. Thus a task has the same execution time on any processor in the system.

A directed graph $G$ is a *directed out-tree* if there is a vertex $u$ in $G$ such that there is exactly one directed path from $u$ to any other vertex. The vertex $u$ is the *root*. Each vertex in $G$ having no outgoing edge is a *leaf*. By reversing the direction of all edges in a directed out-tree, we obtain a *directed in-tree*. See Fig. 1.

In this paper, we consider the case when $PC(J)$ is a directed in-tree or out-tree. Let $e(t_i) = e_i$ and $c(t_i) = c_i$ be the execution and communication time of the task $t_i$, respectively. For convenience, we define the *difference* of task $t_i$ to be $d_i = e_i - c_i$. We schedule $J$ on uniform processors $P_0, P_1, \ldots, P_r$ with a system I/O latency, $L$. Note that $r \leq n$. In this model, task $t_i$ takes $e_i$ time to finish its computation and after its completion (might not be immediately) transmits data to the processor on which task $t_j$ is allocated if there is a precedence relation from $t_i$ to $t_j$. Task $t_j$ cannot start executing unless it has received all data from $t_i$. We assume that the communication time is zero between two tasks allocated to the same processor. If $t_i$ and $t_j$ are allocated to different processors, then the sending time for $t_i$ is $c_j$ and the receiving time for $t_j$ is also $c_j$. All data streams are transmitted in a pipelined fashion, i.e., after $t_i$ starts sending, all data arrive at $t_j$ in $c_j + L$ units of time. If a task needs to send or receive two data elements at the same time, the two I/O operations must take place in sequence. An example of a timing diagram for executing tasks in this model is shown in Fig. 2.

**Realization of a Scheduling** A scheduling, $S$, for $J$ is an assignment of tasks to processors. A *legal realization* for $S$ is the assignment of starting times for all tasks allocated to each processor such that it satisfies the precedence constraints and the I/O latency requirement. Given a realization, let $s(t_i)$ and $f(t_i)$ be, respectively, the start and finish execution times for $t_i$ on the processor to which it is allocated. Let $s(c_i)$ and $f(c_i)$ be the start and finish times to send data to the processor $t_i$ is located on. The *makespan* of a processor $P_i$ for a realization is the time at which the processor $P_i$ finishes all tasks allocated to it. The makespan of a legal realization is the largest makespan among all processors. A legal realization with the smallest makespan is a *best* realization. The makespan of a scheduling $S$ is the makespan of its best realization and is denoted as $M(S)$. An *optimal scheduling* $J$ is a scheduling with the smallest possible makespan.

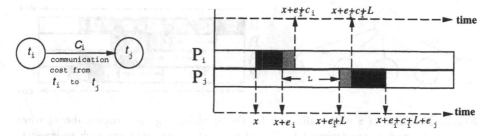

**Fig. 2.** Task $t_i$ is allocated to processor $P_i$ and $t_j$ is allocated to $P_j$. During time $x$-$(x + e_i)$, $t_i$ is executed on $P_i$. The sending time from $t_i$ to $t_j$ is $c_i$. $L$, the system latency, is the units of time from when $P_i$ starts to send data until $P_j$ starts to receive the data from $P_i$, and the receiving time is $c_i$.

For convenience, we assume that $t_0$ is allocated to $P_0$. We now state a property which can be easily verified.

**Lemma 2.1** *Let* $J = \{t_0, t_1, \ldots, t_n\}$ *be a set of tasks whose* $\mathrm{PC}(J)$ *forms a directed one-level tree with the root* $t_0$. *When scheduling* $J$ *on an unlimited number of identical processors, there is an optimal scheduling where every processor, except the one on which* $t_0$ *is located, is allocated no more than one task.*  □

**Lemma 2.2** *Let* $J = \{t_0, t_1, \ldots, t_n\}$ *be a set of* $n+1$ *tasks whose* $\mathrm{PC}(J)$ *forms a directed one-level out-tree with the root* $t_0$ *and whose execution times of tasks other than* $t_0$ *satisfy the condition* $e_i \geq e_{i+1}$, $1 \leq i < n$. *Given a scheduling for* $J$, *let* $t_0, t_{v_1}, \ldots, t_{v_{n-w}}$ *be tasks allocated to* $P_0$ *and let* $t_{u_1}, t_{u_2}, \ldots, t_{u_w}$ *be tasks not allocated to* $P_0$. *There exists a best realization for the given scheduling with the following properties: (1)* $s(t_0) = 0$; *(2)* $u_i < u_{i+1}$, $1 \leq i < w$; *(3)* $s(c_{u_i}) = e_0 + \sum_{j=1}^{i-1} c_{u_j}$, *for all* $1 \leq i \leq w$; *(4)* $s(t_{v_i}) = e_0 + \sum_{j=1}^{w} c_{u_j} + \sum_{j=w+1}^{i-1} e_{v_j}$, *for all* $1 \leq i \leq n - w$; *(5)* $t_{u_i}$ *is allocated on* $P_i$ *with* $s(t_{u_i}) = s(c_{u_i}) + L + c_{u_i}$, $1 \leq i \leq w$.  □

An example for a best realization of a scheduling as described in Lemma 2.2 is illustrated in Fig. 3.

**The Symmetric Property** In the following lemma, let $r(G)$ be the resulting graph obtained from a directed graph $G$ by reversing the direction of each edge in $G$. The weights on nodes and edges remain the same. Note that if $G$ is a directed tree, then $r(G)$ is also a directed tree.

**Lemma 2.3** *Let* $J$ *be a set of tasks whose* $\mathrm{PC}(J)$ *is a directed tree. Let* $J'$ *be the same set of tasks except that* $\mathrm{PC}(J') = r(\mathrm{PC}(J))$. *If* $J$ *has a scheduling with makespan* $M$, *then* $J'$ *also has a scheduling with makespan* $M$.  □

**The Positive Difference Property** Let $J = \{t_0, t_1, \ldots, t_n\}$ be a set of tasks whose $\mathrm{PC}(J)$ is a directed tree rooted at $t_0$. The next lemma shows that a task

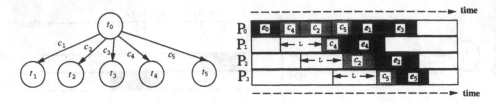

**Fig. 3.** The form of a best realization for the precedence graph (shown above) when tasks $t_0$, $t_1$, and $t_3$ are assigned to $P_0$ and the rest of the tasks are each assigned to another processor. Note that $e_i$ is the execution time for task $t_i$. In this given set of task, $L = 4$, $c_2 = 3$, $c_4 = 2$, $c_5 = 2$, $e_1 = 3$, $e_2 = 4$, $e_3 = 3$, $e_4 = 4$, and $e_5 = 3$. Since $e_4 \geq e_2 \geq e_5$, this is a best realization *for the above task assignment* according to Lemma 2.2.

$t_i$ whose difference (i.e., $e_i$ - $c_i$) is non-positive can be allocated on a processor with its parent to have an optimal scheduling.

**Lemma 2.4** *Let $S$ be a scheduling for a set of tasks whose* $\mathrm{PC}(J)$ *is a directed one-level tree. By re-allocating all tasks with non-positive differences to $P_0$, the resulting scheduling has equal or better makespan than that of $S$.* □

## 3 Algorithms for Scheduling One-Level Task Trees

In this section, we consider the cases of scheduling a directed one-level task tree on both an unlimited number of processors and a fixed number of processors. We have proven the NP-Hardness of both problems via a complex reduction(omitted because of space limitations) from the well-known knapsack problem. Therefore, approximation algorithms must be considered. We will first show an approximation algorithm that finds a scheduling on an unlimited number of processors with a makespan which is less than three times the optimal makespan. By Lemma 2.3, we need only to consider task graphs that are directed one-level out-trees. We also show an approximation algorithm for finding a scheduling on a fixed number of $k$ ($k > 1$) processors with the approximation ratio $3 + \frac{k-2}{k}$. We also consider those cases when the execution times of non-root tasks are equal.

Let $J = \{t_0, t_1, \ldots, t_n\}$ be a set of tasks whose $\mathrm{PC}(J)$ is a directed one-level out-tree rooted at $t_0$. Let $e_i$ and $c_i$ be the execution and communication time of task $t_i$, respectively. Let $L$ be the system I/O latency. We schedule $J$ on $h$ identical processors which are denoted as $P_0, P_1, \ldots, P_{h-1}$.

### 3.1 Scheduling with Arbitrary Task Execution Times

In this section, we describe two approximation algorithms for scheduling directed one-level task out-trees on an unlimited number of processors and on a fixed number of processors.

**Using an Unlimited Number of Available Processors** We show an approximation algorithm for finding a scheduling on an unlimited number of processors whose makespan is less than three times the optimal makespan. The following notation is used:

- $\text{OPT}(J)$ is the makespan of an optimal scheduling for $J$ on an unlimited number of identical processors.
- $J' = \{t_i \mid 1 \leq i \leq n \text{ and } e_i \leq c_i\}$, i.e., the set of tasks other than the root task, each with an execution time no larger than its communication time.
- $J'' = J \setminus (J' \cup \{t_0\})$, i.e., the set of tasks other than the root task, each with an execution time greater than its communication time.
- $E' = \sum_{t_i \in J'} e_i$ and $C'' = \sum_{t_i \in J''} c_i$.

Without loss of generality, assume that $t_0$ is allocated on processor $P_0$ in all schedulings studied in this section. We first give a lemma to help bound from below the value of $\text{OPT}(J)$.

**Lemma 3.1** (i) *An optimal scheduling for $J$ is to schedule all tasks on $P_0$ if and only if for all tasks $t_i$ with $i > 0$ and $e_i > c_i$, $\sum_{i=1}^{n} e_i \leq c_i + L + e_i$. (ii) If scheduling all tasks on $P_0$ is not an optimal scheduling, then $\text{OPT}(J) > e_0 + L$. (iii) $\text{OPT}(J) \geq e_0 + E' + C''$. (iv) $\text{OPT}(J) \geq e_i$, $0 \leq i \leq n$.*

*Proof.* (i) The "only if" part of the proof is trivial since putting a task on another processor in this case only increases the makespan. We now prove the "if" part.

Let $S$ be an optimal scheduling with all tasks allocated to $P_0$. Thus the makespan of $S$ is $e_0 + \sum_{i=1}^{n} e_i$. Assume that there is a scheduling $S'$ with at least one task $t_w$ with $1 \leq w \leq n$ and $e_w > c_w$ such that $\sum_{i=1}^{n} e_i > c_w + L + e_w$. We know that $e_0 + \sum_{i=1}^{n} e_i - e_w + c_w < e_0 + \sum_{i=1}^{n} e_i$ since $e_w > c_w$ and that $e_0 + c_w + L + e_w < e_0 + \sum_{i=1}^{n} e_i$ by our assumption. This implies that $M(S') < M(S)$ which is a contradiction since $S$ was an optimal scheduling. The conclusion follows.

(ii) If scheduling all tasks on $P_0$ is not an optimal scheduling, then we must at least schedule one task $t_i$, $i > 0$, on processor $P_i$. The makespan of $P_i$ is at least $e_0 + c_i + L + e_i$. Thus this part of the lemma holds.

(iii) By Lemma 2.4, we know that scheduling tasks with $e_i > c_i$ on a processor other than $P_0$ does not improve the makespan. Thus all such tasks can be scheduled on $P_0$. The minimum makespan on $P_0$ for any scheduling is at least equal to $e_0 + E' + C''$.

(iv) This part is trivial. □

We now give an approximation algorithm to find a scheduling and prove that its makespan will always be less than three times the optimal makespan.

---

Algorithm A /* a scheduling on at most $n + 1$ processors. */
    1. Check whether scheduling all tasks on $P_0$ is an optimal
    scheduling(Lemma 3.1).
    2. Otherwise, allocate a task $t_i$, $i \neq 0$, with $e_i \geq c_i$ to $P_i$ by itself,
    and the rest of the tasks to $P_0$;

**Lemma 3.2** $(i)$ *Algorithm A runs in $O(n)$ time.* $(ii)$ *The makespan of the scheduling produced by Algorithm A is less than three times the optimal makespan.*

*Proof.* Part $(i)$ is trivial. We prove part $(ii)$. Note that if the condition in Step 1 holds, then Algorithm A finds an optimal scheduling by part $(i)$ in Lemma 3.1. Thus we look at the case where the condition in Step 1 fails. Let $S$ be the scheduling produced in Step 2. The makespan of $P_0$ in $S$ is $e_0 + E' + C''$ which is at most $\mathrm{OPT}(J)$ by part $(iii)$ in Lemma 3.1. The makespan of $P_i$, $i > 0$ and $e_i > c_i$, is less than or equal to $e_0 + C'' + L + e_i$. Therefore, we note that $e_0 + C''$ is no more than $\mathrm{OPT}(J)$ by part $(iii)$ in Lemma 3.1, $L$ is less than $\mathrm{OPT}(J)$ by part $(ii)$ in Lemma 3.1, and $e_i$ is also no more than $\mathrm{OPT}(J)$ by part $(iv)$ in Lemma 3.1. Thus the makespan of any processor is less than $3 \cdot \mathrm{OPT}(J)$. □

**Using a Fixed Number of Available Processors** In this section, we give an approximation algorithm to find a scheduling for a fixed number $k$ of processors, $k > 1$. Our algorithm is used to prove the following lemma.

**Lemma 3.3** *Let $A = \{a_1, \ldots, a_n\}$ be a set of $n$ items, where $w(a_i)$ is the weight of item $a_i$ and let $k$ be a given integer, $k \le n$. Let $W = \sum_{i=1}^n w(a_i)$ and let $U = \max\{\max_{i=1}^n w(a_i), \frac{W}{k}\}$. We can find a disjoint $k$-partition, $k > 1$, $A_1, \ldots, A_k$ for $A$ such that, $(i)$ $\sum_{a_i \in A_j} w(a_i)/U \le 3 - 2/k$, and $(ii)$ for all $2 \le j \le k$, $\sum_{a_i \in A_j} w(a_i)/U \le 2 - 2/k$.*

*Proof.* Assume, without loss of generality, that $w(a_i) \ge w(a_{i+1})$, $1 \le i < n$. Our algorithm uses the following best fit descending order heuristic and runs as follows.

---

Algorithm P /* Partition $n$ items into $k$ subsets, $B_1, B_2, \ldots, B_k$. */
    1. assume, without loss of generality, that $w(a_i) \ge w(a_{i+1})$, $1 \le i < n$;
    2. $B_1 = B_2 = \cdots = B_k = \emptyset$;
    3. $r_1 = -W/k$; $r_2 = \cdots = r_k = 0$;
    4. *for* $i = 1 \cdots n$ *do*
        place $a_i$ into $B_x$ such that $r_x = \min_{j=1}^k r_j$; if there are several
        $x$'s satisfying the above, pick the smallest one; $r_x = r_x + w(a_i)$;

---

Note that our algorithm always places $a_1$ into $B_1$. Let $MB$ be the largest sum of weights in a subset produced by the above algorithm and $B_t$ a subset with sum of weights $MB$. Note that $MB \ge w(a_1)$. We have the following three cases.

*Case 1:* $B_t$ contains one item. Thus $MB \le w(a_1)$. Hence $MB/U \le 1$.

*Case 2:* $B_t$ contains more than one item and for all $2 \le i \le k$, $B_i$ contains at most one item. Thus $t = 1$ and $MB \le w(a_1) + \frac{W}{k}$. Since $U \ge w(a_1)$ and $U \ge \frac{W}{k}$, $MB/U \le 2$.

*Case 3:* $B_t$ contains more than one item and there exists some $i$, $2 \le i \le k$, such that $B_i$ contains two items. Note that our algorithm would never place two items in some $B_i$, $2 \le i \le k$, before placing at least one item to each $B_j$, $1 \le j \le k$. We let $w_i(B_j)$, $j \ne 1$, be the sum of weights in subset $B_j$ right after our algorithm has placed $a_1, \ldots, a_i$ and before it places $a_{i+1}$. The value $w_i(B_1)$ is the sum of weights in subset $B_1$ minus $\frac{W}{k}$ right after our algorithm has placed $a_1, \ldots, a_i$ and before it places $a_{i+1}$. We are interested only in those allocations which increase $MB$. In a worst case scenario, this would mean placing the largest item possible in the minimum-weight bin and producing a new $MB$. Assume that our algorithm places item $a_i$ into subset $B_{r_i}$ and $MB_i = \max_{j=1}^{k} w_i(B_j)$. Let $x$ be the smallest integer such that (1) each $B_i$ contains some item (2) $B_{r_x}$ contains more than one item and (3) $w_x(B_{r_x}) = MB_x$. That is, after the moment each subset contains at least one item, placing $a_x$ creates the first subset with more than one item such that $w_x(B_{r_x}) = MB_x$. By our assumption, we know that such an $x$ could exist and that $x > k$. Also, it can be shown by induction on the value of $z$ that $MB_z - w_z(B_i) \leq w(a_{k+1})$ for all $z \geq x$ and for all $i$. Using these observations, if the sum of weights in $B_1$ is not $MB$, then $W \geq MB + (k-1) \cdot (MB - w(a_{k+1})) + \frac{W}{k}$. This implies $MB \leq \frac{(k-1) \cdot W}{k^2} + \frac{k-1}{k} \cdot w(a_{k+1})$. Note that $W > k \cdot w(a_{k+1})$ and $U \geq W/k$. Thus

$$\frac{MB}{U} \leq \frac{k-1}{k} + \frac{(k-1) \cdot w(a_{k+1})}{W} < \frac{k-1}{k} + \frac{(k-1) \cdot w(a_{k+1})}{k \cdot w(a_{k+1})} = 2 - \frac{2}{k}$$

If the sum of weights in $B_1$ is $MB$, then $MB > \frac{W}{k}$. Let $MB' = MB - \frac{W}{k}$. Then $W \geq MB' + (k-1) \cdot (MB - w(a_{k+1})) + \frac{W}{k}$. Using a similar approach, we can prove that $\frac{MB'}{U} \leq 2 - \frac{2}{k}$. Thus $\frac{MB}{U} \leq 3 - \frac{2}{k}$. $\qquad \square$

---

Algorithm A' /* Scheduling one-level task out-trees using $k$ PE's. */
1. Check whether scheduling all tasks on $P_0$ is an optimal scheduling;
2. Otherwise, use Algorithm P to partition tasks with $e_i > c_i$ into $k$ disjoint subsets using $e_i$ as their weights; Allocate $t_0$, tasks in the first subset, and tasks with $e_i \leq c_i$ on $P_0$;
Allocate tasks in the $(i+1)$th subset to $P_i$, $1 \leq i < k$;

---

**Lemma 3.4** (i) Algorithm A' runs in $O(n \cdot \log n)$ time. (ii) The makespan of the scheduling produced by Algorithm A' is less than $3 + \frac{k-2}{k}$, $k > 1$, times the optimal makespan.

*Proof.* Part (i) is trivial. To prove part (ii), assume that the realization of a scheduling produced by Algorithm A' is for $P_0$ to execute $t_0$ first, then to send data to tasks allocated on other processors, and then finally to execute tasks allocated on itself. Processor $P_i$, $1 \leq i < k$, receives all needed data for all tasks allocated on itself, then executes tasks. Let $OPT(J)$ be the makespan of an optimal scheduling of tasks in $J$ on at most $k$ processors. The proof is similar to the proof of Lemma 3.2. Observe that parts (i) and (iii) in Lemma 3.1 still hold. Let $E''$ be the sum of execution times over tasks with $e_i > c_i$. Let $h$ be the largest execution time among these tasks. Part (ii) in Lemma 3.1 becomes $OPT(J) \geq \max\{h, \frac{E''}{k}\}$. Using part (i) in Lemma 3.3, we know that the sum of task execution time on $P_0$ is at most $2 + \frac{k-2}{k}$ times $OPT(J)$. Using part (ii) in Lemma 3.3, we know that the sum of task execution time in a processor other than $P_0$ is at most $1 + \frac{k-2}{k}$ times $\text{OPT}(J)$. Let $h_i$, $0 \le i < k$, be the sum of execution times for tasks allocated on $P_i$ and $e_i > c_i$. The makespan of $P_i$, $1 \le i < k$, is less than or equal to $e_0 + C'' + L + h_i$. Note that $h_i \le (1 + \frac{k-2}{k}) \cdot \text{OPT}(J)$. Thus the makespan of $P_i$, $i > 0$, is less than $(3 + \frac{k-2}{k}) \cdot \text{OPT}(J)$. The makespan of $P_0$ is also less than $(3 + \frac{k-2}{k}) \cdot \text{OPT}(J)$. Hence this lemma holds. $\square$

## 3.2 Scheduling with Equal Task Execution Times

In this section, we consider the problem of finding an optimal scheduling for directed one-level task out-trees when the execution times of non-root tasks are equal. We can prove the following:

**Theorem 3.5** *An optimal solution to the directed one-level precedence tree scheduling problem can be found in linear time if the execution times of all non-root tasks are equal and tasks are sorted according to their communication costs.* $\square$

**Theorem 3.6** *The problem of finding an optimal scheduling for a directed one-level task tree on a fixed number $k$ of processors is NP-hard even when non-rooted tasks have equal execution times.* $\square$

**Theorem 3.7** *A scheduling of a one-level task out-tree using $k$ processing elements where non-root tasks have equal execution times can be produced in linear time with a performance bound of at most three times the optimal.* $\square$

## 4 Concluding Remarks

Our studies have recently focused on send-receive graphs. Send-receive graphs have a source node which sends tasks to intermediate nodes to be executed and then final results are collected at the sink node. These graphs are frequently encountered in applications where jobs can be partitioned into disjoint unions of parallel tasks. For instance, a workstation could parcel out jobs to free processors in a local area network of workstations and then collect back the results after execution of the jobs. We have discovered that the algorithms and proofs found in this paper can also be adapted to the send-receive graph model.

Our work is a starting point for finding even more tractable algorithms under less stringent conditions. Such work can eventually be used by a compiler to allocate the tasks of a general algorithm to execute in parallel efficiently.

## References

1. F. D. Anger, J. J. Hwang, and Y. C. Chow. Scheduling with sufficient loosely coupled processors. *Journal of Parallel and Distributed Comput.*, 9:87–92, 1990.
2. T. C. E. Cheng and C. C. S. Sin. A state-of-the-art review of parallel-machine scheduling research. *European J. Operational Research*, 47:271–292, 1990.

3. P. Chrétienne. A polynomial algorithm to optimally schedule tasks on a virtual distributed system under tree-like precedence constraints. *European J. Operational Research*, 43:225–230, 1989.

4. P. Chrétienne. Task scheduling with interprocessor communication delays. *European J. Operational Research*, 57:348–354, 1992.

5. P. Chrétienne. Tree scheduling with communication delays. *Discrete Applied Math.*, 49:129–141, 1994.

6. P. Chrétienne, Jr. E. G. Coffman, J. K. Lenstra, and Z. Liu, editors. *Scheduling Theory and its Applications*. John Wiley & Sons Ltd, 1995.

7. J. Y. Colin and P. Chrétienne. C.P.M. scheduling with small communication delays and task duplication. *Oper. Res.*, 39(3):680–684, 1991.

8. D. Culler, R. Karp, D. Patterson, A. Sahay, K. E. Schauser, E. Santos, R. Subramonian, and T. von Eicken. LogP: Towards a realistic model of parallel computation. In *Proc. 4th ACM SIGPLAN Symp. on Principles and Practices of Parallel Programming*, pages 1–12, 1993.

9. O. El-Dissouki and W. Huen. Distributed enumeration on network computers. *IEEE Trans. on Computers*, C-29(9):818–825, 1980.

10. A. Geist, A. Beguelin, J. Dongarra, W. Jiang, R. Manchek, and V. Sunderam. *PVM 3 User's Guide and Reference Manual*. Oak Ridge National Laboratory, Oak Ridge, Tennessee 37831, USA, May 1993.

11. P. B. Gibbons, Y. Matias, and V. Ramachandran. The QRQW PRAM: Accounting for contention in parallel algorithms. In *Proc. 5th ACM-SIAM Symp. on Discrete Algorithms*, pages 638–648, 1994. SIAM J. Comput., to appear.

12. J. J. Hwang, Y. C. Chow, F. D. Anger, and C. Y. Lee. Scheduling precedence graphs in systems with interprocessor communication times. *SIAM Journal on Computing*, 18(2):244–257, 1989.

13. V. M. Lo. *Task Assignment in Distributed Systems*. PhD thesis, Univ. of Illinois at Urbana-Champaign, USA, October 1983.

14. V. M. Lo. Heuristic algorithms for task assignment in distributed systems. *IEEE Trans. on Computers*, 37(11):1284–1397, 1988.

15. D. R. Lopez. *Models and Algorithms for Task Allocation in a Parallel Environment*. PhD thesis, Texas A&M University, Texas, USA, December 1992.

16. P. R. Ma, E. Y. Lee, and M. Tsuchiya. A task allocation model for distributed computing systems. *IEEE Trans. on Computers*, C 31(1):41–47, 1982.

17. M. G. Norman and P. Thanisch. Models of machines and computation for mapping in multicomputers. *ACM Computing Surveys*, 25(3):263–302, 1993.

18. C. Papadimitriou and M. Yannakakis. Towards on an architecture-independent analysis of parallel algorithms. *SIAM Journal on Computing*, 19:322–328, 1990.

19. H. S. Stone. Multiprocessor scheduling with the aid of network flow algorithms. *IEEE Trans. on Software Eng.*, SE-3(1):85–93, 1977.

20. L. G. Valiant. A bridging model for parallel computation. *Communications of the ACM*, pages 103–111, 1990.

21. L. G. Valiant. General purpose parallel architectures. In J. van Leeuwen, editor, *Handbook of Theoretical Computer Science*, pages 944–971. North Holland, 1990.

# Scheduling Algorithms for Strict Multithreaded Computations *

Panagiota Fatourou and Paul Spirakis

Computer Technology Institute
P.O. Box 1122, GR-26110, Patras, Greece
E-mail: {*faturu, spirakis*}@cti.gr

**Abstract.** In this paper, two scheduling algorithms that solve the problem of efficient scheduling $k$-strict multithreaded computations are presented. Both of them use the work-stealing technique, in which a processor that doesn't have work to do, steals threads from other processors. Our results achieve linear speedup and linear expansion of memory, for constant $k$ (that is $k \in O(1)$). However, they are general and hold for all values of $k$.

## 1 Introduction

Dynamically growing "multithreaded" computations are quite common for parallel computers. To execute them efficiently, a scheduling technique is usually employed to ensure that enough threads remain concurrently active to keep the processors busy. At the same time, the concurrently active threads must be within limits in order to control the (dynamic) memory needed. In addition, in order to reduce communication among processors, one should try to maintain many related threads on the same processor. It is not an easy task to design a scheduler to achieve all of the above goals.

Two scheduling paradigms have been considered in the past, work sharing and work stealing. In work sharing, whenever a processor generates new threads, the scheduler attempts to migrate some of them to other (hopefully underutilized at that time) processors. In work stealing however, underutilized processors attempt to "steal" threads from other processors.

Leiserson and Blumofe [2] examined the case of fully strict multithreaded computations, in which all data dependency edges from a thread go to the thread's parent. They introduced a work stealing scheduler and they proved that the expected time $T_P$ to execute a fully strict computation (that is a $k$-strict computation with $k = 1$) on $P$ processors using their scheduler is $T_P = O(T_1/P + T_\infty)$, where $T_1$ is the minimum serial execution time of the multithreaded computation and $T_\infty$ is the minimum execution time with an infinite number of processors. Moreover they proved that the space $S_P$ required

---

* This work was partially supported by the EU ESPRIT Long Term Research Projects ALCOM-IT (Proj. Nr 20244) and GEPPCOM (Proj. Nr 9072 ) and the Greek Ministry of Education.

by the execution satisfies $S_P \leq S_1 P$, while the expected total communication of the algorithm is at most $O(T_\infty S_{max} P)$, where $S_{max}$ is the largest activation record of any thread.

Our paper extends their analysis and proves efficient bounds for $k$-strict multithreaded computations. Our analysis shows that the expected execution time $T_P$ of a $k$-strict multithreaded computation is $O(T_1/P + kT_\infty)$, while space $S_P = O((S_1 + k)P)$ is needed by the execution. We also show that the expected total communication of the algorithm is at most $O(kT_\infty S_{max} P)$. For constant $k$ ($k \in O(1)$), our results achieve linear speedup and linear expansion of memory and also all three of these bounds are existentially optimal to within a constant factor.

## 2   A Model for Multithreaded Computation

A multithreaded computation is composed of a set of threads, each of which contains unit-time tasks.The tasks of a thread executes in a sequential order. Tasks are connected to each other via *continue edges*, that determine the order in which the tasks of a thread executes. During its execution, a thread can *spawn* other threads. The newly created threads are children of the thread that creates them. A thread can create as many threads it likes to. If a task $u_1$ of a thread A spawns a thread B, then a *spawn edge* begins from $u_1$ of A and ends at the first task of thread B. A thread terminates, when it executes its last task.

When a task spawns a thread it allocates an *activation frame* for use by the newly spawned thread. The activation frames hold all the values used by the computation. Thus, there is no global storage available to the computation outside the frames. If we collapse each thread into a single node and consider just the spawn edges, the multithreaded computation becomes a rooted tree with the spawn edges as child pointers. We call this tree the *activation tree*. At a given time t during the execution of a computation, the *activation subtree* at time t is the portion of the activation tree consisting of just those threads that are alive at time t. The space used at a given time in executing a computation is the total size of all frames used by all threads in the activation subtree at that time and the total space is the maximum such value over the course of the execution.

A thread A is located at the $k$th level of the activation tree, if the path connecting A with the root thread is of length $k$. A thread A is located at a higher level than another thread B in the activation tree, if the path connecting A with the root thread is of shortest length than the one of B.

An *execution schedule* for a multithreaded computation determines which processors of a parallel computer execute which tasks at each step. An execution schedule is valid, if the order in which tasks are executed obeys the ordering constraints given by the directed edges of the computation. Another kind of dependency, that an execution schedule should satisfy, in order to be valid, is the *data dependency*. Consider that a task $u$ (consuming task) of a thread A, needs a data item produced by some task $u'$ (producing task) of another thread

B, in order to continue its execution. Data dependency edges are used to enforce such orderings.

A thread *stalls* if the execution of the thread arrives at a consuming task before the producing task has executed. Once the producing task executes, the data dependency is satisfied, which enables the consuming thread to proceed with its execution.

It is obvious from the preceding, that every multithreaded computation can be represented by a directed acyclic graph (DAG) of bounded degree. In such a DAG, every node is a task. The edges of the DAG are spawn, continue and data dependency edges.

Leiserson and Blumofe [3] have proved that general multithreaded computations (that is with arbitrary data dependencies) are impossible to be scheduled efficiently. In this paper $k$-strict multithreaded computations are studied, in which the kind of dependencies are restricted.

**Definition 1.** A k-strict multithreaded computation is one in which all data dependencies from a thread A, go to ancestors of the thread that they are located in at most k higher levels than the level of A, in the activation tree.

There are three main parameters that characterize the time spent and the space used during the execution of a mutlithreaded computation: *activation depth, work* and *dag depth*. The *activation depth* of a thread is the sum of the sizes of all its ancestors, inclusive. The *activation depth* of a multithreaded computation is the maximum activation depth of any of its threads. The minimum amount of space possible for any 1-processor execution of a $k$-strict computation is denoted by $S_1$ and it is equal to the activation depth of the computation. On the other hand, $S_P$ denotes the space used by a P-processor execution schedule of a $k$-strict multithreaded computation.

*Work* of the computation is the total number of tasks of the computation. Work is denoted by $T_1$, while $T_P$ denotes the time spent during the execution of a $P$-processor execution schedule of a $k$-strict multithreaded computation. *Dag depth* of a task is the length of the longest path that terminates at the task, while *dag depth* of the computation is the maximum dag depth of any task of the computation. Let $T_\infty$ denotes the dag depth of a computation, since even with arbitrarily many processors, each task on a path must execute serially.

Leiserson and Blumofe [3] proved the following theorem for *greedy schedules*, those in which at each step of the execution, if at least $P$ tasks are ready, then $P$ tasks execute and if fewer than $P$ tasks are ready, then all execute.

**Theorem 2.** *For any multithreaded computation with work $T_1$ and dag depth $T_\infty$, for any munber $P$ of processors, any greedy execution schedule $\mathcal{X}$ achieves $T_P(\mathcal{X}) \leq T_1/P + T_\infty$.*

# 3 Two Randomized Work Stealing Algorithms for Scheduling k-Strict Multithreaded Computations

In the following two subsections, two algorithms for scheduling $k$-strict multithreaded computations are presented. Both of them use ready lists, data structures which contain ready threads. Each ready list has two ends, top and bottom. Furthermore, the work stealing technique is used by both algorithms. When a processor's ready list is empty, the processor tries to steal work from a victim processor, which was chosen randomly. If the ready list of the victim processor is empty, the thief processor tries again, selecting another victim at random. When the algorithms begin, the ready lists of all processors except one (let it be processor $p$), are empty. All processors except $p$, begin work stealing.

## 3.1 The Algorithm k-SC WS (Work Stealing Algorithm for k-Strict Computations)

As it is mentioned previously, each processor has its own ready list. While the ready list of a processor contains ready threads, the processor extracts the bottommost one and works on it. If a stalled thread is enabled, a new thread is spawned, a thread stalls or dies, the following occur:

1. If a stalled thread A is enabled by another thread B, then thread A, which is now ready, is placed in the ready list of B's processor, exactly above thread B.
2. If a thread A spawns a child B, then A is placed in the bottommost position of the ready list and the processor begins work on B.
3. If a thread A dies or stalls and the ready list of the processor is not empty, then the processor removes the bottommost thread of the ready list and begins work on it. If however, the ready list is empty, the processor begins work stealing. It steals the topmost thread from the ready list of a randomly chosen processor and begins work on it. The work-stealing strategy is the one described at the introduction of this section.

In the case that a thread A enables another thread B and dies, the actions for enabling occur first, before the actions for dying.

**Lemma 3.** *During the execution of any $k$-strict multithreaded computation by Algorithm k-SC WS, consider any processor $p$ and any given time during which $p$ is working on a thread. Let $A_0$ be the thread that $p$ is working on, let $n$ be the number of threads in $p$'s ready list and let $A_1, A_2, \ldots, A_n$ denote the threads in $p$'s ready list ordered from bottom to top, so that $A_1$ is the bottommost and $A_n$ is the topmost. Then the threads in $p$'s ready list satisfy the following properties:*

1. *For $i = 1, 2, \ldots, n - k$, thread $A_{i-1}$ is a child of $A_i$ in the current activation subtree.*
2. *For $i = 1, 2, \ldots, n - k$, thread $A_i$ has not been worked on since it spawned $A_{i-1}$.*

*Proof.* See full paper [1] for proof.

## 3.2 Algorithm $k$-SC WS with Pointers

In this algorithm, there is an extra pointer (WS_ptr) for each ready list. Consider the ready list of a processor $p$. At any time $t$, the WS_ptr is pointing out the thread in the ready list, that is to be stolen, if a work stealing attempt is initiated at time $t$, with target the processor $p$. If the ready list of a processor $p$ is not empty, $p$ removes the bottommost thread from it and begins work on it. If a thread enables a stalled thread, spawns a newly created thread, stalls or dies, the following actions take place:

1. If a thread A enables a stalled thread B (one of the $k$ ancestors of A), the now ready thread B is placed in the proper position of the ready list of A's processor. This position is the one where all threads on higher levels than that of B in the ready list are ancestors of B, while all threads in a lower level are descendants of B.
2. This step is the same as the one for Algorithm $k$-SC WS.
3. This step is also the same as the corespondent step for Algorithm $k$-SC WS with the difference that in the case of stealing, the thief processor steals the thread that the WS_ptr is pointing to, from the ready list of the victim processor.

The WS_ptr pointer needs updating, every time the thread A, in which it points to, is stolen or is executed. In both of these cases, WS_ptr points to the thread that is above A in the ready list. If there isn't such a thread in the ready list and thread A was stolen, then the WS_ptr points at the thread that is under A in the ready list. At the beginning of the execution, WS_ptr points at the topmost thread of the ready list.

**Lemma 4.** *During the execution of any k-strict multithreaded computation by Algorithm k-SC WS with pointers, consider any processor $p$ and any given time during which $p$ is working on a thread. Let $A_0$ be the thread that $p$ is working on, let $n$ be the number of threads in $p$'s ready list and let $A_1, A_2, \ldots, A_n$ denote the threads in $p$'s ready list ordered from bottom to top, so that $A_1$ is the bottommost and $A_n$ is the topmost. Then the threads in $p$'s ready list satisfy the following properties:*

1. *For $i = 1, 2, \ldots, n - k$, thread $A_{i-1}$ is a child of $A_i$ in the current activation subtree.*
2. *For $i = n - k + 1, n - k + 2, \ldots, n$, thread $A_i$ is an ancestor of thread $A_{i-1}$.*
3. *For $i = 1, 2, \ldots, n - k$, thread $A_i$ has not been worked on since it spawned $A_{i-1}$.*

*Proof.* A proof is also provided in full paper [1].

## 4 Analysis of the Algorithms

Three are the major quantities that need to be bounded during this section: space (memory requirements), execution time and communication cost.

## 4.1 Bounding the space requirements

**Theorem 5.** *The total required space for the execution of a k-strict multithreaded computation with activation depth $S_1$ by Algorithm k-SC WS, in a parallel computer with P processors is at most $(S_1 + k)P$.*

*Proof.* See proof in full paper [1].

**Theorem 6.** *The total required space for the execution of a k-strict multithreaded computation with activation depth $S_1$ by Algorithm k-SC WS with pointers, in a parallel computer with P processors is at most $S_1 P$.*

*Proof.* See proof in full paper [1].

## 4.2 Bounding the execution time

The execution time of a multithreaded computation is the total time spent, from the time the computation starts, until its termination. It is worth pointing out, that although there aren't global structures, contention can occur, due to stealing in the ready lists. Leiserson and Blumofe [2] have proved the following lemma, assuming that concurrent accesses to the same data structure are serially queued by an adversary, as in the atomic message passing model of [4] (assumption which is also made in this paper):

**Lemma 7.** *For any $\epsilon > 0$, the total delay imposed due to contention in the ready lists is $O(M + PlgP + Plg(1/\epsilon))$, where M is the total number of steal attempts (initiated by all processors). The expected total delay is at most M.*

For the rest of the analysis, the following accounting argument is used. Consider that every processor pays one dollar for each time step. The processor places the dollar in one of the following three buckets at the end of each step: Work bucket, Steal bucket, Wait bucket. The processor places its dollar in the Work bucket, if it executes a task at the step. If the processor waits for the satisfaction of a stealing attempt, then it places its dollar in the Wait bucket, while if it initiates a steal attempt at the step , the dollar is placed in the Steal bucket.

**Lemma 8.** *When the execution of a k-strict multithreaded computation with work $T_1$, in a parallel computer with p processors, terminates, the total number of dollars in the Work bucket is exactly $T_1$.*

*Proof.* See proof in full paper [1].

In order the number of dollars in the Steal bucket to be bounded, an augmented DAG $D'$ is defined, which is constructed with the addition of some new edges to the execution DAG $D$ of the computation and used for the definition of critical tasks.

Let $D$ denotes the execution DAG of a computation The augmented DAG $D'$ is comprised by the initial graph $D$ with some additional edges, called the *critical edges*. Let $u, v$ be tasks of a thread $A_n$ such that $(u, v)$ is a continue edge in $D$ and $u$ is a task of $A_n$ that have spawned a new task. Moreover, let $A_{n-1}, A_{n-2}, \ldots, A_{n-k}$ be descendants of $A_n$ such that: For $i = n, n-1, \ldots, n-k$, thread $A_{i-1}$ is a child of the thread $A_i$. If $w_{n-1}, w_{n-2}, \ldots, w_{n-k}$ are the first tasks of each one of the threads $A_{n-1}, A_{n-2}, \ldots, A_{n-k}$ correspondently, the *critical edges* $(v, w_{n-1}), (v, w_{n-2}), \ldots, (v, w_{n-k})$, are added in $D'$.

**Definition 9.** At any time during the execution, a task $v$ is said to be critical if every task that precedes $v$ (either directly or indirectly) in $D'$ has been executed, that is, if for every task $w$ such that there is a directed path from $w$ to $v$ in $D'$, task $w$ has been executed.

**Definition 10.** A round of work-steal attempts is a set of at least $3P$ but fewer than $4P$ consecutive steal attempts, such that if a steal attempt that is initiated at time step $t$ occurs in a particular round, then all other steal attempts initiated at time step $t$ are also in the same round.

All the steal attempts that occur during an execution can be partitioned into rounds as follows. If the $i$th round ends at time step $t_i$, then the $(i+1)$st round begins at time step $t_i + 1$ and ends at the earliest time $t_{i+1} > t_i + 1$ such that at least $3P$ steal attempts were initiated at time steps between $t_i + 1$ and $t_{i+1}$, inclusive.

**Definition 11.** A critical sequence is a 3-tuple $(U, R, \Pi)$ satisfying the following conditions:

- $U = (u_1, u_2, \ldots, u_L)$ is a maximal directed path in $D'$. That is, for $i = 1, 2, \ldots, L - 1$, the edge $(u_i, u_{i+1})$ belongs to $D'$ (task $u_1$ must be the first task of the root thread), and task $u_L$ has no outgoing edges in $D'$ (task $u_L$ must be the last task of the root thread).
- $R$ is a positive integer.
- $\Pi = (\pi_1, \pi_2, \ldots, \pi_L)$ is a partition of the integer R.

The critical sequence $(U, R, \Pi)$ is said to occur during an execution, if for each $i = 1, 2, \ldots, L$, at least $\pi_i$ steal attempt rounds occur while task $u_i$ is critical.

**Lemma 12.** *Consider the execution of a k-strict multithreaded computation with dag depth $T_\infty$ by Algorithm k-SC WS (or k-SC WS with pointers) on a computer with P processors. If at least $4P(2T_\infty + R)$ steal attempts occur during the execution, then some $(U, R, \Pi)$ critical sequence must occur.*

*Proof.* The proof is correspondent to the one of Leiserson and Blumofe [2] and it is not presented in this paper. The reader can see the proof of Lemma 8 in [2], instead.

**Lemma 13.** *At all times during the execution of a k-strict multithreaded computation by either of the two algorithms, each critical task must be the ready task of a thread that is one of the top k+1 in its processor ready list.*

*Proof.* The proof of the lemma is provided in full paper [1].

The following lemmas and theorems are proved for Algorithm $k$-SC WS.

**Lemma 14.** *Consider the execution of any k-strict multithreaded computation by Algorithm k-SC WS on a parallel computer with $P \geq 2$ processors. For any task u and any set of $r \geq c(3 + log12)k$ work steal rounds ($c \in \Theta(1)$), the probability that the r rounds occur while the task is critical is the probability that at most k of the steal attempts initiated in the r rounds choose u's processor, which is at most $e^{-2r}$.*

*Proof.* Since $u$ is critical it is in the ready list of some processor $p$. Let $t_a$ be the first time step in which $u$ becomes critical and $t_e$ the time step in which $u$ is executed. Suppose that $r$ work steal rounds occur, while $u$ is critical. Each round contains at least $3P$ steal attempts, so at least $3rP$ steal attempts occur during the $r$ rounds. The last of these steal attempts occurs in a time step $t_b$ strictly before $t_e$, that is $t_b < t_e$. Since at most $P$ steal attempts can be initiated during one time step, at least $(3r - 1)P$ steal attempts are initiated strictly before $t_b$.

Lemma 6 says that at most $k$ tasks in the ready list of a processor can be simultaneously critical. Furthermore, Algorithm $k$-SC WS guarantees that if a thread A is topmost in the ready list of a processor, then another thread can not become topmost whithout either thread $A$ starts its execution or it is stolen. This is true, because whenever a thread $A$ enables a stalled one B, B is placed exactly above A (and A is the bottommost one) in the processor's ready list, without affecting the position of the topmost thread (which is the next to be stolen). From the preceding, it is obvious that the thread that contains $u$ (call it $A'$) will be stolen, after at most $k$ steal attempts with target processor the processor $p$, in whose ready list, $A'$ resides. If all these steal attempts are initiated at time steps $t_1, t_2, \ldots, t_k$ such that: $t_1 \leq t_2 \leq \ldots \leq t_k < t_b$, then the $k$th steal attempt takes the thread $A'$ and executes it at time step $t_k + 1 \leq t_b < t_e$, which is impossible. In the case that after the $k$th steal attempt with target the processor $p$, thread $A'$ haven't been stolen, some other threads should have been placed above $A'$ in the ready list of $p$, event that can occur only if thread $A'$ have been executed at a time step strictly before $t_b$, which is again impossible. The conclusion that can be made from the preceding is that from the $(3r - 1)P$ steal attempts that are initiated at some time step strictly before $t_b$, only $k$ can have as destination the processor $p$. The probability that at most $k$ of the $(3r - 1)P$ steal attempts chose $p$ as a target processor is:

$$S = \sum_{i=0}^{k} \binom{(3r-1)P}{i} \left(\frac{1}{P}\right)^i \left(1 - \frac{1}{P}\right)^{(3r-1)P-i} . \tag{1}$$

The terms of the sum follow the binomial distribution,

$$b(i; (3r-1)P, 1/P) = \binom{(3r-1)P}{i} \left(\frac{1}{P}\right)^i \left(1-\frac{1}{P}\right)^{(3r-1)P-i} . \tag{2}$$

It has been proved (W.Feller [5] mentions it) that:

$$S \le b(k; (3r-1)P, 1/P) \frac{[(3r-1)P-k+1](1/P)}{[(3r-1)P+1](1/P)-k}$$

$$\le c \binom{(3r-1)P}{k} \left(\frac{1}{P}\right)^k \left(1-\frac{1}{P}\right)^{(3r-1)P-k} \quad c = \text{constant},$$

whenever $k \le 3r-1$, because $\frac{[(3r-1)P-k+1](1/P)}{[(3r-1)P+1](1/P)-k} \in \Theta(1)$ . Thus,

$$S \le c \left(\frac{(3r-1)P}{k}\right)^k \left(\frac{1}{P}\right)^k e^{-3r+k+1} \le e^{-2r} ,$$

whenever $r \ge c(3+log12)k$. Since for $r \ge c(3+log12)k$, the inequality $k \le 3r-1$ holds, the proof is completed.

**Lemma 15.** *Consider the execution of a k-strict multithreaded computation with dag depth $T_\infty$ by Algorithm k-SC WS on a parallel computer with P processors. For any $\varepsilon > 0$ with probability at most $1 - \varepsilon$, at most $O(P(kT_\infty + log(1/\varepsilon)))$ work steal attempts occur. The expected number of steal attempts is $O(kPT_\infty)$.*

*Proof.* See full paper [1] for proof.

**Theorem 16.** *Consider the execution of any k-strict multithreaded computation with dag deprth $T_\infty$ and work $T_1$ by Algorithm k-SC WS on a parallel computer with P processors. For any $\varepsilon > 0$, with probability at least $1 - \varepsilon$, the execution time on P processors is $T_P = O(T_1/P + kT_\infty + logP + log(1/\varepsilon))$. The expected running time, including scheduling overhead, is $O(T_1/P + kT_\infty)$.*

*Proof.* See full paper [1] for proof.

Some of the preceding lemmas and theorems have been proved only for Algorithm k-SC WS. However, the results are similar for the Algorithm k-SC WS with pointers too, although some proofs are slightly different than the ones for the first algorithm. These differences are mentioned in what follows.

The only change needed in the proof of Lemma 7 is the reason that if a thread A is topmost in the ready list of a processor, then another thread can not become topmost without either thread A starts its execution or it is stolen. This property is true for Algorithm k-SC WS with pointers, because of the existence of the WS_ptr pointers and the way they are updated.

Nothing changes in the proof of Lemma 8, while the correspondent to Theorem 4 theorem, for Algorithm k-SC WS with pointers follows:

**Theorem 17.** *Consider the execution of any k-strict multithreaded computation with dag depth $T_\infty$ and work $T_1$ by Algorithm k-SC WS with pointers on a parallel computer with $P$ processors. For any $\varepsilon > 0$, with probability at least $1 - \varepsilon$ the execution time on $P$ processors is $T_P = O(T_1/P + k^2 T_\infty + log P + log(1/\varepsilon))$. The expected running time is $O(T_1/P + k^2 T_\infty)$.*

*Proof.* Proof is provided in full paper.

## 5   Bounding the Communication Cost

**Theorem 18.** *The total amount of bytes communicated, during the execution of a k-strict multithreaded computation with dag depth $T_\infty$, by either algorithm, on a parallel computer with $P$ processors, has expectation $O(kPT_\infty S_{max})$, where $S_{max}$ is the size in bytes of the largest activation record in the computation. Moreover, for any $\varepsilon > 0$ with probability at least $1 - \varepsilon$, the total communication incurred is $O(P(kT_\infty + log(1/\varepsilon))S_{max})$.*

*Proof.* See proof in full paper.

For constant $k$, the communication cost incurred is $O(PT_\infty S_{max})$ and this bound is optimal.

## 6   Conclusions

We are currently trying to improve the bounds provided in this paper, in order to make them better for large $k$ (e.g. $k$ near $T_\infty$).

Additionaly, although the total space required was bounded in this paper, the posibility for a processor to run out of memory exists. A solution for this problem is probably the combination of both work-stealing and work-sharing techniques.

## References

1. Fatourou, P., Spirakis, P.: Scheduling Algorithms for Strict Multithreaded Computations. Computer Technology Institute, T.R. 96.5.13, (1996).
2. Blumofe, R., Leiserson, Ch.: Scheduling Multithreaded Computations by Work Stealing. Proc. of the 35th Ann. Symp. on Foundations of Computer Science (FOCS). (1994), 356–368.
3. Blumofe, R., Leiserson, Ch.: Space-Efficient Scheduling of Multithreaded Computation. Proc. of the 25th Ann. ACM Symposium on the Theory of Computing (STOC). San Diego, California, (May 1993), 362–371.
4. Liu, P., Aielo, W., Bhatt, S.: An atomic model for message passing. Proc. of the Fifth Annual ACM Symposium on Parallel Algorithms and Architectures (SPAA). Velen, Germany, (June 1993), 154–163.
5. Feller, W.: An Introduction to Probability Theory and its Application. Third Edition, vol.1, John Wiley & Sons, Inc., 1968.

# On Multi-threaded Paging*

Esteban Feuerstein[1][2] and Alejandro Strejilevich de Loma[1]

[1] Departamento de Computación, FCEyN, Universidad de Buenos Aires, Argentina.
E-mail: {efeuerst,asdel}@dc.uba.ar
[2] Instituto de Ciencias, Universidad de General Sarmiento, Argentina

**Abstract.** In this paper we introduce a generalization of Paging to the case where there are many threads of requests. This models situations in which the requests come from more than one independent source.

Four different problems arise whether we consider fairness restrictions or not, with finite or infinite input sequences. We study all of them, proving lower and upper bounds for the competitiveness of on-line algorithms. The main results presented in this paper may be summarized as follows. When no fairness restrictions are imposed it is possible to obtain good competitive ratios; on the other hand, for the fair case in general there exist no competitive algorithms.

## 1 Introduction

The *Paging Problem* is the problem of managing a memory consisting in two levels, one of which with limited capacity and fast access (the cache) and the other one with slow access but potentially unlimited capacity. *Paging algorithms* must decide which page from fast memory to replace on a *page fault*, that is, when a reference is made to a page that is not present in fast memory. *On-line Paging algorithms* must decide which page to evict based only on the previous requests, with the goal of minimizing the total number of page faults. Paging has been studied from a *competitive analysis* point of view in [8], comparing the performance of on-line algorithms to that of the optimal off-line algorithm. In that work it is shown that, if the cache can hold $k$ pages, no deterministic on-line algorithm can be better than $k$-competitive, that is, guarantee less than $k$ times the optimal off-line number of page faults on every input.

In this paper we introduce the *Multi-threaded Paging* problem (MTP). MTP generalizes Paging to the case in which there is not just one sequence of requests but possibly many threads. This models situations in which the requests come from more than one independent source. Hence, apart from deciding *how* to serve a request, at each stage it is necessary to decide *which* request to serve among several possibilities. In this case there is no notion of "sequence of requests" but a more complex pattern that is not captured by the most general classes of on-line problems proposed in the literature (like Metrical Task Systems [3]

---

* Research supported in part by EC project DYNDATA under program KIT, and by UBACYT project "Modelos y Técnicas de Optimización Combinatoria"

or Request-Answer Games [1]). This is because the order in which the requests are served may be different for each algorithm. Moreover, as we will see later, in some cases even the set of requests that will eventually be served depends on the particular algorithm that is used.

Besides its theoretical interest, the problem we introduce may be significant from a practical point of view. MTP allows to model the situation arising in multi-tasking systems, where all processes independently present their requests of memory pages to the fast memory manager. At each moment, the system must decide whose request to satisfy, and also (as in normal Paging) which page of fast memory to remove on a page fault. The total number of page faults depends therefore not only on the strategy used to determine how each request is served but on which requests are satisfied, and when (in which order) this is done.

We study the MTP problem under two different approaches. In the first one, we simply try to minimize the number of page faults done while serving a set of $w$ sequences of requests. In the second one, we impose *fairness* restrictions, so all algorithms must guarantee that the next request of each thread will be served within a predetermined finite time. For each one of these two problems we consider two different models, namely the finite and the infinite models. For normal Paging it has been proved that it is equivalent to define competitiveness on finite sequences and in the limit [7]. In the case of multiple threads, however, we will see that considering finite or infinite sequences makes a difference (even if in the last case we measure the performance after finite intervals of time).

Fiat and Karlin [5] have considered a related problem, in which the input corresponds to a multi-pointer walk on an *access graph* [2]. Within that framework, the multiple threads of requests are seen as a unique input sequence corresponding to an interleaved execution of the different threads. The way in which the sequences are interleaved in [5] is decided in an earlier stage of the process. On the contrary, in MTP algorithms are free to decide (up to a certain limit, in the case of fairness restrictions) how to do that. In other words, our algorithms act not only as fast memory managers but also as schedulers, while in the cited work, the scheduling is implicitly supposed to be done somewhere else.

Four different problems arise whether we consider fairness restrictions or not, with finite or infinite input sequences. We study all of them, proving lower and upper bounds for the competitiveness of on-line algorithms. The main results presented in this paper may be summarized as follows. When no fairness restrictions are imposed it is possible to obtain good competitive ratios; on the other hand, for the fair case in general there exist no competitive algorithms. However, there is an interesting exception in a particular case that we treat separately.

The remainder of this paper is organized as follows: In Section 2 we formally present the four different versions of our problem, namely the no-fair and fair versions of MTP, with finite and infinite input sequences. In Section 3 we consider the case in which no fairness restrictions are imposed, proving lower and upper bounds. In Section 4 we analyze the fair case, and show that competitive algorithms can only exist in particular situations. Finally, Section 5 is dedicated to describe conclusions and open problems. Due to space reasons, most of the proofs have been omitted. The interested reader may find them in [4].

# 2 Preliminaries

Imagine that $w$ independent processes simultaneously present their requirements of pages of secondary memory that must be brought into fast memory. At each moment, the scheduler and fast memory manager can see only one request per process, precisely the first unserved request of the sequence of requests that the process presents. Serving the present request of a particular process allows the system to see the next request of that process. The system must decide at each step which request should be served, apart from deciding which page of fast memory to evict to make place to the incoming page if that is the case.

We define two different models for this problem. In the first one of them, called Finite-MTP (FMTP), algorithms are faced to a certain number of *finite* sequences of requests that have to be served *completely*, that is, algorithms have to arrive to the end of each one of the sequences. In the second model, that we call Infinite-MTP (IMTP), the input to an algorithm is a set of *infinite* sequences of requests. In this case, however, the system will be observed after a finite number of steps, comparing the costs incurred by different algorithms till that moment. The conceptual difference between the two models is that in the finite one all algorithms are forced to serve the same set of requests (although possibly in different orders), while in the infinite model different algorithms may (because of their choices) see different sets of requests. In a multi-tasking environment where many infinite processes compete for the system's resources, this model may be more adequate than the finite one.

FMTP is given by the universe or set of pages $U$ and two positive integers $k$ and $w$, the cache-size and the number of sequences respectively. $C = \mathcal{P}_k(U)$ is the set of *configurations*, where $\mathcal{P}_k(U)$ is the power-set of $U$ restricted to subsets of size $k$. $\sigma \in (U^*)^w$ is the *input tuple*, $\sigma = \sigma_1, \sigma_2, \ldots \sigma_w$. We can view $\sigma$ as a set of sequences of requests; each request is an element of the universe $U$.

We can imagine to have a pointer to the current position in each sequence $\sigma_i$. In a certain configuration $c$, the system can advance the pointer of some sequence $\sigma_i$ such that $u_i$, the currently pointed page of $\sigma_i$, is present in $c$. Given an input tuple $\sigma$ and an initial configuration, a *schedule* for $\sigma$ is a sequence of pairs $< i_j, c_j >, 1 \le i_j \le w$ such that the currently pointed request of $\sigma_{i_j}$ may be served in configuration $c_j$. The *cost* of such schedule is the summation over the sequence of the Hamming distance between successive configurations. At any stage of a schedule, any sequence $\sigma_i$ whose last request has not been served is called *active*. Note that normal Paging is a particular case of FMTP if $w = 1$.

An on-line algorithm for FMTP receives a tuple of sequences $\sigma$ as input and produces as output a schedule for $\sigma$, with the restriction that the following configuration can only be determined as a function of the known part of $\sigma$. The known part of a tuple of sequences is given by the current tuple of requests and all the requests already served by the algorithm.

We will now modify the definition of FMTP to get the different (but very related) infinite version of the problem, IMTP. In IMTP the sequences of requests are infinite, that is, the input to an algorithm is $\sigma \in (U^\omega)^w$ instead of $\sigma \in (U^*)^w$, where $\Sigma^\omega$ denotes the set of infinite words over an alphabet $\Sigma$. In this case, even if the sequences are infinite, we will observe the system after a finite number of steps, when we will compare the cost in which it incurred with the minimum cost necessary to any algorithm that advances at least the same number of steps. This will be done by considering also an integer $\ell$ (unknown to the on-line algorithm), representing the number of requests that must be served.

In the previous discussion we did not make any consideration regarding the fairness of the algorithms for MTP. In other words, we were interested in optimizing the number of page faults, even if that was achieved by never serving the requests coming from a particular process. This can be satisfactory in some particular situations, but it is not in the framework of multi-tasking systems.

We model fairness restrictions by considering, as part of the input of the problem, a (big) integer $t$ such that no request can "wait" to be served more than $t$ units of time from the moment the previous request of the sequence has been satisfied. Time is measured in the intuitive way: one unit of time elapses each time the system serves a request. Consequently, we define the notion of $t$-fair schedule. An algorithm for the Fair-MTP problem must produce a $t$-fair schedule for the input tuple of sequences. An on-line algorithm must obviously produce the schedule based only in the known part of the input.

All the algorithms presented in this paper for the different versions of MTP are based in Flush-When-Full (FWF), a very well known $k$-competitive on-line algorithm for Paging, that was introduced in [6]. However, it is worthwhile to note that we could have used any deterministic marking algorithm (like, for example, FIFO and LRU); we have chosen FWF to simplify the analysis.

FWF maintains a set of marked pages. Initially the marked pages are exactly those that are present in the cache. The behavior of FWF on a request to page $p$ can be described as follows. If $p$ is in the cache then $p$ is marked and nothing else is done. Otherwise, a page $p'$ is chosen among the unmarked pages of the cache, $p'$ is evicted to make place for $p$, and $p$ is marked; all the marks are erased when a page must be evicted and there is no unmarked page in the cache. FWF works in *phases*, the first phase starting with the first request of the sequence and each new phase starting with the request that would have caused more than $k$ pages to be marked (when the marks are deleted). It is easy to verify that FWF never faults twice on the same page during any given phase, which implies that its cost is at most $k$ per phase.

## 3 Multi-threaded Paging without Fairness Restrictions

In this section we will consider MTP without fairness restrictions. As we will see, both versions of this problem (FMTP and IMTP) are different in what regards lower and upper bounds. However, we will propose the same on-line algorithm for both models. In the infinite model we will show that the algorithm is strongly competitive (by providing a matching lower bound), but in the finite model we were not able to prove whether it is optimal or not.

## 3.1 The Infinite Model

We said that the infinite model allows each algorithm to serve its own set of requests. We will now use this property to prove a lower bound for IMTP.

**Theorem 1.** *No on-line algorithm for IMTP is c-competitive with $c < wk$.*

*Proof.* Given any on-line algorithm A with a cache-size $k$, we will construct a set of sequences $\sigma_1, \sigma_2, \ldots \sigma_w$ for which A cannot behave better than the bound we are trying to prove.

Let $n$ be any positive integer, and consider a set $U$ of $w(k+1)$ pages, partitioned in $w$ disjoint subsets $U_1, U_2, \ldots U_w$, each one of $k+1$ pages. Since $k$ is the cache-size, in any configuration of A there is at least one page of each $U_i$ missing to A. The first request of each sequence $\sigma_i$ is to a page of $U_i$ that is not present in the initial configuration of A. Each new request (up to the $n$-th request) of $\sigma_i$ is to a page of $U_i$ not in A's cache after the algorithm serves the previous request of the same sequence. In this way, A necessarily faults on every request of this part of the input tuple. Suppose there exists some $i$ such that A never arrives to the $n$-th request of $\sigma_i$. The remaining part of $\sigma_i$ could contain requests only to the page of the $n$-th request of $\sigma_i$. Thus the adversary could serve requests of that sequence forever, and the ratio between the costs would be unbounded. Hence, if A achieves a bounded competitive ratio, it must eventually serve the first $n$ requests of every sequence. Let $\sigma_j$ be the last sequence of which A serves the $n$-th request. Then by the moment A serves the $n$-th request of $\sigma_j$, it has incurred a cost of at least $wn$. Now suppose $\sigma_j$ continues with repeated requests to page $s$ which is the $n$-th request of $\sigma_j$. The adversary can serve requests of $\sigma_j$ forever, with a total cost not exceeding $n/k$. In whatever way the other sequences continue, and whatever does A after serving the $n$-th request of $\sigma_j$, we have for the competitive ratio $\frac{C_A}{C_{ADV}} \geq \frac{wn}{n/k} = wk$. $\qquad\square$

We will now present an on-line algorithm for IMTP. We call the algorithm Alternate–Flush-When-Full (AFWF). The algorithm works in rounds; each round consists in applying a phase of FWF to each sequence $\sigma_1, \sigma_2, \ldots \sigma_w$. The algorithm can also be used in the finite model, if it checks whether the sequences are over. We will see now that AFWF is strongly competitive for IMTP.

**Theorem 2.** *Algorithm AFWF is $wk$-competitive for IMTP.*

*Proof.* By definition of AFWF, the cost of each one of the phases of a round is at most $k$, and thus the cost of each round is at most $wk$. Suppose that after $\ell$ requests AFWF completed $m$ rounds and is currently in the $(m+1)$-st round. Therefore its total cost is $C_{AFWF} \leq (wk)m + (wk)$.

There must be at least one sequence for which the adversary served at least the same number of requests that AFWF in that sequence. Let $\sigma_j$ be such a sequence. Restricted to $\sigma_j$ AFWF behaved as FWF in the sense that if we consider all the requests of a round plus the first request of the following round, a set of $k+1$ distinct pages appear. Hence the adversary must fail at least once per completed round, and then we have $C_{ADV} \geq m$, which implies $C_{AFWF} \leq (wk)C_{ADV} + (wk)$, that is, AFWF is $wk$-competitive. $\qquad \Box$

## 3.2 The Finite Model

Here we are constrained by the fact that every algorithm for FMTP must serve the same set of requests. We now give a lower bound for the finite model that is smaller than the one we have obtained for the infinite version. However, if we restrict to normal Paging ($w = 1$) the lower bound is the best possible ($k$). Then we show that algorithm AFWF (proposed above for IMTP) is $wk$-competitive also for the finite problem.

**Theorem 3.** *No on-line algorithm for FMTP is c-competitive if $c < k+1-1/w$.*

**Theorem 4.** *Algorithm $AFWF$ is $wk$-competitive for FMTP.*

# 4 Fair Multi-threaded Paging

We will now analyze MTP when fairness restrictions are explicitly imposed. As we did in Section 3, we will consider finite and infinite input sequences.

## 4.1 The Fair Infinite Model

Without fairness restrictions, a smart on-line algorithm for IMTP could avoid serving sequences that seem to be "too hard". When an algorithm is forced to behave fairly, sooner or later it must serve all the requests it sees, even if there is another way to serve the sequences that never arrives to that place. The following theorem exploits this fact to prove that in general there is no competitive on-line algorithm for Fair-IMTP.

**Theorem 5.** *There is no competitive on-line algorithm for Fair-IMTP with $t \geq w \geq 2$, even if we consider adversaries with cache-size 1 and restrict the sequences of requests to be formed by at most $k+1$ distinct pages.*

*Proof.* Let A be any on-line algorithm with cache-size $k$, and ADV an off-line adversary with cache of size at least 1. We will show a set of sequences $\sigma_1, \sigma_2, \ldots \sigma_w$, for which $C_A$ is arbitraryly high, while $C_{ADV}$ remains bounded.

Let $m$ be any positive integer. Let $U = \{a, b_1, b_2, \ldots b_k\}$ be a set of $k+1$ distinct pages, where $a$ is a page not in A's cache at the beginning. The first request of every sequence is to page $a$. The algorithm A will serve a request with cost at least 1, and no matter what sequence it chooses, the next request of that sequence is to the same page $a$. This continues until A serves $m$ requests. Let $m_i$ be the number of requests of $\sigma_i$ served by A until that moment (possibly

$m_i = 0$). Let $\sigma_j$ be a sequence such that $(\forall i)\, m_i \le m_j$, i. e., a sequence in which A has advanced most.

We will choose the next requests so as to turn $\sigma_j$ very expensive, maintaining the other sequences cheap. As A must still take care of $\sigma_j$ to be fair, it will have a high cost. The off-line adversary, who knows that $\sigma_j$ turns expensive, may only serve the first (cheap) part of it, bounding its cost by a constant.

After A has served the first $m$ requests, and sees in all the sequences requests to page $a$, we choose each new request of $\sigma_j$ to a page of $U$ not in A's cache after it has served the previous request of $\sigma_j$. The other sequences $\sigma_i$ $(i \ne j)$ continue with requests to page $a$.

Let $\ell = m + (n+1)(t+1)$ be the total number of requests to be served, where $n = \left\lfloor m_j \frac{t-w+1}{t+1} \right\rfloor$. After the first $m$ served requests, A must satisfy a minimum of $n+1$ additional requests of $\sigma_j$, because there still remain $n+1$ groups of $t+1$ requests to be served (any fair algorithm must satisfy at least one request of each sequence in every group of $t+1$ consecutive requests that it serves). It is easy to see that with the requests chosen in the way we have described, A faults at least once for each new request of $\sigma_j$ it serves, except eventually the $(m_j+1)$-st request of $\sigma_j$, which is to page $a$. Since A already has paid at least 1 to serve the first $m$ requests, we have that the total cost for $\ell$ requests is $C_A \ge 1 + n$.

We will consider now the adversary's cost. For each $t+1$ requests, it can serve $t$ of the sequences $\sigma_i$ $(i \ne j)$ and only one of $\sigma_j$. With this behavior the adversary serves the minimum number of requests of $\sigma_j$ that guarantees fairness; this minimum is $p_{\min}(\ell) = \left\lfloor \frac{\ell}{t+1} \right\rfloor$, and it depends on the total number of served requests. Having $m \le w m_j$, and $n \le m_j \frac{t-w+1}{t+1}$, it results $\ell \le (m_j+1)(t+1)$, and so $p_{\min}(\ell) \le m_j + 1$. Since the first $m_j + 1$ requests of $\sigma_j$ are all to page $a$, the inequality $p_{\min}(\ell) \le m_j + 1$ means that the adversary always serves requests to that page, and therefore its cost is $C_{ADV} \le 1$.

As $t \ge w$, it follows that $t - w + 1 > 0$. Since $m$ was any positive integer, we can choose it as big as desired, in such a way to make $m_j$ and $n$ as big as desired. This way $C_A$ can be arbitrarly high, while $C_{ADV}$ never exceeds 1. $\quad\square$

The preceding theorem cannot be extended to cover two cases of Fair-IMTP: $w = 1$ and $t = w - 1$. With $w = 1$ fairness restrictions have no sense, and we are faced to regular Paging.

On the other hand, the case $t = w - 1$ requires further analysis. In this situation the algorithms (on-line or not) must apply round-robin, i. e., serve one request of each sequence in a fixed order which is repeated over and over again. This implies that after each $w$ new requests, the set of requests served by any algorithm is the same. That is the reason why the case $t = w - 1$ cannot be included in Theorem 5: intuitively, none or both A and the adversary will serve the expensive part of $\sigma_j$; in the proof the problem arises when we claim that $n$ increases when $m$ does, which is not true if $t = w - 1$.

This particular case of Fair-IMTP is closely related to normal Paging, since after an algorithm has chosen the order in which to serve the requests, we can think that there is only one sequence to be served, as it is the case in normal Paging. Nevertheless, the following points show that the two problems are different:

- In Fair-IMTP we can say that any algorithm has served the same set of requests only after each $w$ new requests, not at every step as in normal Paging.
- The algorithms (of any type) can choose between $w!$ distinct orders of the requests. However, off-line algorithms can choose with information on the whole input tuple, while on-line algorithms can choose only based in the first requests of each sequence. Moreover, any decision made by an on-line algorithm could be fooled by an adversary, say, with two "rows" of requests to the same page (that is, making the first two requests of all the sequences to the same page).
- The on-line algorithms can see the following $w$ requests to serve, not only one of them. It is well known that this kind of lookahead can be easily neutralized by replacing each request with $w$ requests to the same page. Thus on-line algorithms cannot take advantage of this lookahead.

Relying on the above discussion, we can prove a lower bound for the competitive ratio of on-line algorithms for Fair-IMTP with $t = w - 1$. The result is valid for any value of $w$ (and $k$), but it is only interesting when $w \geq 2$.

**Theorem 6.** *No on-line algorithm for Fair-IMTP with $t = w-1$ is c-competitive if $c < k$, even if we restrict the sequences of requests to be formed by at most $k + 1$ distinct pages.*

We present an on-line algorithm for Fair-IMTP called Round-Robin–Flush-When-Full (RRFWF). The algorithm works in "super-phases"; each super-phase consists in applying a phase of FWF to the sequence formed by taking in turn one request of each sequence $\sigma_1, \sigma_2, \ldots \sigma_w$, and then serving the next request and all the other pending requests in the same row; these additional requests are served by RRFWF in an arbitrary (deterministic) way. We want to use this algorithm also for the finite model; in that case it must check for end-of-sequence. Clearly RRFWF is fair for any legal value of $t$. We will show now that it is $(k + w)$-competitive for Fair-IMTP with $t = w - 1$.

**Theorem 7.** *Algorithm RRFWF is $(k+w)$-competitive for Fair-IMTP with $t = w - 1$.*

*Proof.* By definition of RRFWF, the cost of each super-phase does not exceed $k + 1 + (w - 1) = k + w$. Suppose that after $\ell$ requests RRFWF has completed $m$ super-phases and is currently in the $(m + 1)$-st super-phase. Hence we have for its total cost $C_{RRFWF} \leq (k + w)m + (k + w - 1)$. At the end of each one of the completed super-phases, the adversary must have served the same requests that RRFWF, even choosing a different order for the sequences. Since in each super-phase a minimum of $k + 1$ distinct pages appear, the adversary must fault at least once in each completed super-phase, and then we have $C_{ADV} \geq m$, so $C_{RRFWF} \leq (k + w)C_{ADV} + (k + w - 1)$. $\square$

## 4.2 The Fair Finite Model

We have seen in Theorem 5 that in general there is no competitive on-line algorithm for Fair-IMTP. The proof uses an adversary that does not serve the requests that are expensive for the on-line algorithm, which is permitted in IMTP, but not in FMTP (fair or not). Nevertheless, we are able to prove a similar result for the finite model.

**Theorem 8.** *There is no competitive on-line algorithm for Fair-FMTP with $t \geq w \geq 2$, even if we restrict the sequences of requests to be formed by at most $k+1$ distinct pages.*

As in Fair-IMTP, the above theorem excludes an interesting case, that is, the case $t = w-1$. While all the sequences are active the situation is the same that in the infinite version, and the discussion we have done in the previous subsection is valid here. As soon as a sequence finishes, distinct algorithms can serve the remaining requests (if any) in very different ways. This can be used to extend Theorem 8 to the case $w \geq 3$ and $t = w - 1$. The idea of the proof consists in making the on-line algorithm serve a first sequence of length 1, something that was not possible in the infinite model.

**Corollary 9.** *There is no competitive on-line algorithm for Fair-FMTP with $w \geq 3$ and $t = w - 1$ ($t = w - 1 \geq 2$), even if we restrict the sequences of requests to be formed by at most $k + 1$ distinct pages.*

The remainder of our analysis is devoted to the case $w = 2$ and $t = w-1 = 1$, not covered by Theorem 8 and Corollary 9.

**Theorem 10.** *No on-line algorithm for Fair-FMTP with $t = w-1$ is $c$-competitive if $c < k$, even if we restrict the sequences of requests to be formed by at most $k + 1$ distinct pages.*

**Theorem 11.** *Algorithm RRFWF is $(k + w)$-competitive for Fair-FMTP with $w = 2$ and $t = w - 1 = 1$.*

# 5 Conclusions and Open Problems

Table 1 summarizes the results presented in this paper. When there are not fairness restrictions, the infinite case seems to be more difficult for on-line algorithms than the finite one. This is because in the latter all algorithms are obliged to serve the same set of requests, whereas in the former the adversary may take benefit of serving different requests.

When fair behavior is imposed, things become harder even in the finite case, preventing the existence of competitive on-line algorithms in general. The extra difficulty is given by the fact that even if an algorithm finds a sequence to be very expensive, it cannot avoid dealing with it. In the finite case this may force to serve with high cost requests that could be served more efficiently if delayed.

Table 1. Summary of results.

| | IMTP | | FMTP | |
|---|---|---|---|---|
| | l.b. | u.b. | l.b. | u.b. |
| any $w$ | $wk$ | $wk$ | $k + 1 - 1/w$ | $wk$ |

| | Fair-IMTP | | Fair-FMTP | |
|---|---|---|---|---|
| | l.b. | u.b. | l.b. | u.b. |
| $t \geq w \geq 2$ | $\infty$ | – | $\infty$ | – |
| $w \geq 3$ and $t = w - 1$ | | | $\infty$ | – |
| $w = 2$ and $t = w - 1 = 1$ | $k$ | $k + w$ | $k$ | $k + w$ |

As it may be seen in Table 1, lower and upper bounds do not coincide in every case, so one goal is to close the existing gaps. An interesting research direction consists in modeling fairness restrictions in a different way; one possibility is to strengthen the definition by considering (instead of $t$) an integer $\delta$ with the following meaning: the "distance" between the pointers of any pair of sequences can never exceed $\delta$. Alternative definitions of competitiveness may also be considered for the infinite models, like comparing the performances of the different algorithms in the limit, that is, when the number of served requests tends to infinity. Finally, it would be interesting to analyze randomized algorithms.

**Acknowledgments:** We would like to thank Luis César Maiarú for useful discussions about this work.

# References

1. S. Ben-David, A. Borodin, R. Karp, G. Tardos, and A. Widgerson. On the power of randomization in online algorithms. Technical Report TR-90-023, ICSI, June 1990.
2. A. Borodin, S. Irani, P. Raghavan, and B. Schieber. Competitive paging with locality of reference. In *Proc. of 23rd ACM Symposium on Theory of Computing*, pages 249–259, 1991.
3. A. Borodin, N. Linial, and M. Saks. An optimal on-line algorithm for metrical task system. *J. ACM*, 39(4):745–763, 1992.
4. E. Feuerstein and A. Strejilevich de Loma. On multi-threaded paging. Technical Report TR 96-001, Universidad de Buenos Aires, Departamento de Computación, July 1996. http://www.dc.uba.ar/people/proyinv/tr.html.
5. A. Fiat and A. Karlin. Randomized and multipointer paging with locality of reference. In *Proc. of 27th ACM Symposium on Theory of Computing*, 1995.
6. A. Karlin, M. Manasse, L. Rudolph, and D. Sleator. Competitive snoopy caching. *Algorithmica*, 3:79–119, 1988.
7. P. Raghavan and M. Snir. Memory versus randomization in on-line algorithms. Technical Report RC 15622, IBM, 1990.
8. D.D. Sleator and R.E. Tarjan. Amortized efficiency of list update and paging rules. *Communications of ACM*, 28:202–208, 1985.

# A Fast and Efficient Homophonic Coding Algorithm *

Boris Ryabko and Andrey Fionov

Novosibirsk Telecommunication Institute,
Kirov str. 86, 630102, Novosibirsk, Russia
e-mail: ryabko@neic.nsk.su

**Abstract.** Homophonic coding, introduced in [1, 2], is refered to as a technique that contributes to reliability of the secret-key cipher systems. For the methods of homophonic coding known until now, the redundancy $r$ defined as the difference between a mean codeword length and a source entropy can only be reduced at the expense of exponential growth of the memory size and calculation time required. We suggest a method of perfect homophonic coding for which the size of memory and the time of calculation grow as $O(1/r)$ and $O(\log^2 1/r \log \log 1/r)$, respectively, as $r \to 0$.

## 1 Introduction

It is known in cryptology that the plaintext statistics, namely unequal probabilities and mutual dependence of the symbols the plaintext consists of, can help an attacker in breaking a cipher, thus lowering the strength of a secret-key cipher system. To deprive an attacker of such an opportunity cryptographers use special methods of transforming the plaintext before its being encrypted. Homophonic coding represents one of these transformations. The technique has been known for centuries. Among others, it was employed by Gauss (see [3]). In case of *perfect* homophonic coding, a code sequence becomes indistinguishable from a sequence of equiprobable and independent code letters. According to [2], such code enables constructing unbreakable ciphers. All throughout the paper we shall deal only with perfect schemes.

Consider an example of a homophonic coding scheme. Let there be given a Bernoulli source generating letters over the alphabet $A = \{a, b\}$ with probabilities $p(a) = 1/4$ and $p(b) = 3/4$. We shall construct a binary code for each source letter using the following encoding table.

| Letter | Codeword | Probability |
|--------|----------|-------------|
| $a$ | 00 | 1 |
| $b$ | 01 | 1/3 |
| | 10 | 1/3 |
| | 11 | 1/3 |

* Supported by Russian Foundation of Fundamental Research under Grant 96-01-00052

In this example, the letter *b* may be represented by three different codewords that are chosen at random. One can easily find that the probabilities of all codewords are the same and equal to 1/4, all possible two-bit permutations being used. That is sufficient for the code sequence to be completely random.

The above example shows the *conventional* homophonic coding scheme. This scheme has a disadvantage of provoking considerable data expansion. Besides, it cannot produce random binary sequencies if the source letter probabilities are non-binary-rational numbers.

In 1988 Ch. Günther suggested the method of *variable-length* homophonic coding [1]. A year later, in [2] an information-theoretic treatment of the method was given. In Günther's method, several codewords representing a source letter are replaced by one smaller in length. For the above example, the codewords "10" and "11" may be replaced by "1" chosen with the probability 2/3, since their low-order bits contribute in equal shares to the final bit distribution and thus may be thrown aside. The encoding table becomes as follows.

| Letter | Codewors | Probability |
|--------|----------|-------------|
| a | 00 | 1 |
| b | 01 | 1/3 |
|   | 1 | 2/3 |

The code sequence generated by using such a table also proves to be completely random. Apart from perceptible reduction of the code length, the method described is able to encode the source symbols having arbitrary, not binary-rational only, probabilities. Let, for instance, probabilities of the source letters be $p(a) = 1/3$, $p(b) = 2/3$. Then the encoding table assumes the following form (note that $1/3 = 0.010101\ldots$).

| Letter | Codeword | Probability |
|--------|----------|-------------|
| a | 00 | 3/4 |
|   | 0100 | 3/16 |
|   | 010100 | 3/64 |
|   | ... |  |
| b | ... |  |
|   | 01011 | 3/64 |
|   | 011 | 3/16 |
|   | 1 | 3/4 |

Although the set of all codewords is infinite, it has been shown in [2] that the mean code length for such a code does not exceed $H+2$ bits, where $H$ denotes the entropy of a source letter. In the same work, a method of generating arbitrary distributed random variables given the sequence of independent equiprobable bits was also proposed.

In many cases, there arises the task of constructing homophonic codes whose redundancy can be made as small as desired, this being particularly essential for low entropy sources. The problem may be solved by transition from letterwise encoding to block encoding. The latter may be treated as encoding of symbols over some "superalphabet" comprising all letter blocks of a given length as its "superletters". Cardinality of this "superalphabet", however, grows exponentially as the size of the block increases, which implies corresponding growth of memory and time amounts when conventional encoding methods are being used.

The purpose of the present contribution is to introduce a homophonic coding algorithm whose redundancy can be made arbitrary small without any need of exponentially increasing memory and time. This algorithm combines the ideas of homophonic coding from [1, 2] and fast block coding proposed in [4]. It may be applied for sources with known statistics and requires the memory size and the time of encoding and decoding to increase as $O(1/r)$ and $O(\log^2 1/r \log \log 1/r)$, respectively, as the redundancy $r$ converges to zero.

In the following section we describe the version of the algorithm fit for Bernoulli sources, binary encoding alphabet, and binary-rational probabilities of source symbols. Some generalizations of applying the method to Markovian sources, arbitrary finite encoding alphabets, and rational number representation of probabilities, without deteriorating the asymptotic estimates of memory and time, are easily attainable but their full description goes beyond the scope of the present paper.

## 2  Description of the Algorithm

Let there be given a Bernoulli source generating letters over some finite alphabet $A = \{a_1, a_2, \ldots, a_N\}$, $N \geq 2$, with probabilities $p(a_1), p(a_2), \ldots, p(a_N)$ represented as $p(a_i) = \rho_i/2^t$, $\rho_i$ being a positive integer $t$-bit number, $\rho_i < 2^t$ for all $i$. We shall use the binary encoding alphabet $B = \{0, 1\}$. Our aim is to obtain a code sequence, the letters of which should be equiprobable and independent. To gain the given arbitrary small redundancy we shall encode blocks of symbols. However, for ease of understanding, we begin with the letterwise case. A distinctive feature of the algorithm is that, in contrast with Günther's method, it combines constructing and choosing codewords in one process.

Let there be required to construct the homophonic code for some symbol $u$, $u \in A$. Set the lexicographic order over $A$ (note that we do not presume the probabilities to be ordered) and denote

$$Q(a_1) = 0;$$

$$Q(a_i) = \sum_{j=1}^{i-1} \rho_j, \quad i = 2, 3, \ldots, N;$$

$$\tilde{Q}(a_i) = Q(a_i) + \rho_i - 1, \quad i = 1, 2, \ldots, N,$$

so the symbol $u$ is represented by an interval $[Q(u), \tilde{Q}(u)]$.

*Algorithm A1 (homophonic coding of a single symbol)*

Consider the binary expansions of $Q(u)$ and $\tilde{Q}(u)$:

$$Q(u) = q_1 q_2 \ldots q_t, \quad \tilde{Q}(u) = \tilde{q}_1 \tilde{q}_2 \ldots \tilde{q}_t.$$

Denote the $c$ leading coincident bits of $Q(u)$ and $\tilde{Q}(u)$, if any, by $b_1 b_2 \ldots b_c$,

$$b_i = q_i = \tilde{q}_i, \quad i = 1, 2, \ldots, c, \quad c \geq 0.$$

If it occures that $c = t$ or $q_i = 0$ and $\tilde{q}_i = 1$ for all $i > c$, then the process of encoding is terminated and a codeword for the symbol $u$ is defined as

$$C(u) = b_1 b_2 \ldots b_c.$$

In this case there is only one codeword and no choice is needed. Otherwise we have to choose randomly between several possible codewords representing the symbol $u$.

Before describing this process, notice that after having detected the leading coincident bits, $Q(u)$ and $\tilde{Q}(u)$ may be presented as

$$Q(u) = b_1 b_2 \ldots b_c 0 q_{c+2} \ldots q_t,$$

$$\tilde{Q}(u) = b_1 b_2 \ldots b_c 1 \tilde{q}_{c+2} \ldots \tilde{q}_t,$$

where $q_{c+1} = 0$ and $\tilde{q}_{c+1} = 1$ since $Q(u)$ cannot be greater than $\tilde{Q}(u)$ by definition.

We can further "squeeze" the interval by the observation that in $Q(u)$ and $\tilde{Q}(u)$ there may exist a number $s$ of successive bits such that

$$q_i = 1 \text{ and } \tilde{q}_i = 0, \quad i = (c+1)+1, (c+1)+2, \ldots, (c+1)+s, \quad s \geq 0$$

(as will be shown in the proof of the Theorem 1, this "squeezing" is needed to ensure low random bit consumption). After that, $Q(u)$ and $\tilde{Q}(u)$ may be written down as

$$Q(u) = b_1 b_2 \ldots b_c 0 \, 11 \ldots 1 \, q_{n+1} q_{n+2} \ldots q_t,$$

$$\tilde{Q}(u) = b_1 b_2 \ldots b_c 1 \, \underbrace{00 \ldots 0}_{s} \, \tilde{q}_{n+1} \tilde{q}_{n+2} \ldots \tilde{q}_t,$$

where $n = c + 1 + s$. Now we have to deal only with those bits of $Q(u)$ and $\tilde{Q}(u)$ whose indices are greater than $n$.

Denote by $R^{(x)}$ a sequence of random bits generated by a binary symmetric source, $R^{(x)} = r_0^{(x)} r_1^{(x)} r_2^{(x)} \ldots$ . We shall get different sequencies $R^{(1)}, R^{(2)}, \ldots$, until we encounter the sequence $R = r_0 r_1 \ldots r_k$ that satisfies the following conditions:

$$r_0 = 0 \text{ and } q_i = 0 \text{ for all } i > n \tag{1}$$

or

$$r_0 = 0, r_1 = q_{n+1}, \ldots, r_{k-1} = q_{n+k-1},$$
$$\text{and } (r_k > q_{n+k} \text{ or } r_k = q_{n+k} \text{ and } q_i = 0 \text{ for all } i > n+k) \tag{2}$$

or

$$r_0 = 1 \text{ and } \tilde{q}_i = 1 \text{ for all } i > n \tag{3}$$

or

$$r_0 = 1, r_1 = \tilde{q}_{n+1}, \ldots, r_{k-1} = \tilde{q}_{n+k-1},$$
$$\text{and } (r_k < \tilde{q}_{n+k} \text{ or } r_k = \tilde{q}_{n+k} \text{ and } \tilde{q}_i = 1 \text{ for all } i > n+k). \tag{4}$$

Once any one of these four conditions is satisfied, the algorithm stops and a codeword for the symbol $u$ may be constructed as

$$C(u) = b_1 b_2 \ldots b_c r_0 e_1 e_2 \ldots e_s r_1 r_2 \ldots r_k, \quad (c \geq 0, s \geq 0, k \geq 0), \tag{5}$$

where $e_1 = e_2 = \cdots = e_s = 1 - r_0$.

To illustrate the method consider an example. Let $A = \{a, b, c\}$, $p(a) = 93/256$, $p(b) = 11/256$, $p(c) = 152/256$, $t = 8$. The following table gives all possible codes for the letter $u = b$.

| Codewords for the letter $b$ | | | | | | | | | |
|---|---|---|---|---|---|---|---|---|---|
| | 0 | 1 | 0 | 1 | 1 | 0 | | | $T_1(b)$ |
| | 0 | 1 | 0 | 1 | 1 | 1 | 0 | 0 | $T_2(b)$ |
| $Q(b)$ 0 | 1 | 0 | 1 | 1 | 1 | 0 | 1 | | $C_1(b)$ |
| | 0 | 1 | 0 | 1 | 1 | 1 | 1 | | $C_2(b)$ |
| | 0 | 1 | 1 | 0 | 0 | | | | $C_3(b)$ |
| $\tilde{Q}(b)$ 0 | 1 | 1 | 0 | 0 | 1 | 1 | 1 | | |
| $b_1$ | $b_2$ | $r_0$ | $e_1$ | $e_2$ | $r_1$ | $r_2$ | $r_3$ | | |
| $c = 2, s = 2$ | | | | | | | | | |

The codewords which are able to represent a letter $u$ are denoted by $C_i(u)$. Usually, there may be several codewords for each symbol. Their choice depends on what values the random bits participating in the codeword construct may acquire. It can be seen that the conditions (1) — (4) just check whether a codeword constructed fits into the interval $[Q(u), \tilde{Q}(u)]$.

The table also presents "tentative" codewords, denoted by $T_i(u)$, that may arise in the course of the algorithm owing to certain random bit sequences, but are rejected as they go outside the interval $[Q(u), \tilde{Q}(u)]$.

It is necessary to make some remarks on the ways of obtaining random bit sequences $R^{(x)}$. Splitting the output of the binary symmetric source into such sequences should be treated only as a convenient descriptive method. Actually, these sequences are to be formed incrementally, as new random bits are generated. Consider the set of all sequences $R^{(x)}$ involved in constructing a codeword for the letter $b$ of the above example:

$$R^{(T_1)} = 00,$$
$$R^{(T_2)} = 0100,$$
$$R^{(C_1)} = 0101,$$
$$R^{(C_2)} = 011,$$
$$R^{(C_3)} = 1.$$

It is plain that they are prefix-free and therefore may be easily obtained bit by bit when testing the conditions (1) — (4) is in progress.

Decoding may be carried out as follows. Let the decoder receive a code sequence. Denote the first $t$ bits of the sequence by $Z$ and find the letter $w \in A$ to satisfy the conditions $Q(w) \leq Z \leq \tilde{Q}(w)$. This is the encoded letter. It is necessary now to bypass those bits of the input sequence actually representing the codeword for $w$, in order to detect the beginning of the next symbol codeword. To do that, as $Q(w)$ and $\tilde{Q}(w)$ are known, the encoding process is performed with the only claim that corresponding input bits be used instead of random ones, the reconstructed codeword being then deleted from the input sequence.

The properties of the method are summed up in the following

**Theorem 1.** *Let the algorithm A1 be applied to encode messages generated by a Bernoulli source with known statistics. Then the following propositions are held:*

*1) the code sequence is completely random, i. e., all code letters are statistically independent and equiprobable;*

*2) to construct a codeword not more than 12 random bits on average are needed;*

*3) the mean length of a codeword does not exceed $H(u) + 3$ bits, where $H(u)$ is the entropy of a source letter.*

*Proof.* 1) Correctness of the first proposition can be shown by using the same arguments as in [1, 2].

2) First, find the mean length $N_r$ of the random bit sequence $R$ that either satisfies or does not satisfy the conditions (1) — (4). To provide an upper estimate assume that $t = \infty$ and neither $Q(u)$ nor $\tilde{Q}(u)$ contain tails of zeroes and ones, respectively. Notice that $r_0$ always satisfies (2) or (4). The number of additional random bits will be determined by how long the equality between an immediate bit $r_i$ and corresponding bit $q_j$ (or $\tilde{q}_j$) is held, the probability of this equality being

$$\Pr\{r_i = q_j\} = \Pr\{r_i = 0\} \times \Pr\{q_j = 0\} + \Pr\{r_i = 1\} \times (1 - \Pr\{q_j = 0\}) = 1/2$$

This means that, apart from $r_0$, one more random bit will be needed with the probability $1/2$, after which one more random bit will again be needed with the same probability, and so on. Hence,

$$N_r = 1 + \frac{1}{2} \times 1 + \frac{1}{4} \times 2 + \frac{1}{8} \times 3 + \cdots = 1 + \sum_{n=1}^{\infty} \frac{n}{2^n} = 3.$$

Since, due to "squeezing" the interval, either $q_{n+1} \neq 1$ or $\tilde{q}_{n+1} \neq 0$ the sequence $R$, in the worst case, will satisfy either (2) or (4) with the probability $1/4$. Consequently, there will be required 4 different sequences $R^{(x)}$ on average and the total mean number of random bits will be $4N_r = 12$.

3) Let the codeword constructed for the symbol $u \in A$ be defined as (5). Denote the codeword length by $W(u)$,

$$W(u) = c + s + N_r.$$

Notice that $c + s \leq \lfloor \log p(u) \rfloor$. Hence,

$$W(u) \leq \lfloor \log p(u) \rfloor + 3 < \log p(u) + 3.$$

The latter inequality is strict since if $c + s = \log p(u)$ then not more than one random bit is required to construct a codeword. The expected value of $W(u)$ over the whole set of $u \in A$

$$E[W(u)] = \sum_{u \in A} p(u)W(u) < \sum_{u \in A} p(u)(\log p(u) + 3) =$$

$$\sum_{u \in A} p(u) \log p(u) + 3 \sum_{u \in A} p(u) = H(u) + 3.$$

$\square$

Now we proceed to the description of the block homophonic coding method which allows to obtain arbitrary small redundancy together with considerable economy of random bits.

Let $m$ be the size of a block $U = u_1 u_2 \ldots u_m$, $m > 1$, and every symbol $u$ of $U$ be generated by a Bernoulli source over the alphabet $A$, so $P(U) = p(u_1)p(u_2)\ldots p(u_m)$. Denote the set of all $m$-letter words over the alhpabet $A$ by $A^m$ and set over $A^m$ a lexicographic order. To construct a codeword for $U$ we need the values of $Q(U)$ and $\tilde{Q}(U)$,

$$Q(a_1 a_1 \ldots a_1) = 0,$$

$$Q(U) = 2^{mt} \sum_{V < U, V \in A^m} P(V),$$

$$\tilde{Q}(U) = Q(U) + 2^{mt} P(U) - 1,$$

where the factor $2^{mt}$ ensures that $Q(U)$ and $\tilde{Q}(U)$ be integer.

We shall calculate $Q(U)$ and $\tilde{Q}(U)$ using the fast block encoding method proposed in [4].

For ease of designation, assume that $m = 2^\sigma$, where $\sigma$ is integer, $\sigma \geq 1$. Define

$$p(u_1) = \rho_1^0/2^t, \ p(u_2) = \rho_2^0/2^t, \ \ldots, \ p(u_m) = \rho_m^0/2^t;$$
$$Q(u_1) = \lambda_1^0, \ Q(u_2) = \lambda_2^0, \ \ldots, \ Q(u_m) = \lambda_m^0.$$

Define also $d_0 = 2^t$, $d_1 = 2^{2t}$, $d_2 = 2^{4t}$, $\ldots$, $d_\sigma = 2^{mt}$.
*Algorithm A2 (Homophonic coding of a block of symbols)*
Calculate

$$\lambda_k^i = \lambda_{2k-1}^{i-1} d_{i-1} + \rho_{2k-1}^{i-1} \lambda_{2k}^{i-1},$$
$$\rho_k^i = \rho_{2k-1}^{i-1} \rho_{2k}^{i-1}, \quad k = 1, 2, \ldots, m/2^i, \ i = 1, 2, \ldots, \sigma. \tag{6}$$

Then $Q(U) = \lambda_1^\sigma$ and $\tilde{Q}(U) = \lambda_1^\sigma + \rho_1^\sigma - 1$ are represented by $mt$-bit integers and we may apply the algorithm A1 to construct a codeword for $U$, provided that $t$ is replaced by $mt$.

It can be easily seen that the mean per symbol redundancy satisfies the inequality $r < 3/m$. Therefore, to obtain a given arbitrary small redundancy one should take the block length $m$ such that $m > 3/r$.

Decoding is carried out as follows. Let the decoder input receive a code sequence. Denote the word formed out of the first $mt$ bits of the input sequence (supplemented by zeros, if necessary) by $Z_1^\sigma$. Define $Z_1^{\sigma-1} = \lfloor Z_1^\sigma/d_{\sigma-1} \rfloor, \ldots,$ $Z_1^1 = \lfloor Z_1^2/d_1 \rfloor$, $Z_1^0 = \lfloor Z_1^1/d_0 \rfloor$. Find the symbol $w_1$ that satisfies the inequality $Q(w_1) \le Z_1^0 \le \tilde{Q}(w_1)$. It will be the first encoded symbol. Calculate $Z_2^0 = \lfloor (Z_1^1 - \lambda_1^0 d_0)/\rho_1^0 \rfloor$ and find $w_2$ satisfying $Q(w_2) \le Z_2^0 \le \tilde{Q}(w_2)$. Using (6) calculate $\lambda_1^1$ and $\rho_1^1$. Calculate $Z_2^1 = \lfloor (Z_1^2 - \lambda_1^1 d_1)/\rho_1^1 \rfloor$ and from $Z_3^0 = \lfloor Z_2^1/d_0 \rfloor$ find the symbol $w_3$, and from $Z_4^0 = \lfloor (Z_2^1 - \lambda_3^0 d_0)/\rho_3^0 \rfloor$ find $w_4$. As we have known the first 4 symbols, obtain $\lambda_1^2$ and $\rho_1^2$. From $Z_2^2 = \lfloor (Z_1^3 - \lambda_1^2 d_2)/\rho_1^2 \rfloor$, proceeding in the same manner, find $w_5$, $w_6$, $w_7$, $w_8$, $\lambda_2^3$ and $\rho_2^3$. The process going on, we shall eventually obtain $Z_2^{\sigma-1} = \lfloor (Z_1^\sigma - \lambda_1^{\sigma-1} d_{\sigma-1})/\rho_1^{\sigma-1} \rfloor$ that will give us the last $m/2$ encoded symbols, as well as $\lambda_2^{\sigma-1}$ and $\rho_2^{\sigma-1}$. After having decoded the whole block, calculate $\lambda_1^\sigma$ and $\rho_1^\sigma$ and reconstruct the codeword using corresponding input bits instead of random ones and delete it from the input sequence.

The complexity of the method is given by the following

**Theorem 2.** *Let the algorithm A2 be applied to encoding blocks of symbols of the length $m$, $m \ge 1$, generated by a Bernoulli source with known statistics. Then the memory size of the encoder and decoder increases as $O(m)$, and the mean per symbol time of encoding and decoding grows as $O(\log^2 m \log\log m)$ as $m \to \infty$.*

*Proof.* The estimate of the memory size required is based on the observation that in (6), for every "tier" $i$ formed by all $\lambda^i$ and $\rho^i$, the memory amount of $2mt$ bit is needed and all values of the $i$th tier depend only on those of the $(i-1)$th tier, so only two tiers are to be stored at any instant. On decoding, the additional memory for $Z$ values is also confined to $2mt$ bits, since at any instant only one value of every tier suffices to be stored.

The estimate of the algorithm's time consumption is obtained by summing up complexities of the operations involved, these complexities, in their turn, measured in bit operations. We assume that Schönhage-Strassen's method should be used for multiplication and division (see [5]). This method requires $O(n \log n \log\log n)$ bit operations for multiplying two $n$-bit numbers or dividing $2n$-bit number by $n$-bit one. According to [4], it causes the complexity of calculating $\rho_1^\sigma$, $\lambda_1^\sigma$, and $Z_k^i$ to be determined as $O(m \log^2 m \log\log m)$, as $m \to \infty$, while the algorithm A1, given $Q(U)$ and $\tilde{Q}(U)$, requires only $O(m)$ operations, from which the proposition of the theorem immediately follows. $\quad\square$

**Corollary 3.** *The estimates of the memory size and mean per symbol time considered as functions of the redundancy $r$, are determined as $O(1/r)$ and $O(\log^2 1/r \log\log 1/r)$, respectively, as $r \to 0$.*

# References

1. Günther, Ch. G.: A universal algorithm for homophonic coding. Proc. of Eurocrypt-89, Springer-Verlag (1990) 382–394
2. Jendal, H. N., Kuhn, Y. J. B., Massey, J. L.: An information-theoretic treatment of homophonic substitution. Proc. of Eurocrypt-89, Springer-Verlag (1990) 382–394
3. Massey, J. L.: An introduction to contemporary cryptology. Proc. of the IEEE **76** (1988) 533–549
4. Ryabko, B. Y.: Fast and effective coding of information sources. IEEE Trans. Inform. Theory **IT-40, 1** (1994) 96–99
5. Aho, A. V., Hopcroft, L. E., Ullman, J. D.: The design and analysis of computer algorithms. Addison-Wesley Publishing Company (1976)

# An Improvement of the Digital Cash Protocol of Okamoto and Ohta

Osamu Watanabe and Osamu Yamashita

Department of Computer Science
Tokyo Institute of Technology
(watanabe@cs.titech.ac.jp)

**Abstract.** Okamoto and Ohta [OO91, OO93] proposed a digital cash protocol and showed that it satisfies some useful properties for digital cash as well as basic requirements such as security. One of its features is "divisibility"; i.e., a property that enables us to subdivide a piece of given digital cash into several pieces of smaller value. There is, however, some problem in their implementation of this feature: the amount of data transfer per payment may increase depending on the amount of the payment and how the coins have been used. Here we propose a new protocol by which we can fix the amount of data transfer per payment, which is much smaller than the average amount necessary in Okamoto-Ohta protocol.

## 1 Introduction

Recently, "digital cash system" has received considerable attention from both practical and theoretical view points. Roughly speaking, "digital cash" or "electronic cash" is money that can be transferred through computer networks electronically. More specifically, Okamoto and Ohta [OO91, OO93] listed the following six properties for an *ideal* digital cash protocol. (Below the properties are listed according to the order of importance from the authors' point of view.)

(1) *Security:* The digital cash cannot be copied and reused.
(2) *Independence:* The digital cash protocol can be purely implemented on computer networks; that is, it (in particular, its security) is not dependent on any specific physical device.
(3) *Privacy (Untraceability):* The privacy of users is protected; that is, no one can trace the relationship between users and their purchases.
(4) *Off-line payment:* When a user pays for a purchase with the digital cash, the protocol between the user and the merchant is executed off-line; that is, the merchant does not need to consult a bank (i.e., the host of the system) for processing the user's payment.
(5) *Divisibility:* A piece of digital cash can be subdivided and used as several pieces of smaller amounts. For example, $100 digital cash can be used as ten pieces of $10 digital cash.
(6) *Transferability:* The digital cash can be transferred electronically to other users.

Several digital cash protocols have been proposed, and indeed some have been implemented and used in an experimental system (see, e.g., [Sch94, OS96]). Among them, we review here those satisfying the properties (1) ∼ (3). First Chaum [Cha85] proposed a protocol satisfying all propoerties except (4). He used "blind signature" to achieve (3), which became the basis of the following protocols. The first off-line protocol that satisfies (1) ∼ (4) was proposed in [CFN88]. Okamoto and Ohta [OO89] proposed a protocol satisfying (1) ∼ (4) and (5), and later [OO91, OO93], they obtained one satsifying all (1) ∼ (6).

The last protocol proposed by Okamoto and Ohta satisfies all the above properties. Furtheremore, according to their estimation [OO91, OO93], the protocol is feasible; that is, the protocol can be implemented by our current technology to make a reasonably secure digital cash system. Nevertheless, it still has some problem for practical use; the total data transfer for one payment increases when the issued digital cash is subdivided, or it is transferred to another user. While the total data transfer for one payment is, say, $c$ kilobytes when one piece of digital cash is used normally, it becomes (i) $m \times c$ kilobytes when the cash has to be used as $m$ pieces of smaller amounts (see Section 2 for more detail), or (ii) $(n+1) \times c$ kilobytes after the cash has been transferred (i.e., passed on) $n$ times.

It has been shown [CP92, Yam96] that the amount of data per payment needs to grow, at least, proportional to the number of transfers; otherwise, the bank cannot identify a cheating user. Thus, the latter increase is necessary. On the other hand, it would be better if we can keep the amount of data per payment constant no matter how the cash is subdivided. This paper proposes one digital cash protocol that satisfies this requirement. Our protocol, which is based on Okamoto-Ohta protocol, satisfies the above properties (1) ∼ (5); furthermore, no matter how the cash is subdivided, the amount of data per payment is kept constant, which is comparable with the minimum amount in the Okamoto-Ohta protocol. (In order to simpify our discussion, the property (6) is not considered in this paper. However, almost in the same way as the Okamoto-Ohta protocol, we can extend our protocol to the one for transferable digital cash.)

## 2 Okamoto-Ohta Protocol

Here we first give the outline of the basic digital cash protocol [CFN88], and show one important feature of the protocol proposed by Okamoto and Ohta in [OO91, OO93]. We then explain the problem when one piece of the digital cash is used as small pieces.

In the following, we assume some bank that hosts digital cash protocols. We use $U$ to denote a user or a customer, and $V$ to denote a merchant.

The basic digital cash protocol [CFN88] consists of the following four parts. (Note that in this protocol, each piece of digital cash is not divisible; that is, it should be used as a whole.)

**Part 1:** User $U$ opens an account at the bank, and obtain, from the bank, an authorized license $L_U$ to use the digital cash. This licence will be used to authorize issued digital cash.

**Part 2:** $U$ asks the bank to issue one piece of digital cash (in the following, we simply call it a *digital coin*) for, say $1000. The bank issues a coin $C$ for $1000, and deducts $1000 from $U$'s account. This $C$ is valid only when it is used together with the authorized license $L_U$.

**Part 3:** In order to pay $C$ to merchant $V$, user $U$ sends certain information $X$ created from $C$, $L_U$, and a message from $V$. Then $V$ verifies that (i) $C$ is valid w.r.t. $L_U$, and (ii) $X$ is legitimate. If so, $V$ accepts the payment.

**Part 4:** $V$ sends $C$, $L_U$, and $X$ to the bank and asks for $1000. The bank first checks whether (i) $C$ is valid w.r.t. $L_U$, and (ii) $X$ is legitimate. It also checks whether (iii) $C$ has not been claimed before. If all three tests are passed, then the bank pays $1000 to $V$. Otherwise, if either $\neg$ (i) or $\neg$ (ii) occurs, then the bank refuses to pay money to $V$, i.e., rejects $C$. On the other hand, there are the following two cases when $\neg$ (iii) occurs, i.e., $C$ has been claimed before. First, $C$ has been used with the same $X$. This means that $V$ claims money twice for the same $C$; thus, the bank rejects $V$'s claim. Second, $C$ has been used with some different $X'$. This means that $U$ used the same $C$ at two different occasions, and in this case, the bank can identify cheating user $U$ from $X$ and $X'$; thus, the bank can give some penalty to $U$.

This protocol has the following properties: So long as $U$ uses the issued cash $C$ normally, the bank cannot identify $U$ from $C$, $L_U$, and $X$. On the other hand, if $U$ uses the same $C$ twice, then the bank receives two different $X$ and $X'$, and it can identify $U$ from them (together with $C$ and $L_U$). That is, for honest users, their privacy is protected; but the bank can reveal cheating users' identity. This is the key point for achieving the properties (1) and (3) of Section 1. (*Remark.* For the sake of the property (4) (i.e., off-line payment), we need to give up to check whether the same cash is used twice at the payment, i.e., Part 3. Thus, the best we can hope for the property (1) is to identify cheating users at Part 4.)

The protocol proposed by Okamoto and Ohta essentially follows this outline (though there are several technical improvements). One big difference is that an issued digital coin can be subdivided into small pieces in their protocol. For this, they introduce "hierarchical structural table".

Conceptually, a hierarchical structural table is a binary tree of several levels. For example, Figure 1 shows a hierarchical structural table for $1024 up to 4 levels. Each node of the $i$th level corresponds to $1024/2^{i-1}$. A tree like this is one digital cash in Okamoto-Ohta protocol, and some of the nodes in the tree are used for each payment. Clearly, we need the following restrictions: (i) each node cannot be used more than once, and

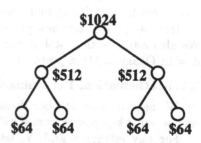

Fig. 1: Hierarchical Structure Table

(ii) if one node is used, then all descendant nodes and all ancestor nodes of this node cannot be used. Okamoto-Ohta protocol is designed so that the bank can identify cheating users who did not follow one of these restrictions. Intuitively, we can satisfy this requirement by considering each node as one digital coin $C$ in the above basic protocol. Then as above, the bank can identify cheating users who break (i). With one more clever idea, Okamoto-Ohta protocol enables the bank to identify cheating users who break (ii).

Now let us see the problem of this protocol mentioned in Section 1 In Okamoto-Ohta protocol, each node in a hierarchical structure table corresponds to one digital coin in the basic protocol. Hence, if one needs to use several, say $m$ nodes for payment, he has to send data $m$ times larger than the basic protocol. Suppose, for example, that user $U$ has a hierarchical structure table for $1024 with 11 levels. (Then the value of nodes in the 11th level is exactly $1.) Let $c$ (kilobytes) be the total amount of data transfer for using one node. Thus, for paying $1024, $U$ needs to send $c$ kilobytes because only one node, i.e., the top level node, is used. On the other hand, for paying $65, $U$ needs to send $2c$ kilobytes. Furthremore, for paying $65 next time, $U$ needs to send $7c$ kilobytes. (There is a way to perform the second $65 payment by sending only $2c$ kilobytes. But then many small nodes are left, and they would cause much bigger data transfer later.) That is, the total amount of data transfer per one payment may increase depending on the amount of payment and the way how nodes have been used. Though this increase may not be so serious in some cases, it would be better if the amount of data transfer is kept constant.

## 3 New Protocol

In Ohta-Okamoto protocol, one digital cash (i.e., one hierarchical structural table) is, intuitively, a collection of small coins. A user selects some of those coins and send them for each payment. On the other hand, one digital cash in our new protocol is considered as a check book with a certain limit on the total amount of money. That is, for each payment, a user writes a figure on one page of his/her digital check book, and pass it to a merchant. Thus, any amount of payment (of course, within the limit) is possible with the same amount of data transfer. Clearly, the important points here are (i) how to prevent cheating users from writing a figure exceeding the limit, and (ii) how to prevent cheating merchants from claiming more money than they can receive.

Here we present our new protocol and explain mainly the above (i) and (ii). We also estimate the total amount of transfer data per payment, and compare it with Okamoto-Ohta protocol.

### 3.1. Mathematical Preliminaries

We use the same number theoretic tool as Okamoto-Ohta protocol. Here we review some key points from [OO91, OO93].

For any integer $x$ and $N$, let $(x/N)$ denote the Jacobi symbol, when $N$ is a composite number, and denote the Legendre symbol, when $N$ is a prime. Define $QR_N$ to be the set of quadratic residue integers in $Z_N^*$, i.e., the set of $x$, $1 \leq x \leq N - 1$, that has a square root under mod $N$. (In this paper, $Z_N^*$ is also used to denote the set of integers relatively prime to $N$.)

An integer $N$ is called a *Williams integer* if $N = PQ$ for some primes $P$ and $Q$ such that $P = 3 \, (\mathrm{mod}\, 8)$ and $Q = 7 \, (\mathrm{mod}\, 8)$. In the protocol, we use Williams integers with large prime factors. Define $W_0 = \{\, N : N$ is a Williams integer with two prime factors of $L_1$ bits $\}$, where $L_1$ is some large number, e.g., $L_1 = 512$.

Williams integers are sepcial case of Blum integers, and thus they have all properties of Blum integers. By using them, the following properties are provable.

**Proposition 3.1.** Let $N$ be a Williams integer. Then for any $x \in Z_N^*$, there is a unique $a \in \{-2, -1, 1, 2\}$ such that $ax \in QR_N$.

**Notation.** For any $x \in Z_N^*$, we use $\langle x \rangle_{QR_N}$ to denote the unique $z \in QR_N$ such that $z = ax \, (\mathrm{mod}\, N)$ for some $a \in \{-2, -1, 1, 2\}$.

**Proposition 3.2.** Let $N$ be a Williams integer. Then for any $x \in QR_N$, there are the following two integers $y_1$ and $y_2$: (i) $y_1^2 = y_2^2 = x \, (\mathrm{mod}\, N)$, (ii) $1 \le y_1, y_2 < N/2$, and (iii) $(y_1/N) = 1$, and $(y_2/N) = -1$. Furthermore, these $y_1$ and $y_2$ are uniquely determined from $x$.

**Notation.** We use $[x^{1/2} \, \mathrm{mod}\, N]_{+1}$ (resp., $[x^{1/2} \, \mathrm{mod}\, N]_{-1}$) to denote the above $y_1$ (resp., $y_2$).

The second proposition says that every $x \in QR_N$ has two (and only two) square roots under mod $N$ satisfying both (ii) and (iii). For any $x \in QR_N$, let $y_1$ and $y_2$ be such square roots of $x$. Notice that we have $(y_1 + y_2)(y_1 - y_2) = 0 \, (\mathrm{mod}\, N)$ since $y_1^2 = y_2^2 = x \, (\mathrm{mod}\, N)$. That is, both $y_1 + y_2$ and $y_1 - y_2$ are prime factors of $N$. Thus, from $y_1$ and $y_2$, we can compute $N$'s prime factorization easily. This plays a key role in Okamoto-Ohta protocol and ours.

In the protocol, we assume some sorts of functions that have been proposed and used in computational cryptography and computational complexity.

For mapping binary sequences of length $l$ into $Z_N^*$, a *hash function* is used. That is, for given $l$ and $N$, we assume a function mapping every binary string of $\{0, 1\}^l$ to some integer in $Z_N^*$ in such a way that the function values are distributed uniformly. For example, an appropriate instance of linear hash functions of [CW79] satisfies our purpose. In order to simplify notation, we will simply denote this type of hash functions by $h$, and assume that they are *appropriately* defined in each context. That is, depending on the context, $h$ may represent different hash functions.

A function $f$ is called *one-way* if computing $f$ is "easy", but computing $f^{-1}$ is "hard on average". If it is also "hard" to control its output, then $f$ is called a *one-way hash function*. (See Remark 1 below for our assumption on the notion of "easy" and "hard".)

In this paper, as typical one-way functions, we consider encryption functions used in the RSA scheme (see [RSA78]). Let $f$ be any encryption function of the RSA scheme. More specifically, $f$ is defined by $f(x) = x^e \, (\mathrm{mod}\, N)$ for some appropriate pair of integers $e$ and $N$, which we call an *encoding key pair*. Then $f$ has some useful properties besides one-wayness. First, for some $d$, we have $f^{-1}(y) = y^d \, (\mathrm{mod}\, N)$. That is, $f^{-1}$ is easy to compute for those who know its *decoding key* $d$. Second, for any $y$ and $r$ in $Z_N^*$, the following relation holds:

$$f^{-1}(y) = f^{-1}(r^e y \pmod{N}))/r \pmod{N}. \tag{3.1}$$

That is, from $f^{-1}(r^e y \pmod{N}))$, we can extract $f^{-1}(y)$. This property has been used [Cha85] to implement blind signatures.

**Remark 1.** (Assumption on Computational Complexity)
From computational complexity theory, some problems, e.g., factorizing a product of two large prime numbers, are believed to be intractable even on average. For example, it is believed that every algorithm with reasonable time bound can factorize numbers on only a small portion of the set $W_0$. (Recall $W_0$ is the set of Williams integers with large prime factors.) Our protocol depends on such assumptions. Furthermore, for the sake of simplicity, we will ingnore small portions. That is, although "hard on average" does not mean "impossible at all", we discuss in the following by assuming *conceptually* that no algorithm with reasonable time bound can factorize numbers *at all* on $W_0$. More specifically, we will make use of the following assumptions:

(a) Consider any Williams number $N$ in $W_0$ (thus, the size of prime factors of $N$ is $L_1$). Then for any $x \in QR_N$, no feasible algorithm can compute any $x^{1/2} \pmod{N}$. (*Cf.* It is easy to compute if the prime factorization of $N$ is also given.) Also given $N \in W_0$, $x \in QR_N$, and any one of square roots, $[x^{1/2} \bmod N]_{+1}$ and $[x^{1/2} \bmod N]_{-1}$, no feasible algorithm can compute the other square root. (If this is possible, then by knowing any one of $x^{1/2} \pmod{N}$, one can factorize $N$.)

(b) Consider any RSA's encoding key pair $(e, N)$ such that $|N| = L_2$ for some large $L_2$, e.g., $L_2 = 512$. Then for any $y \in Z_N^*$, no feasible algorithm can compute $x \in Z_N^*$ such that $y = x^e \pmod{N}$. (*Cf.* It is easy to compute if its decoding key $d$ is also given.)

(c) There is some one-way hash function $g_0$ on $\{0,1\}^{L_3}$ (where $L_3$ is some large number, e.g., $L_3 = 512$) such that for any "small & simple" subset $S$ of $\{0,1\}^{L_3}$, no feasible algorithm can produce $x$ such that $g_0(x) \in S$ with probability larger/smaller than $\|S\|/2^{L_3}$. (In the literature, many one-way hash functions have been proposed. Theoretically, the one proposed in [NY89] is a good candidate for $g_0$. On the other hand, some practical one-way hash functions, e.g., SHA, could be also used though there is no theoretical support for its security. See, e.g., [Sch94] for the survey on one-way hash functions.)

Clearly, in practice, we must estimate the probability that some feasible algorithm can compute, e.g., $x^{1/2} \pmod{N}$, and then determine security parameters $L_1$, $L_2$, and $L_3$ accordingly. Such analysis will be left to the interested reader; see, e.g., [Gol95, Sch94].

### 3.2. Description of the Protocol and its Correctness

Here, following the outline presented in Section 2, we state our new protocol. Again we assume some bank that hosts the protocol, and use $U$ and $V$ to denote a digital cash user and a merchant.

We also assume that every participant $Y$ of the protocol has a unique id, which is denoted as $ID_Y$. For example, $ID_U$ and $ID_V$ are respectively user $U$'s id and merchant $V$'s id.

For the protocol, the bank needs to prepare a trapdoor one-way function. Here we use an encryption function $f_0$ of the RSA scheme; let $(e_0, N_0)$ be its encoding key pair of reasonable size, and let $d_0$ be its decoding key. We assume that only the bank knows $d_0$ while $e_0$ and $N_0$ are public to the participants of the protocol. Also let $h$ denote any appropriate hash function for mapping a given binary string to the domain of $f_0$.

**Part 1:** Issue of the authorized license $L_U$
(Since this part is the same as the one in [OO91, OO93], we state it briefly. In the following, the range of index $i$ is $\{1, ..., K\}$.)

(1) User $U$ generates $K$ William numbers $N_1, ..., N_K$ from $W_0$; let $P_i$ and $Q_i$ be prime factors of $N_i$. Next, for each $i$, $U$ selects random number $r_i$, and computes $w_i = r_i^{e_0} h(I_i \| N_i) \pmod{N_0}$, where $\|$ is a concatination of two binary sequences. Here $U$'s id $ID_U$ is encrypted in $I_i$ in such a way that $ID_U$ is not computable for those who does not know the factorization of $N_i$ (see [OO91, OO93] for detail). Then $U$ sends $w_1, ..., w_K$ to the bank.

(2) The bank chooses $K/2$ random indices $i_{j_1}, ..., i_{j_{K/2}}$, and asks $U$ to open $w_i$'s with these indices. (Here for simplifying notation, we assume that $\{K/2 + 1, ..., K\}$ is selected for $\{i_{j_1}, ..., i_{j_{K/2}}\}$.)

(3) For each $i \in \{K/2 + 1, ..., K\}$, $U$ reveals $I_i, P_i, Q_i$, and $r_i$ to the bank; i.e., $U$ opens $w_i$.

(4) For each received $I_i, P_i, Q_i$, and $r_i$, the bank checks whether (i) $w_i$ is valid w.r.t. them, and (ii) $ID_U$ is correctly encoded in $I_i$. If no problem is found, then the bank sends the following value $B'$ to $U$.

$$B' = \prod_{i=1}^{K/2} f_0^{-1}(w_i) \pmod{N_0}.$$

(5) $U$ will use the following $B$ to certify $U$'s license:

$$B = \prod_{i=1}^{K/2} f_0^{-1}(h(I_i \| N_i)) \pmod{N_0}. \qquad (3.2)$$

Recall that $w_i = r_i^{e_0} h(I_i \| N_i) \pmod{N_0}$. Hence, it follows from (3.1) that $f_0^{-1}(h(I_i \| N_i)) = f_0^{-1}(w_i)/r_i \pmod{N_0}$; thus, $B$ is computable from $B'$. Now a tuple $(\{I_i\}, \{N_i\}, B)$ is $U$'s authorized licence $L_U$. (From now on, the range of the index $i$ is $\{1, ..., K/2\}$, and we omit specifying it.)

We say that $L_U = (\{I_i\}, \{N_i\}, B)$ is $k$-*fake* if it satisfies (3.2), and some $k$ $I_i$'s do not encode $ID_U$ correctly. Since we assumed that no one but the bank can compute $f_0^{-1}$, only the way to make a $k$-fake $L_U$ is to put $k$ wrong $(I_i, N_i)$ pairs into $\{(I_{i'}, N_{i'})\}_{1 \le i' \le K}$. But the probability that the bank asks to open one of such pairs is $1 - 2^{-k}$. Thus, we have the following claim.

**Claim 1.** For any $k$, $1 \le k \le K/2$, the probability that a cheating user $U$ obtains $k$-fake $L_U$ is at most $2^{-k}$.

**Part 2:** Issue of a Digital Check Book
(Here we fix the limit value of each digital check book to $\$Q_0$. Check books with different limit values are realizable by using the same idea in [OO91, OO93].)
(1) For a randomly chosen number $r \in Z_{N_0}^*$, $U$ sends $w = rB \,(\mathrm{mod}\, N_0)$ to the bank and requests for one digital check book.
(2) The bank computes $C' = f_0^{-1}(rB)$. Then the bank sends it back to $U$, and deducts $\$Q_0$ from $U$'s account.
(3) Again by using (3.1), $U$ extracts the following $C$ from $C'$:

$$C = f_0^{-1}(B). \tag{3.3}$$

**Part 3:** Payment of $\$P$
(Here we consider the case that $U$ wants to pay $\$P$ (where $P < Q_0$) to $V$ by using $j$th page of the check book $C$. In some of the following steps, the case that $j = 1$ is treated in a slightly different way.)
(1) For each $i$, $U$ computes the following $\Sigma_{j,i}$:

$$\Sigma_{j,i} = \langle h(j\|C)\rangle_{QR_{N_i}}. \tag{3.4}$$

Then $U$ sends $L_U$, $C$, $j$, and $\Sigma_{j,1}, ..., \Sigma_{j,K/2}$ to $V$. For the case $j \geq 2$, $X_{j-1}$ is also sent.
(2) $V$ checks the validity of $L_U$ (3.2), $C$ (3.3), and $\{\Sigma_{j,i}\}$ (3.4). For the case $j \geq 2$, $V$ also checks the validity of $X_{j-1}$ (3.6), and whether $P < Q_{j-1}$. If no problem is found, then for some randomly generated number $r$, $V$ sends $r$, current time $T$, and the following $b_j$ to $U$:

$$b_j = h_0(g_0(j\|r\|T\|f_0(h(T\|ID_V))\|Q_{j-1}\|Q_j)). \tag{3.5}$$

Where $Q_j = Q_{j-1} - P$ is the balance after this payment. We assume that $h_0$ is a hash function from $\{0,1\}^{L_{g0}}$ to $\{0,1\}^{K/2}$; hence, $b_j$ is a binary string of length $K/2$.
(3) After checking the validity of $b_j$ (3.5), for each $i$, $U$ computes the following $Z_{j,i}$ and send it to $V$:

$$Z_{j,i} = [\Sigma_{j,i}^{1/2} \,\mathrm{mod}\, N_i]_{(-1)^{b_{j,i}}}. \tag{3.6}$$

Where $b_{j,i}$ is the $i$th bit of $b_j$.
(4) $V$ checks the validity of $Z_{j,i}$ (3.6) for each $i$, and accepts the payment if no problem is found. Then both $U$ and $V$ keep the following $X_j$ (as well as $L_U$, $C$, $X_{j-1}$) for their record: $X_j = (j, r, T, f_0(h(T\|ID_V)), Q_{j-1}, Q_j, \{Z_{j,i}\})$.

Suppose that a cheating user $U$ might want to use the $j$th check twice; then $U$ has to use two different $b_j$ and $b'_j$ for the same $\{\Sigma_{j,i}\}$. The situation is the same even if $U$ conspires with merchant $V$ because $f_0(h(T\|ID_V))$ is used to compute $b_j$. (If they use $ID_{V'}$, the id of the previous merchant $V'$, then $V$ will be in trouble for getting money from the bank.) The situation is similar when $U$ tries to change the balance $Q_{j-1}$. On the other hand, as we assumed (Remark 1), no one can control the output of $g_0$ (except using the same input); therefore, the following claim holds.

**Claim 2.** Suppose that $U$ has a $k$-fake license $L_U$, and $U$ tries to use $C$ illegally at the $j$th step. Then the probability that $U$ produces two $b_j$ and $b'_j$ such that $b_{j,i} = b'_{j,i}$ for every correct index $i$ is at most $2^{-K/2+k}$. (Here by "correct index" we mean an index $i$ such that $I_i$ encodes $U$'s id correctly.)

**Part 4:** Claim for a \$$P$ Check

(Here we consider the case that $V$ claims to the bank one piece of \$$P$ check that $U$ paid to $V$.)

(1)  $V$ sends $L_U$, $C$, $X_{j-1}$, and $X_j$ to the bank and asks for $P = Q_{j-1} - Q_j$ dollars. (For the case $j = 1$, $X_{j-1}$ is omitted.)

(2)  The bank checks the validity of $L_U$ (3.2), $C$ (3.3), $X_{j-1}$ (3.6), and $X_j$. For the validity of $X_j$, besides checking (3.6), the bank checks whether $ID_V$ encoded in $f_0(h(T\|ID_V))$ is $V$'s id. It is also checked whether $V$ has claimed the same $(L_U, C, X_{j-1}, X_j)$ before. If no problem is found, then the bank gives \$$P$ credit to $V$'s account.

(3)  It is possible that the bank receives two different $(L_U, C, X_{j-1}, X_j)$ and $(L_U, C, X'_{j-1}, X'_j)$. Such a situation is caused by $U$'s illegal usage of $C$; that is, either $U$ uses the $j$th page of the check $C$ at two occasions, or $U$ changes the balance $Q_{j-1}$. In each case, we have two different $b_j$ and $b'_j$ associated with $X_j$ and $X'_j$ respectively. Suppose that $b_j$ differs from $b'_j$ on the $i$th bit; i.e., $b_{j,i} \neq b'_{j,i}$. Then the bank has two different square roots $Z_{j,i}$ and $Z'_{j,i}$ of $\Sigma_{j,i}$, from which $N_i$ is factorizable. Furthermore, if $ID_U$ is correctly encoded in $I_i$, then the bank, who now knows the factorization of $N_i$, can identify the cheating user $U$.

First consider the possibility that $V$ claims more money than the amount that $U$ has paid to him. To do this, $V$ needs to change $Q_j$, which almost always causes to change some bits of $b_j$. (With very small probability, i.e., $2^{-K/2}$, $V$ can still use the same $b_j$ even if $Q_j$ is changed.) Hence, $V$ has to compute some of $Z_{j,i}$'s by himself, which is impossible from our assumption (Remark 1). Thus, the following claim holds.

**Claim 3.** There is the case that cheating $V$ can claim more money than the amount that he has been paid, and such a claim is accepted by the bank in the above protocol. However, the probability that such case occurs is at most $2^{-K/2}$.

On the other hand, the following claim follows from Claim 1 and Claim 2.

**Claim 4.** Suppose that $U$ has $k$-fake license. Then the probability that $U$ uses an issued digital check book illegally and yet cannot be identified by the bank is at most $2^{-K/2+k}$. Thus, the total probability that $U$ succeeds to obtain $k$-fake license and use a digital check book illegally without being identified is at most $2^{-k} \times 2^{-K/2+k} = 2^{-K/2}$.

Thus, for a sufficiently large $K$, e.g., $K = 40$, our new protocol is reasonably secure; that is, in the worst case, the bank can at least detect cheating users with very high probability. Therefore, the property (1) of Section 1 is satisfied. It is almost clear that the protocol satisfies the other properties as well.

### 3.3. Performance Estimation

We use the following security parameters proposed by Okamoto and Ohta [OO91, OO93]: $K = 40$, $|N_i| = 64$ bytes, and $|N_0| = 100$ bytes. Similarly, we may assume that $g_0$ outputs 100 byte sequence. With these parameters, we estimate the amount of data transfer at Part 3 of our protocol. There user $U$ and merchant $V$ need to exchange $L_U$, $C$, $\{\Sigma_{j,i}\}$, $X_{j-1}$, and $X_j$. Here the major term is the length of $\{\Sigma_{j,i}\}$, $\{Z_{j-1,i}\}$, and $\{Z_{j,i}\}$, where $\{Z_{j-1,i}\}$ and $\{Z_{j,i}\}$ are contained in $X_{j-1}$ and $X_j$ respectively. This length is bounded in total by $3 \cdot (K/2) \cdot |N_i|$ = 3840 bytes. Hence, we may assume that the total amount of data transfer per payment is at most a few kilobytes, which is much smaller than 20 kilobytes (on average) of Okamoto-Ohta protocol.

# References

[CW79]    J.L. Carter and M.N. Wegman, Universal classes of hash functions, *J. Comput. System Sci.*, 18 (1979), 143–154.

[Cha85]   D. Chaum, Security without identification: transaction systems to make big brothers obsolete, *Comm. of the ACM*, 28 (1985), 1030–1044.

[CP92]    D. Chaum and T. Pedersen, Transferred cash grows in size, in *Proc. EUROCRYPTO'92*, Lecture Notes in Computer Science 58 (1993), 390–407.

[CFN88]   D. Chaum, A. Fiat, and M. Naor, Untraceable electronic cash, in *Advances in Cryptology: Proc. of CRYPTO'88*, Lecture Notes in Computer Science 403 (1990), 319–327.

[Gol95]   O. Goldreich, *Foundations of Cryptography*, in Electric Colloquim on Computational Complexity (http://www.informatik.uni-trier.de:80/eccc/), 1995.

[NY89]    M. Naor and M. Yung, Universal one-way hash functions and their cryptographic application, in *Proc. 21st ACM Sympos. on Theory of Computing* (1989), 203–210.

[OO89]    T. Okamoto and K. Ohta, Disposable zero-knowledge authentication and their applications to untraceable electronic cash, in *Advances in Cryptology: Proc. of CRYPTO'89*, Lecture Notes in Computer Science 435 (1990), 134–149.

[OO91]    T. Okamoto and K. Ohta, Universal electronic cash, in *Advances in Cryptology: Proc. of CRYPTO'91*, Lecture Notes in Computer Science 576 (1992), 324–337.

[OO93]    T. Okamoto and K. Ohta, Universal electronic cash scheme (in Japanese), IEICE Trans. D-I, J76-D-I (1993), 315–323.

[OS96]    A. Otsuka and T. Shinohara, On electronic cash (in Japanese), *Journal of Info. Process. Soc. Japan*, 37 (1996), 303–310.

[RSA78]   R.L. Rivest, A. Shamir, L. Adleman, A method for obtaining digital signatures and public-key cryptosystems, *Comm. of the ACM*, 21 (1978), 120–126.

[Sch94]   B. Schneier, *Applied Cryptography*, John Wiley & Sons, Inc., 1994.

[Yam96]   O. Yamashita, *Some Technical Problems on Digital Cash* (in Japanese), Master's Thesis, Tokyo Institute of Technology, 1996.

# Author Index

# Springer
# and the
# environment

At Springer we firmly believe that an international science publisher has a special obligation to the environment, and our corporate policies consistently reflect this conviction.

We also expect our business partners – paper mills, printers, packaging manufacturers, etc. – to commit themselves to using materials and production processes that do not harm the environment. The paper in this book is made from low- or no-chlorine pulp and is acid free, in conformance with international standards for paper permanency.

Springer

# Lecture Notes in Computer Science

For information about Vols. 1–1104

please contact your bookseller or Springer-Verlag